MAGNETISM

Fundamentals

Front Cover Photo:
Courtesy of Pierre Molho, Laboratoire Louis Néel du CNRS, Grenoble, France.
A computer hard disk drive, a device combining many state-of-the-art magnetic technologies (courtesy of Seagate Corporation). The unit is set against a magnetic domain pattern in a garnet film, imaged through the mageneto-optical Faraday effect; the domain width is about 7 um (courtesy Pierre Molho, Laboratoire Louis Néel du CNRS, Grenoble, France).

MAGNETISM

Fundamentals

Edited by

**Étienne du TRÉMOLET de LACHEISSERIE
Damien GIGNOUX
Michel SCHLENKER**

 Springer

Library of Congress Cataloging-in-Publication Data

A C.I.P. Catalogue record for this book is available
from the Library of Congress.

ISBN 1-4020-7222-8 (hardcover)   Printed on acid-free paper.
© 2003 Kluwer Academic Publishers

First Springer Science+Business Media, Inc. softcover printing, 2005
ISBN 0-387-22967-1   E-book ISBN 0-387-23062-9

All rights reserved. This work may not be translated or copied in whole or in part without the written permission of the publisher (Springer Science+Business Media, Inc., 233 Spring Street, New York, NY 10013, USA), except for brief excerpts in connection with reviews or scholarly analysis. Use in connection with any form of information storage and retrieval, electronic adaptation, computer software, or by similar or dissimilar methodology now know or hereafter developed is forbidden.
The use in this publication of trade names, trademarks, service marks and similar terms, even if the are not identified as such, is not to be taken as an expression of opinion as to whether or not they are subject to proprietary rights.

Printed in the United States of America.

9 8 7 6 5 4 3 2 1   SPIN 11313915

springeronline.com

AUTHORS

Michel CYROT - Professor at the Joseph Fourier University of Grenoble, France

Michel DÉCORPS - Senior Researcher, INSERM (French Institute of Health and Medical Research), Bioclinical Magnetic Nuclear Resonance Unit, Grenoble

Bernard DIÉNY - Researcher and group leader at the CEA (French Atomic Energy Center), Grenoble

Etienne du TRÉMOLET de LACHEISSERIE - Senior Researcher, CNRS, Laboratoire Louis Néel, Grenoble

Olivier GEOFFROY - Assistant Professor at the Joseph Fourier University of Grenoble

Damien GIGNOUX - Professor at the Joseph Fourier University of Grenoble

Ian HEDLEY - Researcher at the University of Geneva, Switzerland

Claudine LACROIX - Senior Researcher, CNRS, Laboratoire Louis Néel, Grenoble

Jean LAFOREST - Research Engineer, CNRS, Laboratoire Louis Néel, Grenoble

Philippe LETHUILLIER - Engineer at the Joseph Fourier University of Grenoble

Pierre MOLHO - Researcher, CNRS, Laboratoire Louis Néel, Grenoble

Jean-Claude PEUZIN - Senior Researcher, CNRS, Laboratoire Louis Néel, Grenoble

Jacques PIERRE - Senior Researcher, CNRS, Laboratoire Louis Néel, Grenoble

Jean-Louis PORTESEIL - Professor at the Joseph Fourier University of Grenoble

Pierre ROCHETTE - Professor at the University of Aix-Marseille 3, France

Michel-François ROSSIGNOL - Professor at the Institut National Polytechnique de Grenoble (Technical University)

Yves SAMSON - Researcher and group leader at the CEA (French Atomic Energy Center) in Grenoble

Michel SCHLENKER - Professor at the Institut National Polytechnique de Grenoble (Technical University)

Christoph SEGEBARTH - Senior Researcher, INSERM (French Institute of Health and Medical Research), Bioclinical Magnetic Nuclear Resonance Unit, Grenoble

Yves SOUCHE - Research Engineer, CNRS, Laboratoire Louis Néel, Grenoble

Jean-Paul YONNET - Senior Researcher, CNRS, Electrical Engineering Laboratory, Institut National Polytechnique de Grenoble (Technical University)

Grenoble Sciences

"Grenoble Sciences" was created ten years ago by the Joseph Fourier University of Grenoble, France (Science, Technology and Medicine) to select and publish original projects. Anonymous referees choose the best projects and then a Reading Committee interacts with the authors as long as necessary to improve the quality of the manuscript.

(Contact : Tél. : (33)4 76 51 46 95 - E-mail : Grenoble.Sciences@ujf-grenoble.fr)

The "Magnetism" Reading Committee included the following members :

- **V. Archambault**, Engineer - Rhodia-Recherche, Aubervilliers, France
- **E. Burzo**, Professor at the University of Cluj, Rumania
- **I. Campbell**, Senoir Researcher - CNRS, Orsay, France
- **F. Claeyssen**, Engineer - CEDRAT, Grenoble, France
- **G. Couderchon**, Engineer - Imphy Ugine Précision, Imphy, France
- **J.M.D. Coey**, Professor - Trinity College, Dublin, Ireland
- **A. Fert**, Professor - INSA, Toulouse, France
- **D. Givord**, Senior Researcher - Laboratoire Louis Néel, Grenoble, France
- **L. Néel**, Professor, Nobel Laureate in Physics, Member of the French Academy of Science
- **B. Raquet**, Assistant Professor - INSA, Toulouse, France
- **A. Rudi**, Engineer - ECIA, Audincourt, France
- **Ph. Tenaud**, Engineer - UGIMAG, St-Pierre d'Allevard, France

"Magnetism" (2 volumes) is an improved version of the French book published by "Grenoble Sciences" in partnership with EDP Sciences with support from the French Ministry of Higher Education and Research and the "Région Rhône-Alpes".

BRIEF CONTENTS

I - FUNDAMENTALS

Foreword

Phenomenological approach to magnetism
1. Magnetism, from the dawn of civilization to today - *E. du Trémolet de Lacheisserie*
2. Magnetostatics - *D. Gignoux, J.C. Peuzin*
3. Phenomenology of magnetism at the macroscopic scale - *D. Gignoux*
4. Phenomenology of magnetism at the microscopic scale - *D. Gignoux*
5. Ferromagnetism of an ideal system - *M. Rossignol, M. Schlenker, Y. Samson*
6. Irreversibility of magnetization processes, and hysteresis in real ferromagnetic materials: the role of defects - *M. Rossignol*

Theoretical approach to magnetism
7. Magnetism in the localised electron model - *D. Gignoux*
8. Magnetism in the itinerant electron model - *M. Cyrot*
9. Exchange interactions - *C. Lacroix, M. Cyrot*
10. Thermodynamic aspects of magnetism - *M. Schlenker, E. du T. de Lacheisserie*

Coupling phenomena
11. Magnetocaloric coupling and related effects - *E. du Trémolet de Lacheisserie, M. Schlenker*
12. Magnetoelastic effects - *E. du Trémolet de Lacheisserie*
13. Magneto-optical effects - *M. Schlenker, Y. Souche*
14. Magnetic resistivity, magnetoresistance, and the Hall effect - *J. Pierre*

Appendices
1. Symbols used in the text
2. Units and universal constants
3. Periodic table of the elements
4. Magnetic susceptibilities
5. Ferromagnetic materials
6. Special functions
7. Maxwell's equations

General references

Index by material and by subject

II - MATERIALS AND APPLICATIONS

Foreword

Magnetic materials and their applications

15 Permanent magnets - *M. Rossignol, J.P. Yonnet*
16 Soft materials for electrical engineering and low frequency electronics - *O. Geoffroy, J.L. Porteseil*
17 Soft materials for high frequency electronics - *J.C. Peuzin*
18 Magnetostrictive materials - *E. du Trémolet de Lacheisserie*
19 Superconductivity - *M. Cyrot*
20 Magnetic thin films and multilayers - *B. Dieny*
21 Principles of magnetic recording - *J.C. Peuzin*
22 Ferrofluids - *P. Molho*

Other aspects of magnetism

23 Magnetic resonance imaging - *M. Décorps, C. Segebarth*
24 Magnetism of earth materials and geomagnetism - *P. Rochette, I. Hedley*
25 Magnetism and the Life Sciences - *E. du Trémolet de Lacheisserie, P. Rochette*
26 Practical magnetism and instrumentation - *Ph. Lethuillier*

Appendices

1 Symbols used in the text
2 Units and universal constants
3 Periodic table of the elements
4 Magnetic susceptibilities
5 Ferromagnetic materials
6 Economic aspects of magnetic materials - *J. Laforest*

General references

Index by material and by subject

TABLE OF CONTENTS
I - FUNDAMENTALS

Foreword *by Professor Louis Néel* XXI
Preface XXIII
Acknowledgements XXIV

PHENOMENOLOGICAL APPROACH TO MAGNETISM

1 - Magnetism, from the dawn of civilization to today 3
1. The discovery of lodestone and the observation of magnetic phenomena through the ages 3
   1.1. Objects made from iron, and lodestone in ancient times 3
   1.2. The first texts dealing with lodestone. Origin of the name "magnetism" 4
   1.3. First observations of magnetic phenomena 5
      1.3.1. Magnetic attraction 5
      1.3.2. Magnetic shielding 5
      1.3.3. The compass 5
      1.3.4. Knowledge of magnetism in 1779 6
2. The contribution of the 19th century 7
3. Magnetism in the 20th century 9
4. Magnetism, and technology 11
5. New research lines 14
   5.1. Magnetic nanostructures 15
   5.2. Magnetic multilayers 15
   5.3. Molecular magnetism 16
   5.4. Other lines of research 16
   References 17

2 - Magnetostatics 19
1. Magnetostatics of currents and materials 19
   1.1. Magnetostatics of currents in vacuum 19
      1.1.1. Fundamental laws of magnetostatics: magnetic induction 20
      1.1.2. Ampère's law. Magnetic field 22
      1.1.3. Flux conservation. The vector potential 23
      1.1.4. Boundary conditions for B and A across a current sheet 24
      1.1.5. Induction and field produced by a thread-like circuit.
             The field coefficient of a coil 24
      1.1.6. Some simple current distributions 24
      1.1.7. Induction and field produced at a large distance
             by currents confined to a finite volume:
             magnetic moment, magnetic dipole, and magnetic charge 27
      1.1.8. The fundamental relationships in the Coulombian approach 29

    1.2. Magnetostatics of matter 30
        1.2.1. Magnetic moment, magnetization, and induction associated with matter 30
        1.2.2. The determination of B (and A) from bound currents:
               the Ampérian approach 31
        1.2.3. The magnetic field H in magnetised material 32
        1.2.4. The determination of H from equivalent magnetic charges:
               the Coulombian approach 32
        1.2.5. Application to two simple cases 34
        1.2.6. The demagnetising field H_d 35
        1.2.7. Demagnetising field coefficient 36
    1.3. Response of a material to a magnetic field 37
        1.3.1. The magnetic field H as the independent magnetic variable 37
        1.3.2. Different type of magnetic behavior of a substance 38
        1.3.3. Anisotropic materials 39
        1.3.4. Note 40
        1.3.5. Demagnetising field correction. The external susceptibility 40
    1.4. The general problem of determining B and M 41
    1.5. Magnetic circuit approximation 42
2. Energy, forces, and torques in magnetic systems 45
    2.1. Electromagnetic coupling in vacuum 45
        2.1.1. Lorentz force, and Lorentz field 46
        2.1.2. Electric field created in a moving conductor by a magnetic induction B 46
        2.1.3. Electric field created by a variable magnetic induction 48
        2.1.4. Magnetic induction produced by a variable electric field 49
        2.1.5. Maxwell's equations in vacuum 50
        2.1.6. Electromagnetic coupling in the limit of slow phenomena; magnetic energy 50
        2.1.7. Forces derived from the energy. The nature of magnetic energy 52
        2.1.8. Self induction 52
        2.1.9. Mutual induction 53
        2.1.10. Spontaneous and induced magnetic moments 54
        2.1.11. Energy of a rigid moment in a given induction field 55
        2.1.12. Dipolar interaction energy 55
    2.2. Energy, forces, and torques in magnetic systems 57
        2.2.1. Direct calculation of torques, and forces in a magnetic system 57
        2.2.2. Magnetization energy for matter, and demagnetising field energy 58
        2.2.3. Zeeman energy, and magnetostatic energy 61
        2.2.4. Magnetostatic torque acting on the magnetization 61
        2.2.5. Energy of a complete magnetic system 62
        2.2.6. Forces between magnetic parts in simple situations 63
3. Appendices 64
    3.1. Calculation of B_m outside matter 64
    3.2. Calculation of B_m inside matter 65
    3.3. Calculation of the integral Σ 66
Exercises 67
Solutions to the exercises 73
References 78

3 - Phenomenology of magnetism at the macroscopic scale 79
1. Presentation of some types of magnetic behaviors 79
    1.1. Diamagnetism 79
    1.2. Paramagnetism 80
    1.3. Antiferromagnetism 81

    1.4. Ferromagnetism 82
    1.5. Ferrimagnetism 83
    1.6. Magnetic properties of pure elements in the atomic state 84
    1.7. Magnetic properties of polyatomic materials 84
2. Phenomenology of strongly magnetic materials 86
    2.1. Isothermal magnetization curves 86
    2.2. Weiss domains, and Bloch walls 87
    2.3. Magnetic anisotropy 89
        2.3.1. Cubic symmetry 89
        2.3.2. Tetragonal symmetry 90
        2.3.3. Quadratic symmetry 91
        2.3.4. Uniaxial symmetry 91
        2.3.5. Some remarks on the magnetic anisotropy 91
    2.4. Effects associated with anisotropy - Anisotropy field - Phase rule 92
        2.4.1. Uniaxial material with tetragonal or hexagonal symmetry 92
        2.4.2. Magnetocrystalline anisotropy in cubic symmetry 96
        2.4.3. Magnetic anisotropy in the paramagnetic state 97
    2.5. Time dependent phenomena 97
        2.5.1. Thermal fluctuation after-effect 98
        2.5.2. Diffusion after-effect 99
3. Physical phenomena associated with magnetism 99
Exercises 101
Solutions to the exercises 102
References 103

4 - Phenomenology of magnetism at the microscopic scale 105

1. The classical model of diamagnetism: case of localised electrons 105
2. Systems with non-interacting localised magnetic moments 108
    2.1. Effect of a uniform field on a magnetic moment: precession 108
    2.2. Assembly of localised magnetic moments without interaction:
        Curie paramagnetism 110
    2.3. Superparamagnetism 115
3. Exchange interactions 116
    3.1. General 116
    3.2. Exchange interaction and molecular field approximation 116
    3.3. Experimental 118
4. Ferromagnetism in the molecular field model 118
5. Antiferromagnetism in the molecular field model 121
    5.1. Néel temperature 122
    5.2. Paramagnetic susceptibility 124
    5.3. Perpendicular susceptibility 124
    5.4. Parallel susceptibility 125
    5.5. Metamagnetism 125
6. Ferrimagnetism in the molecular field model 127
7. Other types of magnetic arrangements (helimagnetism,
    sine wave modulated structures...) in the molecular field model 129
    7.1. Helimagnetic structure 129
    7.2. Sine wave modulated structure 130
    7.3. Magnetic structures observed in amorphous substances 132
8. The two main families of magnetic materials 133

9. Arrott plots..133
   9.1. Material without spontaneous magnetization...134
   9.2. Material with a weak spontaneous magnetization...135
10. Conclusions..137
Appendix: determination of magnetic structures using neutron diffraction..........................137
Exercices...140
Solutions to exercices...141
References...141

5 - Ferromagnetism of an ideal system...143

1. Introduction...143
2. The principle of ferromagnetic configurations or the art of compromise........................144
   2.1. Exchange interaction...144
   2.2. Magnetic dipolar interaction..145
   2.3. Competition between exchange and dipolar interactions..146
   2.4. The role of magnetocrystalline anisotropy..146
   2.5. Compromise solutions...147
3. The domain walls..148
   3.1. Domain walls in ordinary systems (Bloch walls)..148
      3.1.1. The geometry of Bloch walls..148
      3.1.2. Evaluation of the domain wall energy, and width..150
      3.1.3. Exact calculation of γ and δ..152
   3.2. Walls in highly anisotropic materials (narrow walls)...155
   3.3. Domain walls in very thin films..157
4. Domain configurations..160
   4.1. Domain configurations in uniaxial crystals...160
   4.2. Closure domains..163
5. Magnetic configurations in small particles. Single domain particles..............................164
   5.1. The most likely stable configurations...165
   5.2. Ellipsoidal particles and shape anisotropy..167
6. Observation of domains and domain walls...168
   6.1. Ferro- and ferrimagnetic domains...168
      6.1.1. The Bitter method...168
      6.1.2. Observation of magnetic domains by magneto-optical effects....................169
      6.1.3. X-ray observation of magnetic domains..170
      6.1.4. Neutron observation of magnetic domains..172
      6.1.5. Electron observation of magnetic domains..174
      6.1.6. Observation of magnetic domains by precipitation anisotropy...................175
      6.1.7. Magnetic Force Microscopy...175
   6.2. Observation of antiferromagnetic domains...180
      6.2.1. Observation of antiferromagnetic domains by optical microscopy.............180
      6.2.2. Observation of antiferromagnetic domains by X-rays and electrons..........181
      6.2.3. Observation of antiferromagnetic domains by neutrons..............................181
7. From the macroscopically demagnetised state to the saturated state:
   magnetization processes under the effect of an external field..183
      7.1. Effect of a field applied parallel to the easy axis in a uniaxial system:
           magnetization by domain wall displacement...184
      7.1.1. Results of observation: growth of domains parallel to the field,
             at the expense of others..184
      7.1.2. The law of behavior: the demagnetising field straight line..........................184

   7.1.3. The mechanism of wall displacement in perfect systems
     where exchange is dominant...187
   7.1.4. Wall displacement mechanism in perfect systems
     with very strong intrinsic anisotropy...189
  7.2. Effect of the field applied normal to the easy axis in a uniaxial system:
   magnetization through moment rotation..190
  7.3. Field applied at an angle to the easy axis...191
  7.4. Cubic single crystal systems (Néel's phase rule)...194
  7.5. Magnetization process in polycrystalline systems...196
 8. Magnetization reversal from the saturated state and coercivity...198
  8.1. Reversibility or irreversibility?..199
  8.2. Reversal by uniform collective rotation of moments
   (rotation of saturation magnetization): the Stoner-Wohlfarth model..........................199
   8.2.1. Presentation and formulation of the model system...199
   8.2.2. Field applied antiparallel to M_r..200
   8.2.3. Inverse field applied at an angle to the direction of M_r....................................202
  8.3. Are other modes of collective reversal possible?...204
   8.3.1. Systems with strong magnetocrystalline anisotropy: Brown's inequality............204
   8.3.2. Systems without magnetocrystalline anisotropy...205
References...207

6 - Irreversibility of magnetization processes, and hysteresis in real ferromagnetic materials: the role of defects 209

1. Introduction: the perfect material model is inadequate...209
2. The role of defects in the irreversibility of the magnetization process
 by domain wall displacement..211
  2.1. Principle of the model: introducing an irregular oscillating potential..........................212
  2.2. Mechanisms of domain wall pinning to specific defects..214
   2.2.1 Inhomogeneous stress field: induced uniaxial anisotropy
     that locally modifies the domain wall energy..214
   2.2.2. Large non-magnetic voids or inclusions:
     local reductions of the wall area and of magnetostatic energy............................216
  2.3. The critical "unpinning" field associated with hindered domain wall displacement....218
   2.3.1. Coercive field associated with the oscillating potential model..........................218
   2.3.2. The critical field in the case of a deformable domain wall................................219
3. The role of defects in magnetization reversal starting
 from the saturated state..220
  3.1. Brown's paradox...220
  3.2. Defects and nucleation..222
   3.2.1. Defects giving rise to reversed domain nucleation in zero field..........................222
   3.2.2. Defects giving rise to nucleation only in a strong inverse field..........................224
4. Hysteresis and irreversibility: experiment and simple models..225
  4.1. Observed behaviors..225
   4.1.1. The initial magnetization curve, starting from a demagnetized state..................225
   4.1.2. The magnetization of a given substance can take all the values
     contained in its major hysteresis loop..227
   4.1.3. The anhysteretic magnetization curve...229
  4.2. Modelling of the behavior observed in Rayleigh's domain..230
   4.2.1. Preisach's representation..230
   4.2.2. Theory of hysteresis in Rayleigh's domain (Néel's random potential)...............232
  4.3. From soft to hard materials...234

5. Magnetic after effects: effects delayed and amplified at the same time..................236
   5.1. Thermal activation as the origin of both types of magnetic after effect..................236
   5.2. The fluctuation magnetic after-effect.
        Example of hard materials near the coercive field..................237
        5.2.1. Foundation of the analysis:
               application to magnetization reversal in permanent magnets..................238
        5.2.2. Variation of magnetization with time or with field: the fluctuation field..........241
   5.3. Diffusion after-effect..................243
        5.3.1. A localised, thermally activated uniaxial anisotropy..................243
        5.3.2. Disaccommodation..................244
References..................246

THEORETICAL APPROACH TO MAGNETISM

7 - Magnetism in the localised electron model..................251

1. Magnetism of a free atom or ion..................251
   1.1. A single electron..................251
        1.1.1. Orbital magnetic moment..................251
        1.1.2. Spin magnetic moment..................253
        1.1.3. States of individual electrons or hydrogen like atoms..................254
   1.2. Many electron atoms..................256
        1.2.1. Hartree's method - The central field approximation: configurations..................256
        1.2.2. Terms..................258
        1.2.3. Spin-orbit coupling..................258
        1.2.4. Multiplets..................259
2. Magnetism of bound atoms..................262
   2.1. Localised and itinerant magnetism..................262
   2.2. Non magnetic materials..................262
   2.3. Which substances display significant magnetism?..................263
   2.4. The two fundamental series of magnetic elements..................264
        2.4.1. The spatial distribution of electronic orbitals..................264
        2.4.2. The influence of neighbouring atoms: crystal field effects..................264
3. Some examples of localised magnetic moments..................266
   3.1. Iron oxides..................266
   3.2. Other ionic compounds of 3d metals..................266
   3.3. Ionic compounds of the rare earth elements..................267
   3.4. Intermetallic rare earth compounds..................268
   3.5. Rare earth metals..................269
4. Magnetocrystalline anisotropy: the crystalline electric field..................269
   4.1. A single d electron in a uniaxial electrostatic potential due to its surroundings..........270
   4.2. Orders of magnitude of the crystal field..................272
   4.3. The effect of the crystal field on the magnetic anisotropy of the 3d elements..................272
        4.3.1. Quenching of the orbital moment..................272
        4.3.2. L = 2 states..................273
        4.3.3. Fe^{3+} and Mn^{2+} ions in insulators (e.g. ferrites)..................273
        4.3.4. Low spin states..................273
        4.3.5. The influence of symmetry..................273
   4.4. The effects of the crystal field upon rare earth ions..................274
        4.4.1. A J = 4 multiplet with $B_2^0 < 0$..................275
        4.4.2. A J = 4 multiplet with $B_2^0 > 0$..................275
        4.4.3. A J = 5/2 multiplet with $B_2^0 > 0$..................276

      4.4.4. Kramers, non-Kramers ions 276
   4.5. The anisotropy of uniaxial 4f compounds with hexagonal or tetragonal symmetry 277

Appendix: detailed treatment of a magnetic moment calculation 279

Exercises 280

Solutions to the exercices 281

References 281

8 - Magnetism in the itinerant electron model 283

1. Generalities 283
2. Special properties of magnetic metals 284
   2.1. Very weak ferromagnets 285
   2.2. Transition metals, and their alloys 285
   2.3. Rare earth metals 285
   2.4. Rare earth - transition metal compounds 285
3. Magnetism of completely free electrons 286
   3.1. Simple description of a metal 286
   3.2. Density of states 287
   3.3. Pauli paramagnetism 287
   3.4. Landau diamagnetism 288
4. Stoner's model of itinerant ferromagnetism 288
   4.1. Criterion for ferromagnetic instability 288
   4.2. Magnetic susceptibility of a metal with interactions 289
   4.3. Ferromagnetic solution 290
   4.4. Applications 292
5. Generalization of Stoner's criterion 292
   5.1. Stoner's theory and molecular field theory 292
   5.2. Generalized susceptibility 292
   5.3. The local molecular field 293
   5.4. Dynamic susceptibility 293
   5.5. The instantaneous local molecular field 294
6. Transition metals 295
   6.1. The d band of the transition metals 295
   6.2. Origin of magnetism 297
   6.3. Crystalline electric field 299
   6.4. Magnetocrystalline anisotropy 300
7. Localized magnetic moment in itinerant magnetism 301
   7.1. Magnetism of impurities 301
      7.1.1. Local density of states 302
      7.1.2. Magnetism of an impurity 302
   7.2. The Kondo effect 303
   7.3. Spin glasses, and frustration 303
   7.4. Magnetic moments and the transition metals 304
8. Magnetism and local environment 305
9. Magnetism of transition metal alloys 306
10. Conclusion 309

References 309

9 - Exchange interactions 311

1. Many-electron wave functions 311

 2. Exchange interactions in insulators 313
  2.1. Superexchange 313
  2.2. Antisymmetric exchange 314
 3. Exchange interactions in metals 315
  3.1. Indirect interactions between 4f moments (the RKKY interaction) 315
  3.2. Exchange interaction in the 3d metals 317
  3.3. Double exchange 318
 4. Summary 319
 References 319

10 - Thermodynamic aspects of magnetism 321
 1. Basic thermodynamics: outside magnetism 321
 2. Thermodynamic potentials for an undeformable magnetic system 323
 3. Maxwell relations and inequalities 325
 4. Situation of a deformable magnetic solid 326
 5. Coupling phenomena 327
 6. Landau-type free energy 328
 7. Critical exponents and scaling laws 330
 8. Magnetic anomalies near T_c 332
 9. The molecular field model under experimental test 333
 Exercises 334
 Solutions to the exercises 334
 References 336

COUPLING PHENOMENA

11 - Magnetocaloric coupling and related effects 339
 1. Thermal variations of magnetization and specific heat 339
 2. The magnetocaloric effect 341
 3. Irreversible thermal effects 343
 4. Size effects associated with the magnetocaloric effect 345
 5. Applications of magnetothermal effects 346
 Exercises 346
 Solutions to the exercises 348
 References 349

12 - Magnetoelastic effects 351
 1. The main magnetoelastic effects 351
  1.1. Anomalous expansion. Spontaneous and forced exchange magnetostriction 351
  1.2. Influence of hydrostatic pressure on the magnetic properties 352
  1.3. Anomalies in the elastic constants: the morphic effect 352
  1.4. Anisotropic, highly magnetic field sensitive, magnetostriction 352
  1.5. Effect of a uniaxial stress on the magnetic properties 353
  1.6. Other magnetoelastic effects 354
 2. Microscopic origin of the magnetoelastic coupling 354
  2.1. Isotropic exchange magnetoelastic coupling 355

- 2.2. Anisotropic magnetoelastic coupling ... 356
  - 2.2.1. Principle of the coupling energy calculation ... 356
  - 2.2.2. The case of cubic symmetry ... 356
- 3. Callen's symmetrised notation ... 357
- 4. Magnetostriction ... 359
  - 4.1. Exchange magnetostriction in cubic symmetry ... 359
  - 4.2. Anisotropic magnetostriction in cubic symmetry ... 361
    - 4.2.1. Measuring magnetostriction coefficients in cubic symmetry ... 363
    - 4.2.2. Thermal variation of magnetostriction ... 364
    - 4.2.3. Variation of anisotropic magnetostriction
      under magnetic field in cubic symmetry ... 366
  - 4.3. Magnetostriction in hexagonal crystals ... 367
  - 4.4. Magnetostriction of isotropic materials ... 368
    - 4.4.1. Magnetostriction curves for an intrinsically isotropic material ... 369
    - 4.4.2. Influence of the demagnetised state on the magnetostriction curve ... 370
    - 4.4.3. Effect of temperature on magnetostriction ... 372
  - 4.5. Magnetostriction in polycrystalline materials ... 372
  - 4.6. Magnetoelastic effects observed on surfaces and in thin films ... 374
  - 4.7. Dipolar magnetostriction, or form effect ... 376
  - 4.8. Notes on the measurement of magnetostriction ... 378
    - 4.8.1. Internal stresses related to magnetostriction ... 378
    - 4.8.2. Accurate determination of Joule magnetostriction ... 378
- 5. Twisting of a magnetised wire carrying a current: the Wiedemann effect ... 379
- 6. Inverse magnetoelastic effects and the ΔE effect ... 380
  - 6.1. Effect of hydrostatic pressure on a magnetic material ... 380
  - 6.2. Effect of a uniaxial stress on the magnetization curve ... 381
    - 6.2.1. Material with cubic symmetry ... 381
    - 6.2.2. Isotropic material ... 382
    - 6.2.3. A polycrystalline material ... 384
  - 6.3. Effect of twist on a magnetised bar
    (the inverse Wiedemann and the Matteuci effects) ... 385
  - 6.4. Deviation from Hooke's law for ferromagnetic solids: the ΔE effect ... 386
- 7. Modelling magnetostrictive transducers ... 387
  - 7.1. Magnetomechanical coupling coefficient ... 388
  - 7.2. Dynamic elastic energy away from resonance ... 390
  - 7.3. Magnetoelastic coupling at resonance ... 390

Appendix: a refresher on linear elasticity ... 391

Exercises ... 393

Solutions to the exercises ... 394

References ... 396

13 - Magneto-optical effects ... 399

- 1. Introduction to the magneto-optical effects ... 399
  - 1.1. Transmission effects ... 400
    - 1.1.1. The Faraday effect ... 400
    - 1.1.2. Magnetic quadratic rectilinear birefringence ... 402
    - 1.1.3. Magnetic linear rectilinear birefringence ... 402
  - 1.2. Magneto-optics in reflection geometry ... 403
    - 1.2.1. The polar Kerr effect ... 403
    - 1.2.2. The longitudinal Kerr effect ... 403
    - 1.2.3. The transverse Kerr effect ... 403

    1.3. Non-linear magneto-optical effects..404
    1.4. Magnetic diffraction and magnetic dichroism in the X-ray range 404
2. Phenomenological treatment..405
    2.1. Propagation equation. Eigenmodes 405
    2.2. Effects in transmission geometry 407
        2.2.1. The Faraday effect..407
        2.2.2. The Voigt effect..408
    2.3. The Kerr effects...409
        2.3.1. The transverse Kerr configuration..410
        2.3.2. The longitudinal Kerr configuration 410
        2.3.3. The polar Kerr configuration 410
    2.4. Thin film systems 411
3. Physical presentation...411
4. Measurement and order of magnitude of the magneto-optical effects.....................415
    4.1. Modulation methods..416
        4.1.1. Amplitude (or intensity) modulation..416
        4.1.2. Phase (and polarization direction) modulation......................................416
    4.2. Magneto-optical determination of the easy magnetization directions..............417
    4.3. Measuring the rotation Θ of a rectilinear polarization direction
        (Faraday effect, polar and longitudinal Kerr effects) 418
    4.4. Measuring ellipticity..419
    4.5. Measuring an intensity variation (transverse Kerr effect,
        circular and rectilinear dichroism)..420
    4.6. Measuring the phase difference and its variation 420
    4.7. Some orders of magnitude...421
        4.7.1. Faraday effect..421
        4.7.2. Polar Kerr effect..422
        4.7.3. Longitudinal Kerr effect...423
        4.7.4. Transverse Kerr effect 424
5. Uses and applications of the magneto-optical effects..424
    5.1. Uses in physics 425
        5.1.1. Magnetic characterization. Hysteresis loop determination 425
        5.1.2. Variation of magnetization in homogeneous systems. Phase transitions....425
        5.1.3. Magneto-optics in diffraction...426
        5.1.4. Resonances..426
        5.1.5. Magnetic domain imaging 426
        5.1.6. Magnetic field maps 428
    5.2. Applications...429
        5.2.1. Optical isolators..429
        5.2.2. Magnetic field sensors 430
        5.2.3. Non destructive testing...431
        5.2.4. Magneto-optical recording..431
Appendix: optics of non-magnetic materials 432
References...439

14 - Magnetic resistivity, magnetoresistance, and the Hall effect 443

1. Definitions..443
    1.1. Electrical resistivity 443
    1.2. Magnetoresistance 443
    1.3. The Hall effect...444
    1.4. The planar Hall effect..444

- 2. Transport in magnetic metals ... 445
  - 2.1. Magnetic resistivity ... 445
  - 2.2. Localised moments ... 445
  - 2.3. Itinerant magnetism, alloys, and transition metals ... 446
    - 2.3.1. A magnetically ordered metal ... 446
    - 2.3.2. Spin fluctuations, and the Kondo effect ... 446
  - 2.4. Magnetoresistance ... 448
    - 2.4.1. Cyclotron effect ... 448
    - 2.4.2. Magnetization dependent terms - Spin disorder ... 449
    - 2.4.3. Anisotropy of magnetoresistance ... 449
  - 2.5. The Hall effect in a magnetic material ... 450
- 3. Magnetotransport in semiconductors ... 450
  - 3.1. Non magnetic semiconductors ... 450
  - 3.2. Magnetic semiconductors ... 451
- 4. Metal-insulator transitions in magnetic materials ... 451
  - 4.1. Semiconducting transition metal compounds ... 452
  - 4.2. Types of metal-insulator transitions ... 452
    - 4.2.1. The "classic" Mott transition ... 452
    - 4.2.2. Mott-Hubbard transitions ... 452
    - 4.2.3. Band crossing transitions (EuO at the Curie point) ... 452
    - 4.2.4. Anderson transitions in disordered systems ... 453
    - 4.2.5. An example: giant magnetoresistance in manganites ... 454
  - 4.3. Half-metals ... 454
- 5. Quantum effects ... 455
  - 5.1. The Shubnikov-de Haas effect ... 455
  - 5.2. The quantum Hall effect ... 456
- 6. Applications ... 457
  - 6.1. Hall effect field sensors ... 457
  - 6.2. Magnetoresistive sensors ... 457

Final note ... 458

References ... 458

APPENDICES

1. **Symbols used in the text** ... 463
2. **Units and universal constants** ... 467
   1. Conversion of MKSA units into the CGS system, and other unit systems of common use ... 467
   2. Some fundamental physical constants ... 468
3. **Periodic table of the elements** ... 469
4. **Magnetic susceptibilities** ... 471
5. **Ferromagnetic materials** ... 475
6. **Special functions** ... 477
   1. Spherical harmonics ... 477
   2. Legendre polynomials ... 479

		3.	The Langevin function	479
		4.	Brillouin functions	480
		5.	Modified Bessel functions	481

7. Maxwell's equations 483

General references 485

Index by material 487

Index by subject 493

FOREWORD

Thousands of years before our time, our ancestors already knew about the amazing properties of lodestone, or magnetite. Ever since, man has been fascinated by magnetic phenomena, especially because of their action at a distance. They are found everywhere in our daily lives: in refrigerator doors, cars, cellphones, suspension systems for high speed trains etc. In pure science they are present at all scales, from elementary particles through to galaxy clusters, not forgetting their role in the structure and history of our Earth.

The last thirty years have seen considerable progress in most of these fields, whether fundamental or technological. The purpose of this book is to present this progress. It is the collective work of faculty members and researchers, most of whom work in laboratories in Grenoble (Universities, CNRS, CEA), often in close cooperation with local industry, and the large international organizations established in the Grenoble area: Institut Laue-Langevin, ESRF (large European synchrotron), etc. This is no surprise, since activities concerning Magnetism have consistently been supported in Grenoble ever since the beginning of the 20th century.

Most of the chapters are accessible to the University graduate in science. Those notions which require a little more maturity do not need to be fully mastered to be able to understand what comes next. This treatise should be read by all who intend to work in the field of magnetism, such an open-ended field, rich in potential for further development.

New magnets, with higher performance and lower cost, will surely be found. The magnetic properties of materials containing unfilled electronic shells are not yet fully understood. Hysteresis plays a key role in irreversible effects. While its behavior is fairly well understood both in magnetic fields which are small with respect to the coercive field, and in very strong fields near saturation, the processes occurring within the major loop have not yet been very well described. When hysteresis depends on the combined action of two variables, such as magnetic field and very high pressure, we know nothing. How are we, for instance, to predict the magnetic state of a submarine cruising at great depth, depending on its diving course?

French scientists, with Pierre Curie, Paul Langevin and Pierre Weiss, played a pioneering role in magnetism. They will certainly have worthy successors, notably in biomagnetism in a broad sense.

This work includes interesting features: exercises with solutions, references fortunately restricted to the best papers and books, and various appendices: lists of symbols, special functions, properties of various materials, economic aspects, and, last but not least, a very necessary summary of units, which the dual coulombic-amperian presentation made so unnecessarily complicated and unpalatable in the past.

I believe this book should satisfy a broad readership, and be a valuable document to students, researchers, and engineers. I wish it a lot of success.

Louis NEEL
Nobel Laureate in Physics,
Member of the French Academy of Science

PREFACE

Magnetic materials are all around us, and understanding their properties underlies much of today's engineering efforts. The range of applications in which they are centrally involved includes audio, video and computer technology, telecommunications, automotive sensors, electric motors at all scales, medical imaging, energy supply and transportation, as well as the design of stealthy airplanes.

This book deals with the basic phenomena that govern the magnetic properties of matter, with magnetic materials, and with the applications of magnetism in science, technology and medicine.

It is the collective work of twenty one scientists, most of them from Laboratoire Louis Néel in Grenoble, France. The original version, in French, was edited by Etienne du Trémolet de Lacheisserie, and published in 1999. The present version involves, beyond the translation, many corrections and complements.

This book is meant for students at the undergraduate and graduate levels in physics and engineering, and for practicing engineers and scientists. Most chapters include exercises with solutions.

Although an in-depth understanding of magnetism requires a quantum mechanical approach, a phenomenological description of the mechanisms involved has been deliberately chosen in most chapters in order for the book to be useful to a wide readership. The emphasis is placed, in the part devoted to the atomic aspects of magnetism, on explaining, rather than attempting to calculate, the mechanisms underlying the exchange interaction and magnetocrystalline anisotropy, which lead to magnetic order, hence to useful materials. This theoretical part is placed, in volume I, between a phenomenological part, introducing magnetic effects at the atomic, mesoscopic and macroscopic levels, and a presentation of magneto-caloric, magneto-elastic, magneto-optical and magneto-transport coupling effects. Volume II, dedicated to magnetic materials and applications of magnetism, deals with permanent magnet (hard) materials, magnetically soft materials for low-frequency applications, then for high-frequency electronics, magnetostrictive materials, superconductors, magnetic thin films and multilayers, and ferrofluids. A chapter is dedicated to magnetic recording. The role of magnetism in magnetic resonance imaging (MRI), and in the earth and the life sciences, is discussed. Finally, a chapter deals with instrumentation for magnetic measurements. Appendices provide tables of magnetic properties, unit conversions, useful formulas, and some figures on the economic place of magnetic materials.

We will appreciate constructive comments and indications on errors from readers, via the web site *http://lab-neel.grenoble.cnrs.fr/magnetism-book*

Acknowledgments

We are grateful for their helpful suggestions to the members of the Reading Committee who worked on the original French edition: V. ARCHAMBAULT (Rhodia-Recherche), E. BURZO (University of Cluj-Napoca, Rumania), I. CAMPBELL (Laboratoire de Physique des Solides, Orsay), F. CLAEYSSEN (CEDRAT, Grenoble), J.M.D. COEY (Trinity College, Dublin), G. COUDERCHON (Imphy Ugine Précision, Imphy), A. FERT (INSA, Toulouse), D. GIVORD (Laboratoire Louis Néel), L. NEEL, Nobel Laureate in Physics (who passed away at the end of 2000), B. RAQUET (INSA, Toulouse), A. RUDI (ECIA, Audincourt), and P. TENAUD (UGIMAG, St-Pierre d'Allevard). The input of many colleagues in Laboratoire Louis Néel or Laboratoire d'Electrotechnique de Grenoble was also invaluable: we are in particular grateful to R. BALLOU, B. CANALS, J. CLEDIERE, O. CUGAT, W. WERNSDORFER. Critical reading of various chapters by A. FONTAINE, R.M. GALERA, P.O. JUBERT, K. MACKAY, C. MEYER, P. MOLLARD, J.P. REBOUILLAT, D. SCHMITT and J. VOIRON helped considerably. Zhang FENG-YUN kindly translated a document from the Chinese, J. TROCCAZ gave helpful advice in biomagnetism, D. FRUCHART, M. HASSLER and P. WOLFERS provided figures, and P. AVERBUCH gave useful advice on the appendix dealing with the economic aspects.

We also would like to thank all our fellow authors for their flawless cooperation in checking the translated version, and often making substantial improvements with respect to the original edition. We are happy to acknowledge the colleagues who, along with the two of us, took part in the translation work: Elisabeth ANNE, Nora DEMPSEY, Ian HEDLEY, Trefor ROBERTS, Ahmet TARI, and Andrew WILLS.

We enjoyed cooperating with Jean BORNAREL, Nicole SAUVAL, Sylvie BORDAGE and Julie RIDARD at Grenoble Sciences, who published the French edition and prepared the present version.

Damien GIGNOUX - Michel SCHLENKER

PHENOMENOLOGICAL APPROACH TO MAGNETISM

CHAPTER 1

MAGNETISM, FROM THE DAWN OF CIVILIZATION TO TODAY

The historical approach of this chapter introduces qualitative ideas which will be reconsidered and analysed in later chapters. The great adventure of magnetism has progressed, slowly at first, following the pioneers who lived long ago at Sumer, in China or in Greece, while the last two centuries have witnessed an explosion in knowledge, investigative techniques, and industrial applications in this model domain of Science. In 1779, the Encyclopædia of Diderot and d'Alembert [1] still said about magnetite:

> *"It is in this metal married with salt and oil*
> *rather than in stony substances*
> *that resides true magnetism"*

We have come a long way in the years that followed... !

1. THE DISCOVERY OF LODESTONE AND THE OBSERVATION OF MAGNETIC PHENOMENA THROUGH THE AGES

The oldest manuscript that mentions the existence of lodestone is the work of a Chinese writer, Guanzhong (died 645 BC), but objects made from magnetic materials have been found in archaeological sites dating from much further back.

1.1. OBJECTS MADE FROM IRON AND LODESTONE IN ANCIENT TIMES

Small tubular beads made from iron of meteoritic origin (containing at least 7.5% nickel) were discovered in many Sumerian and pre-dynastic Egyptian tombs (fourth millenium BC). These appear to be the oldest traces of ferromagnetic objects wrought by human hands. The question remains, however, as to whether their "attractive properties" had been discovered in these distant times. The ancient Egyptians, who called iron *bia–n–pet* (= metal from the sky), did not study the metallurgy of iron until relatively late, and certainly after the Hittites, who did so about 1500 BC; the tomb of Tutankhamon (1340 BC) already contained a dagger and various other objects made of iron and iron ores [2]. Much later, in the temple of Edfu, an inscription mentions "living metal": this was the expression of the ancient Egyptians for lodestone [3]. In Crete, at the palace of Knossos (2000 BC - 1300 BC), the throne room of Minos is paved in the centre with a rectangular flagstone made of iron oxide, consisting mainly of magnetite. Perhaps this was not a random choice, it is possible that the ancient Cretans understood its magnetic properties, which could justify its presence in such a central part of the palace.

1.2. THE FIRST TEXTS DEALING WITH LODESTONE. ORIGIN OF THE NAME "MAGNETISM"

All our information on the origins of magnetism in China were taken from the Chinese work *The History of Electromagnetism* [4] as well as from a recent work by M. Soutif on the origins of Science and Technology in the East [5]. We know from Guanzhong that, in ancient Chinese civilization, lodestones were called "**soft stones**". This is the general name of all strongly magnetic oxides existing in nature and having a permanent magnetic character. They are the magnetic minerals based on iron: γ-Fe_2O_3, FeO-TiO_2-Fe_2O_3, sometimes also FeS_{1+x}, and especially *magnetite* $\mathbf{Fe_3O_4}$. The name "soft stones" is due to their attraction for ferrous metals, in analogy with the tenderness that a mother shows her child.

At the time of the Eastern Han dynasty (25 - 220), Gaoyiu wrote: "Soft Stone is the mother of iron, it can therefore attract its child" [4]. This text leads us to think that iron had already been extracted from magnetic ores. Habits have changed today, magnets are known as "**hard**" magnetic materials as opposed to "**soft**" magnetic materials which demagnetise spontaneously!

In Greece, Aristotle reported that Thales of Miletus (625 BC - 547 BC) knew lodestone, and Onomacritus provides us with the most ancient name known, *magnetes* which evolved into *magnitis*, from which derives the modern term magnetite. Sophocles (495 BC - 406 BC) called lodestone "Lydian rock" while Plato (427 BC - 347 BC) called it "Heraclitian rock" in his work *Tinaeus*. These various names suggest that, in Greco-Roman antiquity, the first magnets were made from ores found at mount Sipylus, close to a town in Asia Minor named "Magnesia ad Sipylum". This town name is the origin of the words "magnetism" and "magnetite". Lucretius confirms in *De Natura Rerum* that the name for lodestone comes from the region where it was extracted. The Latins used the word *sideritis* [1], derived from the Greek word for iron, *sidèros*, which seems to have the same root as the Latin word *sider* meaning *heavenly body*: once again, we come across the belief that iron is of celestial origin, a belief also held by the ancient Egyptians.

1.3. FIRST OBSERVATIONS OF MAGNETIC PHENOMENA

1.3.1. Magnetic attraction

It seems that magnetic attraction was the first magnetic phenomenon to draw the attention of man to "lodestone", since the Chinese name "soft stone" comes from this attraction, as does the name that Hippocrates (460 BC - 377 BC) gave it in *Lib. de sterilib. mulier: the rock that attracts iron*. At the same time, in China, Gui Guzi (400 BC) noticed that rock magnets attracted needles, while Liu An (120 BC) wrote in *Huai Nanzi*: "It is impossible for lodestone to attract tiles as it attracts iron; and the same for copper". Thus, from the very beginning, this mutual attraction was considered as a specific property of iron and its ores.

1.3.2. Magnetic shielding

Saint Augustine (354 - 430) noted that a silver plate could not stop lodestone from attracting iron. Later the question arose whether there existed a material capable of preventing the attraction of a magnet for iron; the *Notes of Guang Yang* gave the reply: "A Ru, son of Liu XianTing (1648 - 1695), had the intuition, and he then experimentally verified, that only iron could act as a shield effective against this attraction". This is the first mention of magnetic shielding.

1.3.3. The compass

The Chinese writings of Gui Guzi and Han Fei (280 BC - 233 BC), showed that the orientation of natural lodestone towards the earth's geographical poles had also been known for a very long time. The oldest "directional tool" known was made in China: it was a natural lodestone sculpted and polished into the form of a spoon. This "directional spoon", the ancestor of the compass, was described by Wang Chong (27 - 97) in *Lun Heng*: "This instrument resembles a spoon, and when it is placed on a plate on the ground, the handle points to the south". On an etching dating from the Han period, we see a spoon placed on a small square tray (fig. 1.1-a), very similar to the trays made from iron-copper alloys, and painted wood which were recently exhumed by archaeologists. It is commonly accepted that this is a painting of a directional instrument.

The interest in these early directional instruments was limited by the rarity in supply of the materials, the need to polish them, and their directional inaccuracy. It was in response to the needs of "geomancers" (commissioned to select the orientation of palaces and towns), soldiers and navigators that artificially magnetised needles, the directional fish, and then compasses evolved.

Zeng Gongliang described the fabrication of a *directional fish* in his *Wu Jing Zong Yao* (1044): "We cut a very thin piece of iron into the shape of a fish... We make it red hot in a coal fire, and then retrieve it with tongs. The tail remaining oriented towards the north, we quench it in water for a few minutes". This process constitutes a quench under magnetic field (the earth's field), which reinforces the magnetism and magnetic remanence of the directional instrument –i.e. the memory effect which allows it to remain magnetised for years. The long and pointed shape increases the directional precision and decreases the risk of demagnetization (loss of magnetism). The slightly raised edges allow the instrument to float (fig. 1.1-b).

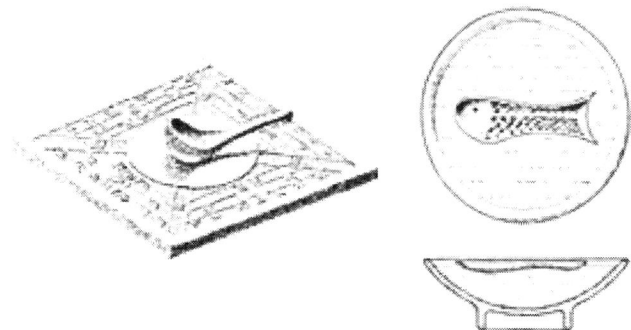

Figure 1.1 - (a) A "directional spoon" - (b) A "directional fish"
(drawing by L. Mouveau)

Meng Xi Bi Tan, written in 1086 by Shen Kuo, contains the description of the *first real compasses*: "When we polish the tip of a needle (made of iron) with a lodestone, it points towards the south but slightly towards the east, not really south". This discovery, and the realization that the magnetic pole is slightly misoriented with respect to the geographic pole, were essential for the use of compasses. The first compasses were suspended with silk wires. Finally, it was the "floating compass" (a magnetised needle carried by a floater on water) that was developed for maritime navigation applications (only from 1099 - 1102).

The first Westerners to publish anything but myths in the field of magnetism were Pierre Pelerin de Maricourt, who revealed his knowledge in a celebrated letter entitled *De Magnete* (1269) in which he introduced the notion of magnetic poles. He was followed much later by an English physician, William Gilbert, who wrote a book also entitled *De Magnete* in 1600.

1.3.4. *Knowledge of magnetism in 1779*

The Encyclopaedia of Denis Diderot and Jean Lerond d'Alembert summarised the knowledge of magnetism of that time under three entries: magnetic poles, attraction, and transmission of magnetism; the absence of any serious theory of magnetism was also noted [1].

Poles

It is possible to observe magnetic field lines with the aid of iron filings: *a magnetised bar has two poles* where the field lines concentrate (fig. 1.2). The poles of a magnet may be modified with a more powerful magnet. A broken magnet may remain magnetised. It should be noted that the geographic north pole (boreal pole) is magnetically a south pole, and that the geographic south pole (austral pole) is magnetically a north pole. Strictly speaking, however, the axis of the geographic poles does not exactly coincide with that of the magnetic poles: their shift, called magnetic *declination*, varies with time (see chap. 24, § 6).

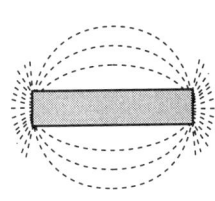
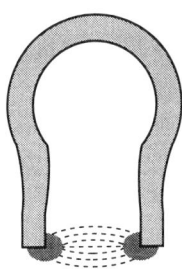

Figure 1.2
The magnetic field lines of a magnet

Attraction

Opposite poles attract, and like poles repel. Mitchell thought that he had observed, with small magnets, that the mutual repulsive force between two like poles is less than the attractive force between two opposite poles separated by the same distance. The attraction between iron and a magnet is greater than the attraction between two lodestones. "Excessive heat" is detrimental to the "properties" of a magnet, which are restored upon cooling. Lodestones can be used more efficiently if they are fitted with an appropriate iron structure.

Transmission of magnetism

It has been known since antiquity that lodestone can transmit its magnetism to a piece of iron just by bringing the two materials close to each other: this can lead to permanent magnetism. Iron rods become magnetised when they remain a long time in a vertical position, e.g. the belfry crosses of Chartres, Delft and Marseilles have been magnetised by the vertical component of the earth's magnetic field. Lightning magnetises steel; a nail can also be magnetised by firstly making it red hot in a fire, and then cooling it when oriented in the north-south direction: this agrees with the Chinese observations of the 11th century. In the same way, impacts can modify the magnetic state of a piece of iron.

Theory of magnetism

The Encyclopaedists had to face the fact that all attempts at interpreting magnetism (including the efforts of the mathematician Bernoulli) were unsuccessful, but they concluded with optimism that their children or grandchildren will no doubt be able to explain all these phenomena, and *perhaps even relate them to the phenomena of electricity*. Good foresight!

2. THE CONTRIBUTION OF THE 19TH CENTURY

Repeating the experiment of Mitchell, Charles de Coulomb demonstrated his famous $1/r^2$ interaction law between "magnetic charges" by using very long magnets, the further magnetic poles of which were sufficiently separated that they could be neglected. This discovery, in 1795, led to a great leap forward in the knowledge of magnetism. Shortly afterwards, Denis Poisson (1781 - 1840) introduced the notion of a *magnetic field*, as the force acting on a unit magnetic charge at all points in space. However, in contrast to electric charges, magnetic charges seem to only exist in pairs, constituting "magnetic dipoles", i.e. pairs of equal magnetic charges with opposite polarity.

Note - *Paul Dirac showed that the existence of "magnetic monopoles" could explain the quantization of electric charge, which always appears as an integer multiple of the charge of the electron, but to this day, all attempts to isolate a magnetic monopole have failed* [6].

In 1820, the Danish physicist H. C. Oersted showed that a magnetic field could be created by the circulation of an electric current in a conducting wire (fig. 1.3). In the weeks that followed, André-Marie Ampère repeated this historic experiment, and succeeded in giving it an elegant formulation. He showed that a current carrying circular coil generates a magnetic field identical to that created by a magnetic dipole, and suggested that *the magnetism of the matter could be due to small electrical current loops circulating on the molecular scale.*

Then began the famous controversy between the advocates of the Coulombian approach (description based on magnetic poles) and the Ampérian approach (description based on currents).

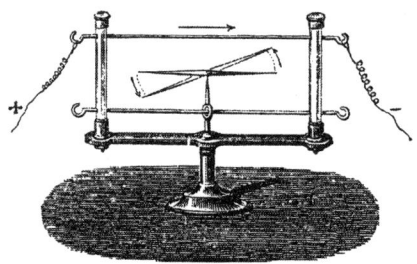

Figure 1.3 - The historic experiment of Oersted after [7]

Today, these are taken as complementary approaches: the Amperian approach is especially useful in atomic physics, while the Coulombian approach is useful in electrical engineering. Moreover, in both cases, the *magnetic property* of a body is defined by the same quantity, its *magnetic moment*, which is proportional to the torque (the *moment* of the mechanical forces) which it experiences in a uniform magnetic field.

In 1821, the phenomenon of induction was discovered by the English physicist Michael Faraday (1791 - 1867), who constructed the first dynamo. Pierre-Simon de Laplace (1749 - 1827) established the expression for the force acting on a current carrying conductor in a magnetic field.

These discoveries caused a storm in the world of physics by unifying the electric and magnetic interactions, and they gave rise to endless discussions on the choice of unit systems. The system of "electrostatic" CGS units or e.s.u. (which defines the unit of electric charge), and the system of "electromagnetic" CGS units or e.m.u. (which defines the unit of magnetic charge) were compatible as long as electric and magnetic phenomena were totally independent. This no longer held when it became clear that a magnetic field could be created either by a magnetic dipole or by moving electric charges.

The equations of electromagnetism, deduced by the Scottish physicist James Clerk Maxwell in 1864, crowned the efforts of half a century, and remain today the analytical basis of magnetism.

Faraday was the first to exploit the motive action of a magnetic field gradient to *measure the magnetism of substances* when he constructed the balance which bears his name to study the magnetic properties of materials. He observed that the application of a magnetic field could induce magnetic effects in all substances, but that, very often, these effects were smaller by many orders of magnitude than those observed in iron.

For certain so called "diamagnetic" substances this very weak *induced magnetism* opposes the magnetic field which creates it, according to Lenz's law (the magnet repels the substance), whereas for "paramagnetic" substances, the situation is reversed: the magnet attracts the substance as in the case of iron. This last result appeared paradoxical, until it was realised that spontaneous magnetism could exist in matter.

William Sturgeon (1783 - 1850) built the first electromagnet in 1824. By adding an iron core to a solenoid, he was able to create a "very intense" magnetic field, exploiting the strong magnetism induced in the core, without having to pass a very high electric current: electrical engineering was born.

3. MAGNETISM IN THE 20TH CENTURY

One would have to wait until the end of the 19th and beginning of the 20th centuries before successful theoretical studies of magnetic material would appear. Pierre Curie (1859 - 1906) introduced or clarified the ideas on diamagnetism, paramagnetism and ferromagnetism, and Paul Langevin (1872 - 1946) dealt with induced and permanent magnetism; the latter also established the classical statistical theory of paramagnetism.

Decisive progress was made in the classical description of the magnetism of solids by Pierre Weiss (1865 - 1940), an unrivalled experimentalist and the creator of the *molecular field* hypotheses, as well as by Louis Néel, who was awarded the Nobel prize in physics in 1970 especially for his theory of *antiferromagnetism* (1936) and *ferrimagnetism* (1948). We note that the first unequivocal antiferromagnet, MnO, was discovered by Bizette and Tsaï one year after Louis Néel had published his theory of antiferromagnetism (the name antiferromagnetism was proposed by Francis Bitter in 1939, i.e. three years after Néel proposed his interpretation of *constant paramagnetism*).

It remained to explain the origin of the giant "molecular fields" observed in strongly magnetic substances. The magnetic dipole interaction between atomic magnetic moments was quite insufficient to explain their existence, and could explain magnetic ordering at very low temperatures only. At this stage quantum mechanics made a fundamental contribution to magnetism with the discovery in 1925, by George Uhlenbeck and Samuel Goudsmit, of an "intrinsic angular momentum" of the electron which was called *spin*. In extending the theoretical studies of the electron, Paul Dirac confirmed the existence of spin in 1927, then in 1929 Werner Heisenberg showed that the strong magnetic interactions responsible for magnetic order were of electrostatic origin, and could be interpreted in terms of coupling between two neighbouring spins.

Thus a solid basis had been created for the study of the magnetic properties of matter. It resulted in an explosive output of scientific literature in this area, supported by the spectacular progress made in experimental methods, most notably in neutron diffraction.

Finally, let us note that atomic nuclei also carry magnetic moments: this nuclear magnetism is three orders of magnitude smaller than electronic magnetism, and can thus be neglected as long as we are interested only in the static and quasi-static properties of condensed matter at not very low temperature. However, one very important application is based on nuclear magnetism: this is magnetic resonance imaging (MRI), which will be covered in chapter 23.

Superconducting materials have zero electrical resistance below a so-called critical temperature. They deserve to be mentioned here because a very strong diamagnetic character (*magnetic susceptibility* $\chi = -1$) is associated with this very strange electrical transport behavior (which allows very intense magnetic fields to be created). Their magnetic behavior is opposite to that of ferromagnetic materials, as can be easily observed: a superconducting needle placed in the airgap of a magnet orients perpendicular to the flux lines, while a ferromagnetic needle ($\chi \gg 1$) orients parallel to the flux lines. Similarly, a magnet suspended on a wire is repelled by a superconducting plate and is attracted by a ferromagnetic plate, while a copper plate (also diamagnetic, but with a very low susceptibility, $\chi \sim -10^{-5}$) does not influence the magnet in any perceptible way, as shown in figure 1.4.

The discovery, in 1986, of high temperature superconductors rejuvenated the research into this family of materials which had started with the discovery by Kamerling Onnes, in 1911, of the superconductivity of mercury. The present record for critical temperature is held by $HgBaCa_2Cu_3O_{8+\delta}$ (T_c = 133 K). It is not possible to discuss superconductors without mentioning *SQUIDS*, devices which exploit quantum interference in superconducting junctions, and are used more and more for ultra-sensitive magnetometry.

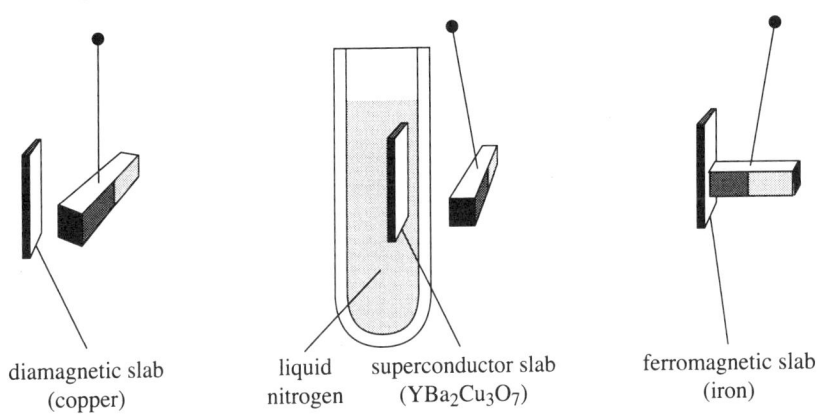

Figure 1.4 - Reaction of a magnet in the presence of diamagnetic, superconducting and ferromagnetic materials

We conclude here this historic introduction to the science of Magnetism; the interested reader should consult the excellent study by Mattis who, in his first chapter, adds a philosophical analysis of Greek, medieval and modern thoughts on science and especially on magnetism to the purely historical facts [8].

4. MAGNETISM AND TECHNOLOGY

Parallel to fundamental research –the objective of which was to explain the origin and basic mechanisms of magnetism– a very active *applied research* was developed to optimise the performance of magnetic materials for industrial applications:
- magnets, which are made of "hard magnetic" materials, the magnetic state of which is difficult to alter, and which are characterised by a reasonable value of remanence (they conserve most of the *magnetic induction **B*** conferred on them by the application of an intense magnetic field). The *magnetic field **H*** needed to demagnetise a magnet (the "coercive field") should also be as high as possible: we thus speak of *high coercivity* materials;
- transformer cores ("soft magnetic" materials) for which, contrary to hard magnetic materials, the lowest possible coercivity is sought;
- magnetic memory materials which must have a high remanence with a moderate coercivity in order to minimise the energy required for writing. Without doubt, this is the area of magnetism in which most progress has been made in the second half of the 20th century;
- materials for electronics which must operate at very high frequencies;

- materials for robotics (magnetic sensors and actuators);
- other applications which should also be mentioned, which are of military interest, include magnetic mines and their corollaries, ship demagnetisers, sonar emitters-receivers: all of these were intensely studied during the second World War.

Technological research is expanding to include new concepts, new applications as well as new working methods; e.g. numerical simulations are used more and more to optimise devices or electrical and magnetic parameters, as well as mechanical and thermal parameters.

Some salient facts highlight the technical progress made by magneticians during the second half of the 20th century.

Magnetic fields

It is possible to create static magnetic fields of 3×10^7 A.m^{-1} in a diameter of 3.2 cm with the use of hybrid coils using superconducting materials and copper conducting coils, though admittedly these installations consume very much electrical energy. Transient fields reaching 5×10^8 A.m^{-1} during 4 μs have been produced in a diameter of 1 cm by implosion techniques [9], which are, however, not easy to develop in a laboratory. Aimé Cotton had already passed 5 MA.m^{-1} (7 teslas) in 1928 with a large electromagnet; a very impressive feat for the time.

Spectacular progress has been recently made in devices for the production of high magnetic fields with low energy consumption, which are easy to operate in standard laboratories: thus, it is possible to create fields of the order of 3×10^7 A.m^{-1} in minuscule coils by simply using the discharge from a bank of capacitors (fig. 1.5). Such coils are suitable for making measurements on the micrometer scale, a scale which is compatible with today's nanostructured materials.

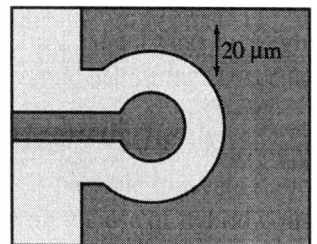

A single coil loop which allows high magnetic fields to be created over a small volume in a laboratory. The duration of the pulse field is relatively short (30 ns), but sufficient to characterise the properties of micron-sized magnetic samples by magneto-optical techniques. The energy needed to operate such a microcoil does not require the large installations needed to generate high static fields: the discharge of a single capacitor into such a coil is sufficient to supply the required energy.

Figure 1.5 - Microcoil capable of creating a 40 T magnetic field

At the other end of the scale, the SQUIDs used in magneto-encephalography can detect transient magnetic fields as low as 5×10^{-9} A.m^{-1} emitted by the brain (an induction of 7 fT in a band width of 1 Hz could be detected at Chiba in Japan in 1993) [10].

Thus man works with magnetic field intensities spanning a scale of 17 decades (fig. 1.6); for reference, the earth's magnetic field is of the order of a few tens of amperes per meter.

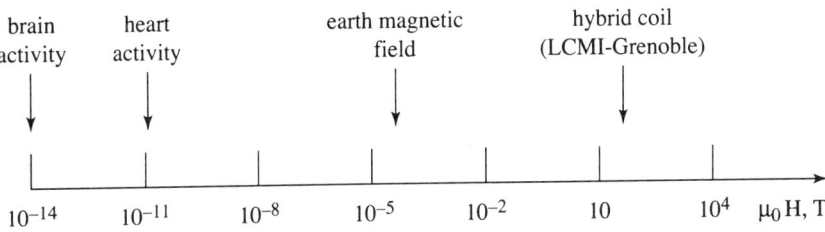

Figure 1.6 - Range of experimentally accessible magnetic fields

Permanent magnets

Our grandmothers used only horse-shoe shaped magnets to collect their sewing needles, as magnets of the day had a tendency to spontaneously demagnetise, and this specific shape led to maximum "flux line" closure, and thus minimised the risk of demagnetization (see fig. 1.2).

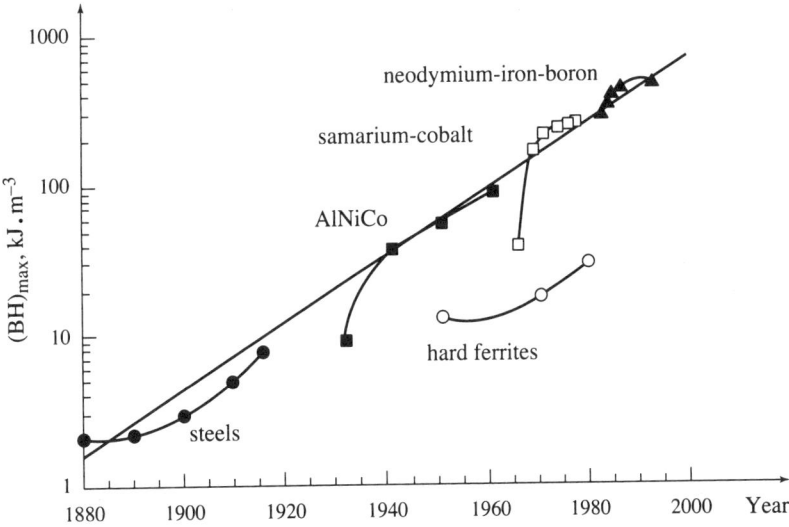

Figure 1.7 - Evolution in the performance of permanent magnets as characterised by their energy products

Logarithm of the maximum energy product ($[BH]_{max}$) expressed in $kJ.m^{-3}$ and plotted as a function of the date of their arrival on the permanent magnet market for different materials belonging to the five principal industrial magnet families, after [11]. The concept of induction, magnetic field and energy product will be discussed in the following chapter.

Today's Nd-Fe-B magnets can remain magnetised in thin sheet form, even when magnetised perpendicular to the plane of the sheet. The progress made over 150 years in permanent magnet fabrication is spectacular (see fig. 1.7), and can be quantified by the growth in energy product: Nd-Fe-B magnets have an energy product which is 1,000 greater than that of ancient magnetite magnets!

New applications

Classical electrical engineering devices (soft iron cores, permanent magnets) have progressed much, and new applications have also appeared, e.g. radio- and ultra high frequency applications for ferrites, as well as sensors and actuators. Nevertheless, it is magnetic recording which has caused the greatest stir: in fifty years, this domain of activity has hoisted itself up to be the number one user of magnetic materials, with a market share surpassing that of all other applications together.

The race towards ultra high density recording has a long way to go as shown in figure 1.8, where two curves are shown. One is a plot of commercial IBM products while the second is a plot of IBM demonstration prototypes. Technological research is flourishing in the domain of storage media (hard disks, mini-disks, super-mini-disks, high density magnetic tapes...) as well as in the area of read-write heads (magnetoresistive heads, giant magnetoresistive heads...).

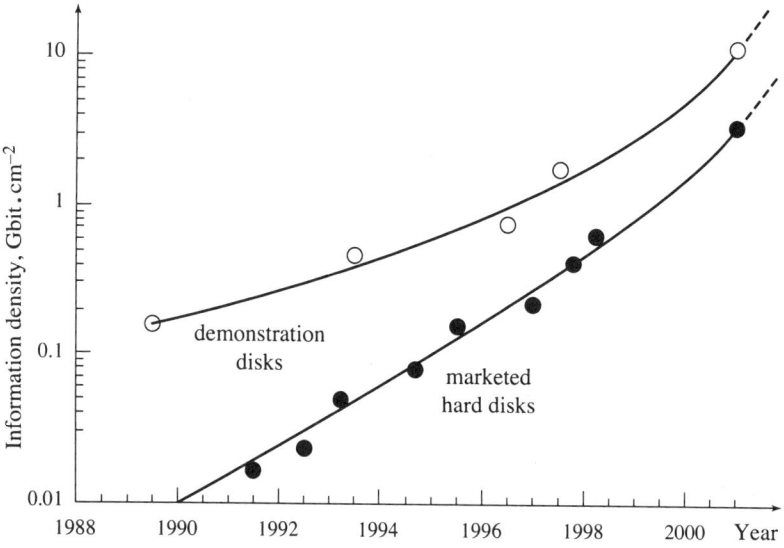

Figure 1.8 - Progress made over the last decade in hard disk magnetic recording density (IBM products)

5. NEW RESEARCH LINES

Magnetism is a very diversified field of study, covering at the same time paleomagnetism, i.e. the magnetism of rocks, biomagnetism (note that in ancient China magnetic rocks were used to treat ear-aches!), and now encompassing the study of new materials which are more ordered (single crystals) or more disordered (fine particles, amorphous ribbons and wires, quasi-crystals, ferrofluids, spin glasses), low dimensionality systems (thin films), etc.

5.1. MAGNETIC NANOSTRUCTURES

The study of magnetic nanostructures is a multi-disciplinary domain, involving a large variety of preparation techniques: mesostructures are prepared by lithography, which allows control of the geometry and organization of particles on the micron scale; on a finer scale, new fabrication techniques have been introduced: point probe techniques (lithography under a scanning microscope, nanoindentation, growth under a point probe, atom displacement), creation of nanowires by the bombardment of a matrix with heavy ions, nanostructure deposition with a focused atom jet, sol-gel precipitation. The smallest nanoparticles are constituted of a collection of a few tens of atoms. The magnetic properties of these nanoparticles can be quite surprising, as it is possible to observe finite size effects, both classical and quantum. The study of these systems has been made possible by the very recent development of microSQUID sensors, an example of which is shown in figure 1.9.

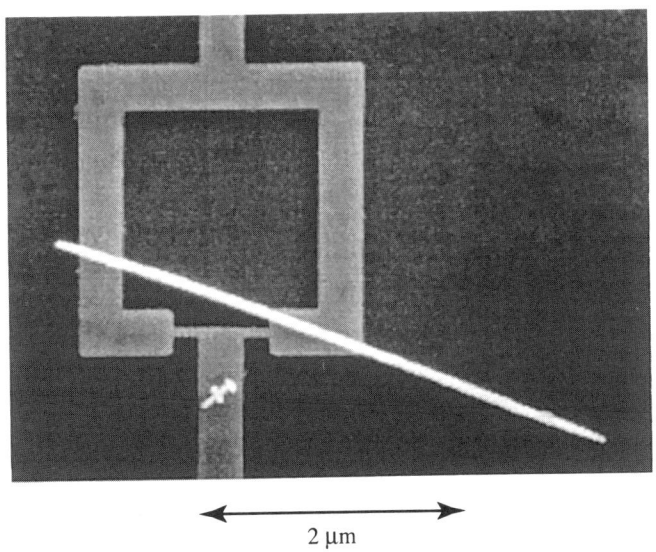

2 µm

Figure 1.9 - MicroSQUID prepared at L2M-CNRS, Bagneux by D. Mailly

The outer dimension of the square is 2 µm. The sample to be measured is a nickel wire 60 nm in diameter. Courtesy of W. Wernsdorfer (thesis, 1996).

5.2. MAGNETIC MULTILAYERS

The possibility to prepare very thin and perfectly structured films has given a great boost to the study of a new type of magnetic structure, namely "multilayer" structures. An example of such a new structure is the "sandwich" structure consisting of hard and soft magnetic layers: these "spring magnets" have original properties which cannot be induced in traditional materials. This is a very promising line of research.

5.3. MOLECULAR MAGNETISM

Molecular magnetism involves the study of not only simple molecules but complex molecular systems also: bistable spin transition systems, high spin molecules, molecular magnets as well as magnets made from molecular precursors, supramolecular magnetic structures. This area is also necessarily multi-disciplinary, since it relies on to close collaboration between magneticians and organic chemists.

5.4. OTHER LINES OF RESEARCH

Finally we mention the revival of old topics, due to recent technological progress in preparation and characterization techniques (e.g. magnetic force microscopy, magneto-optics, dichroism, synchrotron radiation, very high magnetic field facilities).

This revival of interest is particularly impressive in the area of magnetic oxides, and was stimulated by the discovery of high temperature superconducting oxides. Similarly, the discovery of giant magnetoresistance in (La-RE)MnO_3-type perovskites has opened the door to applications in the totally new domain of *spin electronics*.

And it is not finished. This is why we hope that this book will become a daily reference for those researchers and engineers who will tomorrow carry the torch and live the fascinating experience of developing new magnetic materials or probing the fundamentals of magnetism.

Nevertheless, writing an exhaustive book on magnetism is totally unthinkable today. We have thus chosen to present in the following chapters an *introduction* to the *general phenomena* as well as to *today's most used materials,* and invite the reader to refer to other textbooks of magnetism for those subjects which we have not developed. There is an impressive selection of works available, starting from Livingston, who presents the physics of magnetism without any formulae [12], and culminating in the encyclopædic 10 volume series edited by Wohlfarth and Buschow [13]. Between these two extremes we recommend the works of Morrish [14], Cullity (emphasis on magnetic materials) [15], Watson (emphasis on applications of magnetism) [16], and Chikazumi (the original 1964 edition was republished in 1997) [17]. A general bibliography on magnetism is presented at the end of the book.

REFERENCES

[1] D. DIDEROT, J. LEROND D'ALEMBERT, Encyclopédie des Sciences, des Arts et des Métiers (1779) Imprimerie de la Société, Livourne.

[2] J. LECLANT, Le fer dans l'Egypte ancienne, le Soudan et l'Afrique, *in* Colloque International "Le fer à travers les âges", Nancy, 3-6 octobre 1955 (Ann. de l'Est, Fac. des Lettres de Nancy, mémoire n°16, 1956).

[3] S. AUFRÈRE, L'univers minéral dans la pensée égyptienne, Vol. 1 (1991) 235, Institut français d'Archéologie Orientale, Le Caire.

[4] SONG DESHENG, LI GUODONG, History of Electromagnetism: Observation and Utilization of electrical and magnetic Phenomena (1987) Popular Press Publishers, Guang Xi, China (*in Chinese*).

[5] M. SOUTIF, L'Orient, Source de Sciences et de Techniques (1995) Presses Universitaires de Grenoble.

[6] P.A.M. DIRAC, *Phys. Rev.* **74** (1948) 7817.

[7] R. RADAU, Le Magnétisme (1881) 219, Hachette, Paris.

[8] C. D. MATTIS, Theory of Magnetism (1965) Harper & Row, New York.

[9] G. AUBERT, G. MARTINEZ, P. WYDER, Proc. 5th Intern. Conf. on Large Magnetic Field Facilities in Physics (1994) 265-272, Jacob & Schopper, Lausanne.

[10] K. OKAJIMA, T. KOMURO, N. HARADA, A. ADACHI, M. UEDA, A. KANDORI, G. UEHARA, H. KADO, *Proc. IEEE Conf. Record Nuclear Science Symposium & Medical Imaging Conf.* **3** (1993) 194.

[11] J. F. HERBST, J. J. CROAT, *J. Magn. Magn. Mater.* **100** (1991) 57-78.

[12] J.D. LIVINGSTON, Driving Force (1996) Harvard University Press.

[13] E.P. WOHLFARTH, K.H.J. BUSCHOW Eds, Ferromagnetic materials 10 volumes to date (1985, ...) North Holland, Amsterdam.

[14] A.H. MORRISH, The physical principles of magnetism (1965) J. Wiley & Sons, New York.

[15] B.D. CULLITY, Introduction to magnetic materials (1972) Addison-Wesley, Reading.

[16] J.K. WATSON, Applications of magnetism (1980) J. Wiley & Sons, New York.

[17] S. CHIKAZUMI, Physics of Ferromagnetism (1997) Clarendon Press, Oxford.
New revised and corrected edition of the celebrated "Physics of Magnetism" by the same author, first published in 1964.

CHAPTER 2

MAGNETOSTATICS

*This chapter is split into two sections. Only the first of these is required in order to understand the rest of the material in this book as it is here that the fundamental concepts of magnetism (magnetic induction **B**, magnetic moment **m**, magnetization **M**, magnetic field **H**, and magnetic susceptibility χ) are introduced. This section also deals with the determination of magnetic induction occurring in systems where both currents, and materials with a known magnetization **M(H)** are present.*

The second section tackles two delicate problems concerning energy, forces, and magnetic torque. The magnetic energy associated with a current is accounted for by the concept of electromagnetic coupling, whereas that which is associated with the magnetization of the material has the peculiarity of depending upon the shape of the material. The total energy of the most general magnetic system (consisting of both currents, and magnetised material) is the sum of two independent terms, the energy associated with the magnetization of the material (in the absence of currents), and the energy associated with the currents in vacuum (in the absence of magnetization). The forces and the torques within a magnetic system can be calculated either by applying Laplace's equation (or equations deriving from it) directly, or by calculating the energy involved using the theorem of virtual work –without forgetting the work done by the induced electric field.

1. MAGNETOSTATICS OF CURRENTS AND MATERIALS

1.1. MAGNETOSTATICS OF CURRENTS IN VACUUM

As described in the first chapter, the historical description of magnets was based on the idea of magnetic charges. Contrary to this *Coulombian approach*, we introduce here magnetostatics from the point of view of electric currents, the *Ampérian approach*, which corresponds better to the modern perspective on magnetism. We make the assumption that the reader is familiar with electric current, and we will only recall some of the key results, whose derivations can be found in one of the many text books on electromagnetism and magnetostatics [1, 2, 3].

1.1.1. *Fundamental laws of magnetostatics: magnetic induction*

In the literature one often comes across two terms, magnetic induction, and magnetic field, to describe the same quantity **B**. The International Electrotechnical Commission recommends calling **B** the magnetic induction, and **H** the magnetic field. In this text, we also use the terms "**B** – field", and "**H** – field" to denote these two quantities. We will see later that the difference between the two, apart from the trivial difference in dimensional character (Tesla *vs* $A.m^{-1}$), is only important within magnetised material.

The idea of magnetic induction arises naturally when one studies the forces between current carrying conductors. An example of this is the origin of the force and torque that are applied to each other by two closed and rigid wire circuits, C_1 and C_2, which pass respectively the currents I_1 and I_2. It is handy to imagine that each of these circuits creates a vector field **B(r)** that is called the magnetic induction. The force and torque experienced by one of these circuits are considered to be the effects of the induction created by the other, and not the result of a direct interaction between the two circuits.

This step is completely analogous to electrostatics where the electric field vector, **E**, is defined as the intermediary for the forces of attraction or repulsion between charges. Numerous arguments, but notably the fact that the same electric field or magnetic induction can be created using different methods, and still produce the same effects, justify this statement.

With this in mind, and to explain the phenomena observed in magnetostatics, we will state two fundamental laws. These laws of magnetostatics are then compared to the analogous laws of electrostatics.

The Biot-Savart law

This gives the expression of the magnetic induction due to a wire-like circuit (C) carrying a current I:

$$\mathbf{B} = \frac{\mu_0}{4\pi} \int_C \frac{I d\mathbf{l}}{r^2} \times \frac{\mathbf{r}}{r} \tag{2.1}$$

μ_0, the permeability of vacuum, is a universal constant equal to $4\pi \times 10^{-7}$ $N.A^{-2}$; **r** is the vector between a point P in the circuit (C) and the point M where the induction **B** is observed, and r is its modulus. d**l** is an infinitesimal element of the circuit (C) at point P, and the vector d**l** points in the direction of the current I at this point.

This law applies in general not only to threadlike circuits, but also to continuous volume distributions of current density **j** (expressed in $A.m^{-2}$): in this case the expression for the magnetic induction becomes:

$$\mathbf{B} = \frac{\mu_0}{4\pi} \int_v \frac{\mathbf{j} dV}{r^2} \times \frac{\mathbf{r}}{r} \tag{2.2}$$

2 - MAGNETOSTATICS

Laplace's law

This expresses the force, d**F**, suffered by a line element of circuit, d**l**, carrying a current I, in the presence of an induction **B**:

$$d\mathbf{F} = I\, d\mathbf{l} \times \mathbf{B} \tag{2.3}$$

The magnetic induction is expressed in tesla (T) (1T = 1 N.A^{-1}.m^{-1}).

Comparison with electrostatics

The equivalent of equation (2.2) in electrostatics is that for the electric field created by a distribution of charges with density ρ:

$$\mathbf{E} = (4\pi\varepsilon_0)^{-1} \int_V \rho dV\, \mathbf{r}/r^3$$

with $\varepsilon_0 = 1/\mu_0 c^2$, and c being the speed of light in vacuum. equation (2.3) is similar to the expression for the force applied by an electric field on a charge dq: d**F** = dq **E**.

These equations lead one to thinking in terms of the current element I d**l** (or **j**dV) as the elementary object of magnetostatics, in analogy with the point charge dq = ρdV, the elementary object of electrostatics. Despite its convenience in the formalism, an isolated current element has (unlike the electric charge) no physical reality because it does not satisfy the law of conservation of charge, at least not in the DC regime.

Note on the nature of the vector B - *The definition of the induction (eq. 2.1 or 2.2) shows that it does not obey the same rules of symmetry as those for the vector current density j, force F or electric field E, which are polar vectors. In fact B behaves in the same way as the vector angular velocity ω defining the rotation of an object. Such vectors are known as axial vectors or pseudo-vectors, and are in actual fact tensors. This observation is important when performing calculations of the induction created by a given current distribution.*

It is often handy to make use of the symmetry properties of current distributions. A practical rule to bear in mind is that **B** transforms like an oriented closed loop (fig. 2.1). **B** will remain invariant under inversion with respect to a centre. Mirror symmetry leaves the component of **B** perpendicular to the plane unchanged, whereas the tangential component will change sign.

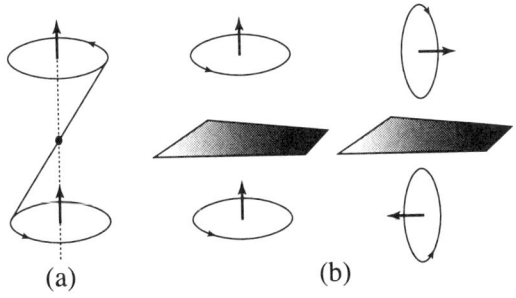

Figure 2.1 - Symmetry transformations of the vector B on inversion with respect to a point (a), and in mirror reflection with respect to a plane (b)

1.1.2. Ampère's law. Magnetic field

Ampère's name remains associated with the elegant mathematical interpretation that he gave to the historic experiment of Oersted. We consider a surface (S) bounded by a closed curve (Γ), and a closed threadlike circuit (C) carrying an electric current I that gives rise to an induction **B**. Starting from the definition in (2.1) it can be shown that two cases arise:

- the circuit C does not cross the surface (S), then:

$$\oint_\Gamma \mathbf{B}\,d\mathbf{l} = 0 \tag{2.4}$$

- the circuit C crosses (S) once. In this case the line integral of **B** is written as:

$$\oint_\Gamma \mathbf{B}\,d\mathbf{l} = \mu_0 I \tag{2.5}$$

Having chosen the positive sense of the circulation along Γ, the electric current is counted as being positive if it runs in the direction of the normal **n**, oriented conforming to the usual rule. An observer standing on the surface, and seeing the vector **n** coming out from his feet towards his head will see the circulation turning in the trigonometrically direct sense (counterclockwise) (fig. 2.2).

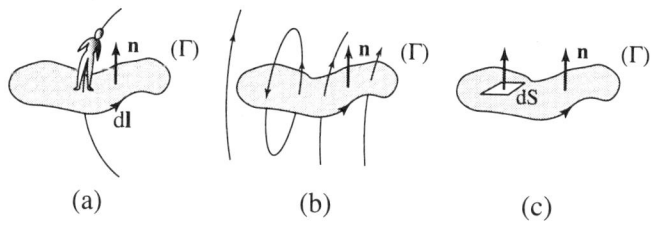

(a) (b) (c)

Figure 2.2 - Ampère's law

*(a) A single threadlike circuit - (b) Several threadlike circuits. The arrows indicate the sense of the currents - (c) Volume distribution of current density **j**.*

Let's define a new vector $\mathbf{H} = \mathbf{B}/\mu_0$ which we will call the magnetic field. equation (2.5) now becomes:

$$\oint_\Gamma \mathbf{H}\,d\mathbf{l} = I \tag{2.6}$$

which represents the standard form of Ampère's law. This definition of magnetic field is valid only in vacuum. It will be generalised later on for the case of magnetised material, and then it will become really useful. The unit of magnetic field is an ampere per metre ($A.m^{-1}$).

Equation (2.6) is valid in general for any type of wire-like current. I is simply the algebraic sum of all of the currents that thread (S). The field created by a continuous distribution of volume currents, with density **j**, is obtained similarly:

$$\oint_\Gamma \mathbf{H}\,d\mathbf{l} = \int_S \mathbf{j}\,\hat{\mathbf{n}}\,dS \tag{2.7}$$

2 - MAGNETOSTATICS

where $\hat{\mathbf{n}}$ is the unit vector normal to the surface (S) following the same sign convention as before. The integral in equation (2.7) can be written in differential form as equation (2.8) using Stokes' theorem. This form is often referred to as the Ampère-Maxwell equation:

$$\mathbf{curl\ H} = \mathbf{j}, \quad \text{or} \quad \mathbf{curl\ B} = \mu_0 \mathbf{j} \tag{2.8}$$

1.1.3. Flux conservation. The vector potential

Consider a closed volume V bounded by a surface (S). It can be shown, using equations (2.1), and (2.2), that the flux of **B** coming out of (V) through (S), Φ, is zero.

$$\int_S \mathbf{B}\hat{\mathbf{n}} dS = 0 \tag{2.9}$$

Equation (2.9) also implies that the flux entering into (V) is equal to the flux leaving (V), in other words that the flux of **B** is conserved. Equation (2.9) can also be expressed in differential form:

$$\text{div } \mathbf{B} = 0 \tag{2.10}$$

Note - *This property is already true for the induction $d\mathbf{B} = \mu_0 I\, d\mathbf{l} \times \mathbf{r}/4\pi r^3$ created by a current element. It also remains true for circuits that are not closed (in contrast to Ampère's law).*

Equation (2.10) signifies that **B** derives from a vector potential **A** such that:

$$\mathbf{B} = \mathbf{curl\ A} \tag{2.11}$$

where **A** can be determined in two different ways in the presence of a given set of currents,

• from the expression deriving directly from the definition (2.2) of **B**:

$$\mathbf{A} = \frac{\mu_0}{4\pi} \int_V \frac{\mathbf{j}}{r} dV \tag{2.12}$$

A is thus a polar vector in the same way as **j**.

• from the following equation deduced from (2.8):

$$\mathbf{curl\,(curl\ A)} = \mu_0 \mathbf{j} \tag{2.13}$$

In case (b), **A** is obtained from partial differential equations which must be solved taking into account the boundary conditions (discussed in § 1.4).

In practice, the choice between the two different approaches given above depends upon the geometry of the problem. As an example, in the case where the symmetry allows a circuit to be chosen (generally circular) along which |**B**| and **B.dl** are constant, one should use Ampère's law. In the case of a circular current loop, or a cylindrical set of windings, one should use the Biot-Savart law. In other situations one should integrate equation (2.13).

1.1.4. Boundary conditions for B and A across a current sheet

A current sheet is a surface upon which a surface current density $\mathbf{j_s}$ exists (expressed in A.m^{-1}). Such a surface corresponds to a singularity of the volume current density $\mathbf{j}$. We can easily check by taking equations (2.12), and (2.5) to the limit of localised distributions, or more directly by equation (2.10), that **A** and **B** satisfy the following conditions on a current sheet:
- the vector potential **A** is continuous,
- the component of **B** perpendicular to the surface is continuous, which can be written as:

$$\hat{\mathbf{n}}_{12} \mathbf{B}_1 = \hat{\mathbf{n}}_{12} \mathbf{B}_2 \qquad (2.14)$$

where $\mathbf{B}_1$ and $\mathbf{B}_2$ are the inductions in the media 1 and 2 respectively, and $\hat{\mathbf{n}}_{12}$ is the unit vector perpendicular to the surface which points from medium 1 to medium 2,
- the tangential component of **B** is discontinuous such that:

$$\mathbf{B}_{T2} - \mathbf{B}_{T1} = \mu_0 (\mathbf{j_s} \times \mathbf{n}_{12}) \qquad (2.15)$$

where $\mathbf{B}_{T1}$ and $\mathbf{B}_{T2}$ are the components of **B** tangential to the surface.

1.1.5. Induction and field produced by a thread-like circuit. The field coefficient of a coil

The induction and field created at a point P by the most general coil, in which a current I is flowing, can be written (as a result of eq. 2.1):

$$\mathbf{B} = \mathbf{C_B} I, \quad \text{and} \quad \mathbf{H} = \mathbf{C_H} I \qquad (2.16)$$

where $\mathbf{C_B}$, and $\mathbf{C_H}$ depend only on the geometry of the circuit and the position of the point P relative to it. $\mathbf{C_H}$ is often called the (vectorial) field coefficient. These two coefficients will be useful in the following section.

1.1.6. Some simple current distributions

After the very general presentation given in the preceding sections, we will now tackle some concrete and useful examples of field and induction calculations for some simple current distributions.

An infinite, straight, wire-like current

The current I is along the z-axis in the positive sense. Let r, θ, and z be the cylindrical polar coordinates of a point. The system is invariant with respect to translation along z, and rotation around the z-axis. There is also a symmetry plane through the origin, perpendicular to the z-axis. This results, taking the transformation rules of **B** (which are also valid for **H**) into account, in $B_z = B_r = 0$, and in **B** depending only upon r. The only non-zero component of **B** is B_θ, and it is constant on a circle of radius r around the z-axis (fig. 2.3).

2 - MAGNETOSTATICS

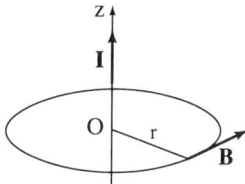

Figure 2.3 - A thread-like current

Applying Ampère's law immediately gives:

$$H_\theta = I/2\pi r \quad (B_\theta = \mu_0 I/2\pi r, \text{ in vacuum}) \tag{2.17}$$

Note : The force between two straight, infinitely long current carrying wires, and the definition of the ampère - *Direct application of equation (2.17) and Laplace's equation gives the force F' per unit length between two straight, rigid, wires carrying the same current I: $|F'| = \mu_0 I^2/2\pi r$. This force is attractive if the currents are in the same sense, and repulsive if they are in opposite senses.*

If $r = 1$ m, $F' = \mu_0 I^2/2\pi$. In the international system we have seen that $\mu_0 = 4\pi \times 10^{-7}$. This relationship defines the ampère: $I = 1$ A if $F' = 2 \times 10^{-7}$ N.m^{-1}.

An infinite, uniform, planar current sheet

In the right-handed coordinate system Oxyz, we suppose that the $z = 0$ plane contains the surface current density j_s parallel to Ox. Translational invariance along Ox and Oy tells us that the induction **B** is independent of x and y. We can thus calculate B for example at point (0, 0, z) by summing up all of the elementary inductions produced by the straight currents j_s dy (fig. 2.4-a).

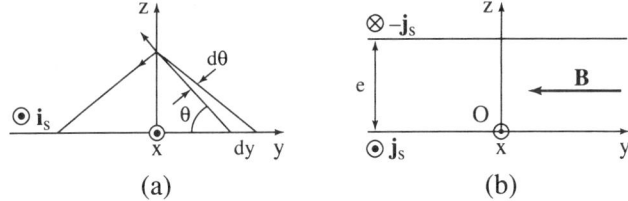

Figure 2.4 - (a) Current sheet $i_s x$ - (b) Double current sheet

Combining elements that are symmetric with respect to the $y = 0$ plane, we observe that the only non-zero component of **B** is B_y. We find that:
$dB_y = 2(\mu_0/\pi r)j_s dy \sin\theta = (\mu_0/\pi)j_s d\theta$, and finally for $z > 0$:

$$\mathbf{B} = \int_{\pi/2}^{0} \frac{\mu_0 j_s}{\pi} d\theta = -\mu_0 j_s/2 \tag{2.18}$$

The plane of the sheet being a symmetry plane, we have for $z < 0$: $B_y = +\mu_0 j_s/2$. The induction is thus independent of the distance from the plane of the sheet, but undergoes a discontinuity on the surface of this plane. One can verify that $B(+0) - B(-0) = -\mu_0 j_s$, conforming to equation (2.15), and that the induction does not cancel out at infinity.

Double current sheet

We associate with the sheet of density $\mathbf{j}_s$ at $z = 0$ a second sheet with density $-\mathbf{j}_s$ at $z = e$. When we superpose the fields from the two sheets, we obtain for $0 < z < e$:

$$B = -\mu_0 j_s \qquad (2.19)$$

in the sense of the arrow representing B in figure 2.4-b. Outside of this interval the field is zero.

This model is reminiscent of a plane capacitor in electrostatics, and we will come back to it later on.

A circular current loop of radius R

When we limit ourselves to calculating the induction on the axis of a circular current loop, it is advantageous to use directly equation (2.1). One can show that at a point P on the axis the induction **B** is along the axis with magnitude (fig. 2.5):

$$B = (\mu_0 I / 2R) \sin^3 \theta \qquad (2.20)$$

where θ is the angle between the vector joining P to the loop, and the axis of the loop.

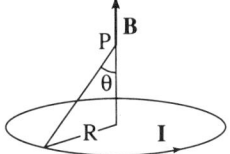

Figure 2.5 - Current loop

Infinite Solenoid

On the surface of a cylinder with axis Oz and radius R, we suppose that transverse surface current densities j_s are present. A single wire of negligible diameter wound around a cylindrical former is a good approximation to this theoretical model. We thus have $j_s = n'I$ where I is the current in the coil, and n' is the number of windings per unit length. If the solenoid is assumed to be infinitely long, then there is translational invariance along Oz if we neglect the small effect of the pitch of the windings.

The system is furthermore invariant in any rotation around Oz, and with respect to mirror symmetry through any plane perpendicular to Oz if we again ignore the small effect of the helicity of the windings. This results in the induction being parallel to Oz, and only depending on the distance to the axis r. The application of Ampère's law to a rectangle constructed in a plane containing Oz (and thus having two sides parallel to Oz) shows that the induction is uniform in each of the two regions $r < R$, and $r > R$. On the other hand **B** displays a discontinuity on the cylinder for $r = R$ (fig. 2.6).

2 - MAGNETOSTATICS

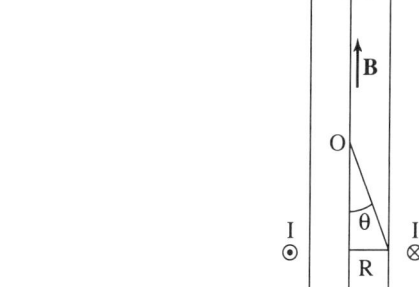

Figure 2.6 - Solenoid

In order to calculate the induction inside the solenoid, the simplest approach is to calculate it on the axis using equation (2.20), and replacing I by n'Idz. We can then either integrate z from $-\infty$ to $+\infty$, or θ from 0 to π. Inside the solenoid we obtain:

$$\mathbf{B} = \mu_0 n'I \,\hat{\mathbf{z}} \qquad (2.21)$$

where $\hat{\mathbf{z}}$ is the unit vector along Oz. When the discontinuity on the cylinder is taken into account, the induction becomes zero outside of the cylinder. This equation is valid for a cylindrical solenoid whatever the shape of its cross-section.

1.1.7. Induction and field produced at a large distance by currents confined to a finite volume: magnetic moment, magnetic dipole, and magnetic charge

Consider a current distribution confined to the neighbourhood of the origin O, inside a finite volume (V). Starting from equation (2.12) it can be shown that the vector potential at a large distance from the origin reduces to:

$$\mathbf{A} = \frac{\mu_0}{4\pi} \frac{\mathbf{m} \times \mathbf{r}}{r^3} \qquad (2.22)$$

where the vector $\mathbf{m}$ is the *magnetic moment*, expressed as a function of the current density $\mathbf{j}$:

$$\mathbf{m} = \frac{1}{2} \int_V \mathbf{r} \times \mathbf{j}(\mathbf{r}) dV \qquad (2.23)$$

The magnetic moment is the analogue of the angular momentum in mechanics: we will come back later to the relationship between these two quantities. Starting from equation (2.22), we can calculate the induction far from the origin:

$$\mathbf{B} = \frac{\mu_0}{4\pi} \left[3\frac{(\mathbf{m}\cdot\mathbf{r})\mathbf{r}}{r^5} - \frac{\mathbf{m}}{r^3} \right] \qquad (2.24)$$

The associated magnetic field can be deduced as being: $\mathbf{H} = \mathbf{B}/\mu_0$. In the case of a flat current loop (C), carrying a current I, the expression for the magnetic moment becomes (using eq. 2.18):

$$\mathbf{m} = IS\,\hat{\mathbf{n}} \qquad (2.25)$$

where S is the area of the loop, and $\hat{\mathbf{n}}$ is the unit normal vector oriented following the usual rule (fig. 2.2-c) with the sense of circulation being that of the current. From equation (2.25), we note that the magnetic moment is expressed in $A \cdot m^2$. We also point out the similarity between the expressions for the field **H** or the induction **B**, and that of the electric field produced by an electric dipole moment $\mathbf{p} = q.\boldsymbol{\delta}$ ($\boldsymbol{\delta}$ being the vector joining the charge $-q$ to the charge $+q$): $\mathbf{E} = (4\pi\varepsilon_0)^{-1}[3(\mathbf{p}.\mathbf{r})\mathbf{r}/r^5 - \mathbf{p}/r^3]$. As a result, the field lines of a magnetic moment are the same as those for an electric dipole, but only at distances from the sources large with respect to their size (fig. 2.7). In the region of the sources the field lines of **B** and **H** are very different from those of **E**. In particular, right at the heart of the sources, they are opposite to one another. This observation is important in what follows.

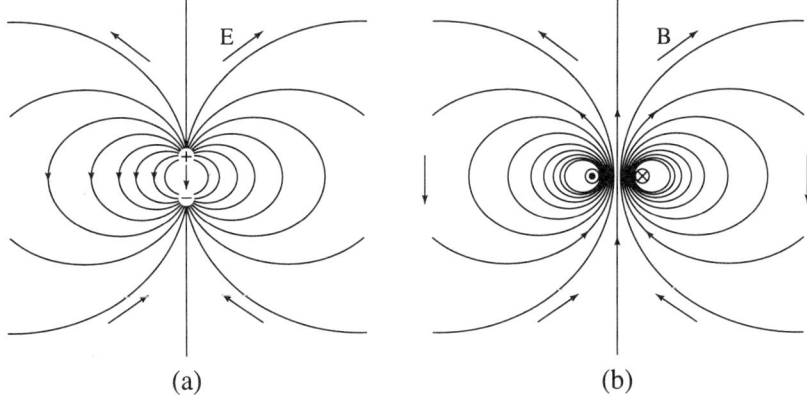

(a) (b)

Figure 2.7 - (a) The electric field produced by a pair of equal and opposite charges - (b) The magnetic induction produced by a current loop

At a large distance the field lines become identical, and so we speak in terms of a dipole field.

The similarity between **B** or **H** and **E** far from the sources allows us to replace a current loop by a pair of point-like magnetic charges $-q_m$ and $+q_m$ placed a distance $\boldsymbol{\delta}$ apart, with $q_m \boldsymbol{\delta} = SI\hat{\mathbf{n}}$. This entity is known as a magnetic dipole, the magnetic moment of which being given by $\mathfrak{m} = q_m \boldsymbol{\delta}$. In this *Coulombian* approach we consider that **H** and **B** are produced by the magnetic charges according to Coulomb's law. In this way we write the field from a point-like magnetic charge q_m as being:

$$\mathbf{H} = \frac{1}{4\pi} \frac{q_m \mathbf{r}}{r^3} \qquad (2.26)$$

The usefulness of this approach –in particular why we emphasise the magnetic field– will become apparent when we move on to study magnetised materials.

The elementary vortex, and the point-like dipole: If one lets the dimensions of a loop go to zero, and the current to infinity (keeping $\mathfrak{m} = IS$ constant), one obtains an object known as an elementary vortex [1]. If the origin is on the loop then equation (2.22) and those following it are exact for all non zero values of r. One can also start from the dipole, and let $\boldsymbol{\delta}$ tend to zero and q_m to infinity (keeping $\mathfrak{m} = q_m \boldsymbol{\delta}$

constant). The object obtained in this process is a point-like magnetic dipole. The elementary vortex and the magnetic dipole are two equivalent mathematical representations of magnetic effects. They can be used in the calculation of fields produced by circuits, or more generally by current distributions. We will come back to this point during the study of magnetised material, and now illustrate it by a very simple example. Figure 2.8 shows that a single current loop of finite dimensions can be decomposed (in many different ways) into an equivalent set of smaller loops. Going to the limit, one obtains a distribution of elementary vortices that can in turn be replaced by dipoles. The object obtained is known as a magnetic sheet –the coulombian representation of a current loop.

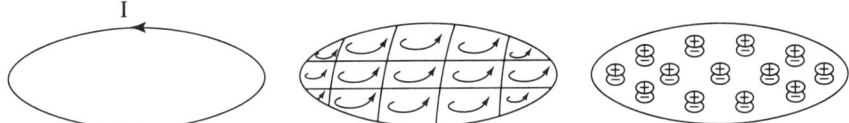

Figure 2.8 - Current loop, and magnetic sheet

The induction and field at any point can be calculated either by using the Biot-Savart law (2.1), or by integrating the contributions from the elementary dipoles over the magnetic sheet. The sheet itself must however be excluded because it does not satisfy the condition of validity of the dipole formalism. In the example studied, this does not represent a severe constraint since the sheet has zero thickness, and its geometry is largely arbitrary (it must only be bounded by the loop). In other more general cases this limitation is less innocent.

1.1.8. The fundamental relationships in the Coulombian approach

By integrating the Laplace force, and its moment around a current loop, we show that the torque Γ and the force $\mathbf{F}$ applied by an induction $\mathbf{B}$ or a field $\mathbf{H}$ on a moment $\mathfrak{m}$ is given by:

$$\Gamma = \mathfrak{m} \times \mathbf{B} \tag{2.27}$$

$$\mathbf{F} = (\mathfrak{m} \cdot \mathbf{grad})\mathbf{B} \tag{2.28}$$

These expressions show that:
- a magnetic moment tends to align parallel to a field, in the same direction as the field,
- a force only exists if the field is inhomogeneous,
- if $\mathfrak{m}$ is parallel or anti-parallel to $\mathbf{B}$, the moment is attracted towards the region of maximum or minimum field respectively.

One can easily verify that equations (2.27) and (2.28) are also valid using the magnetic charge approach if one assumes that a magnetic charge q_m in a field $\mathbf{H}$, or induction $\mathbf{B} = \mu_0 \mathbf{H}$ experiences the force:

$$\mathbf{F} = \mu_0 q_m \mathbf{H} = q_m \mathbf{B} \tag{2.29}$$

Equations (2.26) and (2.29) are the fundamental relationships in the Coulombian approach to magnetostatics.

1.2. MAGNETOSTATICS OF MATTER

The magnetic effects of some materials were identified well before those of electric currents (see chap. 1). On reading what follows it will become apparent that in terms of pedagogical approach it is preferable to deal with the magnetostatics of currents before that of matter.

1.2.1. Magnetic moment, magnetization, and induction associated with matter

A piece of magnetised material shows the same characteristics as the magnetic moment associated with a current loop. It creates the same magnetic induction **B**, it experiences the same torque and the same force. The origin of this magnetic moment will be discussed in detail in chapter 7. Letting d**m** be the magnetic moment of a volume element of matter dV, by definition the local magnetization is the magnetic moment per unit volume:

$$\mathbf{M} = d\mathbf{m}/dV \qquad (2.30)$$

If we consider that the material consists of atoms of type i with magnetic moment $\mathbf{m}_i$ with a number density N_i per unit volume, the magnetization can also be written as:

$$\mathbf{M} = \sum_i N_i \mathbf{m}_i \qquad (2.31)$$

Outside of the material, the induction **B** is created not only from the macroscopic currents, but also from the elementary current loops comprising the material. One can always make use of the dipole approximation, even in the immediate vicinity of the material, due to the fact that each loop is considered to be pointlike (so long as one does not consider distances down to the atomic scale). While there is no difficulty in defining the induction **B** outside of the material, this is not the case for the induction inside the material. Here one is in the middle of a distribution of micro-currents, and the local induction (generally written as **b**) fluctuates enormously in space, in particular from atom to atom. It is difficult to determine **b** (without using neutron diffraction), and as far as magnetostatics is concerned its determination does not bring much insight. With this in mind we define the induction **B** as being the spatial average of the local induction **b**, such that **B** = <**b**>. It is this last quantity which will be of interest in what follows.

<**b**> can be measured from the deviation produced by the Lorentz force (eq. 2.57) on a beam of charged particles crossing the material. This is exactly what is performed in Lorentz microscopy [4].

At this stage the problem that we are interested in is the determination of the induction **B** created at all points in space by a known distribution of currents and magnetization **M**. For the moment, we are not interested in how the magnetization came about.

1.2.2. The determination of B (and A) from bound currents: the Ampérian approach

In the presence of matter, one can show (see appendix § 3 at the end of this chapter) that the determination of the induction **B** and the vector potential **A** at all points in space can be reduced to a problem of electromagnetism in vacuum where two types of current have to be considered:
- on the one hand the real *free currents* with density $\mathbf{j}_0$, and
- on the other hand the currents associated with the magnetised material, or bound currents, with volume and surface densities, $\mathbf{j}_m$ and $\mathbf{j}_{ms}$ respectively, given by:

$$\mathbf{j}_m = \text{curl } \mathbf{M} \quad (2.32)$$

$$\mathbf{j}_{ms} = \mathbf{M} \times \hat{\mathbf{n}} \quad (2.33)$$

where $\hat{\mathbf{n}}$ is the unit vector normal to the surface of the material, and pointing outwards. equation (2.32) reduces to (2.33) when one only considers the surface as the transition between the material ($\mathbf{M} \neq 0$ inside) and the vacuum ($\mathbf{M} = 0$ outside), and going to the limit where δ (the thickness) goes to zero. One can thus determine **B** using the method given in section 1.1.3 taking $\mathbf{j} = \mathbf{j}_0 + \mathbf{j}_m$.

Notes

- *When the magnetization is uniform, which is often the case, $j_m = 0$, and only the surface current density j_{ms} is left. An example, the **B** field produced everywhere (inside and outside of the cylinder) by a cylinder uniformly magnetised along its cylindrical axis is the same as from a solenoid of current density (current per unit length along the solenoid) $|j_{ms}| = |\mathbf{M}|$ (see fig. 2.9).*

- *The conditions at the interface between two materials are the same as those given in section 1.1.4, taking into account the fact that i (or rather i_s in eq. 2.15) is the sum of the free and bound current densities.*

- *One can always write that:*

$$\mathbf{B} = \mathbf{B'}_0 + \mathbf{B}_m \quad (2.34)$$

where $\mathbf{B'}_0$ and $\mathbf{B}_m$ are the contributions from the free currents and magnetised material respectively.

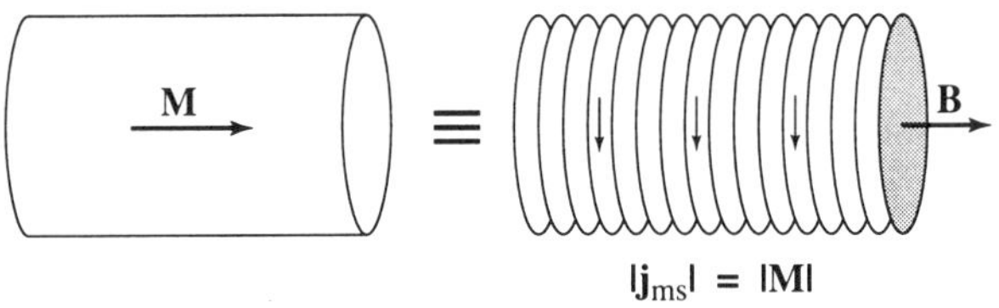

Figure 2.9 - *The equivalence between magnetization and surface current density in the case of a cylinder uniformly magnetised parallel to its cylindrical axis*

1.2.3. The magnetic field H in magnetised material

From equation (2.8), **curl B** = $\mu_0 \mathbf{j}$ = $\mu_0 (\mathbf{j}_0 + \mathbf{j}_m)$ = $\mu_0 (\mathbf{j}_0 + \text{curl } \mathbf{M})$, whence:

$$\text{curl } [(\mathbf{B}/\mu_0) - \mathbf{M}] = \text{curl } \mathbf{H} = \mathbf{j}_0 \tag{2.35}$$

In this way the vector $\mathbf{H} = \mathbf{B}/\mu_0 - \mathbf{M}$, called the magnetic field, keeps alive the notion of free (or real) currents from Ampère's law, despite this no longer being true for the induction **B**. This is the usefulness of the concept of magnetic field.

We note that equation (2.35) leads to a more general definition of **H** than that given in section 1.1.2, but reduces to the latter when the magnetization is zero (M = 0). The definition of **H** is:

$$\mathbf{B} = \mu_0 (\mathbf{H} + \mathbf{M}) = \mu_0 \mathbf{H} + \mathbf{J} \tag{2.36}$$

where $\mathbf{J} = \mu_0 \mathbf{M}$ is called the magnetic polarization. Some authors consider the vector **J** as the magnetization, and denote it as **M**, which can lead to confusion [1].

We will now explain the physical significance of **H**. Consider an element of material where a uniform induction **B** exists, and then imagine making a cylindrical cavity in this element that is elongated along the direction of **M**. Due to the superposition theorem, the new value of the induction $\mathbf{B}_i$ in the cavity is equal to the old **B** minus the induction $\Delta \mathbf{B}$ produced by the solenoid equivalent to the part that has been removed. $\Delta \mathbf{B}$ is given by equation (2.21) in which n'I = i_{ms} = M thus $\Delta \mathbf{B} = \mu_0 \mathbf{M}$. It turns out that:

$$\mathbf{B}_i = \mathbf{B} - \mu_0 \mathbf{M} = \mu_0 \mathbf{H} \tag{2.37}$$

where **H** is the magnetic field in the material. In this way the field **H** is none other (to within the factor $1/\mu_0$) than the magnetic induction measured in a cylindrical cavity whose axis is aligned parallel to the magnetization.

Note - *H is the "total" field. In a material it is called the "internal" field (also written H_i).*

1.2.4. The determination of H from equivalent magnetic charges: the Coulombian approach

The force acting on a point magnetic charge q_m located at point P due to a charge q'_m at P' can be written as:

$$\frac{\mu_0}{4\pi} \frac{q_m q'_m}{r_{p'p}^3} \mathbf{r}_{p'p}$$

Equation (2.35) by no means signifies that only free currents produce **H**. The fact that an isolated magnet produces in its neighbourhood an induction **B**, and thus a field $\mathbf{H} = \mathbf{B}/\mu_0$ is enough to convince oneself of that. We can thus write that:

$$\mathbf{H} = \mathbf{H}_m + \mathbf{H'}_0 \tag{2.38}$$

where $\mathbf{H'}_0$ is due to the free currents, and $\mathbf{H}_m$ is due to magnetised material. Knowing that $\mathbf{curl}\,\mathbf{H'}_0 = \mathbf{j}_0$, equation (2.35) implies that $\mathbf{curl}\,\mathbf{H}_m = 0$. $\mathbf{H}_m$ thus can be associated with a potential which we will call V_m such that $\mathbf{H}_m = -\,\mathbf{grad}\,V_m$.

Furthermore since $\mathrm{div}\,\mathbf{B} = \mathrm{div}\,\mathbf{B'}_0 = \mathrm{div}\,\mathbf{B}_m = \mathrm{div}\,\mathbf{H'}_0 = 0$, equation (2.36) leads us to:

$$\mathrm{div}\,\mathbf{H}_m = -\,\mathrm{div}\,\mathbf{M} \qquad (2.39)$$

By analogy with the equations for electrostatics:

$$\mathbf{curl}\,\mathbf{E} = 0 \quad ; \quad \mathrm{div}\,\mathbf{E} = \rho/\varepsilon_0 \qquad (2.40)$$

where $\mathbf{E}$ is the field, and ρ the electric charge density, one can transpose (give or take a factor of ε_0) these results to magnetostatics by introducing fictitious magnetic charges with volume and surface densities, ρ_m and σ_m respectively, given by:

$$\rho_m = -\,\mathrm{div}\,\mathbf{M} \qquad (2.41)$$

$$\sigma_m = \mathbf{M}.\hat{\mathbf{n}} \qquad (2.42)$$

where $\hat{\mathbf{n}}$ is the unit vector normal to the surface, and pointing outwards from the magnetised region. This last relationship derives from equation (2.41) by going to the limit analogous to that used to get from equation (2.32) to equation (2.33).

Once the charges equivalent to the material are known, the different methods of determining $\mathbf{H}_m$ (or V_m) inside and outside of the material are the same as in electrostatics:

♦
$$\mathbf{H}_m = \frac{1}{4\pi}\int \frac{\rho_m \mathbf{r}}{r^3}\,dV \qquad (2.43)$$

♦
$$V_m = \frac{1}{4\pi}\int \frac{\rho_m}{r}\,dV \qquad (2.44)$$

♦ Gauss' theorem (the flux of $\mathbf{H}_m$ leaving a closed surface is equal to the total magnetic charge inside that surface) which is the method to apply if the symmetry is appropriate.

♦ Poisson's equation: $\qquad \Delta V_m + \rho_m = 0 \qquad (2.45)$

(or Laplace's equation $\Delta V_m = 0$ where $\rho_m = 0$). The constants arising from the integration of this partial differential equation are obtained from the boundary conditions as discussed below.

Notes

♦ *The nature of the vectors H, M, and the scalars ρ_m and q_m.*
H and M are pseudo-vectors (axial vectors) in the same way as B, and the scalar $\rho_m = -\,\mathrm{div}\,M$ is in fact what is known as a pseudo-scalar. This comes about because its sign depends on the convention adopted for the positive sense of circulation. The transformation rules of magnetic charges during a symmetry operation are deduced from those of M (identical to those of B) as is shown in figure 2.10.

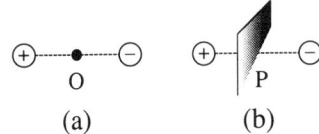

Figure 2.10 - The transformation of a point magnetic charge (a) on inversion with respect to a centre of symmetry, and (b) in mirror reflection with respect to a plane

- When the magnetization is uniform, which is often the case, $\rho_m = 0$, and only the surface density of charges σ_m, remains. For example the field $\mathbf{H}_m$ produced everywhere (both inside, and outside of the material) by a cylinder uniformly magnetised parallel to its axis is the same as that produced by two discs with uniform surface density $\sigma_m = \pm |\mathbf{M}|$ placed parallel to the plane faces of the cylinder (fig. 2.11).

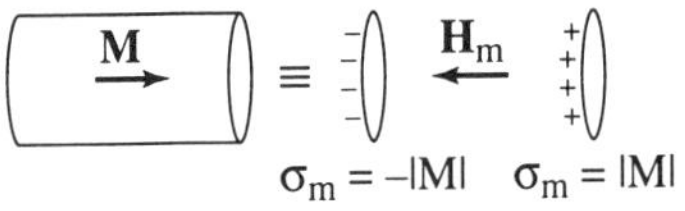

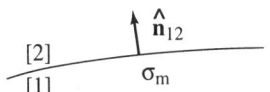

Figure 2.11 - Equivalence between magnetization and magnetic charge density in the case of a cylinder uniformly magnetised along its axis

- The continuity conditions across a surface, in particular between one region and another, are the following:
  - those concerning $\mathbf{H'}_0$ are those imposed on $\mathbf{B'}_0$ (and $\mathbf{A'}_0$), and are given in section 1.1.4.
  - those concerning $\mathbf{H}_m$ are the same as those in electrostatics, viz:
    - V_m is continuous, and thus so is the tangential component of $\mathbf{H}_m$,
    - The perpendicular component of $\mathbf{H}_m$ is discontinuous with $(H_{m2} - H_{m1}) = \hat{n}_{12}\, \sigma_m$ (see fig. 2.12). If the regions 1 and 2 are magnetised material with magnetization $\mathbf{M}$ and vacuum respectively, $\sigma_m = \mathbf{M} \cdot \hat{n}_{12}$. If one considers two regions with magnetization $\mathbf{M}_1$ and $\mathbf{M}_2$, equation (2.39) leads to $\sigma_m = \hat{n}_{12}\,(\mathbf{M}_1 - \mathbf{M}_2)$.

Figure 2.12 - The interface between a magnetic region [1] and a different, non magnetic region [2]

1.2.5. Application to two simple cases

Real current density $\mathbf{j}_0$ and magnetization M known at all points

In summary, the determination at all points in space of **B** and **H** produced by known arrangements of current $\mathbf{j}_0$ and magnetization **M** can proceed in two different ways. If one of the fields is known, then the other can be deduced using equation (2.36). In both approaches, one turns the question into a problem of magnetostatics in vacuum.

- *The ampérian approach*: the determination of $\mathbf{B} = \mathbf{B'}_0 + \mathbf{B}_m$

  $\mathbf{B}_0$ is produced by the free currents $\mathbf{j}_0$ whereas $\mathbf{B}_m$ is produced by the volume and surface densities of bound currents given by equations (2.32) and (2.33), $\mathbf{j}_m = \mathbf{curl}\,\mathbf{M}$, and $\mathbf{j}_{ms} = \mathbf{M} \times \hat{n}$. Thence there are four ways to determine **B** (or **A**):
  - by using the Biot-Savart law directly (eq. 2.1),

2 - MAGNETOSTATICS

- by applying Ampère's law to **B**: $\oint_\Gamma \mathbf{B} d\mathbf{l} = \mu_0 \iint (\mathbf{j}_0 + \mathbf{j}_m) dS$,
- by calculating directly $\mathbf{A} = \dfrac{\mu_0}{4\pi} \int_V \dfrac{\mathbf{j}}{r} dV$, where $\mathbf{j} = \mathbf{j}_0 + \mathbf{j}_m$, and finally
- by integrating the partial equations obtained from $\mathbf{curl}\,(\mathbf{curl}\,\mathbf{A}) = \mu_0 \mathbf{j}$.

♦ *Coulombian approach*: the determination of $\mathbf{H} = \mathbf{H'}_0 + \mathbf{H}_m$

$\mathbf{H'}_0 = \mathbf{B'}_0/\mu_0$ is produced by the free currents, and $\mathbf{H}_m = -\mathbf{grad}\,V_m$ is produced by the equivalent volume and surface magnetic charges given by equations (2.41) and (2.42): $\rho_m = -\,\text{div}\,\mathbf{M}$, and $\sigma_m = \mathbf{M}\cdot\hat{\mathbf{n}}$. There are four ways to determine $\mathbf{H}_m$ (or V_m):

- by using Coulomb's law directly (eq. 2.43),
- by applying Gauss' theorem $\left(\iint \mathbf{H}_m d\mathbf{S} = \iiint \rho_m dV\right)$,
- by calculating directly $V_m = \dfrac{1}{4\pi} \int \dfrac{\rho_m}{r} dV$ (eq. 2.44),
- by integrating the partial differential equation $\Delta V_m + \rho_m = 0$ (eq. 2.45).

If one obtains **A** or V_m by integration of a partial differential equation, the constants are determined from the boundary conditions, in particular those associated with going from one medium to another:

♦ the continuity of **A**, B_N (normal component), V_m, and $\mathbf{H}_{mT}$ (tangential component),
♦ the discontinuity of H_{0T} (tangential component): $\mathbf{H}_{0T2} - \mathbf{H}_{0T1} = \mathbf{j}_{0S} \times \hat{\mathbf{n}}_{12}$, where $\hat{\mathbf{n}}_{12}$ is the unit vector normal to the interface between 1 and 2.

Known current density $\mathbf{j}_0$, and a linear, homogenous, and isotropic material

In a Linear, Homogenous, and Isotropic material (LHI, see § 1.3.2), $\mathbf{M} = \chi \mathbf{H}$, $\mathbf{B} = \mu_0 (1 + \chi)\,\mathbf{H} = \mu_0 \mu_r \mathbf{H}$, where χ, and μ are scalars characterising the material. It follows that:

♦ $\mathbf{j}_m = \mathbf{curl}\,\mathbf{M} = \chi\,\mathbf{curl}\,\mathbf{H} = \chi\,\mathbf{j}_0$. Wherever there are no free currents, $\mathbf{j}_m = 0$; and so only equivalent surface currents are left. In each region we thus have to solve the differential equation: $\mathbf{curl}\,(\mathbf{curl}\,\mathbf{A}) = 0$.
♦ $\rho_m = -\,\text{div}\,\mathbf{M} = -(\chi/\mu_r\mu_0)\,\text{div}\,\mathbf{B} = 0$. In each region we have to solve Laplace's equation: $\Delta V_m = 0$.

Starting from either of these approaches, we can arrive at the other using equation (2.36).

1.2.6. The demagnetising field H_d

We introduce the concept of the *demagnetising field* by considering the case where the only source of magnetic field is a known distribution of magnetization in a material. As shown in figure 2.13, for the case of a uniformly magnetised cylinder, the field $\mathbf{H}_m$ in the material is generally opposed to the magnetization (more precisely, at a given point, the projection of $\mathbf{H}_m$ onto the direction of the magnetization is opposite to **M**). This is why **H**ₘ is called the demagnetising field, and is denoted by **H**_d. Some authors extend the use of the term "demagnetising field" to the field produced by magnetised matter outside of the material itself. Unfortunately certain authors also call **H**ₘ or **H**_d the *dipole field*, or even *dipole interaction field*. We will explore the difference between the dipole and demagnetising field energies in the second part of this chapter.

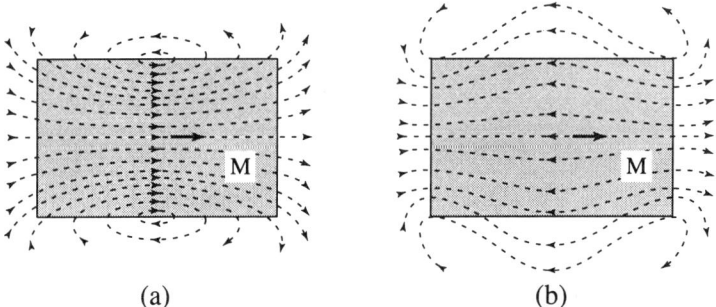

Figure 2.13 - Field lines (a) of the induction B, and (b) of the field H, in the presence of a uniformly magnetised cylinder. At all points in the material, $H = (B/\mu_0) - M$

It is easy to calculate the demagnetising field for uniformly magnetised material with various geometries [5] by using the formalism expressed in equations (2.33) to (2.39):
- very elongated cylinders where the magnetization is parallel to the axis: $\mathbf{H}_d = 0$ (except near the extremities);
- disks or plane thin films with thickness negligible compared to their other dimensions, and magnetization perpendicular to their faces: $\mathbf{H}_d = -\mathbf{M}$ far from the edges;
- at the centre of an infinite cylinder, magnetised along its axis (fig. 2.14), the demagnetising field becomes: $\mathbf{H}_d = -(1 - \cos\theta)\mathbf{M}$.
From this formula the demagnetising field in cases (a), and (b) can easily be deduced.

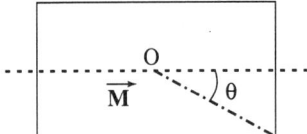

Figure 2.14
A cylinder magnetised along its axis

1.2.7. Demagnetising field coefficient

Although non trivial, it is possible to show that for a uniformly magnetised ellipsoid, **H**_d, and in consequence **B**ₘ are also uniform in the material. The demagnetising field in a uniformly magnetised ellipsoid can thus be written:

$$\mathbf{H}_d = -N\mathbf{M} \quad (2.46)$$

where **N** is the tensor of the coefficients of the demagnetising field. Note that for axes coincident with the symmetry axes of an ellipsoid, the matrix representing this tensor is diagonal, and such that:

$$N_{xx} + N_{yy} + N_{zz} = 1 \quad (2.47)$$

If the magnetization is along one of these axes, $\vec{H}_d$ is collinear to it. For a sphere it is easy to deduce that $N_{xx} = N_{yy} = N_{zz} = 1/3$. A special case that often occurs is that of an *ellipsoid of revolution* of length c along the axis of revolution z, and diameter a perpendicular to this axis. If $r = c/a > 1$ (an elongated, or prolate, ellipsoid) then:

$$N_{zz} = N_{\parallel} = \frac{1}{r^2 - 1}\left[\frac{r}{\sqrt{r^2-1}}\cosh^{-1}(r) - 1\right] \quad (2.48)$$

If $r < 1$ (flattened, or oblate, ellipsoid), then:

$$N_{zz} = N_{\parallel} = \frac{1}{1 - r^2}\left[1 - \frac{r}{\sqrt{1-r^2}}\cos^{-1}(r)\right] \quad (2.49)$$

In both cases: $N_{xx} = N_{yy} = N_{\perp} = (1 - N_{\parallel})/2$.

Notes

- The internal field $\vec{H}$ (or $\vec{H}_i$) in a specimen is usually written as $\vec{H} = \vec{H}_0 + \vec{H}_d$ where $\vec{H}_0$ is the "applied" or "external" field, i.e. that produced by all field sources (free currents and magnetised materials) except for those belonging to the specimen itself.
- With the above definition, the field outside the sample is not the external field.

1.3. RESPONSE OF A MATERIAL TO A MAGNETIC FIELD

In the same way as for the electric polarization and elastic deformation, the magnetization **M** is a configuration variable of matter. One of the major problems of magnetic materials is to determine (either experimentally of theoretically) the relationship between **M** and the different variables upon which it depends. The magnetic field **H** and the induction **B** are without doubt the most important, but since **H** and **B** are linked by equation (2.36), one can consider either one of them as the independent magnetic variable.

1.3.1. The magnetic field H as the independent magnetic variable

This traditional choice is only justified by practical considerations. It is easier to control the internal magnetic field of a sample than its internal induction (particularly in static measurements). As previously stated the main interest of the vector **H** is to verify Ampère's law also inside magnetised material. Figure 2.15-a shows how to make use of this property to set up a given field inside a toroidal sample. The toroid carries a dense winding of n turns regularly distributed around its perimeter, and a current I is passed through the winding. If the system, including the material, conserves the symmetry of revolution around the axis Oz of the toroid, the field will only comprise the azimuthal component H which is constant over a circle with axis Oz such that:

$$H = nI/l \qquad (2.50)$$

where l is the average perimeter of the toroid. The magnetization is thus azimuthal, and constant along a circle around Oz, and the demagnetising field is zero (Oz is the symmetry axis of the toroid, perpendicular to the plane of figure 2.15-a).

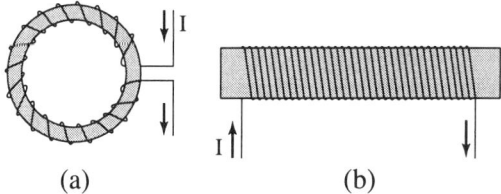

Figure 2.15 - Two configurations where one can impose a given value of H inside a sample: (a) A toroid - (b) An elongated rod

Another useful configuration is a rod, highly elongated in the direction of its magnetization (fig. 2.15-b). The demagnetising field is also zero in this case.

1.3.2. Different type of magnetic behavior of a substance

The magnetic characterization of a material ideally consists of measuring **M** or **B** as a function of **H**. In practice one has to be content with measuring M_H or B_H as a function of H, where M_H and B_H are the projections of **M** and **B** onto **H**. The curves $M_H(H)$ and $B_H(H)$ are known as magnetization curves. When there is no ambiguity over the direction of the projection, these classical relationships are written more simply as M(H) and B(H).

Depending on the type of material, very different behaviors are observed which depend on external parameters such as temperature or pressure, the history of the material, and the direction of the applied field (in the case of anisotropic materials). As an example, figure 2.16 shows typical responses of material to a magnetic field.

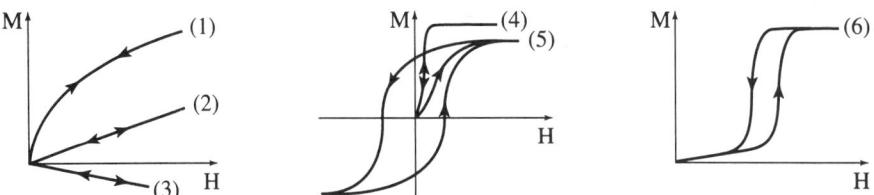

Figure 2.16 - Schematic representation of the different types of response observed depending upon the type of substance considered

2 - MAGNETOSTATICS

Some materials have a linear response up to high fields (cases 2, and 3). If in addition they are isotropic, and homogenous (LHI – *linear, homogeneous, and isotropic*), the magnetization is collinear with the field, and can be expressed as:

$$\mathbf{M} = \chi \mathbf{H} \qquad (2.51)$$

where χ, a dimensionless scalar in MKSA units, is the magnetic *susceptibility*.

For other materials, linear behavior is only observed for small changes in field around a point (H_0, M_0), which is not necessarily the origin (cases 4, and 6 in figure 2.16). In this case, $\chi_i = (dM/dH)_{H_0}$ characterises the response in this region of field, and is called the *differential magnetic susceptibility*. The *initial susceptibility* is defined as $\chi_i = (dM/dH)_0$, i.e. the slope at the origin. Values of the susceptibility range from around -10^{-5} in very weak magnetic materials (diamagnets such as copper, and carbon) up to values of around $+10^6$ (ultra-soft ferromagnets such as certain amorphous or nanocrystalline alloys).

The *permeability* μ in an LHI region is defined as:

$$\mathbf{B} = \mu \mathbf{H} = \mu_r \mu_0 \mathbf{H} = \mu_0 (\mathbf{H} + \mathbf{M}) \qquad (2.52)$$

It follows immediately that $\mu_r = 1 + \chi$. At this point attention needs to be paid to the notation: as many authors do, we will often write equation (2.52) in the form $\mathbf{B} = \mu \mu_0 \mathbf{H}$, where μ is now the *relative permeability*.

Some materials show a hysteresis loop (curve 5) such that when zero field is reached again after the application of a strong field a non zero magnetization, known as the *remanent magnetization* (M_r), is left over: This is the case for permanent magnets (for example some rare earth alloys).

In general, the permeability, and magnetic induction are more interesting for engineers whereas the susceptibility, and magnetization are the favourite parameters for physicists.

1.3.3. Anisotropic materials

For materials with macroscopic anisotropy (e.g. single crystals), the magnetization is not in general collinear with the field. If the behavior is linear, one can write the following tensor equation:

$$\mathbf{M} = \chi \mathbf{H} \qquad (2.53)$$

where χ is the susceptibility tensor. Equation (2.53) projects, under a given coordinate system, to become the matrix equation $\mathbf{M} = [\chi] \mathbf{H}$ where $[\chi]$ is a 3×3 matrix which represents χ in the coordinate system used.

In the time-independent regime we are considering, it can be shown that the representative matrix in a general coordinate system is symmetric with respect to its diagonal, and thus has at most 6 independent components.

A particular system of axes then exists, called the *principal axes*, where the matrix is diagonal. The principal axes are always linked to symmetry elements when these exist. The scalar or tensorial susceptibilities (or permeabilities) that we have defined thus far are the *static* susceptibilities (or permeabilities), and are valid for the case of a DC or slowly varying field. In chapter 17 we will study the susceptibility (or permeability) for fields that vary rapidly with time.

1.3.4. Note

Engineers write equation (2.36) in the form: $\mathbf{B} = \mu_0 \mathbf{H} + \mathbf{J}$ whereas physicists write it as: $\mathbf{B} = \mu_0 (\mathbf{H} + \mathbf{M})$ in the rationalised MKSA system of units (SI). Most of the older literature on magnetism use the c.g.s. unit system where this equation is written as: $\mathbf{B} = \mathbf{H} + 4\pi\mathbf{M}$, where B is expressed in gauss, **H** in oersted, and **M** in uem.cm^{-3}. One gauss or oersted is $(250/\pi)$ A.m^{-1}, and one uem/cm^3 is 10^3 A.m^{-1}.

1.3.5. Demagnetising field correction. The external susceptibility

That the intrinsic response of a magnetic material can only be directly obtained for samples having the form of a very elongated rod, or a toroid has already been emphasized. Nevertheless, samples having ellipsoidal shape can still be used if a correction for the demagnetising field is made. This consists of determining M(H), where H is the internal field in the material, from M(H$_0$) where H$_0$ is the applied field. Knowing the demagnetising field coefficient N in the direction under consideration (necessarily a principal axis of the ellipsoid), one obtains M(H) = M(H$_0$ – NM). This correction is applicable to any type of material, although LHI materials are highlighted here as their intrinsic response is characterised by a single parameter, the susceptibility χ.

One can write $M = \chi H$, with $H = H_0 + H_d$, and $H_d = -NM$, and deduce that:

$$M = [\chi/(1 + N\chi)] H_0 \qquad (2.54)$$

The slope of the curve is thus lower when one plots a graph of the magnetization as a function of the external field (fig. 2.17). This slope $\chi/(1 + N\chi)$ is known as the *external susceptibility*. It is important to note that the external susceptibility characterises not the material itself, but a sample of the material with a specific geometry. In particular, if χ is infinite (such as for ideal ferromagnets in a weak field) the external susceptibility is equal to 1/N, i.e. the reciprocal of the demagnetising field coefficient (fig. 2.18). The change of magnetization as a function of applied field (where this field is weak) is linear, and depends only on the geometry (the *demagnetising field straight line*). demagnetising field line

These considerations, as well as magnetization measurements, will be discussed in depth in chapter 26 on experimental methods. Unless otherwise stated, the curves presented in this work for ferromagnetic substances will be plotted as a function of the internal field H (Am^{-1}) or μ_0H (tesla).

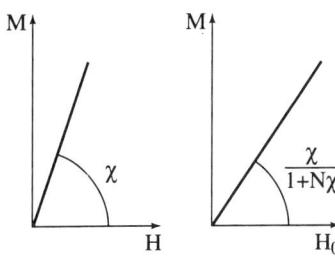
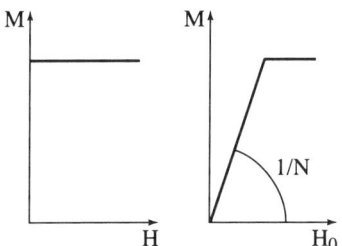

Figure 2.17
Schematic magnetization curves as a function of internal field (H) and external field (H₀), for a substance having finite susceptibility

Figure 2.18
Schematic magnetization curves as a function of internal field (H) and external field (H₀), for a substance with infinite susceptibility

1.4. THE GENERAL PROBLEM OF DETERMINING B AND M

In section 1.2.5 we assumed, in order to obtain **H** (or **B**) everywhere, that the magnetization is known at all points in the material. As we have seen in the previous section the magnetization depends on the field. The most general problem is the following: in the presence of known free currents (if these exist), and perfectly characterised material, we seek to determine at all points the magnetization **M**, and the field **H** (or **B**). The difficulty comes from the fact that locally the magnetization depends on the total field **H** which is the sum of **H'**$_0$ (created by the free currents), and **H**$_m$ (or **B**$_m$) created by the magnetization. We thus have a coupled problem, which most of the time can only be solved numerically in a self-consistent manner. The steps of such a calculation are the following:

1. The determination of **H'**$_0$ (or **B'**$_0$) from the current distribution if this exists.
2. Choice of an initial distribution of the magnetization **M**$_i$ (or of polarization **J**$_i$).
3. The determination of **H**$_m$ (or **B**$_m$) by one or other of the approaches presented in section 1.2.
4. The calculation of the distribution of magnetization from the response of the material.
5. Comparison between the result of step 4, and the starting distribution.
6. If these are the same then the problem has been solved. If not then replace the initial distribution with the calculated one, and go back to step 3.

Thanks to more and more powerful calculation techniques, increasingly complicated problems are being solved through numerical simulation, a field that is flourishing. An example is the finite element method where one chooses a mesh where the magnetization of each element is considered to be uniform. Amongst the difficulties encountered are the choice of mesh and boundary conditions required. The simulations are obviously more accurate the finer the mesh.

Amongst the rare situations where analytical approaches are available are:
- the case of a sphere or spherical shell of LHI material in a uniform external field,
- and the case of an infinite cylinder or cylindrical shell of LHI material in a uniform external field perpendicular to the axis of the cylinder.

These cases (either from the point of view of bound currents or equivalent charges) are the subject of exercises proposed at the end of this chapter.

Note: Micromagnetism - *In the general problem treated up to now we assume that the response of the material is perfectly described by a relation of the form M (H) or B (H) which has local significance irrespective of the scale. In reality, in a magnetic material, there are always one or more characteristic length scales that break the scale invariance.*

The most obvious length scales are related to the microstructure in the case of polycrystalline material, or composites. For example, for slow spatial variations on the scale of the grains in a polycrystal, it is sufficient to describe the response of a material by a macroscopic equation of state M = M(H) where M and H are averages, calculated over a large volume on the scale of characteristic lengths. M (H) defined in this way does not take into account the heterogeneous character of the material. For this reason it is often known as the homogenization law. If the spatial variations are fast, homogenization is no longer possible. The step size in the calculation must then be much smaller than the size of the grains, each of which must be characterised by its own equation of state. If however the grains are quite large – and we will quickly define what is meant by that– this equation of state can retain its local character, and the nature of the problem will not fundamentally change.

On the other hand, the problem will become radically different if the spatial variations are fast at another length scale linked to the exchange interaction. We will see in subsequent chapters that the exchange interaction is behind magnetic ordering, and is not magnetostatic in origin. This brings one to a *non-local* description of the magnetic response of a material: we are no longer dealing with a problem of magnetostatics in the sense of this section. This is called micromagnetism because the characteristic length scales of exchange are typically smaller or even much smaller than 1 µm.

Classical examples of problems of static micromagnetism include determining the distribution of magnetization inside a domain wall, and the calculation of critical switching fields in magnets or magnetic recording media.

1.5. MAGNETIC CIRCUIT APPROXIMATION

At the interface between two materials, in the absence of surface currents, the normal component of **B** and the tangential component of **H** are continuous.

If the materials are LHI, it is easily shown that the field lines (of either **B** or **H**) undergo refraction such that $\tan\theta_2 / \tan\theta_1 = \mu_2/\mu_1$ where θ_1 and θ_2 are the angles between the interface normal and the field lines in mediums 1 and 2 respectively (fig. 2.19). In other words, in a region of high permeability (soft ferromagnet), the field lines tend to lie tangentially to the external surface, except where they are strictly perpendicular to it.

Figure 2.19 - Boundary between two magnetic materials

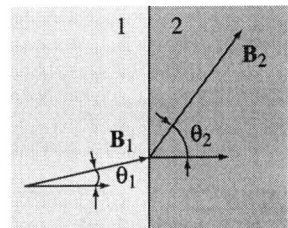

This property is used in what is known as a *magnetic circuit*. The most simple example is shown in figure 2.20: the flux produced by a exciting coil is channelled by a toroid of high permeability in the same way as an electric circuit channels the current produced by a generator. This is where the term magnetic circuit comes from. In the set-up shown in figure 2.20, the windings consist of n turns carrying a current I. S is the cross-section (assumed to be constant) of the toroid, and p is its average perimeter.

Figure 2.20 - Magnetic flux lines due to a coil (a) in vacuum, and (b) in the presence of a toroid of magnetic material

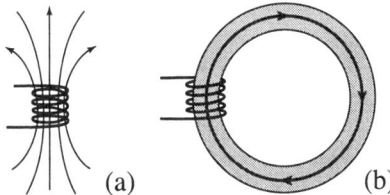

The magnetic circuit approximation is that the field lines are parallel to one another, and to the central line of the toroid. From conservation of flux it is apparent that B and H are invariant along a given field line. In the regime where $p/2\pi \gg S^{1/2}$ the field lines all have lengths close to p. Ampère's law allows one to write $H = nI/p$, and:

$$\Phi = BS = \mu_r \mu_0 H = nI/\mathcal{R} \qquad (2.55)$$

where $\mathcal{R} = (1/\mu_r\mu_0)\, p/S$ is called the *reluctance*.

The similarity between equation (2.55), and Ohm's law $V = RI$ is immediately apparent. The analogue of V is the quantity $\mathcal{E} = nI$ which is called the *magnetomotive force*. The analogue of current is obviously the flux, and that of resistance is the reluctance. This analogy with electric circuits can be generalised to more complicated set-ups having several branches in series or parallel, for which similar laws to those of Kirchhoff can be established.

In figure 2.21 the main circuit branches into two channels, 1 and 2. Along a field line that goes through the main circuit and branch 1, Ampère's law gives:

$$nI = \mathcal{R}_0 \Phi_0 + \mathcal{R}_1 \Phi_1$$

Similarly for a line going through branch 2:

$$nI = \mathcal{R}_0 \Phi_0 + \mathcal{R}_2 \Phi_2.$$

Conservation of flux gives $\Phi_0 = \Phi_1 + \Phi_2$. From this we obtain:

$$nI = (\mathcal{R}_0 + \mathcal{R}') \Phi_0 = \mathcal{R} \Phi_0$$

with $1/\mathcal{R}' = \Lambda' = 1/\mathcal{R}_1 + 1/\mathcal{R}_2 = \Lambda_1 + \Lambda_2$ and $\mathcal{R} = \mathcal{R}' + \mathcal{R}_0$.

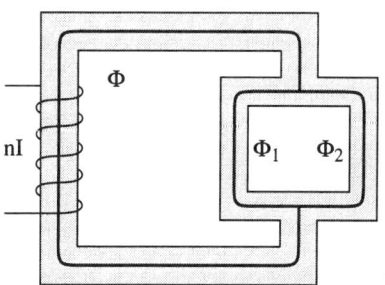

Figure 2.21 - Magnetic circuit with two arms in parallel

In the above the reciprocal of the reluctance, known as the *permeance,* which is the analogue of conductance, appears. As for electric circuits, the reluctances of parts of a circuit that are in series add up whereas the permeances add up for branches in parallel.

The method of magnetic circuits will be illustrated by two important practical examples:

♦ The example in figure 2.22-a corresponds to a simplified representation of an electromagnet. It arises from the circuit in figure 2.20 after introducing what is known as an *air gap*. This is a sheet of air which is perpendicular to the average line of the circuit, and whose thickness is much less than the average perimeter p of the toroid. If the additional condition that $g \ll S^{1/2}$ is imposed, where S is still the cross-section of the toroid, the edge effects at the gap can be ignored, and one can assume that the field lines remain parallel to the average line of the toroid over most of the air gap. Then the flux Φ through the circuit, and particularly the gap, can be written as: $\Phi = \mu_0 H_g S$, hence $\Phi = nI/(\mathcal{R}_i + \mathcal{R}_g)$ where $\mathcal{R}_i = (1/\mu_r \mu_0)(p - g)/S$, and $\mathcal{R}_g = (1/\mu_0) g/S$ are respectively the reluctance of the soft material (iron), and that of the air gap (where $\mu = 1$). H_g is the field in the air gap. Knowing that $g \ll p$, we have: $H_g = nI/[(p/\mu) + g]$. If μ is much larger than p/g, one obtains: $H_g = nI/g$. This formula shows the possibility of *producing a large field* by concentrating the field produced by a much larger coil into a small air gap.

♦ Figure 2.22-b shows the same situation, but where this time the source of flux is a permanent magnet with rigid magnetization **M** instead of a coil. A more general situation in terms of the circuit than the previous is as follows: The cross-sectional area changes from S_m at the magnet to S_g at the air gap. Defining m as the thickness of the magnet, and e as that of the air gap, B_m and H_m are the induction and the field in the magnet respectively (these are assumed to be uniform), B_g, and H_g are the induction and field in the air gap. The field in the iron is neglected because the permeability of iron is huge.

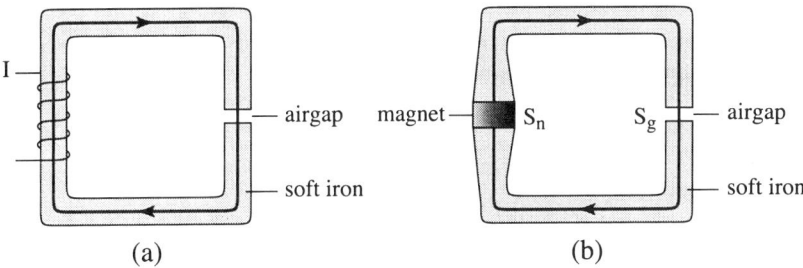

Figure 2.22 - Magnetic circuits with air gaps

Flux produced (a) by a solenoid (b) by a permanent magnet.

Ampère's law gives $m\,H_m + g\,H_g = 0$, and the conservation of flux gives $S_m B_m = S_g B_g$. As $B_m = \mu_0(H_m + M)$ and $B_g = \mu_0 H_g$, one arrives at a linear system of equations: $m\,H_m + g\,H_g = 0$; $S_m H_m - S_g H_g = -S_m M$, giving $H_g = M/(g/m + S_g/S_m)$.

This result can be obtained more directly by using the idea of equivalent currents for the magnetization. The magnet being a plate of thickness m, and cross-sectional area S_m (with $m \ll S_m^{1/2}$), the system behaves as if there were a winding of n turns carrying a current I such that $nI/m = M$, where M is the magnetization of the magnet (assumed rigid). This equivalent coil must be considered as wound, not around the soft magnetic material, but around an air gap of length m, and cross-sectional area S_m. The flux is obtained using the law for adding reluctances neglecting that of iron (due to the fact that the field in the iron is not considered):

$$\Phi = \mu_0 S_g H_g = \mu_0 M\,m/(m/S_m + g/S_g), \text{ and thus, } H_g = M/(S_g/S_m + g/m).$$

Being able to concentrate the effects of a large magnet into a smaller volume illustrates the usefulness of magnetic circuits. Further applications are considered in chapters 15, 16, and 17.

2. ENERGY, FORCES, AND TORQUES IN MAGNETIC SYSTEMS

This section is more detailed than the former, and may be omitted in a first reading.

2.1. ELECTROMAGNETIC COUPLING IN VACUUM

Electrostatics deals only with systems of fixed electric charge while magnetostatics deals exclusively with distributions of steady currents. Strictly speaking, these are independent disciplines. When one is interested in time varying phenomena, one must take into consideration electromagnetic coupling, as formulated by, among others, Faraday, Lenz, Ampère, Maxwell, and Lorentz.

Though this chapter is mostly concerned with magnetostatics, electromagnetic coupling cannot be neglected, because on the one hand it is needed to explain the energy associated with a distribution of steady currents while on the other hand it forms the basis of many important applications.

2.1.1. Lorentz force and Lorentz field

Apply the Laplace equation, introduced at the beginning of this chapter as one of the two fundamental laws of magnetostatics, to a conducting wire in which the mobile charge carriers have a velocity **v**, creating a current $I = N'q\mathbf{v} \cdot \hat{\mathbf{u}}$. Here q is the charge of a carrier, N' the number of charge carriers per unit length, and $\hat{\mathbf{u}}$ is the unit vector along the wire at the point considered. The velocity vector **v** is parallel to $\hat{\mathbf{u}}$ because the charge carriers are confined within the wire, which is itself stationary.

In an induction field **B**, a current element $Idl\,\hat{\mathbf{u}}$ experiences a Laplace force $Idl\,\hat{\mathbf{u}} \times \mathbf{B} = dN\,q\,\mathbf{v} \times \mathbf{B}$ where dN is the number of mobile charge carriers in the element dl. Thus, the Laplace force experienced by a single charge carrier is given by:

$$\mathbf{F} = q\mathbf{v} \times \mathbf{B} \tag{2.56}$$

This equation is valid not only for the case of a current travelling in a wire but in all circumstances, for example, a singly charged particle travelling in vacuum. In such a case, **v** is measured relative to the induction sources (we will return to this point later).

If an induction field **B** and an electric field **E** are experienced simultaneously, one writes:

$$\mathbf{F} = q(\mathbf{E} + \mathbf{v} \times \mathbf{B}) \tag{2.57}$$

This equation, known as the Lorentz equation, has been directly verified with electron beam experiments, and is thus taken to be more than just a microscopic interpretation of the Laplace equation. Note that in equation (2.57), the term $\mathbf{v} \times \mathbf{B}$ plays the same role as an electric field. Henceforth, for convenience, it will be called the *Lorentz field*, but it must be remembered that this term is also used in a very different context (when referring to local fields in dielectrics).

2.1.2. Electric field created in a moving conductor by a magnetic induction B

Consider a rigid conducting wire, with finite length l, oriented along a direction defined by a unit vector $\hat{\mathbf{u}}$.

Now consider moving the conductor at a velocity **v** in a uniform induction field **B** created by fixed sources which are invariant in time (fig. 2.23). At equilibrium, the charges which create the current in the conductor are stationary with respect to the conductor since the current is zero (the conductor is isolated). The velocity of each of these charges in this fixed reference frame is thus reduced to the drift velocity **v**. Therefore, in the wire there exists a Lorentz field $\mathbf{v} \times \mathbf{B}$, with a component $\hat{\mathbf{u}} \cdot (\mathbf{v} \times \mathbf{B}) = \mathbf{B} \cdot (\hat{\mathbf{u}} \times \mathbf{v})$ along the wire axis.

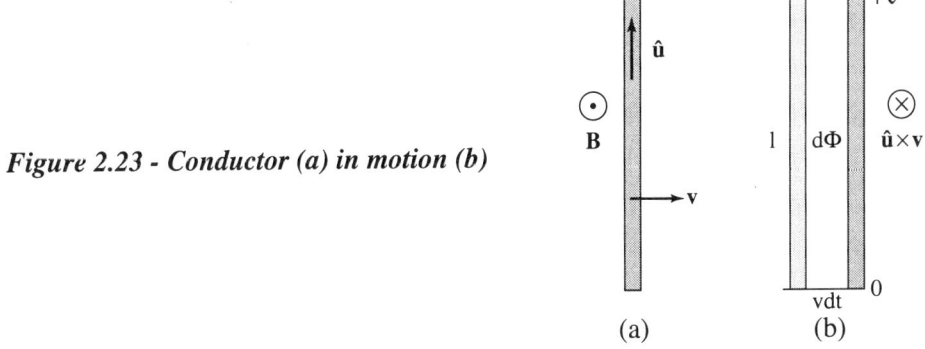

Figure 2.23 - Conductor (a) in motion (b)

The charges must experience another field, which compensates the Lorentz field, if they are to remain stationary with respect to the wire. This question forms the basis of exercise 2.11.

The magnetically induced electric voltage, more commonly known as the induced *electromotive force* (emf) e, a term which is unfortunately less explicit, is defined as the line integral of the Lorentz field along the conductor:

$$e = l\mathbf{B}(\hat{\mathbf{u}} \times \mathbf{v}) = \mathbf{BS'} \qquad (2.58)$$

where **S'** is a vector normal to the plane ($\hat{\mathbf{u}}$, **v**), the modulus of which is the area swept out by the conductor in unit time. |**BS'**| is therefore the flux swept out by the segment in unit time. Taking into consideration the sign convention used in figure 2.23, e may then be written as:

$$e = -d\Phi/dt \qquad (2.59)$$

where, at a given instant, Φ is the flux swept over by the wire during its displacement (note that it is irrelevant that the flux is only defined to within a constant). The value of e has a practical significance as it is the voltage measured between the two ends of the wire by a fixed voltmeter (see ex. 2.11).

Equation (2.59) can be generalised to the case of a non-rigid *closed* wire circuit C which is distorted in an induction field, which may be inhomogeneous, but is invariant in time. At all times t we have:

$$e = \int_C \mathbf{E}_L \, d\mathbf{l} = -d\Phi/dt \qquad (2.60)$$

where Φ is the flux which crosses any surface bounded by the circuit C(t), at time t, and $\mathbf{E}_L$ is the Lorentz field. We recall here that Φ does not depend on the choice of surface, because the flux of **B** is conserved. In this case also, the voltage directly measurable at the extremities of the circuit is equal to e.

It is remarkable that the emf depends only on the rate of change of the flux crossing the circuit C, as seen from equation (2.60). We can also vary the flux by carrying out a different experiment, i.e. not by moving the circuit in a steady induction field, but by varying the induction field itself, the circuit remaining fixed. This looks, at first sight, like a different experiment

2.1.3. *Electric field created by a variable magnetic induction*

Lenz formulated a law to explain the creation of an electric field by a time varying induction field, an effect which had been experimentally demonstrated by Faraday. Lenz's law states that an emf is induced in a stationary closed circuit placed in a varying induction field:

$$e = \int_C \mathbf{E}_{em} d\mathbf{l} = -d\Phi/dt \qquad (2.61)$$

where Φ is the flux crossing the surface bounded by the circuit C. Here we use the normal sign convention: the positive direction of the normal to the surface is chosen arbitrarily, the positive direction of circulation on C is defined by the right-hand-rule.

Conventionally, the field $\mathbf{E}_{em}$, defined in (2.61), is termed an *electromotive field*, even though *magnetically induced field* would again be more explicit if not more expressive. Nevertheless, a specification is for the moment necessary to distinguish the field in question from the Lorentz field as the two are, at first sight, distinct.

Lenz's formulation could lead one to believe that the electromotive field only exists in the presence of a circuit. Of course this is not the case, the conductor should be considered as just a reservoir of mobile charges, serving to reveal the existence of this electromotive field.

Imagine placing a single electric charge q in a varying induction field $\mathbf{B}$ in vacuum. The force $\mathbf{F} = q\mathbf{E}_{em}$ which acts on the charge is defined and measurable, and therefore we can determine the electromotive field $\mathbf{E}_{em}$ at each point in space.

Equation (2.61) is applicable to any closed curve, not necessarily a circuit, and taking the explicit form of Φ, it may be written as:

$$e = \int_C \mathbf{E}_{em} d\mathbf{l} = -\int_S (\partial \mathbf{B}/\partial t) d\mathbf{S} \qquad (2.62)$$

which, according to vector analysis, leads to:

$$\mathbf{curl}\ \mathbf{E}_{em} = -(\partial \mathbf{B}/\partial t) \qquad (2.63)$$

Speculating on the fundamental reason for the similarity between equations (2.60) and (2.61), we see that a key question is the definition of velocity $\mathbf{v}$ which appears in equation (2.58). We have already written that it is the velocity with respect to the sources creating the induction field $\mathbf{B}$. Yet, it is necessary to specify which source we are talking about.

If we create a variable magnetic induction field by changing the current in a solenoid, as in Faraday's experiment, $\mathbf{v}$ must be zero if the current carriers in the solenoid are

2 - MAGNETOSTATICS

taken to be the relevant field sources. We must then consider that this situation is completely different from that examined in the previous section, and we cannot understand the analogy between (2.60), and (2.61).

However, if we assume that the relevant sources are the magnetic poles introduced in section 1.1.7, and that they obey the conservation principle, then we are led to the conclusion that Lorentz type (moving conductor), and Faraday type experiments are not distinct.

In fact, the only way to vary an induction field produced by magnetic poles which are conserved is to move these poles with respect to the circuit. Therefore we naturally recover the idea of relative velocity in Faraday's experiment, and we verify that the electromotive field is equivalent to the Lorentz field.

Of course this interpretation raises a problem, as it is based on the idea of fictitious magnetic poles, when in fact the only real physical sources which have been identified in magnetostatics are currents.

Moreover, it is not very satisfying to have to assume that a constant homogeneous induction field, created for example by two parallel sheet conductors carrying opposite surface current densities, produces different electrical effects depending on whether they are due to stationary or moving sheets.

These difficulties disappear in the special theory of relativity where the only elementary sources for all fields are electric charges. Readers are referred to reference [6] for a more detailed explanation.

2.1.4. Magnetic induction produced by a variable electric field

The two electromagnetic coupling effects which we have seen above concern magnetically induced electrical effects. Conversely, a varying electric field induces magnetic effects. This, at any rate, is Maxwell's assumption, and it is fully supported by its consequences. It can be formulated as:

$$\int_C \mathbf{B}_e \cdot d\mathbf{l} = \mu_0 \varepsilon_0 \int_S (\partial \mathbf{E}/\partial t) d\mathbf{S} = \frac{1}{c^2} \int_S (\partial \mathbf{E}/\partial t) d\mathbf{S} \qquad (2.64)$$

where $\mathbf{B}_e$ is the magnetic induction created by a variable electric field $\mathbf{E}$. C is any closed curve bounding an arbitrary surface S. Note that this equation is a formulation of Ampère's law (2.5), with

$$\mathbf{j} = \mathbf{j}_D = \varepsilon_0 \partial \mathbf{E}/\partial t \qquad (2.65)$$

The vector $\varepsilon_0 \mathbf{E} = \mathbf{D}$ is called the electric displacement. We see that a varying electric displacement in vacuum plays the same role as a current of charged particles of density $\mathbf{j}_D$ equal to $\partial \mathbf{D}/\partial t$. It is termed the displacement current, to denote its special character.

2.1.5. *Maxwell's equations in vacuum*

It is interesting to try to establish the equations that relate only the *observables*, i.e. the *total fields*.

For this we assume that the total electric field **E** is the sum of an electrostatic field, the curl of which is zero, and the divergence of which equals ρ/ε_0 (where ρ is the instantaneous volume density of charge), and an electromotive field, the curl of which is given by equation (2.63), with **B** being the *total* induction, the divergence of which is zero.

In the same way, the total magnetic induction **B** is the sum of a *magnetostatic* induction, satisfying equations (2.8) and (2.10), and an electrically induced induction which satisfies equation (2.64) (or its differential form $\mathbf{curl\,B}_e = \mu_0 \varepsilon_0 \partial \mathbf{E}/\partial t$), where **E** is again the *total* electric field. We are thus led to Maxwell's famous equations:

$$\mathbf{curl\,E} = -\partial \mathbf{B}/\partial t$$

$$\mathbf{curl\,B} = \mu_0 (\mathbf{j} + \varepsilon_0 \partial \mathbf{E}/\partial t)$$

$$\mathrm{div}\,\mathbf{E} = \rho/\varepsilon_0 \quad ; \quad \mathrm{div}\,\mathbf{B} = 0 \tag{2.66}$$

In the absence of sources, these reduce to:

$$\mathbf{curl\,E} = -\partial \mathbf{B}/\partial t$$

$$\mathbf{curl\,B} = \mu_0 \varepsilon_0 \partial \mathbf{E}/\partial t$$

$$\mathrm{div}\,\mathbf{E} = \mathrm{div}\,\mathbf{B} = 0 \tag{2.67}$$

The most spectacular consequence of these equations is the phenomenon of coupled propagation of electric, and magnetic fields at a finite speed c, the speed of light, equal to $(\mu_0 \varepsilon_0)^{-1/2}$. Propagation effects will not be dealt with in more detail in this work, but the interested reader is referred to [6] and [7].

2.1.6. *Electromagnetic coupling in the limit of slow phenomena; magnetic energy*

We are now ready to tackle the problem of magnetic energy in vacuum. For this we consider once again the model of a double current sheet (fig. 2.4-b), and we ask the following question: what energy must be given to the mobile charges in the sheets to go from $j_s = 0$, and thus $B = 0$ everywhere, to $j_s = \bar{j}_s \neq 0$, and thus $B = \mu_0 \bar{j}_s$, just between the sheets? If we exclude the kinetic energy of the charge carriers of non-zero mass which create the current $\bar{j}_s$, and if we neglect energy losses due to friction, the energy supplied to the system is the *magnetic energy* associated with the creation of magnetic induction **B** in vacuum.

Trace a rectangular loop, of height e, and unit length, in the plane Oxz, which contains the normal to the two parallel sheets and the current density vector $\mathbf{j}_s$, so that it is bounded by the two sheets (fig. 2.24, where e is not the emf $\mathcal{e}$). The flux through this loop is $-\mu_0 j_s e$, and when j_s increases by dj_s in a time dt, the change in flux is $-\mu_0 e\, dj_s$.

Figure 2.24 - Double sheet

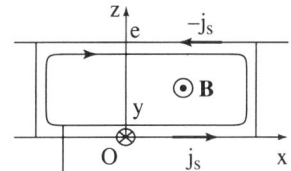

loop for calculation of the electromotive field

According to Lenz's law, an electromotive field is created, with a line integral around the loop equal to $+\mu_0 e\, dj_s/dt$. It is important to note that
- taking account of the invariance of the system under all translations parallel to the plane Oxy, the contribution of the vertical branches of the loop (parallel to Oz) to the line integral of the electromotive field is zero;
- for the same reason, this electromotive field is uniform on each sheet;
- the system is also invariant under a rotation of π around any axis parallel to Oy, and contained in the median plane (at $z = e/2$). As a consequence, if we denote the tangential component of the electromotive field on the lower sheet as E_{em}, the same component on the upper sheet is $-E_{em}$, and the line integral along the loop is given by $-2\, E_{em}$.

Therefore, we have $E_{em} = -1/2\, \mu_0 e\, dj_s/dt$, which shows that the electromotive field *opposes* the increase in current. Thus, to increase j_s by dj_s in a time dt, the surroundings must do work *against* the electromotive field. We calculate this work for a unit length of the sheet in the direction Oy, and thus per unit area of the two parallel sheets. It is, to first order, the work needed to move a line charge density $j_s dt$ around the loop, i.e. $(\mu_0 e\, dj_s/dt)\, j_s\, dt = \mu_0 e\, j_s\, dj_s$. Integrating this work from zero to $\bar{j}_s$, we find per unit area of the double sheet the magnetic energy $(\mu_0/2)\, e\, j_s^2$ which can be expressed in terms of the magnetic induction as $e\, B^2/2\mu_0$. Normalised to unit volume in the region where B exists (between the parallel sheets), the magnetic energy in vacuum may be written, as a function of the induction B or the field $H = B/\mu_0$, as:

$$U' = B^2/2\mu_0 = \mu_0 H^2/2 \qquad (2.68)$$

It can be shown that this result, derived for a particular configuration here, is general. The energy which must be given to mobile charges in any circuit to create the current I, hence the induction $\mathbf{B(r)}$ or the field $\mathbf{H(r)}$, is expressed as an integral over all space of a volume density given by equation (2.68).

We emphasise again that this integral represents only the magnetic part of the energy needed to create induction. The *kinetic* contribution associated with charge carriers of non-zero mass is however negligible in macroscopic circuits. We will see that it is no longer negligible at the atomic level.

Finally, we must define what we mean by the expression "in the limit of slow phenomena", which we used in the title of this section. In the case of the double sheet model, it can be we shown, with the help of Maxwell's equations, that if the final current $\bar{j}_s$ is reached in a time t comparable to e/c, i.e. the propagation time between sheets, the energy given to the charges is greater than that given by equation (2.68). The excess energy corresponds to an energy radiating out of the double sheet. It tends to zero as t tends to infinity.

Generally, we arrive at equation (2.68) only if the process of moving the charges occurs over a time scale which is long compared to the propagation time relevant to the interior of the system considered. In fact, this is a question of thermodynamic reversibility. The energy density of equation (2.68) is often termed a *magnetostatic* energy density.

2.1.7. Forces derived from the energy. The nature of magnetic energy

Let us return once again to the double sheet model (fig. 2.24). The direct application of the Laplace equation gives the pressure P exerted on one sheet by the other:

$$P = j_s B/2 = B^2/2\mu_0 \qquad (2.69)$$

This pressure is repulsive. We may be tempted to calculate the energy per unit area by taking the derivative of $(B^2/2\mu_0)e$ with respect to e. However, we see that this method would give a pressure of the same absolute value but of opposite sign!

The reason for this discrepancy comes from the fact that we have neglected the electromagnetic coupling. In fact, in the displacement δe, there is a variation in the flux concentrated between the sheets. To keep the induction, and thus the current, constant, an additional energy must be supplied from outside the system.

If P' is the pressure *that is applied to* the sheet as it moves by δe, the expression for the energy per unit area supplied from outside the system becomes:

$$P' \delta e + B j_s \delta e = P' \delta e + (B^2/\mu_0) \delta e.$$

The second term corresponds to the work done against the electromotive field. It is this sum which is equal to $(B^2/2\mu_0) \delta e$, therefore: $P' = - B^2/2\mu_0$, and $P = - P' = B^2/2\mu_0$ in accordance with equation (2.69).

Generally, when possible, it is wiser to calculate the forces between the currents using the Laplace equation directly. If we want to work with energy, we must include the electromagnetic coupling, or else use the Lagrange formalism in treating the magnetic energy associated with macroscopic currents as a *kinetic* energy [1].

2.1.8. Self induction

The magnetic energy stored in a long solenoid of n turns, with a cross-sectional area S, and a length l, is written, after equations (2.21), and (2.68), as:

$$W = \frac{1}{2\mu_0} \left(\mu_0 \frac{n}{l}\right)^2 I^2 Sl = LI^2/2 \qquad (2.70)$$

This relation defines the self-inductance:

$$L = \mu_0 n^2 S/l \qquad (2.71)$$

2 - MAGNETOSTATICS

This definition can be generalised to any coil, on condition that the diameter of the wire remains finite, so as to avoid field and energy density singularities. Even in this case, another difficulty may appear: the field values will depend on the exact current density distributions in the conductors; this point is discussed in detail by Durand [1].

Nevertheless, in practice, these considerations are not very important, and the self-inductance is well defined by the relation:

$$\frac{1}{2}LI^2 = \frac{\mu_0}{2}\int H^2 dV = \frac{1}{2\mu_0}\int B^2 dV \qquad (2.72)$$

where the volume of the conductors is excluded from the integral. Introducing the field coefficient $\mathbf{C_H}$ defined in section 1.1.5, this becomes:

$$L = \mu_0 \int C_H^2 dV \qquad (2.73)$$

The self-inductance L can also be defined in terms of the work, $(d\Phi/dt) I\, dt = I\, d\Phi$, which must be done against the emf to increase I by dI in a time dt. Here Φ is the flux generated across the coil by its own current (from which comes the term *self-inductance*).

Therefore, as a consequence $dW = LI\, dI = I\, d\Phi$, and, since Φ vanishes with I:

$$\Phi = LI \qquad (2.74)$$

Equations (2.73) and (2.74) are taken to be of course equivalent, but are not always equally convenient.

2.1.9. Mutual induction

Consider two coils C_1 and C_2, in which currents I_1 and I_2 are circulating, respectively. If $\mathbf{C_{H1}}$ and $\mathbf{C_{H2}}$ are the vectorial field coefficients of each of these circuits at any point, the total magnetic energy is written:

$$W = \frac{\mu_0}{2}\int (C_{H1} I_1 + C_{H2} I_2)^2 dV = \frac{1}{2}L_1 I_1^2 + \frac{1}{2}L_2 I_2^2 + MI_1 I_2 \qquad (2.75)$$

with

$$L_1 = \mu_0 \int H'_1{}^2 dV \ ; \ L_2 = \mu_0 \int H'_2{}^2 dV \ ; \ M = \mu_0 \int C_{H1} C_{H2} dV \qquad (2.76)$$

L_1 and L_2 are the self-inductances of the respective circuits, when isolated, and M is a new coefficient called the *mutual inductance*. If Φ_1 and Φ_2 are the fluxes crossing C_1 and C_2 respectively, taking into consideration the linearity of the equations of magnetostatics, we can write:

$$\Phi_1 = L_1 I_1 + M_{12} I_2 \ ; \ \Phi_2 = L_2 I_2 + M_{21} I_1 \qquad (2.77)$$

When I_1, and I_2 vary by dI_1 and dI_2, respectively, the work done against the electromotive field is, after equations (2.59), (2.75), and (2.77):

$$dW = I_1 d\Phi_1 + I_2 d\Phi_2 = I_1 (L_1 dI_1 + M_{12} dI_2) + I_2 (L_2 dI_2 + M_{21} dI_1)$$
$$= L_1 I_1 dI_1 + L_2 I_2 dI_2 + M(I_1 dI_2 + I_2 dI_1) \qquad (2.78)$$

This identity proves the reciprocal character of mutual inductance:

$$M_{12} = M_{21} = M \qquad (2.79)$$

2.1.10. Spontaneous and induced magnetic moments

As a simple model for a microscopic magnetic moment, consider a cylindrical tube with axis Oz, of cross-sectional area S and length l, which carries a spontaneous circumferential current I_0. Suppose that I_0 is due to charge carriers of zero mass which move without friction along the surface of the cylinder (perpendicularly to Oz). If $l \gg S^{1/2}$, the self energy of the moment is reduced to $W = (\mu_0/2)(S/l)I_0^2$, or $W = (\mu_0/2)(m_0^2/V)$ where $m_0 = SI_0$ is the modulus of the spontaneous moment associated with I_0. Note that W tends to infinity when, for constant m_0, we reduce the volume $V = Sl$ of the cylinder towards zero.

Now apply an induction field B parallel to the moment. Applying Lenz's law to our model gives $\mu_0 \Delta I / l = -B$ where ΔI is the variation in current induced by B. The relative variation in current, and thus that of the moment, is $\Delta I / I_0 = Bl/\mu_0 I_0 = BV/\mu_0 m_0$. We see that the relative variation in current induced by B tends to zero at the same time as V, at a given spontaneous moment m_0. The moment m remains practically equal to m_0, independent of the induction field it experiences. For this reason it is considered *rigid*.

Here we have not taken into consideration the kinetic energy of charge carriers with non-zero mass. However, we can make the model more complex, and even introduce various other forms of internal energy (potential energy).

Taking into account these additional energies does not qualitatively change the essential conclusion of this section: the existence of rigid spontaneous moments on the atomic scale can be understood, in accordance with the law of induction, as a result of the presence of a reservoir of internal energy –of large capacity– which absorbs or supplies, practically keeping the moment constant, the work of the electromotive field.

We will see that the spontaneous atomic moments are at the origin of many properties of strongly magnetic materials. In particular, we will see that the magnetization processes in these materials basically involve the rotation of these spontaneous moments.

If we set $I_0 = 0$ in the simple model defined at the beginning of this section, there are no longer spontaneous moments, and the only moment that appears in the presence of an applied induction field B is $m = -BV/\mu_0 = -VH$, where H is the applied field. We then speak of an induced moment, and we can introduce the magnetic polarisability $\alpha = m/H$, in analogy with electrostatics. We already explained above (§ 1.3.1) why we chose H rather than B to measure the excitation field.

We then note that α reduces to $-V$: in particular the induced moment opposes the induction or the applied field. This is the reason why we talk here of diamagnetism.

Nevertheless, atomic diamagnetic polarization given by this simple model is largely overestimated because we have not taken into account the kinetic and potential energies of the electrons. If we consider these, we find that α, while remaining negative, is reduced by a factor of the order of 10^4 to 10^5 with respect to the previous value.

In the rest of the book, we will deal mostly with strongly magnetic materials, the magnetization of which is due, as we stated above, to the ordering of the spontaneous magnetic moments, and in which diamagnetism is negligible.

2.1.11. Energy of a rigid moment in a given induction field

The energy of a moment m in an induction field $\mathbf{B}$ is, by definition, the work supplied (in the algebraic sense of the word) by the moment to the external system as it is removed to infinity, where $\mathbf{B}$ is assumed to be zero. However, conservation of the modulus of a rigid moment m does not require exchange of energy with the external system. Therefore, the required energy is confined to the work done by the force given in equation (2.23), which immediately gives:

$$W = -m \cdot \mathbf{B} \tag{2.80}$$

2.1.12. Dipolar interaction energy

Dipolar interactions will be frequently referred to throughout the rest of this work, and we will now precisely define them. Consider two dipoles, dipole 1 of moment m_1 situated at $\mathbf{r}_1$, and dipole 2 of moment m_2 at $\mathbf{r}_2$. The mutual energy, or energy of interaction, of these dipoles is given by the following equivalent expressions:

$$E_D = -m_1 \mathbf{B}_{12} = -m_2 \mathbf{B}_{21} = -(m_1 \mathbf{B}_{12} + m_2 \mathbf{B}_{21})/2 \tag{2.81}$$

Here $\mathbf{B}_{12}$ and $\mathbf{B}_{21}$ are the magnetic induction fields created at $\mathbf{r}_1$ by m_2, and at $\mathbf{r}_2$ by m_1, respectively. We note that the first of these expressions is the magnetic energy of m_1 in the induction field created by m_2, as defined in the last section while the second expression is the magnetic energy of m_2 in the induction field created by m_1. The equivalence of these expressions comes from the fact that removing dipole 2 to infinity, keeping dipole 1 fixed, is equivalent to removing dipole 1 to infinity, keeping dipole 2 fixed. Expression (2.81) is referred to as the dipole-dipole pair interaction energy.

It is instructive to examine the situation when the moments of the pair are parallel, and at a constant distance d from each other. There exist two noteworthy configurations, termed the polar configuration, and the equatorial configuration.

In the polar configuration both moments lie on a single line parallel to their axes. For simplicity, we assume that each moment has the same modulus m, so that the interaction energy is $-2\mu_0 m^2/4\pi d^3$, according to equations (2.81) and (2.24).

In the equatorial configuration both moments are in the same equatorial plane (plane perpendicular to the axes). The interaction energy is then given by $+\mu_0 m^2/4\pi d^3$.

Therefore, the polar configuration has lower energy than the equatorial configuration; more generally, this is the lowest energy configuration of two moments which are free to rotate but are kept at a constant distance.

Now consider the more general case of an assembly of n dipoles, of moment $\mathbf{m}_i$ situated at $\mathbf{r}_i$ (i = 1, 2, ..., n). The dipolar energy of such an assembly is also, by definition, the work which must be done against the interaction forces to assemble the moments in their final positions, each dipole coming from infinitely far away. It is equal to the sum of the energies of all distinct pairs, i.e.:

$$E_D = -\frac{1}{2}\sum_{i,j\neq i} \mathbf{m}_i \mathbf{B}_{ij} = -\frac{1}{2}\sum_i \mathbf{m}_i \mathbf{B}_i \quad ; \quad \mathbf{B}_i = \sum_{j\neq i} \mathbf{B}_{ij} \qquad (2.82)$$

Here $\mathbf{B}_{ij}$ is the induction field created on the site of moment i by moment j < i, while $\mathbf{B}_i$ is the induction field created at site i by all the other moments.

We now consider a situation which we will encounter very often throughout the rest of this book, viz a lattice of moments. Consider a sample consisting of parallel moments, of the same modulus $\mathfrak{m}$, situated at the nodes of a cubic lattice of cell parameter a. One could imagine that, for an infinite sample (in practice of dimensions much greater than a) the notion of a dipolar interaction energy density has a precise meaning. In fact, this is not at all true: we find that the energy density calculated from equation (2.82), where $\mathbf{B}_{ij}$ is given by equation (2.24), depends on the shape of the sample even if its dimensions are much greater than a. In the case of an ellipsoid, this energy density can be calculated analytically.

Here we just illustrate the results for an ellipsoid of revolution with major axis parallel to the moments. The reduced dipolar interaction energy density, $E_D / \{(\mu_0/2)[\mathfrak{m}/a^3]^2\}$, is plotted as a function of the ratio c/b in figure 2.25, where b and c are the lengths of the minor axis and of the major axis of revolution of the ellipsoid, respectively. Note that this dipolar interaction energy density is maximum for c/b = 0, i.e. for a flat ellipsoid, while it decreases as c/b increases, going to zero for c/b = 1, i.e. for a sphere. When c/b $\to \infty$ the dipolar interaction energy density tends towards a negative value.

This behavior is close to that of the pair dipolar energy, discussed above. We see that the energy of an elongated ellipsoid is lower than that of a flat ellipsoid (of equal volume) because of the greater prominence of the low energy polar configuration in the former case.

We will see shortly that it is convenient to consider the dipolar energy of a lattice as the sum of two contributions. An *intrinsic* contribution which is the minimum energy value corresponding to an infinitely elongated ellipsoid (c/b $\to \infty$), and a magnetostatic or "demagnetising field" contribution which is the energy needed to move from an infinitely elongated ellipsoid to the sample shape considered, while maintaining the number of dipoles, and the lattice volume constant.

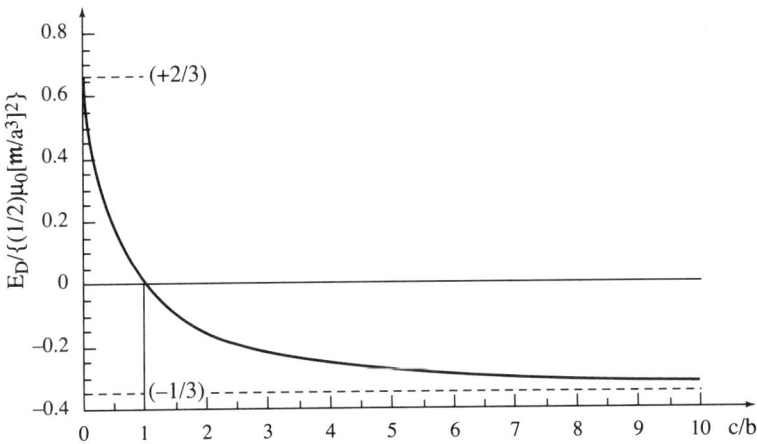

Figure 2.25 - Reduced dipolar interaction energy of a cubic lattice of parallel dipoles with the same moment m

The lattice has cell parameter a, and fills an ellipsoid of revolution with minor axis b, and major axis c parallel to the rotation axis.

2.2. ENERGY, FORCES, AND TORQUES IN MAGNETIC SYSTEMS

In the most general case, by magnetic system we mean an entity which includes both currents (in the form of continuous distributions or wire-like circuits), and magnetised material consisting of spontaneous and induced moments as introduced in section 2.1.10. Energy, forces, and torques in such systems have been dealt with in detail by Durand [1]. Here we shall examine only the situations which are most important in practice, starting with the most obvious.

2.2.1. Direct calculation of torques and forces in a magnetic system

By direct calculation, we mean the application of the Laplace equation (2.3) and equations (2.27) and (2.28), which are derived from the Laplace equation. This method allows the direct determination of torques and forces within a system constituted of rigid circuits and magnetised bodies. For rigid elements, the *force* $\mathbf{F}_i$ and the *torque* $\mathbf{\Gamma}_i$ with respect to a given reference point are directly measurable quantities. For a *rigid circuit* C_i, the basic laws give:

$$\mathbf{F}_i = \int_{C_i} \mathrm{Idl} \times \mathbf{B}_i \quad ; \quad \mathbf{\Gamma}_i = \int_{C_i} (\mathbf{r} - \mathbf{r}_i) \times \mathrm{Idl} \times \mathbf{B}_i \tag{2.83}$$

where $\mathbf{B}_i$ is the induction created by all elements other than the circuit considered. $\mathbf{r}$ defines the point considered, and $\mathbf{r}_i$ the reference point.

For an element V_i, constituted of *solid matter*, equations (2.27), and (2.28) may be directly applied. It is convenient to start from the elementary force and torque which act on the volume of matter dV:

$$d\mathbf{F} = (\mathbf{M}\,dV.\mathbf{grad}).\mathbf{B}_i \quad ; \quad d\mathbf{\Gamma} = \mathbf{M}\,dV \times \mathbf{B}_i \tag{2.84}$$

where $\mathbf{B_i}$ is once again the induction field created by all elements of the system, other than the ith element. We then integrate over the volume of the element considered:

$$\mathbf{F_i} = \int d\mathbf{F_i} \quad ; \quad \mathbf{\Gamma_i} = \int \left[d\mathbf{\Gamma_i} + (\mathbf{r} - \mathbf{r_i}) \times d\mathbf{F_i} \right] \quad (2.85)$$

where $\mathbf{r_i}$ defines the reference point for the ith element.

This force can be equally well written in terms of magnetic pole densities. Recalling that the magnetic pole volume and surface densities equivalent to a given magnetization distribution are $\rho_m = -\text{div}\,\mathbf{M}$, and $\sigma_m = \hat{\mathbf{n}}\cdot\mathbf{M}$, respectively, we may write:

$$\mathbf{F_i} = \int_{V,S} (\rho_m dV + \sigma_m dS)\mathbf{B_i}, \text{ and } \mathbf{\Gamma_i} = \int_{V,S} (\rho_m dV + \sigma_m dS)(\mathbf{r} - \mathbf{r_i}) \times \mathbf{B_i}$$

Here again the induction field $\mathbf{B_i}$ is that produced by all elements of the system, other than the one considered. We see that these direct methods can only be used when we know the fields and magnetization at every point, as well as the respective contributions of all elements of the system. This is illustrated in exercise 14.

The Lorentz force within the material

The Lorentz force can also be directly calculated. In fact, a particle of electric charge q crossing the magnetised material with a velocity $\mathbf{v}$ experiences an instantaneous force $q\,\mathbf{v}\times\mathbf{b}$ where $\mathbf{b}$ is the microscopic induction. If the particle's path length in the material is much longer than the interatomic distance, it is evident that, on average, the particle experiences a force $q\,\mathbf{v}\times\mathbf{B}$. In particular, in a magnetised sample, in the absence of an applied magnetic field, the force is $q\,\mathbf{v}\times\mu_0\mathbf{M}$. We have already mentioned that this effect is exploited in Lorentz electron microscopy [4].

2.2.2. *Magnetization energy for matter, and demagnetising field energy*

Let us consider a sample of a magnetic material, in the shape of an ellipsoid of revolution about the Oz axis, having a demagnetising factor N along this axis, and ask ourselves the following question: *If the sample is initially demagnetised, what energy must be supplied to the sample so that it is uniformly magnetised with a magnetization M_0 along Oz?*

This energy is, by definition, its magnetization energy. We will show that it is convenient to consider it as the sum of two contributions. For this, let us suppose that magnetization results from an ordering of spontaneous atomic moments. Let us virtually cut our sample into elementary magnetised tubes (fig. 2.26) of cross-sectional area dS with their axes parallel to Oz. For example, we can imagine that each tube crosses the sample completely, and has a square cross-section of width da which is much smaller than the major axis of the ellipsoid but much larger than the interatomic spacing. Thus, the sample is composed of a compact array of $S/(da)^2$ elementary tubes, where S is the cross-sectional area of the ellipsoid, taken at the centre of the sample, and perpendicular to the major axis.

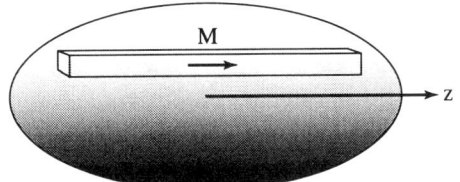

Figure 2.26 - Division of a magnetised sample into elementary magnetised tubes

Now suppose that we section the sample, and remove to infinity a small fraction of the elementary tubes, with total volume dV, taking care to evenly distribute the removed tubes. We thus reduce the macroscopic magnetization of the sample **M** in a uniform manner. If V is the volume of the ellipsoid, we have $d\mathbf{M} = -\mathbf{M}_0(dV/V)$. The work gained (in the algebraic sense of the word) by the operator to section, and then remove to infinity one elementary tube is, according to equation (2.80), $dW_d = -d\mathbf{m}\cdot\mathbf{B}_i$, where $d\mathbf{m}$ is the magnetic moment of the tube, and $\mathbf{B}_i$ is the induction field experienced by the tube which is created by all the other tubes. Remember that here we treat magnetic moments as rigid moments, as discussed in section 2.1.10.

$\mathbf{B}_i$ is thus the magnetic induction in a cylindrical cavity with axis parallel to Oz, thus to $\mathbf{M}_0$ and **M**. We saw in section 1.2.3 that $\mathbf{B}_i = \mu_0\mathbf{H}$, where **H** is the field in a material of magnetization **M**. As **M** is the only field source considered, **H** is also, by definition, the demagnetising field $\mathbf{H}_d$.

If dV is the total volume of tubes taken away to reduce MV by $M_0 dV$, we have:

$$dW_d = -\mu_0\mathbf{M}_0 dV \cdot \mathbf{H}_d = +\mu_0 d\mathbf{M}\cdot\mathbf{H}_d V \qquad (2.86)$$

Since $\mathbf{H}_d = -N\mathbf{M}$, $dW_d = -\mu_0 N\mathbf{M}\cdot d\mathbf{M}\, V$ we finally get, after integration, the energy:

$$W_d = -\mu_0\int_{M_0}^{0} NM\cdot dM\, V = \frac{1}{2}\mu_0 NM_0^2 V = -\frac{1}{2}\mu_0\mathbf{H}_d\cdot\mathbf{M}_0 V \qquad (2.87)$$

Here we get the positive energy W_d which must be supplied to reconstruct the sample by gathering together the tubes, already magnetised, which we call, by definition, the *demagnetising field energy*. The terms *magnetostatic energy* and sometimes, unfortunately, *dipolar energy*, are also used. Note that each isolated magnetised tube is characterised by a zero demagnetising factor along its axis. Therefore it no longer has any demagnetising field energy. The reader can also verify that the definition of this energy given here is the same as that given in section 2.1.12. Each tube contains an assembly of atomic moments, between which there obviously exists an interaction energy.

We saw in section 2.1.12 that there exists an energy contribution, which is dipolar in origin, and which we termed intrinsic. We will see in the following chapters, in particular when dealing with ferromagnetism, that there exist other energy contributions, the two principal ones being exchange energy and anisotropy energy.

It is natural to term the energy which is needed to magnetise an infinitely long bar, from the demagnetised state ($M = 0$), an intrinsic energy. Here we suppose that the state of the system can be described fully by the single variable M, however this is not always true, as we will see below.

The form of the magnetostatic energy density given by equation (2.87) was derived for an isolated, uniformly magnetised ellipsoid. This may be generalised to a sample of any shape which is non-uniformly magnetised, and even to a collection of distinct samples which interact:

$$W'_d = -(\mu_0/2)\, \mathbf{H}_d\, \mathbf{M} \tag{2.88}$$

where $\mathbf{M}$, and $\mathbf{H}_d$ depend on the point considered.

An infinitesimal variation $d\mathbf{M}(\mathbf{r})$ of $\mathbf{M}(\mathbf{r})$ inevitably produces a variation $d\mathbf{H}_d(\mathbf{r})$ in $\mathbf{H}_d(\mathbf{r})$. Thus, we have $dW'_d = -(\mu_0/2)(\mathbf{H}_d\, d\mathbf{M} + \mathbf{M}\, d\mathbf{H}_d)$.

To go from $\mathbf{M}$ to $\mathbf{M} + d\mathbf{M}$ (and thus from $\mathbf{H}_d$ to $\mathbf{H}_d + d\mathbf{H}_d$) we must firstly assemble, at infinity, the moments which create the rigid distribution $d\mathbf{M}$, requiring work which is of second order in $d\mathbf{M}$. Then we can either move the distribution creating $d\mathbf{M}$ so that it superimposes on $\mathbf{M}$ or inversely, we can move $\mathbf{M}$ to superimpose on $d\mathbf{M}$. In the first case, the work needed is $\mu_0 \int_V \mathbf{H}_d\, d\mathbf{M}\, dV$ while in the second case it is $\mu_0 \int_V \mathbf{M}\, d\mathbf{H}_d\, dV$. We thus find:

$$dW_d' = -(\mu_0/2)(\mathbf{H}_d\, d\mathbf{M} + \mathbf{M}\, d\mathbf{H}_d) = -\mu_0\, \mathbf{H}_d\, d\mathbf{M} \tag{2.89}$$

We will later see that the demagnetising field, and the energy associated with it cause the formation of magnetic *domains* in ferromagnetic materials.

Intrinsic magnetization energy

We now recall the significance of the energy density given by equation (2.88): it is the energy which must be supplied to assemble the elementary tubes of material, already having a magnetization $\mathbf{M}$ (parallel to the tube). Thus, this does not include the energy needed to magnetise the tubes, which we have termed the intrinsic magnetization energy. To give a concrete meaning to this energy, consider a sample of magnetic material in the form of an elongated bar on which we wind a coil of n' turns per metre. The field created by a current I passing through the coil is zero outside the coil, and equal to n'I inside the coil (see § 1.1.6.5). As we have already seen, the presence of a very elongated bar does not modify the field. Starting from the situation $I = 0$ and $M = 0$, we reach a magnetization M by applying a current I. For simplicity, we assume the magnetization remains collinear with the field. Consider the elementary reversible processes (from the thermodynamic point of view) used to pass from I to $I + dI$, and from M to $M + dM$; the energy, per unit length of coil, supplied to the system during these processes reduces to: $I\, d\Phi = I\, n'\, S\, dB = H_i\, S\, dB$.

The notation H_i (i for intrinsic) recalls that in a reversible process we must move along the static response curve of the elongated bar, which precisely characterises the intrinsic response of the material. The volume energy density which was supplied to the system (coil and sample) to reach the state with I, H, M, B is finally given by:

$$W' = \int_0^B H_i dB = \frac{1}{2}\mu_0 H_i^2 + \mu_0 \int_0^M H_i dM \qquad (2.90)$$

This energy density must be attributed to the entire system, i.e. vacuum and material: it is composed of a term which corresponds to the creation of field H_i in vacuum, and a term due to the magnetised material. Only the second term can be considered as the intrinsic magnetization energy.

Note that, from the theoretical point of view, it is not necessary to apply a magnetic field to magnetise a sample: we could imagine directly applying, at the microscopic level, the forces and torques needed to rotate the spontaneous magnetic moments. In this case, the energy supplied to the material is, as expected, reduced to:

$$W'_i = \mu_0 \int_0^M H_i dM \qquad (2.91)$$

To summarise, it is convenient to think of the process of magnetising a sample as occurring in two steps:
- the magnetization of isolated elementary tubes, which involves the supply of an energy density given by equation (2.91),
- the reconstruction of the sample by assembling the elementary tubes, which involves the supply of a demagnetising field energy density given by (2.88).

Note that, in this treatment, we have not considered the work of the electromotive field since we have considered the moments to be rigid in the sense given in section 2.1.10.

2.2.3. Zeeman energy, and magnetostatic energy

Take a magnetised sample, in zero external field, and move it into a field $\mathbf{H}_0$, which is not necessarily uniform, keeping the magnetization constant.

We know from equation (2.80) that the work required to do this is $W_z = -\int_V \mu_0 \mathbf{M}\cdot\mathbf{H}_0 dV$ which we call the Zeeman energy. The corresponding energy density is:

$$W'_z = -\mu_0 \mathbf{M}\cdot\mathbf{H}_0 \qquad (2.92)$$

Finally, the energy density supplied to reconstruct a magnetised sample by assembling magnetised tubes in an *external magnetic field* $\mathbf{H}_0$ is given by the sum $W'_m = W'_d + W'_z$. This is often termed the magnetostatic energy.

2.2.4. Magnetostatic torque acting on the magnetization

Consider a sample of magnetised material in a magnetic system, where we think of the sample as a finite lattice of parallel moments of equal modulus, as we did in section 2.1.12.

Contrary to what we did in the previous sections, now we must consider the sample to be formed of two decoupled systems, the lattice which possesses all the degrees of freedom of a solid, and the magnetization which has two rotational degrees of freedom. The torque density that we are interested in is that which acts on the magnetization only, and is thus the torque produced by all other components of the system considered, including the lattice bearing the magnetization. This density, Γ', is calculated by considering the variation in the magnetostatic energy density for an infinitesimal rotation of the magnetization, $d\omega$, the lattice (and thus the material) remaining fixed.

Rotate the magnetization vector $\mathbf{M}$ by an infinitesimal amount $d\omega$ so as to produce a variation in magnetization $d\mathbf{M} = d\omega \times \mathbf{M}$ at every point.

According to equations (2.88), and (2.89), this becomes:

$$dW'_m = \Gamma d\omega = -\mu_0[(1/2)\mathbf{H}_d \cdot d\mathbf{M} + (1/2)\mathbf{M} \cdot d\mathbf{H}_d + \mathbf{H}_0 \cdot d\mathbf{M}]$$
$$= -\mu_0 \mathbf{H} \cdot d\mathbf{M} = -\mu_0 \mathbf{H}(d\omega \times \mathbf{M}) = \mu_0(\mathbf{H} \times \mathbf{M})d\omega \qquad (2.93)$$

where $-\Gamma$ is the torque applied by the operator to produce the rotation $d\omega$, and $\mathbf{H} = \mathbf{H}_d + \mathbf{H}_0$ is the total field acting in (permeating) the material. Thus:

$$\Gamma = \mu_0 \mathbf{M} \times \mathbf{H} \qquad (2.94)$$

This torque can be called the magnetostatic torque. The total torque also has a contribution associated with the intrinsic energy, and we will return to this in the following chapters.

2.2.5. *Energy of a complete magnetic system*

Now we consider a complete system, including currents and magnetised material, where the latter is taken to be made up of rigid atomic moments (rigid in the sense discussed in section 2.1.10). We take the state in which the currents and magnetization are zero everywhere as the zero energy reference state. To reach the considered state of the system by processes which are reversible in the thermodynamic sense of the word, we can proceed in two steps: firstly, we create the magnetization distribution $\mathbf{M}(\mathbf{r})$ without applying a field (by the virtual process we described above, whereby we apply, on the microscopic scale, the torque necessary to rotate each dipole). Thus we supply the intrinsic magnetization energy of the system as well as the demagnetising field energy. After which, the magnetization being blocked (which requires no energy for rigid moments) we increase the currents from zero to their nominal values. In doing this, we supply only the magnetic energy $(\mu_0/2)\int \mathbf{H}_0^2 dV$ where $\mathbf{H}_0$ is the field produced by the *currents alone*, in the absence of magnetised material, the integral being taken over the total volume.

We see that the total energy of the system remains equal to the sum of two *independent* terms, the energy of the currents, and the energy of the magnetic moments. This total energy is independent of the relative position of the sub-system

2 - MAGNETOSTATICS

of the magnetic moments with respect to the sub-system of currents, each sub-system being considered rigid. This can be exploited when calculating forces and torques. Now we will derive a useful theorem.

Reciprocity theorem

Consider a simple system formed of a single coil (which creates an induction $\mathbf{B}_0$ in vacuum), and one sample of magnetised material. Starting with the magnetised sample positioned far from the coil (where $\mathbf{B}_0$ is zero), we move the sample towards the coil, to a given position where $\mathbf{B}_0$ is not zero. To do this, we must supply the energy given by equation (2.92), $-\int \mathbf{MB}_0 dV = -I \int \mathbf{MC}_B dV$ where $\mathbf{C}_B = \mu_0 \mathbf{C}_H$, $\mathbf{C}_H$ being the vectorial field coefficient of the coil, as defined in equation (2.16). But we must consider the phenomenon of induction as well as the work needed to maintain a current I in the coil. This work is equal to $I\Delta\Phi$ where, in these circumstances, $\Delta\Phi$ is the flux created by the sample in the coil. As mentioned above, the energy of the system is independent of the relative positions of the magnetised sample and the coil, so that: $I\Delta\Phi - I\int \mathbf{MC}_B dV = 0$ from which we get the particularly interesting relation, known as the reciprocity relation:

$$\Delta\Phi = \int \mathbf{MC}_B dV \qquad (2.95)$$

In fact, this is just a simplified version of the more general reciprocity theorem which can be found elsewhere, notably in the book by Durand [1].

2.2.6. Forces between magnetic parts in simple situations

We illustrate three approaches.

- *Use of the equivalent charge distributions*
  The magnetised media may in general be replaced by the equivalent charge densities, which are often located predominantly on the surfaces. One can then calculate directly the force between these charges.

- *Attraction between two pole pieces or two magnets close to each other: use of the stored energy*
  Let us assume that the magnetised pole-pieces have a cross-section S large compared to the length g of the air gap, so that the induction B_g can be considered as uniform in the air gap. The magnetostatic energy stored in the air gap is then:

$$W_g = B_g^2 Sg / 2\mu_0$$

  whence the attraction force $F = dW_g / dg$:

$$F = B_g^2 S / 2\mu_0$$

• *A magnet in front of a soft ferromagnetic plate: image approach*
One can show (this is a generalization of the result of exercice E.12.3) that a system consisting of a magnet and a soft ferromagnetic plate ($\mu_r \gg 1$) at a distance g is equivalent to two identical magnets, at a distance 2g from each other, both for the induction distribution between them and for the attraction force between the magnet and the plate.

The basis for this *magnetic image* approach is the fact that the boundary conditions at the surface of the plate are satisfied. The relative orientation of the magnet and its image are determined by the condition that the tangential component of **H** at the plate surface must be zero. The result is identical to the effect of a *mirror plane* on a magnetic moment, as discussed in section 1.1.1 (fig. 2.1) and 1.2.4 (fig. 2.10) of the present chapter. The mirror plane coincides with the surface of the ferromagnetic plate.

3. APPENDICES

3.1. CALCULATION OF B_M OUTSIDE MATTER

Once we are outside the material, we may always consider, however near we are, that we are far from the field sources since the material can be subdivided into pieces as small as we wish. Using the dipolar approximation, we can generalise the expression for the vector potential created by a magnetic moment which is given by equation (2.17):

$$\mathbf{A}_m = \frac{\mu_0}{4\pi} \int_V \frac{\mathbf{M} \times \mathbf{r}}{r^3} dV$$

Now:
$$\frac{\mathbf{r}}{r^3} = -\mathbf{grad}_K\left(\frac{1}{r}\right) = \mathbf{grad}_P\left(\frac{1}{r}\right)$$

where K, and P are defined in figure 2.27.

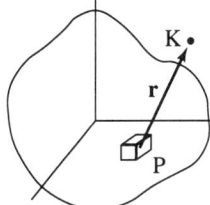

Figure 2.27 - Point K is outside of the magnetic material

Furthermore **curl** (f.**V**) = f. **curl V** – **V** × **grad** f, where f, and **V** are a scalar, and a vector, respectively. Therefore:

$$\mathbf{A}_m = \frac{\mu_0}{4\pi} \int_V \frac{\mathbf{curl\,M}}{r} dV - \frac{\mu_0}{4\pi} \int_V \mathbf{curl}\left(\frac{\mathbf{M}}{r}\right) dV$$

and
$$\int_V \mathbf{curl\,A}\, dV = -\int_S (\mathbf{a} \times \hat{\mathbf{n}})\, dS$$

2 - MAGNETOSTATICS

where S is the surface bounding the volume V, and $\hat{n}$ is the unit vector which is normal to the surface, and directed outwards. It is then deduced that:

$$\mathbf{A}_m = \frac{\mu_0}{4\pi}\int_V \frac{\mathbf{curl\,M}}{r}dV + \frac{\mu_0}{4\pi}\int_S \left(\frac{\mathbf{M}\times\hat{n}}{r}\right)dS \qquad (2.96)$$

Comparing this equation with equation (2.12), we conclude that $\mathbf{A}_m$, and consequently $\mathbf{B}_m$, are created by two equivalent current densities (currents associated with the magnetised material, called bound currents):
- a volume density $\mathbf{j}_m = \mathbf{curl\,M}$,
- a surface density $\mathbf{j}_{ms} = \mathbf{M}\times\hat{n}$.

3.2. CALCULATION OF B_M INSIDE MATTER

In this and the following section, we denote the vector going from A to B by $\mathbf{r}_{AB}$. Locally we can define the microscopic field $\mathbf{b}_m$, which fluctuates greatly in space, in particular from one atom to another. It is not possible to determine it (at least not simply), and from the point of view of magnetostatics, it is not important. As we saw in section 3, we are interested in the field $\mathbf{B}_m$ which is a spatial average of the local field $\mathbf{b}_m$, i.e. $\mathbf{B}_m = \langle\mathbf{b}_m\rangle$. The dipolar approximation is not valid, *a priori*, inside the material.

We must start from the real sources, which are the currents associated with electronic motion, and which can be characterised by the density of bound currents $\mathbf{j}_m$ such that:

$$d\mathbf{b}_m = \frac{\mu_0}{4\pi}\frac{\mathbf{j}_m\times\mathbf{r}}{r^3}dV \qquad (2.97)$$

To calculate $\mathbf{B}_m = \langle\mathbf{b}_m\rangle$ at a point O we do the following. We consider a sphere, centred at O, of volume V large compared with the interatomic distances but small compared with the scale over which magnetization varies, so that the magnetization in this sphere may be considered uniform. We assume we can remove this sphere. We want to calculate $\langle\mathbf{b}_m\rangle_V$ which is the average of $\mathbf{b}_m$ over the volume V. It is the sum of $\langle\mathbf{b}_m\rangle_V$ (int) created by the sources inside the sphere, and $\langle\mathbf{b}_m\rangle_V$ (ext) created by the external sources, including those situated on the surface of the spherical cavity.

First, we are interested in $\langle\mathbf{b}_m\rangle_V$ (int), and we look for the average value in the volume V of the field $\mathbf{b}_{mI}$ created by a source $\mathbf{j}_m\,dV_I$ localised in the neighbourhood of a point I inside the sphere. We have:

$$\langle\mathbf{b}_{mI}\rangle_V = \frac{\mu_0}{4\pi V}\int_V \frac{\mathbf{j}_m dV_I \times \mathbf{r}_{IM}}{r_{IM}^3}dV_M = \frac{\mu_0}{4\pi V}\mathbf{j}_m dV_I \times \int_V \frac{\mathbf{r}_{IM}}{r_{IM}^3}dV_M \qquad (2.98)$$

We will show in section 3.3 that this last integral is equal to: $-(4\pi/3)\mathbf{r}_{OI}$. Therefore:

$$\langle\mathbf{b}_{mI}\rangle_V = -\frac{\mu_0}{3V}\mathbf{j}_m dV_I \times \mathbf{r}_{OI} \qquad (2.99)$$

Now considering all the sources inside the sphere, we obtain:

$$\langle \mathbf{b}_m \rangle_V (\text{int}) = \frac{\mu_0}{3V} \int_V (\mathbf{r}_{OI} \times \mathbf{j}_m) dV_I.$$

Now, the quantity $\frac{1}{2} \int_V (\mathbf{r}_{OI} \times \mathbf{j}_m) dV_I$ is, by definition, the magnetic moment $\mathfrak{m}$ of the complete sphere. This leads to:

$$\langle \mathbf{b}_m \rangle_V (\text{int}) = \frac{2\mu_0}{3V} \mathfrak{m} = \frac{2\mu_0}{3} \mathbf{M} \qquad (2.100)$$

Now let us focus on $\langle \mathbf{b}_m \rangle_V (\text{ext})$. We can take the expression (2.98) for $\langle \mathbf{b}_{mI} \rangle_V$, assuming now that the sources are outside the sphere (and on its surface). Using the corresponding value of Σ (see § 3.3), we get:

$$\langle \mathbf{b}_{mI} \rangle_V = \frac{\mu_0}{4\pi} \mathbf{j}_m dV_I \times \frac{\mathbf{r}_{IO}}{r_{IO}^3} = \mathbf{b}_{mI} (O).$$

Therefore, $\langle \mathbf{b}_m \rangle_V (\text{ext}) = \mathbf{b}_m (O) (\text{ext})$, i.e. the value of $\mathbf{b}_m$ at the centre O and originating from the sources external to the sphere. The latter can be calculated as in the previous section (because point O considered is in a cavity, i.e. outside of the material) from the equivalent current densities. Thus $\langle \mathbf{b}_m \rangle_V (\text{ext})$ is the field at point O arising from three contributions, and may be written as:

$$\langle \mathbf{b}_m \rangle_V (\text{ext}) = -\frac{2\mu_0}{3} \mathbf{M} + \frac{\mu_0}{4\pi} \int \frac{\mathbf{j}_m \times \mathbf{r}}{r^3} dV + \frac{\mu_0}{4\pi} \int \frac{\mathbf{j}_{ms} \times \mathbf{r}}{r_{IM}^3} dS \qquad (2.101)$$

where the three contributions arise respectively from:
- the surface density of bound currents lining the cavity, the calculation of which is treated in exercice n° 4,
- the surface density of bound currents distributed on the outer surface of the material, and
- the volume density of bound currents.

We are thus led to the expression $\mathbf{B}_m = \langle \mathbf{b}_m \rangle_V (\text{int}) + \langle \mathbf{b}_m \rangle_V (\text{ext})$, which is identical to that used for a point outside the material, viz:

$$\mathbf{B}_m = \frac{\mu_0}{4\pi} \int \frac{\mathbf{j}_m \times \mathbf{r}}{r^3} dV + \frac{\mu_0}{4\pi} \int_V \frac{(\mathbf{j}_{ms} \times \mathbf{r})}{r_{IM}^3} dS \qquad (2.102)$$

We can check, from the expressions for the current densities as a function of magnetization (see § 2.3.1), that this expression for $\mathbf{B}_m$ is indeed the curl of the vector potential $\mathbf{A}_m$ defined by equation (2.96).

3.3. CALCULATION OF THE INTEGRAL Σ

Consider a sphere of volume V, centred at O, with M any point in the sphere, and I any arbitrary point. We are interested in the integral:

$$\Sigma = \int_{\text{sphere}} \frac{\mathbf{r}_{IM}}{|\mathbf{r}_{IM}|^3} dV_M = -\int_{\text{sphere}} \frac{\mathbf{r}_{MI}}{|\mathbf{r}_{MI}|^3} dV_M \qquad (2.103)$$

which is, to within a constant, the expression for the electric field **E** created at I by a uniformly charged sphere with charge density ρ (centred at O, and of volume V) i.e.:

$$\mathbf{E} = \frac{\rho}{4\pi\varepsilon_0} \int_{sphere} \frac{\mathbf{r}_{MI}}{|\mathbf{r}_{MI}|^3} dV_M.$$ This field can be easily calculated with Gauss' theorem: $\mathbf{E} = (\rho/4\pi\varepsilon_0)\mathbf{r}_{OI}$ if I is inside the sphere, and $\mathbf{E} = (\rho V/4\pi\varepsilon_0)\mathbf{r}_{OI}/|\mathbf{r}_{OI}|^3$ if I is outside the sphere. It follows that if point I is inside the sphere $\mathbf{\Sigma} = -(4\pi/3)\mathbf{r}_{OI}$, and if I is outside the sphere, $\mathbf{\Sigma} = V\mathbf{r}_{IO}/|\mathbf{r}_{IO}|^3$.

EXERCISES

E.1 - Field along the axis of a cylinder, uniformly magnetised along this axis

A cylinder of length 2L, and radius R is uniformly magnetised along its axis Oz. Derive expressions for the magnetic induction **B** and the magnetic field **H** at any point P on the axis, using successively the ampérian and coulombian methods.

B and **H** should be expressed as functions of the half cone angles α_1 and α_2 (where the bases of the cones coincide with those of the cylinder, and the tops with P (α_1, $\alpha_2 \in [0, \pi/2]$), and of z, with the origin taken at the centre of the cylinder (fig. 2.28). Study the limiting cases where D << R, and D >> R.

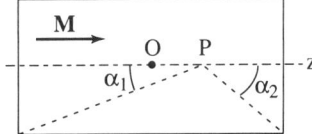

Figure 2.28 - A magnetised cylinder

E.2 - The field in the air gap of a magnet

Two long, coaxial, cylindrical rods of radius R are placed with their ends a distance 2g apart. The rods are uniformly magnetised along their common axis z in the same direction (fig. 2.29).

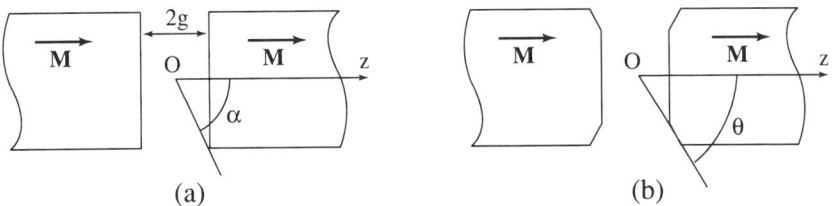

Figure 2.29 - The air gap of a magnet with pole pieces which are (a) cylindrical, and (b) a truncated cone

Calculate using successively the ampérian and coulombian methods the induction **B**₀ at the centre of the air gap O, as a function of the magnetization M and the cone angle α (fig. 2.29-a). Repeat this for truncated conical pole pieces where the new variable θ is the angle of the conical truncation (fig. 2.29-b). Show that B_0 has a maximum value at $θ = θ_0$, and find an expression for this quantity.

E.3 - The field produced by a uniformly magnetised sphere: method one

Two spheres with radius R carry uniformly distributed volume densities of magnetic charge $-ρ_m$ and $+ρ_m$. Their centres N and P are such that $|r_{NP}| \ll R$, and $\mathbf{r}_{NP}$ (the vector joining N and P) is parallel to the z axis. The origin is taken to be at the middle of the line joining N and P.

E.3.1 - Show that the system is practically equivalent to a spherical surface density $σ_m = |r_{NP}| ρ_m \cos θ$ where θ is the angle between Oz, and the vector joining O to the point under consideration.

E.3.2 - Using this equivalence, show that:
- the field $\mathbf{H}_m$ inside the sphere is uniform, and give its direction, and value;
- the potential V_m, and hence the field $\mathbf{H}_m$ outside the sphere, are those of a point-like magnetic dipole located at O, with dipole moment $\mathfrak{m} = \mathbf{r}_{NP} Q_m$, where Q_m is the absolute value of the total magnetic charge of each sphere.

E.3.3 - Using these results show that, for a sphere having uniform magnetization **M**:
- the demagnetising field is uniform, and that $\mathbf{H}_d = -\mathbf{M}/3$. From there show that $\mathbf{B}_m$ is uniform, and that $\mathbf{B}_m = 2μ_0\mathbf{M}/3$;
- the field $\mathbf{H}_m$, and the induction $\mathbf{B}_m$ outside of the sphere are those of a point-like magnetic moment located at the centre of the sphere, having the value $\mathfrak{m} = \mathbf{M}V$, where V is the volume of the sphere.

E.4 - The field within a spherical cavity

Using the results from the previous exercise, give expressions for the field $\mathbf{H}_m$ and the induction $\mathbf{B}_m$ in a spherical cavity within a uniformly magnetised material of infinite dimensions.

Exercises E.5 through E.9 may require results of vector calculus given at the end of appendix 7.

E.5 - The field produced by a uniformly magnetised sphere: method two

Imagine a sphere of radius a, uniformly magnetised with the magnetization **M** parallel to Oz. The goal is to obtain the results in exercise E.3.3 using both the ampérian, and coulombian methods.

E.5.1 - **Ampérian method** (bound currents). By symmetry, $\mathbf{A} = A_φ(θ, r)\hat{\mathbf{e}}_φ$ in spherical coordinates, where $\hat{\mathbf{e}}_φ$ is a unit vector. Show that, in each medium, $A_φ$ is a solution of the equation **curl** (**curl A**) = 0, and can be written as $A_φ = f(r) \sin θ$. Show that $f(r) = C_m r^m + C_n r^n$, and determine m and n. From the boundary conditions at the interface, determine C_m and C_n in each medium. Thus rederive the results of E.3.3.

2 - MAGNETOSTATICS

E.5.2 - Coulombian approach (equivalent charges). Show that, in each of the media, V_m is a solution of Laplace's equation: $\Delta V_m = 0$, and can be written as $V_m = f(r) \cos \theta$. Show that $f(r) = D_m r^m + D_n r^n$. From the boundary conditions at the interface determine D_m and D_n in each of the media. Rederive the results of E.3.3.

E.6 - Sphere with permeability μ subjected to a uniform field

A sphere of radius a, made up of a LHI material having relative permeability μ_r, is subjected to a uniform magnetic induction $\mathbf{B}_0 = \mu_0 \mathbf{H}_0$ parallel to Oz. The aim is to determine at any given point $\mathbf{B}$, $\mathbf{H}$, and $\mathbf{M}$ using both methods.

E.6.1 - Ampérian approach (bound currents). By symmetry, $\mathbf{A} = A_\varphi(\theta, r) \hat{\mathbf{e}}_\varphi$ in spherical coordinates, where $\hat{\mathbf{e}}_\varphi$ is a unit vector. Show that, in each medium, A_φ is a solution of the equation **curl (curl A)** $= 0$, and can be written as $A_\varphi = f(r) \sin \theta$. Show that $f(r) = C_m r^m + C_n r^n$, and determine m, and n. From the boundary conditions at the interface, determine C_m and C_n in each medium. Show that $\mathbf{B}$, $\mathbf{H}$, and $\mathbf{M}$ are uniform in the sphere, and find expressions for them as functions of $\mathbf{H}_0$ and μ_r.

E.6.2 - Coulombian approach (equivalent charges). Show that, in each of the media, V_m is a solution of Laplace's equation: $\Delta V_m = 0$, and can be written as $V_m = f(r) \cos \theta$. Show that $f(r) = D_m r^m + D_n r^n$. From the boundary conditions at the interface determine D_m and D_n in each of the media. From these find expressions for $\mathbf{B}$, $\mathbf{H}$, and $\mathbf{M}$ within the sphere.

E.7 - A cylindrical rod with permeability μ, subjected to a uniform field perpendicular to its axis

A cylindrical rod (radius a) of LHI material (relative permeability μ_r) with infinite length is subjected to a uniform induction $\mathbf{B}_0 = \mu_0 \mathbf{H}_0$ perpendicular to its axis. The aim is to determine $\mathbf{B}$, $\mathbf{H}$, and $\mathbf{M}$ at any point using both methods.

E.7.1 - Ampérian approach (bound currents). By symmetry, $\mathbf{A} = A_z(\theta, r) \hat{\mathbf{e}}_z$ in cylindrical coordinates, where $\hat{\mathbf{e}}_z$ is a unit vector. Show that, in each medium, A_z is a solution of the equation **curl (curl A)** $= 0$, and can be written as $A_z = f(r) \sin \theta$. Show that $f(r) = C_m r^m + C_n r^n$, and determine m and n. From the boundary conditions at the interface, determine C_m and C_n in each medium. Show that $\mathbf{B}$, $\mathbf{H}$, and $\mathbf{M}$ are uniform in the cylinder, and find expressions for them in terms of $\mathbf{H}_0$ and μ_r.

E.7.2 - Coulombian approach (equivalent charges). Show that, in each of the media, V_m is a solution of Laplace's equation: $\Delta V_m = 0$, and can be written as $V_m = f(r) \cos \theta$. Show that $f(r) = D_m r^m + D_n r^n$. From the boundary conditions at the interface determine D_m and D_n in each of the media. From these find expressions for $\mathbf{B}$, $\mathbf{H}$, and $\mathbf{M}$ within the cylinder.

E.8 - A spherical magnetic shield

A hollow sphere (external radius a, and internal radius b) of LHI material with relative permeability μ_r is subjected to a uniform induction $\mathbf{B}_0 = \mu_0 \mathbf{H}_0$.

E.8.1 - Ampérian approach (bound currents). By symmetry, $\mathbf{A} = A_\varphi(\theta, r)\hat{\mathbf{e}}_\varphi$ in spherical coordinates, where $\hat{\mathbf{e}}_\varphi$ is a unit vector. Show that, in each medium, A_φ is a solution of the equation **curl (curl A)** $= 0$, and can be written as $A_\varphi = f(r) \sin\theta$. Show that $f(r) = C_m r^m + C_n r^n$, and determine m and n. From the boundary conditions at the interface, determine C_m and C_n in each medium. Find, as a function of $\mathbf{B}_0$ and μ_r, the induction **B** inside the cavity.

E.8.2 - Coulombian approach (equivalent charges). Show that, in each of the media, V_m is a solution of Laplace's equation: $\Delta V_m = 0$, and can be written as $V_m = f(r)\cos\theta$. Show that $f(r) = D_m r^m + D_n r^n$. From the boundary conditions at the interface determine D_m and D_n in each of the mediums. Find, as a function of $\mathbf{B}_0$ and μ_r, the induction **B** inside the cavity.

By how much is B_0 reduced for $\mu_r = 5{,}000$, $a = 64$ mm, and $b = 60$ mm?

E.9 - A cylindrical magnetic shield

A hollow cylinder (external radius a, and internal radius b) of LHI material with relative permeability μ_r is subjected to a uniform induction $\mathbf{B}_0 = \mu_0 \mathbf{H}_0$ perpendicular to its axis.

E.9.1 - Ampérian approach (bound currents). By symmetry, $\mathbf{A} = A_z(\theta, r)\hat{\mathbf{e}}_z$ in cylindrical coordinates, where $\hat{\mathbf{e}}_z$ is a unit vector. Show that, in each medium, A_z is a solution of the equation **curl (curl A)** $= 0$, and can be written as $A_z = f(r)\sin\theta$. Show that $f(r) = C_m r^m + C_n r^n$, and determine m and n. From the boundary conditions at the interface, determine C_m and C_n in each medium. Find, as a function of $\mathbf{B}_0$ and μ_r, the induction **B** inside the cavity.

E.9.2 - Coulombian approach (equivalent charges). Show that, in each of the mediums, V_m is a solution of Laplace's equation: $\Delta V_m = 0$, and can be written as $V_m = f(r)\cos\theta$. Show that $f(r) = D_m r^m + D_n r^n$. From the boundary conditions at the interface determine D_m and D_n in each of the media. Find, as a function of $\mathbf{B}_0$ and μ_r, the induction **B** inside the cavity.

By how much is B_0 reduced for $\mu_r = 5{,}000$, $a = 64$ mm, and $b = 60$ mm?

E.10 - The magic cylinder

E.10.1 - Let two cylinders, of infinite length and radius R, contain a uniform volume density of magnetic charge. The first cylinder has volume density $-\rho_m$, and its axis intersects the $z = 0$ plane at O_1 with coordinates $x = -a$, and $y = 0$. The second cylinder has current density $+\rho_m$, and its axis cuts the $z = 0$ plane at O_2 having coordinates $x = +a$, and $y = 0$. Show that, if $a \ll R$, this system is equivalent to a cylinder of radius R having surface magnetic charge density at a point A (in cylindrical coordinates R, θ, and z) given by $\sigma_m(\theta) = 2a\rho_m \cos\theta$.

E.10.2 - Using the superposition law, show that the magnetic field created, at a point P inside the cylinder, by a cylindrical surface charge density of the type $\sigma_m = \sigma_0 \cos\theta$ is uniform, and given by $\mathbf{H}_m = -(\sigma_0/2)\hat{\mathbf{e}}_x$, where $\hat{\mathbf{e}}_x$ is a unit vector (hint: calculate the field taking P in the z = 0 plane, and explicitly include the vectors O_1P and O_2P in the contributions from each cylinder).

E.10.3 - A hollow cylinder (internal and external radii R_i and R_e respectively), considered as infinite in length, has magnetization $\mathbf{M}(r, \theta, z)$, with modulus M_0, perpendicular to z, and making an angle 2θ with $\hat{\mathbf{e}}_x$.

- State the densities of surface magnetic charges equivalent to the material, and determine, in the cavity, the magnetic field produced by all of these.
- State the volume magnetic charge density equivalent to the material.
- Using the result of *E.10.1*, and *E.10.2*, give an expression for the field $d\mathbf{H}_m$, in the cavity, produced by a cylindrical layer of material of radius r, and thickness dr.
- Give an expression for the total field $\mathbf{H}_m$ in the cavity.

E.11 - Electromagnetic coupling

Consider a flat capacitor made from two identical circular electrodes A and B with area S. A strandlike conductor of length d takes the shortest path between the two electrodes. The positive sense on the strand is defined by the unit vector $\hat{\mathbf{u}}$ which goes from A to B. A magnetic induction $\mathbf{B}$ is applied parallel to the plates of the capacitor (thus perpendicular to the strand), and the whole set-up moves with velocity $\mathbf{v}$ parallel to the plates.

Imagine a naive model of an electrometer consisting of a mobile point charge q at the centre of a rigid frame, and pulled back towards its resting position by a spring of known stiffness. Assume that the frame and the electrodes have no influence upon the charge q.

E.11.1 - Describe the state of equilibrium in the strand, and calculate the charge acquired by the capacitor.

E.11.2 - The electrometer is placed in the gap between the electrodes, and is moved with the capacitor. What electric field will it indicate? What is the electric field if the electrometer stays between the two electrodes, but does not move with them?

E.11.3 - The electrometer is in vacuum, far from the flat capacitor, and still moves with velocity $\mathbf{v}$. What will be the measured electric field?

E.12 - Magnetic torques

Consider a small ellipsoidal magnetic sample of volume V, having a demagnetising coefficient N along its symmetry axis.

E.12.1 - The sample is made of soft linear and isotropic material with susceptibility χ, and is placed in an induction $\mathbf{B} = \mu_0 \mathbf{H}$ making an angle θ with its symmetry axis. Calculate the torque Γ to which it is subjected, discussing interesting special cases.

E.12.2 - As for the previous question but for a sample having a remanent magnetization M_0 along its axis (χ remains linear).

E.12.3 - The sample still has a fixed magnetization $\mathbf{M}_0$ but now the susceptibility is zero as is the applied induction **B**. The sample is placed a distance d from a sheet of soft iron assumed to have infinite permeability. The distance d is large compared to the size of the sample such that this can be considered pointlike.

Show that the field (and induction) created by this arrangement in vacuum is that of two *dipoles*, the first being the moment under consideration, and the second being an image dipole with the same moment, whose position and orientation are to be specified.

E.12.4 - Using the same geometry as section E.12.3, the sample is now free to rotate around an axis parallel to the iron sheet. Calculate the torque to which it is subjected. What are the stable equilibrium positions?

E.12.5 - We would like to construct a magnetometer based on the measurement of the torque Γ applied to a sample of known fixed moment by an unknown induction **B**. What sort of experimental artefacts does one expect with this principle?

E.13 - Magnetic suction cup

Figure 2.30 is a simplified outline of a *magnetic suction cup*. The magnetic circuit in the form of half a rectangular toroid of perimeter l, and cross-sectional area S is brought near a large piece of soft steel which is assumed to have infinite permeability. Let g be the residual air gap between the suction cup, and the sheet. The magnetic circuit is excited by a coil of length l, having n uniformly spaced windings carrying a current I. In these conditions let $\mathbf{M}(\mathbf{r})$ be the magnetization distribution in the half toroid.

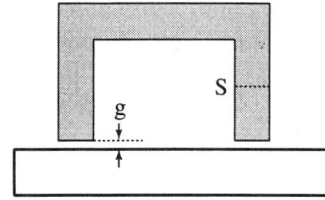

Figure 2.30 - Magnetic suction cup

E.13.1 - Firstly suppose that I = 0, but that by applying the necessary torques at the atomic level the same magnetization **M** is re-created in the half toroid. In the approximation of magnetic circuits, calculate the field H_i in the circuit. Thus deduce the demagnetising field coefficient N.

E.13.2 - Now I ≠ 0, and the toroid is in the linear regime with permeability μ.
Give the expression of the total energy of the magnetic system as a function of I and g. Use the approximations that $g \ll l \ll \mu g$.

E.13.3 - By applying the law of virtual work in two different ways for a change δg of g, calculate the force F exerted by the suction cup on the soft steel piece:
- for g changing with fixed current I,
- for g changing with the flux constant (i.e. the induction is constant in the circuit).

2 - MAGNETOSTATICS

E.13.4 - The circuit is replaced by a permanent magnet having the same fixed magnetization M_0 and of course the windings are removed. Suppose that M_0 stays parallel to the lateral surfaces of the toroid, and that div $M_0 = 0$. Using magnetic charges, and the method of images (see the preceding exercise) describe qualitatively the shape of the field produced by the arrangement. Does this conform to the approximation of magnetic circuits? Why?

E.13.5 - Using the direct method applied to magnetic charges, calculate the force F.

SOLUTIONS TO THE EXERCISES

S.1 Inside: $\mathbf{H} = -M(2 - \cos\alpha_1 - \cos\alpha_2)/2$, $\mathbf{B} = \mu_0 M(\cos\alpha_1 + \cos\alpha_2)/2$.
Outside: $\mathbf{H} = M|\cos\alpha_1 - \cos\alpha_2|/2$, $\mathbf{B} = \mu_0 M|\cos\alpha_1 - \cos\alpha_2|/2$.

For all z, $\mathbf{B} = \dfrac{\mu_0 M}{2}\left(\dfrac{L-z}{\sqrt{R^2+(L-z)^2}} + \dfrac{L+z}{\sqrt{R^2+(L+z)^2}}\right)$

Outside: $\mathbf{H} = \dfrac{\mathbf{B}}{\mu_0}$.

Inside: $\mathbf{H} = -\dfrac{M}{2}\left(2 - \dfrac{L-z}{\sqrt{R^2+(L-z)^2}} + \dfrac{L+z}{\sqrt{R^2+(L+z)^2}}\right)$.

Inside an elongated cylinder sufficiently far from its ends: $\mathbf{H} = 0$, and $\mathbf{B} = \mu_0 M$.
Inside a thin disk: $\mathbf{H} = -M$ and $\mathbf{B} = 0$.

S.2 Before truncation, $B_0 = \mu_0 M(1 - \cos\alpha)$.
After truncation, $B_0 = \mu_0 M[1 - \cos\theta + \sin^2\theta \cos\theta \ln(\tan\alpha/\tan\theta)]$.
$\theta_0 = \cos^{-1}\sqrt{3}/3 = 54.73°$.

S.3 *S.3.2* ♦ $\mathbf{H}_m$ is the vector sum of the fields produced by each sphere, and is calculated using Gauss' theorem.
Thus $\mathbf{H}_m = \rho_m(\mathbf{r}_{PA} - \mathbf{r}_{NA})/3 = -\mathbf{r}_{NP}\rho_m/3$ where A is the point where the field is calculated.
♦ $V_m = Q_m/(4\pi[(1/|\mathbf{r}_{PA}|) - (1/|\mathbf{r}_{NA}|)]$. The first term in the expansion of the distances PA and NA as a function of $r = OA$ and θ gives: $V_m = Q_m |\mathbf{r}_{NP}| \cos\theta/4\pi r^2 = \mathbf{m}\cdot\mathbf{r}/4\pi r^3$, the first term of the expansion for a dipole.

S.3.3 The sphere is equivalent to a surface density $\sigma_m = M\cos\theta$. In the results above, one substitutes $|\mathbf{r}_{NP}|\rho_m$ by M, and $\mathbf{m}$ by MV.

S.4 $\mathbf{H}_m = -M/3$, and $\mathbf{B}_m = 2\mu_0 M/3$.

S.5 *S.5.1* **curl (curl A)** $= \mu_0(\mathbf{j}_0 + \mathbf{j}_m) = 0$, as there are no free currents ($\mathbf{j}_0 = 0$), and $\mathbf{j}_m =$ **curl (M)** $= 0$. m = 1, and n = -2. Boundary conditions: A_φ stays

finite at r = 0, **B** = 0 at infinity, continuity of A_φ (or B_r), and H_θ at r = a.
Outside: $C_m^e = 0$, and $C_n^e = \mu_0 M a^3/3$.
Inside: $C_m^i = \mu_0 M/3$, and $C_n^i = 0$.

S.5.2 $\Delta V_m = 0$ as $\rho_m = -\text{div } \mathbf{M} = 0$. m = 1, and n = – 2.
Boundary conditions: V_m stays finite at the origin, and at infinity, continuity of V_m (or H_θ), and B_r at r = a.
Outside: $D_m^e = 0$, and $D_n^e = Ma^3/3$.
Inside: $D_m^i = M/3$, and $D_n^i = 0$.

S.6 S.6.1 **curl** (**curl A**) = μ_0 ($\mathbf{j}_0 + \mathbf{j}_m$), with $\mathbf{j}_m$ = **curl** (**M**) = χ **curl** (**H**) = χ $\mathbf{j}_0$. Thus **curl** (**curl A**) = 0, where there are no free currents, i.e. in each medium. m = 1, and n = – 2. Boundary conditions: A_φ stays finite at r = 0, **B** = **B**$_0$ at infinity, continuity of A_φ and H_θ at r = a.
Outside: $C_m^e = \mu_0 H_0/2$, and $C_n^e = \mu_0 (\mu_r - 1)a^3 H_0/(\mu_r + 2)$.
Inside: $C_m^i = 3\mu_0\mu_r H_0/2(\mu_r + 2)$, and $C_n^i = 0$.
B = $3\mu_0\mu_r H_0/(\mu_r + 2) = \mu_0\mu_r \mathbf{H} = \mu \mathbf{H}$, and **M** = $3(\mu_r - 1)H_0/(\mu_r + 2)$.

S.6.2 $\Delta V_m + \rho_m = 0$ with $\rho_m = -\text{div } \mathbf{M} = -(\chi/\mu)\text{div } \mathbf{B} = 0$, thus $\Delta V_m = 0$. m = 1, and n = – 2. Boundary conditions: V_m stays finite at the origin and infinity, continuity of V_m and B_r at r = a.
Outside: $D_m^e = 0$, and $D_n^e = (\mu_r - 1)a^3 H_0/(\mu_r + 2)$.
Inside: $D_m^i = (\mu_r - 1)H_0/(\mu_r + 2)$, and $D_n^i = 0$.

S.7 S.7.1 **curl** (**curl A**) = μ_0 ($\mathbf{j}_0 + \mathbf{j}_m$), where $\mathbf{j}_m$ = **curl** (**M**) = χ **curl** (**H**) = χ $\mathbf{j}_0$. Thus **curl** (**curl A**) = 0, where there are no free currents, i.e. in each medium. m = 1, and n = – 1. Boundary conditions: A_z stays finite for r = 0, **B** = **B**$_0$ at infinity, continuity of A_z and H_θ at r = a.
Outside: $C_m^e = \mu_0 H_0$, and $C_n^e = \mu_0 (\mu_r - 1)a^2 H_0/(\mu_r + 1)$.
Inside: $C_m^i = 2\mu H_0/(\mu_r + 1)$ and $C_n^i = 0$.
B = $2\mu H_0/(\mu_r + 1) = \mu \mathbf{H}$, and **M** = $2H_0(\mu_r - 1)/(\mu_r + 1)$.

S.7.2 $\Delta V_m + \rho_m = 0$, with $\rho_m = -\text{div } \mathbf{M} = -(\chi/\mu)\text{div } \mathbf{B} = 0$, thus $\Delta V_m = 0$. m = 1, and n = – 1. Boundary conditions: V_m stays finite at the origin and infinity, continuity of V_m, and B_r at r = a.
Outside: $D_m^e = 0$, and $D_n^e = (\mu_r - 1)a^2 H_0/(\mu_r + 1)$.
Inside: $D_m^i = (\mu_r - 1)H_0/(\mu_r + 1)$, and $D_n^i = 0$.

S.8 **B** = **B**$_0[1 + 2(\mu_r - 1)^2(1 - b^3/a^3)/9\mu_r]^{-1}$, **B** = **B**$_0/196$.

S.9 **B** = **B**$_0[1 + (\mu_r - 1)^2(1 - b^2/a^2)/4\mu_r]^{-1}$, **B** = **B**$_0/152$.

S.10 S.10.3 ♦ $\sigma_m (R_e) = \mathbf{M} \cdot \hat{\mathbf{e}}_r = M_0 \cos\theta$, and $\sigma_m (R_i) = \mathbf{M} \cdot (-\hat{\mathbf{e}}_r) = -M_0 \cos\theta$, where $\hat{\mathbf{e}}_r$ is a unit vector. The result of E.10.1 shows that the two contributions lead to a zero resulting field.

2 - MAGNETOSTATICS

- $\rho_m = -(M_0/r)\cos\theta$.
- Uniform field $d\mathbf{H}_m = (M_0 \, dr/2r)\,\hat{\mathbf{e}}_x$.
- Uniform field $\mathbf{H}_m = \left(\dfrac{M_0}{2}\right)\ln\left(\dfrac{R_e}{R_i}\right)\hat{\mathbf{e}}_x$.

S.11 *S.11.1* The free current carriers are in equilibrium when the field to which they are subjected is zero: this field is the sum of the Lorentz field $\mathbf{E}_L = \mathbf{v}\times\mathbf{B}$, and the electrostatic reaction field $\mathbf{E}_e$ produced by the charges Q_A, and $Q_B = -Q_A$ accumulated on the plates of the capacitor, and is given by:

$$E_e = +Q_A/\varepsilon_0 S = -Q_B/\varepsilon_0 S \text{ where: } Q_B = \varepsilon_0 S\,(\mathbf{v}\times\mathbf{B})\,\hat{\mathbf{u}}.$$

S.11.2 The electrometer placed between the two plates, and moving with them measures the total field to which the strands are subjected, i.e. 0.
The fixed electrometer only measures the electrostatic field, thus $-\mathbf{v}\times\mathbf{B}$, and the corresponding potential difference $V_B - V_A = \hat{\mathbf{u}}\cdot(\mathbf{v}\times\mathbf{B})\,d$ is thus equal to the induced electromotive force: $e = \mathbf{E}_L\cdot\hat{\mathbf{u}}\,d$.

S.11.3 Only the Lorentz field $\mathbf{v}\times\mathbf{B}$.

S.12 *S.12.1* Oxyz is a right handed system of perpendicular axes *linked to the sample* such that Ox is the axis of revolution, and the plane Oxy contains the applied induction **B**. By virtue of symmetry, Oxyz is the *principal system of axes* of the external susceptibility $[\chi_e]$. The coefficients of the demagnetising field along Oy, and Oz are both equal to $(1-N)/2$. Using equation (2.54) the following equation is obtained:

$$[\chi_e] = \begin{bmatrix} \dfrac{\chi}{1+N.\chi} & 0 & 0 \\ 0 & \dfrac{\chi}{1+\dfrac{(1-N)\chi}{2}} & 0 \\ 0 & 0 & \dfrac{\chi}{1+\dfrac{(1-N)\chi}{2}} \end{bmatrix}$$

Knowing the components of field $H_x = H\cos\theta$, $H_y = H\sin\theta$, and $H_z = 0$, this matrix immediately gives M_x, and M_y, and the torque $\mathbf{\Gamma} = \mu_0 \mathbf{M}\times\mathbf{H}\,V$ with component Γ_1 along Oz:

$$\Gamma_1 = (1/2)\,\mu_0 H^2 V \sin 2\theta \,\dfrac{1-3N}{\left(\dfrac{2}{\chi}-N+1\right)\left(\dfrac{1}{\chi}+N\right)}$$

The notable special cases are those which allow the formula to be verified by highlighting obvious results: for the sphere ($N = 1/3$), the torque cancels for all angles; the equilibrium positions are $\theta = 0$ or $\pi/2$ with the stable position at $\theta = 0$ for an elongated ellipsoid ($N < 1/3$), and at $\theta = \pi/2$ for a flattened ellipsoid ($1/3 < N < 1$); $N = 0$ with $\chi \gg 2$, the rod magnetises mainly along its axis, and the torque is written directly as $\Gamma_1 = \mu_0 MVH \sin \theta$, so that

$\Gamma_1 = \mu_0 (\chi H \cos \theta\, V)\, H \sin \theta = (1/2) \mu_0 \chi\, H^2\, V \sin 2\theta$

which is also the result obtained by setting $N = 0$, and $\chi \gg 2$ in the general formula.

S.12.2 The magnetization **M** of the sample is such that $\mathbf{M} = \mathbf{M}_0 + \chi \mathbf{H}_i$ where $\mathbf{H}_i$ is the *internal field*. In the Oxyz system of axes this relationship is written as: $[M] = [M_0] + \chi\,([H] - [N][M])$ where the notation [ ] is used irrespective of whether matrices or vectors are involved.
Hence $([1] + \chi\,[N])\,[M] = [M_0] + \chi\,[H]$. The solution of this linear system gives $M_x = (M_0 + \chi H \cos \theta)/(1 + \chi N)$,
$M_y = \chi H (\sin \theta)/[1 + \chi (1 - N)/2]$, $M_z = 0$, and finally the torque is given as: $\Gamma_z = \Gamma = BM_y \cos \theta - BM_x \sin \theta = \Gamma_1 - BM_0 \sin \theta /(1 + \chi N)$ where Γ_1 is the torque found in question 12.1. Special cases: $M_0 = 0$, in agreement with 12.1; $\chi = 0$, one regains the obvious result $BM_0 \sin \theta$.

S.12.3 The field created by the two dipoles is *by construction* a solution of the magnetostatic equations in the empty half space above the soft iron substrate. In the approximation of infinite susceptibility, the field H is *necessarily zero* in the iron (otherwise the magnetization, and the induction would be infinite). The boundary conditions are, for the tangential component of the field just above the iron, $H_t = 0$, and for the normal component of the induction $B_n = \mu_0 M_n$ where M_n is the component of magnetization in the iron perpendicular to the interface. The first condition is satisfied if the image moment has the same normal component as the source moment, an opposite tangential component, and a position symmetrical with that of the source moment with respect to the interface. The second allows the component M_n and the density of equivalent magnetic charges $\sigma_m = - M_n$ induced on the surface of the iron by the dipole under consideration to be calculated. Knowing that the solution is unique, that which satisfies simultaneously the magnetostatic equations and the boundary conditions is necessarily *the* solution to the problem.

S.12.4 The induction $\mathbf{B}_a$ acting on the dipole of moment $\mathfrak{m}$ ($= \mathbf{M}_0 V$) is that *created by the image*. The direct calculation of the torque thus follows. If α is the angle that $\mathfrak{m}$ makes with the normal to the interface, $\mathbf{B}_a$ has as its *normal* component B_n the axial component from the moment

$\mathfrak{m} \cos \alpha$, and as its *tangential* component B_t the equatorial induction from the moment $\mathfrak{m} \sin \alpha$ (refer to eq. 2.24), giving:
$B_n = (2\mu_0/4\pi)(\mathfrak{m} \cos \alpha/r^3)$, and $B_t = -(\mu_0/4\pi)(\mathfrak{m} \sin \alpha/r^3)$ where $r = 2d$. Hence the torque has a single component Γ_a perpendicular to the plane where the rotation of α is made:
$\Gamma_a = -(1/2)(\mu_0 \mathfrak{m}^2/32\pi d^3) \sin 2\alpha$. This torque cancels for $\alpha = 0$ or π, and $\alpha = \pm\pi/2$. The positions 0 and π are stable (as $d\Gamma/d\alpha < 0$), and only the positions $\alpha = \pm\pi/2$ are unstable. The moment will orient itself perpendicular to the sheet.

S.12.5 For measurements in the vicinity of soft material (e.g. in the air gap of an electromagnet), the result will need correcting for the image effect discussed above if the induction to be measured is the same order of magnitude as $\mu_0 \mathfrak{m}/32\pi d^3$.

S.13 *S.13.1* In the approximation of magnetic circuits, the modulus of the field is uniform, and equal to H_i in the iron, and similarly uniform, and equal to H_g in the air gaps. The fields and the magnetization **M** are everywhere perpendicular to the section of the circuit. Ampère's law, and the conservation of flux gives:
$lH_i + 2g H_g = 0$; $H_i + M - H_g = 0$; where $H_i = -2g M/(l + 2g)$. This is a relation of the form $H = -NM$ with $N = 2g/(2g + l) \cong 2g/l$, if $g \ll l$.

S.13.2 If $M = (\mu - 1) H_i$, it follows that $lH_i + 2g H_g = nI$; $\mu H_i - H_g = 0$, hence $H_i = nI/(l + 2\mu g)$, and $M = (\mu - 1) nI/(l + 2\mu g)$.
If $2\mu g \gg l$, and $l \gg g$, $M = nI/2g$. The same result can be obtained by considering the reluctances.
The energy is written using (§ 2.2.5), knowing that μ is infinite:
$W = (1/2) L_0 I^2 + (1/2) \mu_0 NM^2 lS$, whence:
$W \cong (1/2) L_0 I^2 + (1/2) \mu_0 (2g S) M^2 = (1/2) L_0 I^2 + (1/2) \mu_0 V_g M^2$
where $L_0 = \mu_0 n^2 S/l$ is the self-inductance of the winding in the absence of magnetic material, and V_g is the total volume of the air gaps. With the approximate expression for M, one has
$W = (1/2) I^2 [L_0 + \mu_0 V_g (n^2/4g^2)]$, hence
$W = (1/2) \mu_0 n^2 S I^2 (1/l + 1/2g) \cong (1/2) \mu_0 (n^2 S/2g) I^2 = (1/2) LI^2$
where L is none other than the self inductance of the winding in the presence of the magnetic circuit (in the approximation that $2\mu g \gg l$).

S.13.3 Balancing the magnetostatic force F experienced by the toroid with an applied force $-F$, and allowing the air gap g increase by an infinitesimally small amount (δg) we have:
- at constant current, $\delta W = -F \delta g + I \delta \Phi = (1/2) I^2 \delta L$.
  But $\delta \Phi = I \delta L$, thus $-F \delta g = -(1/2) I^2 \delta L = -(1/2) LI^2(-2\delta g/g)$,
  let: $F \delta g = -(1/2) \mu_0 M^2 S 2\delta g$, where: $F/2S = -(1/2) \mu_0 M^2$.

The magnetostatic pressure of the suction cup is thus negative, which corresponds to an attraction.

♦ at constant flux, the electromotive field is zero,
thus $\delta W = -F \delta e = \delta [1/2) LI^2]$, hence $\delta W = (1/2) I^2 \delta L + LI \delta I$.
At constant flux, and thus constant field in the air-gap, $\delta I = +2\delta e/e$, and finally, $F\delta g = +(1/2) LI^2 2\delta g/g = -(1/2)\mu_0 M^2 2S\delta g$, and so $F/2S = -(1/2)\mu_0 M^2$.

S.13.4 The distribution of magnetic charges equivalent to the magnetization M_0 reduces to a surface density of $+ M_0$ at one extremity of the half toroid, and $- M_0$ at the other. The images are charge distributions of opposite signs, symmetric with respect to the surface of the large plate. The field produced in the upper half space (magnet side), both in vacuum and in the magnet, is thus that of two magnetic sheets of finite thickness (2g). In particular, it is not parallel to the lateral surfaces of the half toroid, and is not negligible outside of these. The field does not conform to the approximation of magnetic circuits, which is not surprising as we are far from having a circuit of high permeability. In such a circuit there would necessarily be charges on the lateral surfaces which would impose on the field the form assumed in the magnetic circuit approximation.

S.13.5 There is no winding, and one is not concerned with induction phenomena. The direct method can be applied. The field produced on a sheet of charges by its image is simply $(1/2) M_0$, and the force is $(1/2) \mu_0 M_0^2 (2S)$, which result is identical to that stated previously.

REFERENCES

[1]   E. DURAND, Magnétostatique (1968) Masson et Cie, Paris.

[2]   L. LANDAU, E. LIFSHITZ, The Classical Theory of Fields (1987) Butterworths-Heinemann, Oxford.

[3]   R.P. FEYNMAN, M.L. SANDS, R.B. LEIGHTON, The Feynman Lectures on Physics, Vol. 1. (1970) Addison-Wesley, Reading, MA.

[4]   P.B. HIRSCH, A HOWIE, R.B. NICHOLSON, D.W. PASHLEY, M.J. WHELAN *in* Electron microscopy of thin crystals, Chap. 16 (1965) 388, Buttersworths, London.

[5]   M. BERTIN, J.P. FAROUX, J. RENAULT, Electromagnétisme 4 (1984) Dunod, Paris.

[6]   E.M. PURCELL, The Berkeley Physics Course: Electricity and Magnetism, Vol. 2 (1973) McGraw-Hill, New York.

[7]   C. VASSALO, Electromagnétisme classique dans la matière (1980) Dunod, Paris.

CHAPTER 3

PHENOMENOLOGY OF MAGNETISM AT THE MACROSCOPIC SCALE

The principal classes of magnetic materials are presented schematically, and are classified according to their response to the magnetic field: dia-, para-, antiferro-, ferro- and ferrimagnetic, etc. On account of the outstanding importance of ferro- and ferrimagnetic substances, their macroscopic magnetic properties are then described in more details: the notions of Weiss domains and magnetic anisotropy are introduced. Finally the magnetic contributions to the other physical properties of these substances are briefly mentioned, which allows us to introduce some coupling phenomena.

1. PRESENTATION OF SOME TYPES OF MAGNETIC BEHAVIORS

An atom is said to be magnetic if it carries a permanent magnetic moment, usually represented by a vector of constant modulus. Every substance is formed from an assembly of atoms which can be either non magnetic or magnetic; in this latter case, the direction, and sometime the modulus of the magnetic moment can depend on the particular environment of each atom (nature and position of the neighbouring atoms, temperature, applied magnetic field).

We will now present very briefly the main types of magnetic behavior, pending a more detailed description which is given in chapter 4. These main types of magnetism are the following: diamagnetism, paramagnetism, antiferromagnetism, ferromagnetism, and ferrimagnetism.

1.1. DIAMAGNETISM

Diamagnetism characterises substances that have only non magnetic atoms: their magnetization, induced by the field, is very weak, and is opposite to the field direction. The susceptibility, virtually independent of the field and temperature, is negative (see § 1 of chap. 4), and is usually of the order of 10^{-5} (fig. 3.1). This magnetism originates from the change in the electronic orbital motion under the effect of the applied magnetic field. According to the Lenz law, the induced currents give rise to an induction flux opposite to the change in the applied field. This magnetism also exists in substances with magnetic atoms, but it is so weak that it is then completely masked by the contribution of these magnetic atoms. However a very strong diamagnetism is present in superconducting substances, with a susceptibility equal to -1: a special chapter is devoted to superconductivity (chap. 19) on account of the very special physical effects which are associated with, and of the increasing number of technological applications, of superconducting materials.

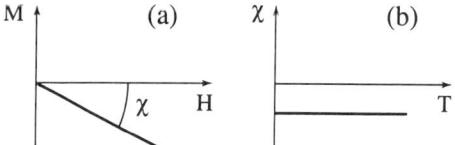

Figure 3.1 - Field dependence of magnetization (a) and thermal variation of the magnetic susceptibility (b) for a diamagnetic substance

The values of the diamagnetic susceptibility of some substances of general use are given in appendix 4.

1.2. PARAMAGNETISM

In a large number of cases the magnetism of paramagnetic substances originates from the permanent magnetic moments of some or all of the constituent atoms or ions. If these moments have negligible interactions with each other, and can orient themseves freely in any direction, this is called paramagnetism of free atoms (fig. 3.2-a).

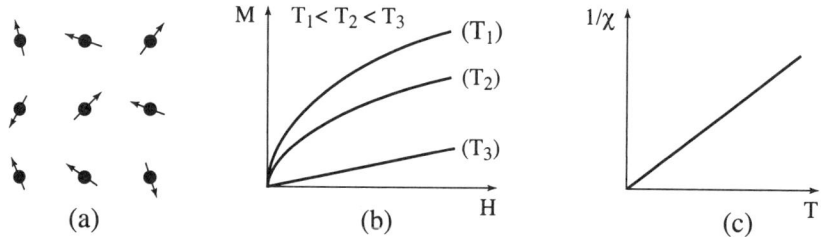

Figure 3.2 - Paramagnetism of free atoms

On applying a magnetic field, the average direction of the moments is modified, and an induced magnetization parallel to the field appears. This magnetization is lower the higher the temperature, i.e. the larger thermal agitation (fig. 3.2-b). When increasing temperature the variations of magnetization as a function of the field become more and more linear. The low field susceptibility is positive, becomes infinite at 0 kelvin, and decreases when temperature is increased. This susceptibility is generally of the order of 10^{-3} to 10^{-5} at room temperature. In the ideal case, the reciprocal susceptibility varies linearly with temperature: this is the *Curie law* (fig. 3.2-c).

In real materials, deviations from Curie's law are often observed, in particular at low temperature. One of the most usual contributions to these deviations is the so called Van Vleck's paramagnetism that will be discussed in chapter 7. In metals, conduction electrons can also give rise to a paramagnetic behavior, called *Pauli paramagnetism*, with a susceptibility that is virtually temperature independent.

The values of the paramagnetic susceptibility of some commonly used substances are given in appendix 4.

1.3. ANTIFERROMAGNETISM

Antiferromagnetism, macroscopically similar to paramagnetism, is a weak form of magnetism, namely with a weak and positive susceptibility. However the thermal variation of the reciprocal susceptibility, measured on a polycrystalline sample, exhibits a minimum at the so called *Néel temperature* T_N (fig. 3.3-c). This maximum in the susceptibility originates from the appearance, below T_N, of an antiparallel arrangement of the magnetic moments. The latter are distributed, in the most simple cases, into two sublattices, with their magnetization equal and opposite, in such a way that, in the absence of magnetic field, the resulting magnetization is zero (fig. 3.3-a). This antiparallel arrangement of the atomic moments results from interactions between neighbouring atoms (called *negative exchange interactions*). These interactions work against the effect of the applied field which would tend to align all the moments parallel. When temperature is reduced down to below T_N, the susceptibility decreases as the thermal agitation, which works against the antiferromagnetic ordering of the moment, decreases. At high temperature, thermal agitation overcomes interaction effects, and one observes again a thermal variation of the susceptibility similar to that of a paramagnet (fig. 3.3-b, and 3.3-c).

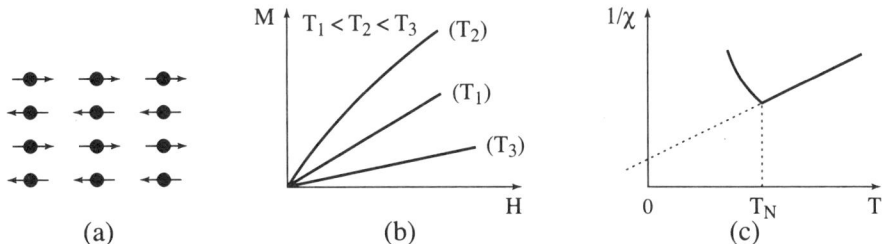

Figure 3.3 - Antiferromagnetism: (a) Spin lattice - (b) M(H) - (c) χ^{-1} (T)

This is the simplest case of an antiferromagnetic material, made of two antiparallel sublattices. In fact many antiferromagnetic substances feature more complex magnetic structures, especially non colinear structures as we will see in chapter 4.

1.4. FERROMAGNETISM

Contrary to the above case, so called *positive exchange interactions* favor, in ferromagnetic substances, a parallel arrangement of magnetic moments in neighbouring atoms (fig. 3.4-a). The effect is then the same as that of a magnetic field, called the *molecular or exchange field*, which would align the moments (this fictitious field will be useful to describe ferromagnetism, but in fact it is not a true magnetic field). As for antiferromagnetism, at high temperature, thermal agitation leads to a susceptibility that is similar to that of a paramagnet: this is the *Curie-Weiss law* that is schematized in figure 3.4-c.

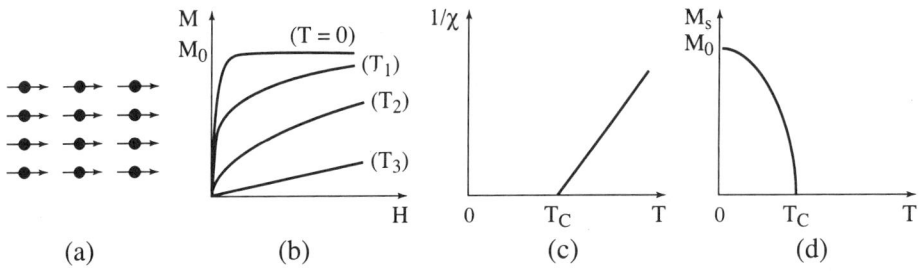

Figure 3.4 - Ferromagnetism: (a) Spin lattice - (b) Field dependence of magnetization ($T_1 < T_c \leq T_2 < T_3$) - (c) Thermal dependence of $1/\chi$ (d) Thermal dependence of the spontaneous magnetization

However, on account of the magnetic interactions, susceptibility –instead of becoming infinite at 0 K as in a paramagnet– becomes infinite at a characteristic temperature, called the *Curie temperature* T_C. Below this temperature, interactions overcome thermal agitation, and a *spontaneous magnetization* (M_s) appears in the absence of an applied magnetic field. This spontaneous magnetization reaches its maximum value, M_0, at 0 K corresponding to parallelism of all the individual moments (fig. 3.4-b and 3.4-d).

In spite of the existence of a spontaneous magnetization below T_C, a piece of ferromagnetic material is not necessarily spontaneously magnetised: its magnetic moment can be zero. The material is then said to be demagnetised. This results from the fact that the material is divided into magnetic domains, called Weiss domains; each domain, which encompasses a large number of atoms, is spontaneously magnetised. From one domain to the other, the moment direction, i.e. the local spontaneous magnetization, varies in orientation in such a way that the resulting magnetic moment of the whole sample is zero. However, under the application of a magnetic field, the distribution of domains is modified, giving rise to the magnetization curve shown in full line in figure 3.5, and called the *initial magnetization curve*: thus, at the macroscopic scale, a ferromagnet is a material in which a strong magnetization is induced by a field.

3 - PHENOMENOLOGY OF MAGNETISM AT THE MACROSCOPIC SCALE

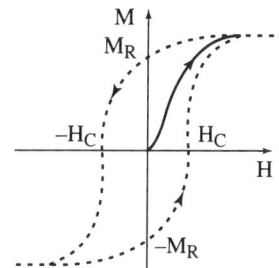

Figure 3.5 - Magnetization curve of a demagnetised material or initial magnetization curve (full line). Hysteresis loop (dashed line)

Here the field scales (1 to 10^6 A.m^{-1}) are quite different from those of figure 3.4-b (from 10^3 to 10^7 A.m^{-1}).

In large enough magnetic fields, magnetization tends to saturate. The laws of the approach to saturation, from which the spontaneous magnetization can theoretically be deduced, will be discussed later in chapter 5 (§ 7.5).

If the applied field is then continuously changed between two extreme values $\pm H_0$, the magnetization process is not reversible, and is described by a *hysteresis loop*. In addition to a strong magnetization, the initial magnetization curve as well as the hysteresis loop are characteristic of ferromagnetic substances. Most technological applications are based on the existence of this loop.

1.5. FERRIMAGNETISM

Ferrimagnetism characterises a material which, microscopically, is antiferromagnetic-like, but in which the magnetizations of the two sublattices are not the same (fig. 3.6-a). The two sublattices no more compensate each other exactly. As a result, below the ordering temperature T_C, a spontaneous magnetization appears in such a way that, as shown in figures 3.6-b, and 3.6-d, the macroscopic properties of a ferrimagnet in this temperature range are close to those of a ferromagnet.

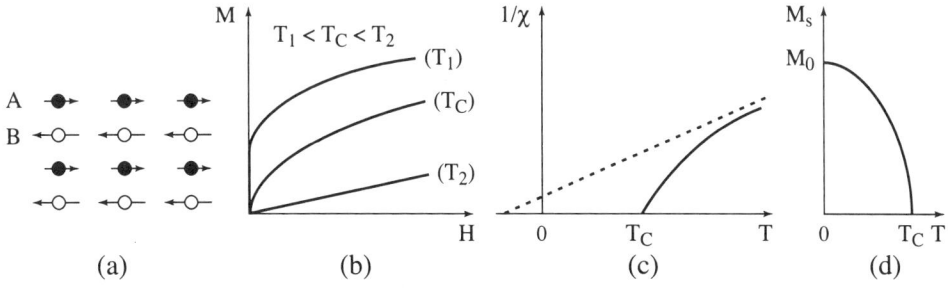

Figure 3.6 - Ferrimagnetism: (a) Spin lattice - (b) Magnetization curves (c) Thermal variation of $1/\chi$ - (d) Thermal variation of spontaneous magnetization

Note however that the spontaneous magnetization of ferrimagnets can exhibit thermal variations much less monotonous than that of figure 3.6-d. In particular it can cancel at a temperature T_{comp} smaller than T_C, on account of the exact compensation of both sublattices: T_{comp} is then called *compensation temperature*. Also, whereas at very high temperature the thermal variation of the reciprocal susceptibility is almost linear, it noticeably shifts away from this linear behavior when approaching the Curie temperature.

Furthermore, the asymptote to the high temperature variation of $1/\chi$ (T) generally crosses the temperature axis in the region of negative values (see fig. 3.6-c), contrary to the case of ferromagnetic substances. These behaviors will be discussed in more details in the next chapter. In the following we will see that a large number of materials which exhibit non collinear magnetic structures, and/or which are made of several types of magnetic atoms with different values of moments, can be classified as ferromagnets or ferrimagnets.

1.6. MAGNETIC PROPERTIES OF PURE ELEMENTS IN THE ATOMIC STATE

79 of the 103 first pure elements carry an atomic moment in the atomic ground state, as shown in table 3.1.

Table 3.1 - Magnetic properties of pure elements in the atomic state

H																	He
Li	**Be**											B	**C**	N	O	F	**Ne**
Na	**Mg**											Al	**Si**	P	S	Cl	**Ar**
K	**Ca**	Sc	Ti	V	Cr	Mn	Fe	Co	Ni	Cu	**Zn**	Ga	**Ge**	As	Se	Br	**Kr**
Rb	**Sr**	Y	Zr	Nb	Mo	*Tc*	Ru	Rh	**Pd**	Ag	**Cd**	In	**Sn**	Sb	Te	I	**Xe**
Cs	**Ba**	La	Hf	Ta	**W**	Re	Os	Ir	Pt	Au	**Hg**	Tl	**Pb**	Bi	Po	At	**Rn**
Fr	*Ra*	*Ac*															
			Ce	Pr	Nd	*Pm*	**Sm**	Eu	Gd	Tb	Dy	Ho	Er	Tm	**Yb**	Lu	
			Th	Pa	U	Np	Pu	Am	Cm	Bk	Cf	Es	Fm	Md	No	Lw	

*In the atomic ground state, the only **non** magnetic elements (bold framed) are those for which J = 0 (see chap. 7): Be..., Zn..., He..., Pd, Yb ($^1S_0$), C... ($^3P_0$), W ($^5D_0$), and Sm ($^7F_0$). The radioactive atoms are shown in italics.*

1.7. MAGNETIC PROPERTIES OF POLYATOMIC MATERIALS

On the contrary, as soon as the elements belong to a polyatomic substance (molecule, pure crystal, compounds and alloys...), only a small number of atoms carry a permanent magnetic moment. This characteristic is illustrated in table 3.2 in the case of pure elements in their solid state. Very few of them are ferromagnetic or

3 - PHENOMENOLOGY OF MAGNETISM AT THE MACROSCOPIC SCALE

ferrimagnetic: it is the case of iron, cobalt, and nickel, as well as several metals of the 4f series of the rare earth elements (Eu to Tm). Furthermore, the magnetic character can depend on the crystallographic structure: in its body centered cubic phase, iron (α-Fe) is ferromagnetic at room temperature and pressure, whereas in its face centered cubic phase (γ-Fe), it is not ferromagnetic at room pressure and, in addition, its moment is then very sensitive to the interatomic distances (tab. 3.3).

Table 3.2 - Low temperature magnetic properties of pure elements in the solid state. In bold framed: the magnetically ordered substances

H																	He
																	dia
Li	Be											B	C	N	**O**	F	Ne
para	dia											dia	dia	dia	**AF**	dia	dia
Na	Mg											Al	Si	P	S	Cl	Ar
para	para											para	dia	dia	dia	dia	dia
K	Ca	Sc	Ti	V	**Cr**	**Mn**	**Fe**	**Co**	**Ni**	Cu	Zn	Ga	Ge	As	Se	Br	Kr
para	para	para	para	para	**AF**	**AF**	**Ferro**	**Ferro**	**Ferro**	dia	dia	dia	dia	dia	dia	dia	dia
Rb	Sr	Y	Zr	Nb	Mo	Tc	Ru	Rh	Pd	Ag	Cd	In	Sn	Sb	Te	I	Xe
para	para	para	para	para	para		para	para	para	dia	dia	dia	*	dia	dia	dia	dia
Cs	Ba	La	Hf	Ta	W	Re	Os	Ir	Pt	Au	Hg	Tl	Pb	Bi	Po	At	Rn
para			para	para	para	para	para	para	para	dia	dia	dia	dia	dia			dia
Fr	Ra	Ac															
			Ce	Pr	**Nd**	Pm	**Sm**	**Eu**	**Gd**	**Tb**	**Dy**	**Ho**	**Er**	**Tm**	Yb	Lu	
			*	para	**AF**		**AF**	**Ferri**	**Ferro**	**Ferro**	**Ferro**	**Ferro**	**Ferri**	**Ferri**	para	para	
			Th	Pa	U	Np	Pu	Am	Cm	Bk	Cf	Es	Fm	Md	No	Lw	
			para		para												

Complex structures (helimagnetic...) are classified as Ferri; AF means antiferromagnetic. The magnetic state (para or dia) of Sn and Ce depends on their crystallographic structure. The substances painted grey are superconductors at very low temperature, and those in italics are radioactive.

In the case of substances consisting of more than one type of atom, the magnetic moment of each type of atom also depends on the considered material, and no simple approach allows one to *a priori* predict whether a given substance will be –or will not be– magnetic. This is well illustrated in table 3.3 in the case of iron.

Table 3.3 - Magnetic moment of iron atoms, at 0 K, and in zero applied field, in several materials

Compound	γ-Fe$_2$O$_3$	α-Fe	YFe$_2$	γ-Fe	YFe$_2$Si$_2$
type	ferri	ferro	ferro	antiferro	–
m$_{Fe}$ (μ_B)	5.0	2.2	1.45	unstable	0

Note - *The Bohr magneton μ_B is a useful unit for magnetic moments at the atomic scale: $1\ \mu_B = 0.9274 \times 10^{-23} A.m^2$ (see § 1 of chap. 7, and app. 2).*

There exist materials in which none of the atoms exhibits ferromagnetic behavior in the pure state, and which however are ferromagnetic: for instance, the Heusler alloy Cu_2MnAl is ferromagnetic at room temperature, whereas copper is diamagnetic, manganese is antiferromagnetic, and aluminium is paramagnetic. The quantum theory of magnetism, presented in chapters 7 and 8, allows one to interpret these results which are *a priori* surprising.

2. PHENOMENOLOGY OF STRONGLY MAGNETIC MATERIALS

On account of the large similarity between the macroscopic magnetic properties of ferrimagnets and ferromagnets at a given temperature, the term "ferromagnetic behavior" is generally used in both cases. More generally, *in the rest of this chapter*, any magnetically ordered substance, in which the moments are neither necessarily parallel nor of the same magnitude, but which features a non-zero magnetization at the *mesoscopic scale* in the absence of applied field, will be classified as ferromagnetic. Mesoscopic scale is considered as intermediate between microscopic (or atomic), and macroscopic scales: we will see that it involves dimensions of the same order of magnitude as those of Weiss domains (thus dependent on the considered material, but typically of the order of the µm).

2.1. ISOTHERMAL MAGNETIZATION CURVES

At the *macroscopic scale*, the field dependence of the magnetization of a ferromagnetic substance, in the case of a bulk and rather pure material which has never been magnetised, generally exhibits the variations reported in figure 3.7.

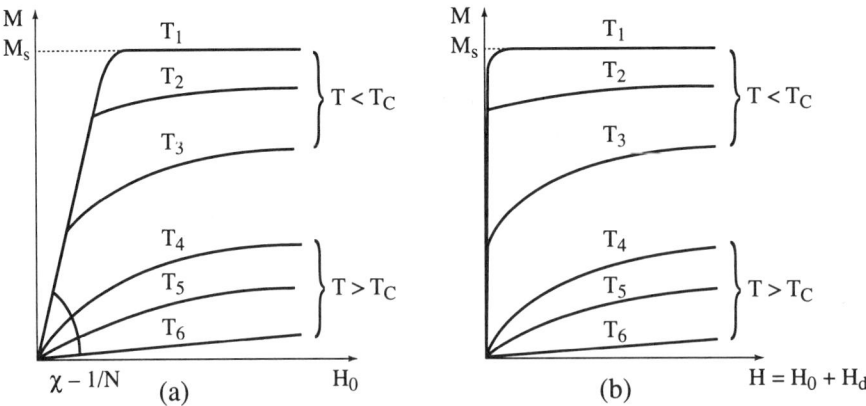

Figure 3.7 - Schematic experimental magnetization curves at different temperatures ($T_1 < T_2 < T_3 < T_4 < T_5 < T_6$) for a ferromagnet without defects as function of the applied field H_0 (a), and of the internal field H (b)

For T < T_C, in low field the susceptibility is large (fig. 3.7-a), and if the magnetization is plotted as a function of the total field (or internal field) H = H_0 + H_d, the initial susceptibility is "nearly" infinite (fig. 3.7-b). As soon as the field is increased, magnetization tends to saturate. The extrapolation to zero internal field of the magnetization curve in the range where magnetization weakly depends on the field defines the spontaneous magnetization $M_s(T)$ (the value which is determined from a neutron diffraction experiment). When T is increased, the magnetization curves are similar, but $M_s(T)$ decreases on account of thermal agitation, and exhibits the variation schematized in figure 3.8. At the critical temperature T_C, called the *Curie temperature*, the spontaneous magnetization vanishes.

For T > T_C, the substance is paramagnetic, and schematic magnetization curves are those shown in figure 3.7 for different temperatures. When the temperature is increased, they become more and more linear, and the initial susceptibility decreases. At high enough temperature, the thermal variation of the reciprocal susceptibility follows the linear *Curie-Weiss law* (fig. 3.8):

$$1/\chi = (T - \theta_p)/\mathscr{C} \tag{3.1}$$

where $\mathscr{C}$ is the *Curie constant, and* θ_p is the *paramagnetic Curie temperature* which is often slightly larger than T_C.

Specialists of magnetism have the habit to show on the same graph these two thermal variations, M_s being zero for T > T_C, and $1/\chi$ being zero for T < T_C. Remember that the curves represent quite different quantities!

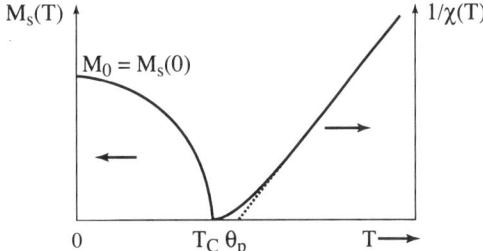

Figure 3.8 - Thermal variations of the spontaneous magnetization $M_s(T)$ and of the reciprocal paramagnetic susceptibility for a ferromagnetic substance

2.2. WEISS DOMAINS AND BLOCH WALLS

Since a ferromagnetic material (in the absence of an applied field) has a spontaneous magnetization, one would expect that the whole sample possesses a permanent moment. In fact, such is not the case, and generally a piece of iron does not attract another piece of iron (except if it has been previously magnetised). Pierre Weiss lifted this contradiction by assuming that at the macroscopic scale (typically for sizes larger than one micrometer) a ferromagnetic material is spontaneously divided into domains. Each domain has a spontaneous magnetization, but from domain to domain the resulting magnetization does not have the same direction. Thus, at the macroscopic level, there is no resulting moment. For example, at 20°C, in a cobalt single crystal with hexagonal structure, magnetization tends to be parallel to the **c** axis: the sample is divided into the same amount of two types of domains corresponding to the two possible orientations of moments along this axis (fig. 3.9-a). Similarly, an iron single crystal, with cubic structure, and in which the magnetic moments are along one of the three fourfold axes [100], [010] or [001] [1], is divided into the same amount of six types of domains (six phases) corresponding to the six possible orientations of the magnetic moments: figure 3.9-b shows four of these phases for a thin slab parallel to the (001) plane.

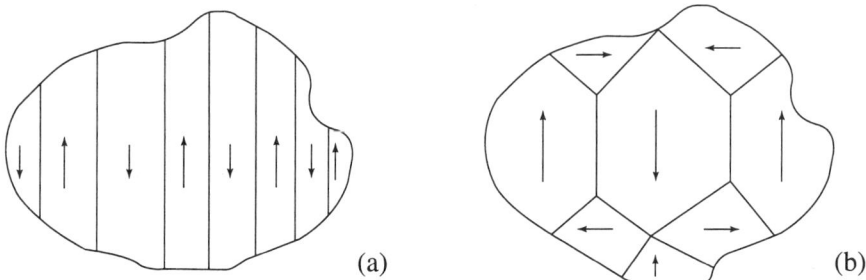

(a) (b)

Figure 3.9 - Schematic representation of domain distributions : (a) Hexagonal crystal in which the 6-fold axis [001] is the easy magnetization direction - (b) Cubic crystal in which the 4-fold axes <100> are the easy magnetization directions (notation <100> means any of the ± [100], [010] or [001] axes)

Magnetic domains are named *Weiss domains*. They are separated by walls, the *Bloch walls*, which consist of a certain number of atomic planes in which the orientation of magnetic moments more or less progressively passes from that of one domain to that of one other. Under the effect of an increasing applied field, H_0, the dominant magnetization process, i.e. $M(H_0)$ in low field, involves wall displacements associated with the volume increase of the domains which are energetically favored, and the progressive disappearance of the others.

In a sample with few defects, walls move nearly freely so that the macroscopic magnetization reaches the spontaneous magnetization for a very weak internal magnetic field: such a material will be termed a *soft ferromagnetic* material. Conversely, in a sample with a large number of defects, the latter constitute obstacles to the wall motion so that the magnetization increases more slowly as a function of the applied field: this is the case of *hard ferromagnetic* materials.

In fact, if the region of the magnetization curve where the internal field is very weak is observed in detail, it turns out that the susceptibility remains finite: the initial part of the magnetization curve strongly depends on the purity of the sample, and on its magnetic, metallurgical, and mechanical histories. This is why magnetization curves below T_C are not systematically similar to that described above, and why it is difficult to characterize a ferromagnet from its magnetization curves alone.

---

1   One speaks of an **n-fold axis** to specify that the system remains invariant under rotations by $2\pi m/n$ where m, and n are integers.

Magnetic domains and the associated walls play a key role in most of the technological applications of magnetism. They will be studied in much more detail in the following, in particular in chapters 5, and 6.

2.3. MAGNETIC ANISOTROPY

We have seen in the previous section that the magnetization of a crystallized material tends to orient preferentially along definite crystallographic directions: this is one of the features of magnetic anisotropy that is explained by the symmetry of the local environment of magnetic atoms. In liquids and amorphous materials, the random distribution of local surroundings usually leads to a macroscopic anisotropy which is zero.

In crystallized magnetic materials, the most frequent case, this anisotropy arising from the local surroundings plays a key role: it is called the *magnetocrystalline anisotropy*, and generally comes from the electrostatic interaction between the charge distribution of the environment, and the magnetic electron orbitals of the atom under investigation: one then uses the term of crystalline electric field (CEF). We will see later that there exist other types of anisotropy such as shape anisotropy, exchange anisotropy, induced anisotropy... but they are generally weaker.

Magnetocrystalline anisotropy has a key influence on the magnetic properties of most materials, especially those with rare earth elements: thus, the field dependence of magnetization and the thermal dependence of susceptibility strongly depend on the orientation of the sample when they are measured on a single crystal. The phenomenological approach to magnetocrystalline anisotropy consists in considering that energy only depends on the orientation of the magnetic moment (actually the crystalline electric field has an effect also on the modulus of the moment).

The free energy density F is therefore expressed in terms of an expansion involving even powers (because it should be invariant under the reversal of magnetization for a given direction) of the direction cosines of magnetization relative to the crystal axes. The terms involved in this polynomial depend on the symmetry of the system in such a way that the energy be invariant under symmetry operations. As a result the number of terms is lower the higher the symmetry. Moreover, experiments show that generally the order of magnitude of these terms very rapidly decreases when their power increases, so that it is enough to consider only the first terms of the polynomial. The expressions of this magnetocrystalline free energy density are given hereafter for the most frequent cases.

2.3.1. Cubic symmetry

Let α_1, α_2, α_3 be the cosines of the angles between magnetization and the x, y, z axes which are parallel to the fourfold axes. The anisotropy energy density can then be written as:

$$F = K_1(\alpha_2^2\alpha_3^2 + \alpha_3^2\alpha_1^2 + \alpha_1^2\alpha_2^2) + K_2\alpha_1^2\alpha_2^2\alpha_3^2 + ... \quad (3.2)$$

The K_i's are *anisotropy constants* which are characteristic of a given material, and which depend on temperature. As an example figure 3.10 shows the variation of F, in a plane parallel to the three main types of symmetry axes, in polar representation, for two simple cases where $K_2 = 0$: $K_1 > 0$, and $K_1 < 0$.

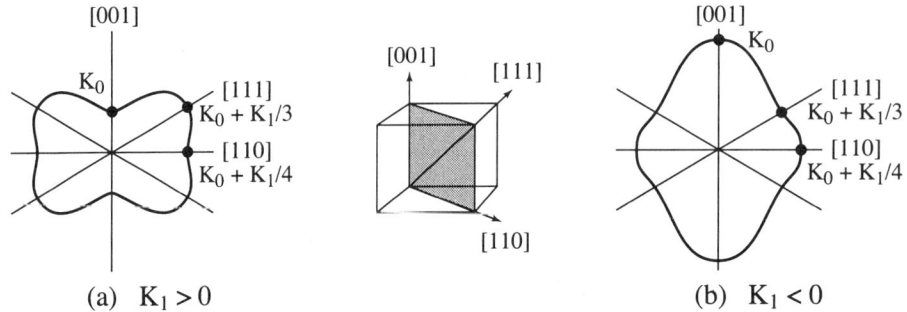

(a) $K_1 > 0$ (b) $K_1 < 0$

Figure 3.10 - Polar representation of magnetocrystalline anisotropy

Case of a cubic crystal. The energy is plotted as a function of the direction of magnetization in a plane (perpendicular to a twofold axis) parallel to the three main symmetry directions. The energy being defined to within a constant, a K_0 term has been added to make the representation easier to read.

It can be easily concluded that, for $K_1 > 0$, the energy is minimum when the moments are directed along a fourfold <001> axis ([001], [100] or [010] because all these directions are equivalent). These axes are then the easy magnetization directions. Conversely, for $K_1 < 0$, the easy magnetization directions are along the threefold axes.

2.3.2. Tetragonal symmetry

θ and ϕ being the angles in polar coordinates, and the z axis being parallel to the sixfold axis [001] (fig. 3.11-a), the free energy density can be written as:

$$F = K_1 \sin^2\theta + K_2 \sin^4\theta + K_3 \sin^6\theta + K_4 \sin^6\theta \cos 6\phi + \ldots \quad (3.3)$$

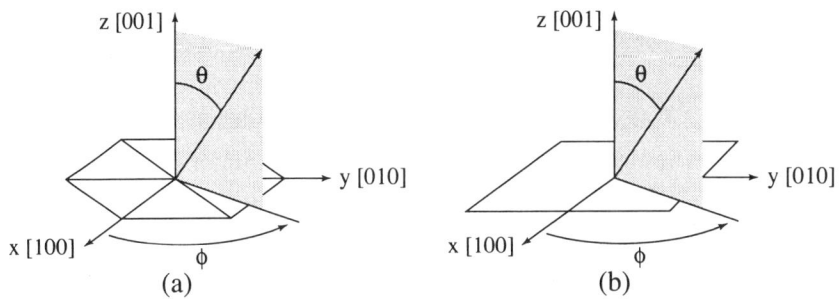

Figure 3.11 - Definition of the θ and ϕ angles in hexagonal (a), and tetragonal (b) symmetries

Note - *In hexagonal symmetry, the [100] and [010] directions make an angle of 120°, whereas in tetragonal symmetry they are perpendicular. It is the [120] direction which is perpendicular to [100] in hexagonal symmetry.*

2.3.3. Quadratic symmetry

In this symmetry (fig. 3.11-b), characterized by a four-fold axis [001] parallel to z, one has the following expression for the magnetic anisotropy energy:

$$F = K_1 \sin^2\theta + K_2 \sin^4\theta + K_3 \sin^4\theta \cos 4\phi + \ldots \tag{3.4}$$

2.3.4. Uniaxial symmetry

In hexagonal and quadratic symmetries, we have seen that the first term is that of second order which only depends on the angle between the magnetic moment, and the axis of highest symmetry. The situation is similar in rhombohedral symmetry, where the axis of highest symmetry is the 3-fold axis. The term of next higher degree is the fourth order one which is generally at least one order of magnitude smaller. This is the reason why, for these three types of symmetry, the expression of the anisotropy energy is often limited to the second order term which involves a unique symmetry axis:

$$F = K_1 \sin^2\theta \tag{3.5}$$

This anisotropy, called *uniaxial*, can also be found in amorphous materials submitted to a stress, or in isotropic materials annealed in the presence of a magnetic field.

2.3.5. Some remarks on the magnetic anisotropy

- Magnetocrystalline anisotropy constants are usually expressed as energy per unit volume. They can range from around $10\,\text{J.m}^{-3}$ in very soft materials to $10^7\text{-}10^8\,\text{J.m}^{-3}$ in some rare earth alloys, passing through values such as $10^3\text{-}10^4\,\text{J.m}^{-3}$ for 3d metals of cubic symmetry (nickel, iron).

- Anisotropy constants generally rapidly decrease when the order of the associated term increases. This is why the expansion is limited to the first terms. The second order term does not exist in cubic symmetry, whereas it appears for all the other symmetries. As a result the anisotropy of a cubic material is generally weaker (for the same type of magnetic atoms). Similarly, in hexagonal (or tetragonal) symmetry, the anisotropy within a plane perpendicular to the six fold (or four fold) axis is much smaller than that between this plane and the six fold (or four fold) axis.

- Let us finally mention that anisotropy constants often vary as a function of temperature much faster that magnetization does. In magnetic insulators, these constants generally vary as $[M_s(T)]^{n(n+1)/2}$ at low temperature ($M_s(T)$ is the spontaneous magnetization, and n the order of the term associated with the anisotropy constant): thus at low temperature in hexagonal symmetry (formula 3.3), K_1 varies as $[M_s]^3$, and K_2 as $[M_s]^{10}$, whereas in cubic symmetry (eq. 3.2), K_1 varies as $[M_s]^{10}$. Around T_C, the thermal dependence is slower, for instance it goes as $[M_s(T)]^2$ for K_1 in hexagonal symmetry. In fact, the situation is usually more complicated, and a detailed thermodynamic study has been published by Callen and coworkers [1], in the case of insulators: these authors show that an analysis of magnetocrystalline anisotropy in terms of harmonic polynomials of cosines of the angles between magnetization and the reference axes would be better than the above presented analysis in terms of homogeneous polynomials (eq. 3.2 to 3.4). For more details the reader is invited to see appendix 6, section A6.5. In the case of metals, no general rule seems to apply, because the details of band structure often play a key role: this is for instance the case for nickel, in which K_1 varies rapidly with magnetization at low temperature, and furthermore in which the higher order anisotropy constants retain noticeable values up to a very high order (n = 34) [2].

2.4. EFFECTS ASSOCIATED WITH ANISOTROPY - ANISOTROPY FIELD - PHASE RULE

In order to understand the effect of magnetocrystalline anisotropy on the magnetic properties of solids, the magnetization curves of some perfect crystals with different symmetries will be studied first. The case of cubic symmetry will then be presented, before looking at the effect of anisotropy on the paramagnetic susceptibility.

2.4.1. *Uniaxial material with tetragonal or hexagonal symmetry*

As pointed out previously, the K_1 (second order) anisotropy constant (see formulae 3.3 and 3.4) is usually dominant compared to those of higher order. We make this assumption hereafter. Let us consider a spherically shaped single crystal, and investigate the low temperature magnetization curves when the applied field is parallel to each of the three main symmetry directions, namely the **c** axis, and the two other directions of the plane perpendicular to **c**, also called *basal plane*. Considering a spherical shape allows one to keep aside the shape anisotropy, and moreover guarantees that the field is uniform in the material (§ 1.2.7 of chap. 2). We will look at two opposite situations depending on whether K_1 is positive or negative.

The "c" axis is the easy magnetization direction

When K_1 is positive, equations (3.3), and (3.4) easily show that **c** is the easy magnetization direction.

- In zero field, the material is divided into the same amount of two types of domains. We can, with no loss in generality, just consider two domains with equal volumes as shown in figure 3.12.

- When the applied field H_0 is parallel to **c**, the free motion of the walls leads to the disappearance of the domains with magnetization opposite to the field, as soon as the latter is larger than $M_s/3$ (i.e. the maximum demagnetising field reached when the sphere has become single domain), to the benefit of domains with magnetization in the same direction as the field. The macroscopic magnetization is then equal to the microscopic magnetization M_s (i.e. the magnetization within one domain).

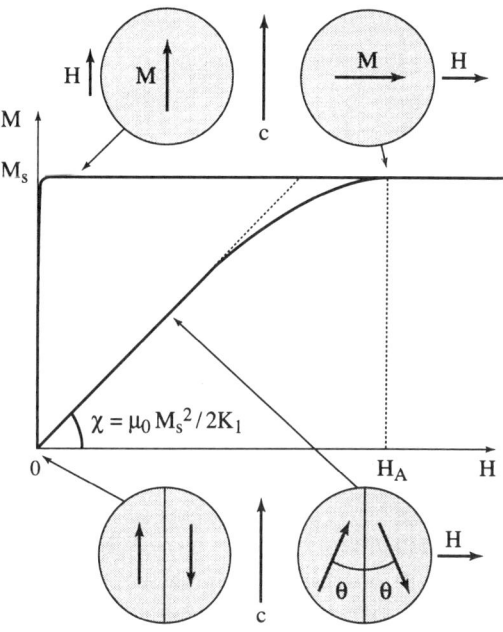

Figure 3.12 - Uniaxial ferromagnetic material in which the high symmetry axis (c) is the easy magnetization direction: low temperature theoretical magnetization curves in terms of internal field when the applied field is parallel or perpendicular to the high symmetry axis

Schematic representation of domain distribution for different fields.

- When the applied field is perpendicular to **c**, none of the domains being favoured, their volume does not change. The magnetization in each of them rotates by the same angle θ. For an applied field H_0, limiting the expansion of the anisotropy to the fourth order term, the free energy density can be written as:

$$F = K_1 \sin^2\theta + K_2 \sin^4\theta - \mu_0 H_0 M_s \sin\theta + \mu_0 N M_s^2 \sin^2\theta / 2 \qquad (3.6)$$

In this equation the next to last term, called the Zeeman term (eq. 2.80) corresponds to the energy of a rigid dipole with moment $M_s V$ (with V the volume of the sample) submitted to an induction $\mu_0 H_0$ which makes a angle $\frac{\pi}{2} - \theta$ with M_s. The last term expresses the demagnetising field energy. Minimizing this energy with respect to θ leads to two solutions:
- $\cos\theta = 0$, i.e. the magnetization in each domain is parallel to the field, and $M = M_s$; this is the solution which takes place when the field is large enough,

♦ and in smaller field, i.e. as long as magnetization M_s within each domain is not parallel to the magnetic field, one obtains the equation:

$$2K_1 \sin\theta + 4K_2 \sin^3\theta = \mu_0 (H_0 - NM_s \sin\theta) M_s = \mu_0 H M_s,$$

which, keeping in mind that $\sin\theta = M/M_s$, leads to:

$$2K_1 M/M_s + 4K_2 (M/M_s)^3 = \mu_0 H M_s \qquad (3.7)$$

M is the macroscopic magnetization, which is parallel to the applied field, and which is nothing but the magnetization component parallel to the field in each domain. equation (3.7) shows that the low field susceptibility is: $\chi = \mu_0 M_s^2 / 2K_1$, and that saturation is reached for a given field H_A called the *anisotropy field* which is given by the equation:

$$H_A = \frac{2K_1 + 4K_2}{\mu_0 M_s} \qquad (3.8)$$

It corresponds to the internal field for which the magnetization of each domain has become parallel to the applied field. It can also be shown that the area enclosed between the curves for $H_0 // c$, and $H_0 \perp c$ is equal to $K_1 + K_2$. Thus measuring M(H) curves parallel and perpendicularly to c allows a determination of the anisotropy constants K_1 and K_2. Note that $\chi \sim M_s / H_A$ if $K_2 \ll K_1$.

Note also that some authors use the definition $H_A = \dfrac{2K_1}{\mu_0 M_s}$.

As an example, M(H) curves measured at 4.2 K on a single crystal of the hexagonal compound YCo_5, in which only cobalt is magnetic, are reported in figure 3.13.

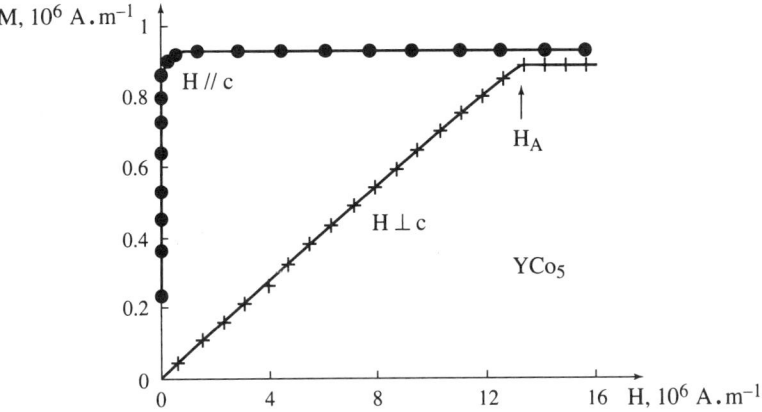

Figure 3.13 - Variations of magnetization as a function of the internal field measured at 4.2 K on a single crystal of the YCo_5 compound, after [3]

Only cobalt bears a magnetic moment. The curve in full line –for $H \perp c$– shows the variation fitted with the theoretical model (rel. 3.7), slightly modified in order to take into account a weak variation of the magnetization magnitude during the process (anisotropy of magnetization).

It can be noticed that, above the anisotropy field, the perpendicular magnetization is slightly smaller that the parallel magnetization. Such a gap, often observed, is called *anisotropy of magnetization*.

Remark on the role of the anisotropy within the plane perpendicular to c

If higher order terms, which correspond to the anisotropy within the basal plane (term in K_4 and K_3 in equations (3.3) and (3.4) respectively), are taken into account, the slope at the origin of the magnetization curve is still equal to $\mu_0 M_s^2 / 2K_1$ for both symmetry directions in the basal plane. On the other hand there will be a slight difference between the anisotropy fields. For instance, in tetragonal symmetry, H_A is given by the equation (3.8) in which K_2 is replaced by $K_2 + K_3$ or $K_2 - K_3$ depending on whether the applied field is along <100> or <110>, respectively. This effect is generally not observed in hexagonal symmetry, on account of the weakness of the K_4 constant.

Planar Anisotropy: "c" is the hard magnetization direction

When K_1 is negative, **c** is the hard magnetization direction. The easy magnetization direction is then one of the symmetry axes of the basal plane, and in low field noticeable differences are observed depending on whether the field is applied along one or the other of these symmetry axes. As an example, the low temperature magnetization curves of a tetragonal crystal are simulated (fig. 3.14) when the field is applied along the three types of symmetry axes, namely <001>, <100>, and <110>. In this example we have considered that $K_3 < 0$ (see eq. 3.4), and $|K_3/K_1| = 10^{-2}$. It can be easily shown that the <100> directions are the easy magnetization ones. In zero applied field, the material is divided into four types of domains in which the spontaneous magnetization is parallel to one of these directions. In a weak field applied along a <100> axis the sample becomes single domain, and the measured magnetization is equal to the spontaneous magnetization M_s. When the field is applied along a <110> axis, in low field there remain only two types of domains, the magnetizations of which are directed at 45° on either side of the field. The magnetization extrapolated to zero internal field is then $M_s \cos 45° = M_s / \sqrt{2}$. This is the so called phase rule (cf § 2.4.2). In increasing field the magnetization increases, and reaches the spontaneous magnetization all the faster the smaller K_3. When the field is applied along <001> none of the domains is favoured, and the four types of domains keep the same volume but the magnetization curve is the same as that of a single domain sample.

In increasing field, magnetization varies from zero up to M_s, a value which is reached for the anisotropy field H_A given by:

$$H_A = 2|K_1|/\mu_0 M_s \qquad (3.9)$$

One could define an anisotropy field when the field is applied along <110> but, in the case of a uniaxial system, the term "anisotropy field" is generally used to characterize the anisotropy between the **c** axis and the basal plane. A deeper analysis of the magnetization curve of the above situation is treated in exercice 2.

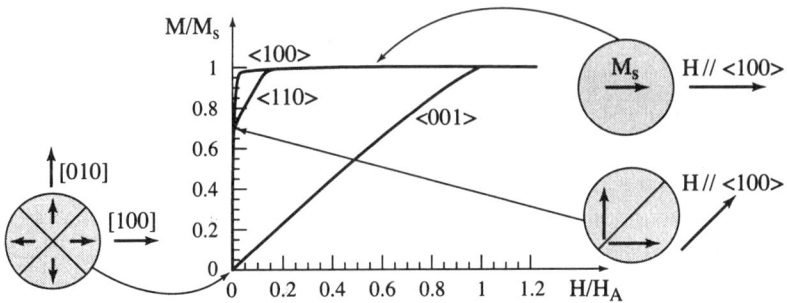

Figure 3.14 - Magnetization curves of a tetragonal single crystal with planar anisotropy

Magnetization curves are plotted for the three main symmetry axes when the <100> axes are the easy magnetization directions (see text). H_A is given by equation (3.9). The calculation has been performed for $K_2 = |K_1|/10$, and $|K_3| = |K_1|/100$. H is the internal field.

In the case of a hexagonal substance, with easy magnetization basal plane, in zero field six types of domains are present. They correspond to the three equivalent easy magnetization directions, and occupy the same volume. When a low field is applied along one of the three equivalent more difficult axes of the basal plane, the *phase rule* leads to a measured magnetization of $M_s \cos 30° = M_s \sqrt{3}/2$.

2.4.2. Magnetocrystalline anisotropy in cubic symmetry

In cubic symmetry, there exist three different types of high symmetry axes, namely the <100>, <110>, and <111> axes (fourfold, twofold, and threefold axes, respectively). The observed magnetic behaviors are different depending on which of these axes is a hard or an easy magnetization direction. As an example, magnetization curves measured at 300 K on an iron single crystal are reported in figure 3.15 after [4].

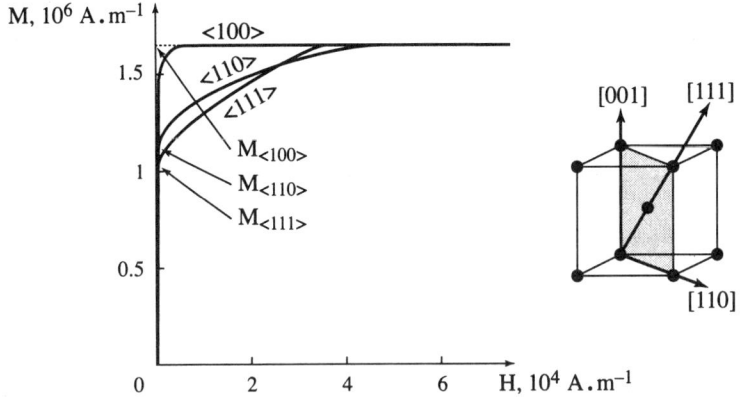

Figure 3.15 - Magnetization curves (in terms of internal field) of an iron single crystal (cubic symmetry) when the field is applied along the three high symmetry directions

Along <100>, magnetization is the highest, and it reaches its maximum value in very low field; it is the easy magnetization direction. In zero field, the material is divided into six types of domains, called *phases*, in which the spontaneous magnetization M_s is directed along one of these directions. When the field is applied along the <110> and <111> axes, the *phase rule* is observed: in very weak field the magnetization remains parallel to <100>, and the magnetizations extrapolated to zero internal field, $M_{<110>}$ and $M_{<111>}$ (fig. 3.15), correspond to the projection of the spontaneous magnetization $M_{<100>} = M_s$ along each of these directions ($M_{<110>} = M_s / \sqrt{2}$, and $M_{<111>} = M_s / \sqrt{3}$).

An analysis similar to that previously developed in the uniaxial situation shows that:
- the area between the curves measured as a function of the internal field H along the <100> and <110> directions is equal to $K_1/4$,
- and the field for which these two curves meet can be written as $H_{A<110>} = 2K_1/\mu_0 M_s$.

2.4.3. Magnetic anisotropy in the paramagnetic state

Let us note that in uniaxial systems, of hexagonal or tetragonal symmetry for instance, in which magnetism originates from rare earth atoms having $L \neq 0$ (see chap. 7), the strong magnetocrystalline anisotropy of the latter leads to an *anisotropy of the magnetic susceptibility*. In particular, above the ordering temperature, two shifted Curie-Weiss laws are observed for the susceptibilities parallel and perpendicular to the high symmetry **c** axis, namely:
$1/\chi_\parallel = (T - \theta_{p\parallel})/\mathscr{C}$ and $1/\chi_\perp = (T - \theta_{p\perp})/\mathscr{C}$. In this case the difference between the parallel ($\theta_{p\parallel}$) and perpendicular ($\theta_{p\perp}$) paramagnetic Curie temperatures is a measure of the main contribution to the anisotropy, namely the second order term.

2.5. TIME DEPENDENT PHENOMENA

Once a magnetic field is applied on a ferromagnetic substance, the magnetization of the latter reaches a value which depends on the intensity of the magnetic field, on the nature of the substance, on the temperature, but also on the history of the considered sample: in particular, magnetization often strongly depends on the order in which *thermal treatments*, *mechanical stresses*, and variations of the *magnetic field* have been made. However another key parameter has also to be considered, namely *time*, because time dependence of magnetic properties is often observed: *a substance submitted to a magnetic field does not immediately reach a final and well defined magnetization.*

Many effects contribute to this time dependence of magnetization:
- *aging* means an evolution of magnetic properties associated with irreversible modifications of the structure of the material under investigation (for instance, formation of precipitates); these structural modifications result from mechanical treatments, chemical effects (oxidation or others) or from atomic diffusion phenomena.

♦ various *relaxation* phenomena can also be observed. We will see in section 2 of chapter 4 that a certain time interval, which can reach 10^{-12} to 10^{-7} s, elapses before an *atomic magnetic moment* becomes parallel to the field which has just been applied; this relaxation phenomenon can be critical in some applications in fast electronics (see chap. 17 and 23). At the scale of the sample, a relaxation of electromagnetic origin can also be observed in conductors. It is due to *eddy currents,* associated with magnetic flux variations in the material which, during some seconds or fractions of a second, oppose the penetration of a rapidly changing field into the metallic material. The speed of the domain wall motion, measured by Sixtus and Tonks inside metallic fibres [5], is –for instance– in agreement with what is predicted by calculations which account for the delay imposed by the eddy currents.

♦ *magnetic after-effect*: once a magnetic field has been applied, magnetization changes only after a certain time, and goes on changing during a time range which allows experiments; there is a delay between the cause and its effect. One distinguishes the thermal fluctuation after-effect (irreversible), of thermodynamic character, and occurring in all ferromagnetic substances, and the diffusion after-effect (reversible) due to motion of particles inside the material. Two major works [6, 7] give an excellent and detailed presentation of these effects, which are briefly presented hereafter, and which are developed in section 5 of chapter 6.

2.5.1. Thermal fluctuation after-effect

We have seen in figure 3.5 that, for the same value of the magnetic field, three different values of magnetization could be obtained: the first corresponds to the initial magnetization curve, the second is observed when the magnetic field decreases from a high and positive value, and the third when the field increases from negative values. Each of these values corresponds to a minimum of the total energy of the system, but they are metastable equilibria. In particular it is the case of the remanent magnetization, that is observed when the magnetic field is reduced down to zero after saturation of the sample. One easily imagines that thermal agitation allows the system to slowly evolve toward a more stable equilibrium state: the remanent magnetization then decreases little by little, and tends more and more slowly toward zero, a value reached for an infinite time. The final equilibrium state ($M = 0$, $H = 0$) will be the same whatever the sign and the intensity of the magnetic field applied before it is reduced to zero.

This point ($M = 0$, $H = 0$) corresponds to the origin of the so called *anhysteretic magnetization curve* of the substance. As a matter of fact, to any non zero value H_0 of the magnetic field applied to a ferromagnetic substance, there also corresponds a unique value of the anhysteretic magnetization, the value which would be observed for an infinite time. However there is a more realistic technique to obtain this anhysteretic magnetization curve for a soft material. It consists in superposing to the static field H_0 an AC magnetic field which is slowly decreased down to zero: the final value is the required magnetization. Figure 3.16 shows the result obtained in this way on an alloy of iron with a small amount of silicon: the anhysteretic curve has been plotted in the first quadrant (H > 0, M > 0). It lies above any experimental curve strarting from the origin (the initial magnetization curve included), and below any curve measured in decreasing field from a field larger than the saturation field.

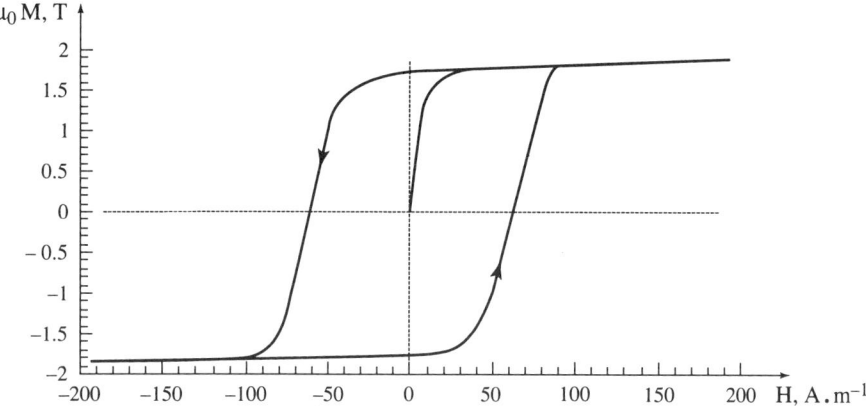

Figure 3.16 - Magnetization curve of a textured Fe-Si alloy: hysteresis loop and anhysteretic curve

The purely magnetic after-effect that has just been discussed only implies magnetic moment reversal. The host crystal or amorphous material is assumed to be perfectly stable: the values of the saturation magnetization and of the magnetocrystalline anisotropy constants remain constant in this process.

2.5.2. Diffusion after-effect

Contrary to the previous situation, diffusion after-effect is associated with the reversible diffusion of atoms, and/or of vacancies between interstitial lattice sites of the magnetic material. This can lead to a variation of its magnetization, and even of its anisotropy constants: the main magnetic characteristics of the substance are then modified. This effect takes place only in substances that can contain small size impurities in interstitial sites (carbon, nitrogen, ...), and it only occurs in the temperature range where diffusion of these impurities is fast enough. It can be used to induce, in an initially isotropic material, a uniaxial magnetic anisotropy called *induced anisotropy*. In this way, for instance, ferrite manufacturers can optimise the performance of their products.

3. PHYSICAL PHENOMENA ASSOCIATED WITH MAGNETISM

So far we have considered the effects of an applied magnetic field on the magnetic response, i.e. magnetization or magnetic induction. In fact magnetism in a substance also gives rise to secondary effects, which affect all physical properties: thermal, elastic, thermo-elastic, electrical, etc.; as an illustration some examples are presented hereafter. All these secondary effects are most dramatic in the vicinity of the critical temperature (T_C, T_N) as shown in following figures.

Figure 3.17 shows the thermal variations of the magnetic contribution to the specific heat of the antiferromagnetic compound $ErGa_2$.

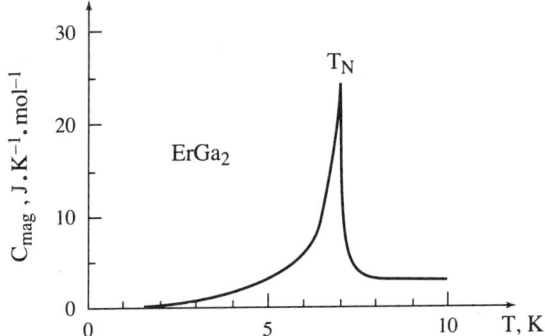

Figure 3.17 - Thermal variation of the specific heat of $ErGa_2$ [8]

The well marked peak observed near the Néel temperature means that the entropy of the magnetic moment system markedly increases at T_N, i.e. at the transition from the antiferromagnetic ordering towards the more disordered paramagnetic state; more detailed explanations on the coupling between thermal and magnetic properties will be given in chapter 11, devoted to the magnetocaloric coupling.

The thermal variations of the reduced resistivity ρ/ρ_0 of $Tb_{75}Gd_{25}$, and of its derivative $d\rho/\rho_0 dT$, are both reported in figure 3.18: when temperature is increased, the absolute value of the magnetic contribution to the electrical resistivity ρ regularly decreases, and vanishes in the vicinity of T_C (see also fig. 14.2 of chap. 14). Its derivative exhibits a well marked lambda type anomaly, centered on T_C.

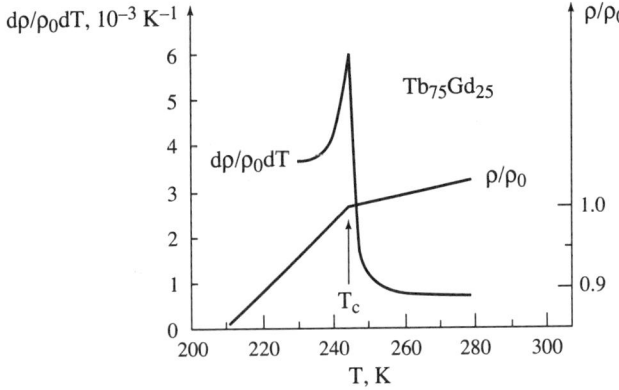

Figure 3.18 - Thermal variations of the electrical resistivity of $Tb_{75}Gd_{25}$ along the b axis, and of its derivative (ρ_0 = resistivity at T_C) [9]

The dotted line shows the behavior expected for a non magnetic substance.

Two additional examples are also presented in figures 3.19 and 3.20: the thermal expansion of $GdAl_2$ exhibits a standard linear behavior above its Curie temperature $T_C = 170$ K. At lower temperature, one observes a contraction of the lattice parameter, of magnetic origin, that we will later identify as due to *exchange magnetostriction*, i.e. a magnetoelastic coupling effect of first order (see chap. 12). Finally, the thermal variation of the c_{44} elastic constant of GdZn also exhibits a positive anomaly below $T_C = 270$ K: this is a magnetoelastic coupling effect of second order. The derivatives of these physical quantities with respect to temperature (i.e. the thermal expansion coefficient, and the thermal coefficient of the elastic constant) also exhibit a well marked lambda-type anomaly at T_C.

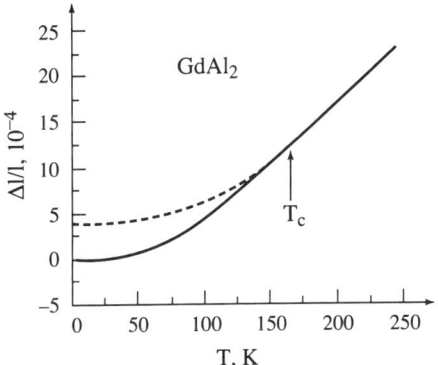
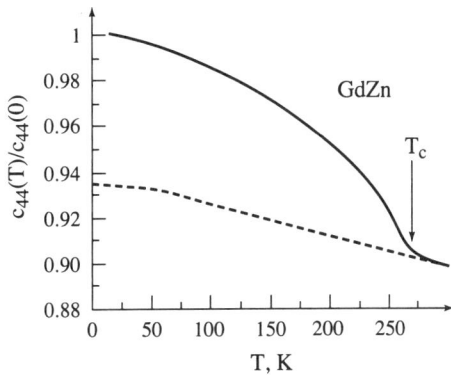

Figure 3.19 - Longitudinal thermal expansion of the compound $GdAl_2$ [10]

Figure 3.20 - Thermal variation of the c_{44} elastic constant of GdZn [11]

The dotted lines are the behaviors expected for non magnetic substances.

All these effects, and many others also, are the manifestation of couplings between the different physical properties of the substance: magnetocaloric, magnetoelastic, magnetoelectric, magneto-optic, etc. ... couplings. Their study is not pure speculation: indeed, on the one hand their knowledge can be valuable to the researcher who wants to investigate a magnetic structure, but also, and especially, these effects give rise to many technological applications which will be treated further in this book.

The thermodynamical approach of magnetism that will be presented in chapter 10 will allow to account for the formalism of these couplings, to predict the associated effects, and to give a qualitative description of them.

EXERCISES

E.1 - Consider a tetragonal ferromagnetic compound in which the **c** axis is the easy magnetization direction. The table hereunder gives some couples of values of the applied field H_0, and of the magnetization M measured at 4 K in a direction perpendicular to the **c** axis.

H_0 (10^4 A.m^{-1})	9.58	43.77	63.42	95.49	127.32
M (10^4 A.m^{-1})	5.83	26.24	37.46	54.57	70.10

E.1.1 - Knowing that the sample is spherical, give for each couple the internal field H.

E.1.2 - From formula (3.7) deduce the expression of H/M as a function of M².

E.1.3 - Plot this experimental curve. Does the shape of this curve correspond to that expected?
Deduce the values of K_1 and K_2, knowing that $M_s = 85.12 \times 10^4$ A.m^{-1}.

E.2 - Consider a crystal with tetragonal symmetry having the characteristics considered in section 2.4.1, in which $K_1 < 0$ and $K_3 < 0$, with K_1 much larger than K_2 and K_3 (see eq. 3.4) in absolute values.

E.2.1 - When the field is applied along the <001> direction, show that the initial slope of the magnetization curve (*vs* internal field) can be written as: $\chi = dM/dH = \mu_0 M_s^2 / (2|K_1| - 4K_2 + 4|K_3|)$, and find again expression (3.9) for the anisotropy field H_A.

E.2.2 - When the field is applied along <110>, show that the initial slope of the magnetization curve (*vs* internal field) can be written: $\chi = dM/dH = \mu_0 M_s^2 / 32|K_3|$. Also show that, in the basal plane (001), the field for which the magnetization curves along <110> and <100> meet is $H'_A = \dfrac{16|K_3|}{\mu_0 M_s}$.

E.3 - Consider a ferromagnetic crystal with cubic symmetry in which the magnetocrystalline anisotropy is such that $K_1 < 0$, and $K_2 = 0$.

E.3.1 - What are the easy magnetization directions? How many types of magnetic domains are present in zero field?

E.3.2 - From an analysis similar to that of section 2.4.3, plot schematically the variations of magnetization as a function of the internal field when H_0 is applied successively along the three main symmetry directions. Discuss the different magnetization processes involved.

SOLUTIONS TO THE EXERCISES

S.1 *S.1.1* $H = H_0 - M/3$.

> *S.1.3* The experimental variation of H/M as a function of M² is linear, as expected. From the intercept of this linear variation with the vertical axis one deduces $K_1 = 5.96 \times 10^5$ J.m^{-3}, and from the slope one obtains $K_2 = 0.59 \times 10^5$ J.m^{-3}.

S.2 *S.2.1* During the process $\phi = 0$ (see fig. 3.11). Minimizing the total energy with respect to θ, one can deduce a relationship between M and H, for H ranging from 0 to H_A.

S.2.2 During the process $\theta = 90°$. Minimizing the total energy with respect to ϕ one finds an expression of H as a function of ϕ. Knowing that $M = M_s \cos(\pi/4 - \phi)$, the slope at the origin can be calculated from the derivatives $dH/d\phi$ and $dM/d\phi$. For the determination of H'_A, an expansion is necessary.

S.3 *S.3.1* <111>. There are 8 types of magnetic domains.

S.3.2 Let M_s be the magnetization of one domain. One obtains variations similar to those of figure 3.15, with the difference that the magnetizations extrapolated to zero internal field are in the present case $M_{<111>} = M_s$, $M_{<110>} = M_s \sqrt{2}/\sqrt{3}$, and $M_{<100>} = M_s/\sqrt{3}$.

REFERENCES

[1]   E.R. CALLEN, H.B. CALLEN, *J. Phys. Chem. Solids* **16** (1960) 310.

[2]   R. GERSDORF, *Phys. Rev. Lett.* **40** (1978) 344.

[3]   J.M. ALAMEDA, D. GIVORD, R. LEMAIRE, Q. LU, *J. Appl. Phys.* **52** (1981) 2079.

[4]   K. HONDA, S. KAYA, *Sci. Reports Tohoku Univ.* **15** (1926) 721.

[5]   K.J. SIXTUS, L. TONKS, *Phys. Rev.* **37** (1931) 930; *ibid.* **42** (1932) 419.

[6]   A. AHARONI, Introduction to the Theory of Ferromagnetism (1996) Clarendon Press, Oxford.

[7]   J.L. DORMANN, D. FIORANI, E. TRONC, Magnetic relaxation in fine-particle systems, *in Advance in Chemical Physics* **XCVIII** (1997) 284-494, I. Prigogine & S.A. Rice Eds, Wiley & Sons, New York.

[8]   D. SCHMITT, private communication.

[9]   J.B. SOUSA, M.M. AMADO, M.E. BRAGA, R.S. PINTO, J.M. MOREIRA, *Communications on Physics* **2** (1977) 95.

[10] E. DU TRÉMOLET DE LACHEISSERIE, *J. Magn. Magn. Mater.* **73** (1988) 289.

[11] J. ROUCHY, P. MORIN, E. DU TRÉMOLET DE LACHEISSERIE, *J. Magn. Magn. Mater.* **23** (1981) 59.

CHAPTER 4

PHENOMENOLOGY OF MAGNETISM AT THE MICROSCOPIC SCALE

*This chapter presents some simple models which account for the three main types of magnetism previously introduced: that of substances without magnetic atoms (**diamagnetism**), that of substances with magnetic atoms without interactions (**paramagnetism**), and finally that of substances with magnetic atoms which strongly interact with their surroundings (**ferro, antiferro, ferrimagnetism**, etc.). The models described in this chapter only apply to substances **in which the electrons responsible for magnetism are well localised**. They do not apply to substances in which magnetism originates from itinerant electrons. For the latter, other models have been proposed (Landau diamagnetism, Pauli paramagnetism, and Stoner-Wohlfarth ferromagnetism): these "itinerant electron" models of magnetism are presented in chapter 8.*

1. THE CLASSICAL MODEL OF DIAMAGNETISM: CASE OF LOCALISED ELECTRONS

We have seen in chapter 3 that a diamagnetic substance is one in which the atoms have no permanent magnetic moment. Correspondingly it exhibits a magnetic susceptibility that is generally weak, and virtually temperature independent.

The simplest way to account for this property is to consider the classical model of an electron moving on a circular orbit. The situation is identical to that of a current loop. When a magnetic field is applied perpendicular to the plane of the orbit, the current is modified in such a way that the flux variation originating from the loop itself is equal and opposite to that due to the applied field (Lenz's rule). The result is an orbital magnetic moment variation opposite to the applied field. This magnetic moment change is the same whatever the sign of the electron displacement on its orbital, i.e. the sign of the initial orbital magnetic moment with respect to the applied field: the variation is alway opposite to the magnetic field (fig. 4.1).

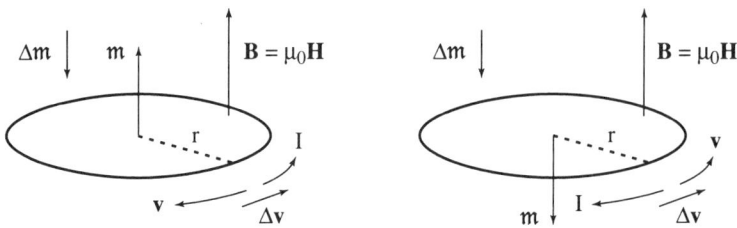

Figure 4.1 - Effect of a magnetic field on two electrons moving in opposite directions on the same circular orbit

As an electron on an orbit is equivalent to a superconductor current loop, i.e. without resistance, the current variation, and accordingly the magnetic moment variation, remain unchanged as long as the applied field is maintained.

Let us assume that two electrons with opposite spins move on the same orbit in opposite directions. The total orbital and spin angular momenta, and accordingly the total magnetic moment, are zero. Under the application of a field, the orbital magnetic moment changes being opposite to the field for both electrons, the system acquires a magnetic moment opposite to the field.

Let us look more quantitatively at the magnetic moment variation in the case of one electron (charge $-e$, and mass m_e) moving at speed v on a circular orbit of radius r (fig. 4.1). We will see farther (eq. 7.3) that the orbital magnetic moment is:

$$m_o = -erv/2 \tag{4.1}$$

According to the Lenz's rule, the electromotive force u resulting from the flux variation Φ in the current loop, when a field **B** is applied perpendicular to the plane of the orbit, can be written as:

$$u = -\frac{d\Phi}{dt} = -\pi r^2 \frac{dB}{dt} \tag{4.2}$$

We assume, as it will be justified farther, that the orbit radius is not modified. Let **E** be the electric field acting on the electron along the orbit ($u = 2\pi r E$). One obtains:

$$E = -\frac{\pi r^2}{2\pi r}\frac{dB}{dt} = -\frac{r}{2}\frac{dB}{dt} \tag{4.3}$$

It results in a force acting on the electron parallel to its velocity:

$$F = -eE = m_e \frac{dv}{dt} \tag{4.4}$$

whence:

$$\frac{dv}{dt} = \frac{er}{2m_e}\frac{dB}{dt} \tag{4.5}$$

Integrating over the total induction variation, from 0 to B, leads to:

$$\Delta v = \frac{er}{2m_e}\int_0^B dB = \frac{er}{2m_e}B \tag{4.6}$$

and, inserting this variation of v into equation (4.1), one deduces the magnetic moment variation:

$$\Delta m = - (e^2 r^2 / 4 m_e) B \qquad (4.7)$$

At this point, let us justify the assumption that the orbit is not modified by the field. When the field is applied, the electron is submitted to two additional radial forces:
- on the one hand, the force exerted by the field **B** on the electron: $F_B = e v B$,
- on the other hand, the variation of the centrifugal force associated with the speed variation: $\Delta F_c = 2 m_e v \Delta v / r$.

Replacing Δv by its expression in (4.6), it appears that these two forces are equal and opposite (fig. 4.2), and that accordingly the orbit radius remains unchanged.

Figure 4.2 - Balance of radial forces when applying a magnetic field

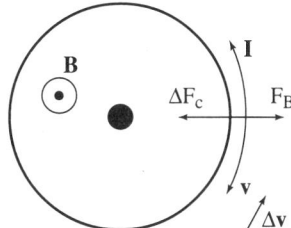

The expression of the magnetic moment variation (4.7) was established in the case of a planar orbit perpendicular to the applied field. Actually one has to consider that the orbit is not planar. In the simplest case, the electron moves on a spherical orbital of average radius $\sqrt{<r^2>}$. Let z be the direction parallel to the field. In expression (4.7), $r^2 = x^2 + y^2$, the average value of which is equal to 2/3 of the average value of the square of the radial distance $<r^2>$ ($<r^2> = <x^2> + <y^2> + <z^2>$).

Expression (4.7) then becomes:

$$\Delta m = - (e^2 r^2 / 6 m_e) B \qquad (4.8)$$

The magnetic moment variation of an atom with Z electrons is:

$$\Delta m_{\text{per atom}} = - (e^2 B / 6 m_e) \Sigma <r_n^2> \qquad (4.9)$$

where $\Sigma <r_n^2>$ is the average value of the square of the radius of the nth orbital. The summation can be replaced by $Z R^2$ where R^2 is the average value of the square of the radius of the different orbitals.

Finally, if N is the number of atoms of the considered element per unit volume, the diamagnetic susceptibility can be written as:

$$\chi = \partial M / \partial H = - \mu_0 N e^2 Z R^2 / 6 m_e \qquad (4.10)$$

The value of R^2 can be estimated thanks to quantum mechanics. The diamagnetic susceptibility calculated with the above model is in rather good agreement with the values measured in insulators, especially in rare gases. These susceptibilities are of the order of 10^{-6} (see app. 4).

The parameters on which the diamagnetic susceptibility, given by expression (4.10) depends, are temperature independent (Z, e, m_e), or very weakly dependent (N, R^2) on temperature. This is in agreement with the fact that the susceptibilities of diamagnetic substances are, with a few exceptions, temperature independent.

In diamagnetic substances, atoms possess only filled electronic orbitals either in the atomic state, or resulting from the formation of chemical bonds. These substances are mainly:

- the monoatomic rare gases such as He, Ne, Ar,
- most of the polyatomic gases such as H_2, N_2,
- ionic solids such as NaCl,
- substances with covalent bonds, in particular organic compounds. Susceptibility measurements performed on the latter supplied valuable information, through the determination of R^2, on the electronic orbital sizes.

The diamagnetism due to conduction electrons in metals will be treated in chapter 8, as mentioned in the introduction. Diamagnetism also exists, obviously, in all substances with permanent magnetic moments, because it results from a universal rule (Lenz's law), but its contribution to the susceptibility is so weak compared to the others that it is, in most cases, neglected.

Finally it is worth noticing that superconductors are perfect diamagnets in which the susceptibility is –1 (Meissner effect). They will be studied in chapter 19.

2. SYSTEMS WITH NON-INTERACTING LOCALISED MAGNETIC MOMENTS

2.1. EFFECT OF A UNIFORM FIELD ON A MAGNETIC MOMENT: PRECESSION

Before looking at the behavior under a magnetic field of an assembly of magnetic moments, we will examine the case of an isolated magnetic moment submitted to a uniform magnetic field. We will see that it behaves like a gyroscope. Indeed, let us first consider a gyroscope of mass m and of angular momentum $\pounds$, both extremities O and M of which are initially maintained fixed in such a way that the rotation axis makes an angle θ with the vertical (fig. 4.3-a), and then let us release the extremity M leaving the gyroscope free to rotate around the point O.

The effect of gravity leads to a torque:

$$\mathbf{\Gamma} = \mathbf{r}_{OG} \times m\mathbf{g} \qquad (4.11)$$

where G is the center of gravity of the gyroscope, and **g** the acceleration of gravity.

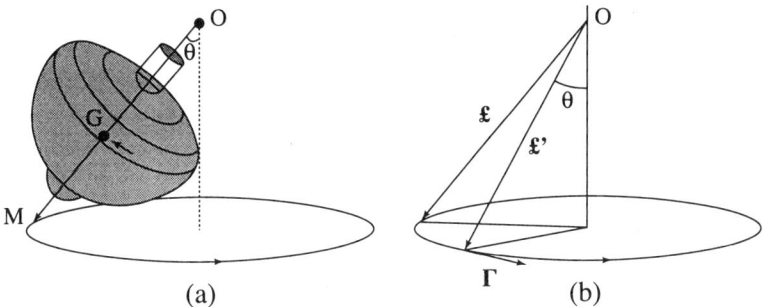

Figure 4.3 - Effect of gravity on a gyroscope

The basic equation of dynamics for a solid under rotation can be written in that case as:

$$d\pounds/dt = \Gamma \tag{4.12}$$

Γ is alway perpendicular to the vertical plane which contains the axis of the gyroscope. The result is that the extremity M will move on a circle of radius $OM \sin\theta$. This is the so called precession motion that each of us has observed in the case of a spinning top (in this case, it is the upper extremity that moves on a circle). Let us assume that, during the time Δt, $\pounds$ becomes $\pounds'$ as shown in figure 4.3-b, with the same angle θ, and let ω_p be the angular precession speed. During this time interval, the variation of the angular momentum $\Delta\pounds$ can be written: $\Delta\pounds = (\pounds \sin\theta)(\omega_p \Delta t)$. Hence $d\pounds/dt = \pounds\, \omega_p \sin\theta$. Inserting this value into (4.12), and keeping in mind that $\Gamma = mgl \sin\theta$ with $l = |r_{OG}|$, one obtains:

$$\omega_p = mgl/\pounds \tag{4.13}$$

It can be seen that the precession frequency is independent of the angle that the gyroscope (or the spinning top) makes with the vertical. Therefore, although submitted to a torque, the gyroscope axis does not move towards the vertical, as would be the case for a solid without angular momentum, but precesses keeping the same angle with the vertical. However, if a friction force opposes the precession motion, the gyroscope progressively yields under the torque due to gravity: the extremity M of the gyroscope describes a spiral of decreasing radius until the gyroscope takes a vertical position that corresponds to the energy minimum. This is what occurs when the extremity M is maintained in contact with a liquid. Let us come back now to magnetism: a magnetic moment behaves like a gyroscope because it is always associated with an angular momentum as it will be seen in chapter 7. When subjected to a torque from a magnetic field B_0, a free magnetic moment m does not align parallel to the latter but precesses (fig. 4.4), like the gyroscope in the gravity field. Equation (4.12) then becomes:

$$d\pounds/dt = \Gamma = m \times B_0 \tag{4.14}$$

Let us consider the angular momentum and the magnetic moment in a classical way, and let us use equation (7.28) of chapter 7. The same approach as in the case of gyroscope leads to: $\omega_0 = |\gamma B_0|$ (Larmor precession). The precession frequency is proportional to B_0, and its measurement allows the determination of the gyromagnetic factor γ.

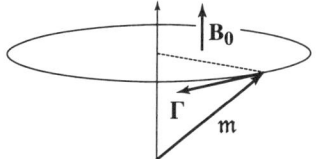

Figure 4.4 - Precession of a magnetic moment m around a magnetic field B_0

This measurement can be performed with different techniques, among others ferromagnetic resonance. It is worth noticing that *if the atom is no more isolated*, the magnetic moment behaves like a damped gyroscope, its damping arising from the interactions with neighbouring atoms: progressively, while precessing, the magnetic moment rotates toward the field. This process is very fast, so that, unless precession is maintained by an energy supply, the magnetic moment almost immediately (10^{-7} to 10^{-11} s) reaches its ground state: parallel to the field. Generally, magnetization has two components, one parallel to the field, and the other perpendicular. The latter is at the origin of the physical phenomenon called *spin waves* (see § 9 of chap. 10), in which precession is maintained thanks to a rotating field perpendicular to the static field.

2.2. ASSEMBLY OF LOCALISED MAGNETIC MOMENTS WITHOUT INTERACTION: CURIE PARAMAGNETISM

In a large number of substances, one observes a magnetic susceptibility that is positive, generally weak at room temperature ($10^{-6} \leq \chi \leq 10^{-4}$), and is proportional to reciprocal temperature over a large temperature range (fig. 4.5):

$$\chi = \mathscr{C}/T \tag{4.15}$$

This is the *Curie law*, which characterises the paramagnetism of free atoms. $\mathscr{C}$ is called the *Curie constant*.

This behavior can be interpreted as follows: a paramagnetic substance contains magnetic atoms the moments of which are free to take any direction. In the absence of field, moments point randomly in all directions, and the magnetization of the substance is zero. When a magnetic field is applied, moments tend to become parallel to the latter, and, if nothing opposes this effect, the system acquires a large magnetization. This is what occurs at zero temperature. On the other hand, at non zero temperature, thermal agitation works against parallelism, so that moment alignment is partial, and that one only observes a weak positive susceptibility. As temperature increases, so does disorder, and susceptibility decreases in accordance with the Curie law.

4 - PHENOMENOLOGY OF MAGNETISM AT THE MICROSCOPIC SCALE

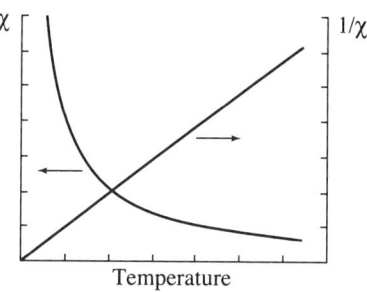

Figure 4.5 - Curie law for magnetic moments without interaction

The first quantitative model, developed by Langevin, uses a classical description in which the magnetic moment of each atom can take any direction. The energy of an atomic magnetic moment m of modulus m_0 in a field $\mathbf{H}$ is: $E(\theta) = -\mu_0 m_0 H \cos\theta$, where θ is the angle between $\mathbf{m}$ and $\mathbf{H}$. At a given temperature T, the probability for the moment energy to be $E(\theta)$ is proportional to the Boltzmann factor: $e^{-E(\theta)/k_B T}$ where k_B is the *Boltzmann constant*. The probability for the moment to make an angle between θ, and $\theta + d\theta$ with respect to the field (fig. 4.6) is proportional to the corresponding solid angle ($2\pi \sin\theta \, d\theta$), and can be written as:

$$dp(\theta) = 2\pi \exp[-E(\theta)/k_B T] \sin\theta \, d\theta / Z \tag{4.16}$$

where Z is the partition function:

$$Z = \int_0^\pi 2\pi \exp[-E(\theta)/k_B T] \sin\theta \, d\theta \tag{4.17}$$

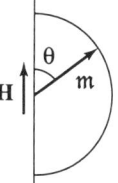

Figure 4.6 - In classical mechanics, the magnetic moment can point in any direction

The thermal average of the magnetic moment component parallel to the field can be expressed as:

$$\langle m \rangle_T = \int_0^\pi m_0 \cos\theta \, dp(\theta) = m_0 \frac{2\pi \int_0^\pi \cos\theta \exp\left(\frac{\mu_0 m_0 H \cos\theta}{k_B T}\right) \sin\theta \, d\theta}{2\pi \int_0^\pi \exp\left(\frac{\mu_0 m_0 H \cos\theta}{k_B T}\right) \sin\theta \, d\theta} \tag{4.18}$$

Defining $x = \mu_0 m_0 H / k_B T$, one easily finds out that $Z = 4\pi \sinh(x)/x$, and that:

$$\langle m \rangle_T = m_0 \frac{1}{Z} \frac{dZ}{dx} \tag{4.19}$$

The magnetization M(T, H) along the field direction is then deduced:

$$M = Nm_0 [\tanh^{-1}(x) - 1/x] = M_0 \mathcal{L}(x) \qquad (4.20)$$

where N is the number of magnetic atoms per unit volume, and $M_0 = M(0,0)$ is the spontaneous magnetization at T = 0. $\mathcal{L}(x)$, called the *Langevin function*, accounts for the reduced magnetization $m = M/M_0$ as a function of the ratio $\mu_0 m_0 H / k_B T$. It is shown in figure A6.3 of appendix 6 at the end of the book.

One sees that $\mathcal{L}(x)$ tends toward 1, i.e. that M tends toward M_0, when H becomes large, and/or T becomes small, conditions which are necessary in order for the alignment effect due to the field to overcome thermal agitation.

At room temperature, and for rather large fields ($\mu_0 H \approx 1$ T), x is of the order of 1.5×10^{-2} for a gadolinium based paramagnetic substance, and M/M_0 only reaches 5×10^{-3}. The temperature must be of the order of 1 K for M/M_0 to reach 0.8 in this field.

So, in particular when dealing with the initial susceptibility, i.e. in weak field, one will be almost always concerned by the initial part of the curve, the range in which the Langevin function has the following expansion:

$$\mathcal{L}(x) = x/3 - x^3/45 + \ldots \qquad (4.21)$$

whence, to first order, one deduces:
$$M = \frac{\mu_0 N m_0^2 H}{3 k_B T} \qquad (4.22)$$

The susceptibility is then:
$$\chi = \frac{\mu_0 N m_0^2}{3 k_B T} = \frac{\mathcal{C}}{T} \qquad (4.23)$$

One finds again the experimental Curie law, with the following expression for the Curie constant:

$$\mathcal{C} = \mu_0 N m_0^2 / 3 k_B \qquad (4.24)$$

The susceptibility of a paramagnet is typically of the order of 10^{-5} to 10^{-4} at room temperature, thus generally much larger than the diamagnetic susceptibility but, as it will be seen farther, markedly smaller than that of a ferromagnet.

Figure 4.7 shows the magnetization curves observed at very low temperature, for three salts containing Cr^{3+}, Fe^{3+}, and Gd^{3+} ions. For each of them, the experimental points plotted as a function of the ratio H/T perfectly fit the same curve, but one remark should be made: the variations of the reduced magnetizations M/M_0 as a function of H/T are not exactly the same for these different salts, and *none of them fits the Langevin curve*. The deviation is larger for smaller magnetic moment per atom. For given temperature, saturation occurs for magnetic fields systematically weaker than predicted by the Langevin theory [$\mathcal{L}(x)$ is plotted in dotted lines].

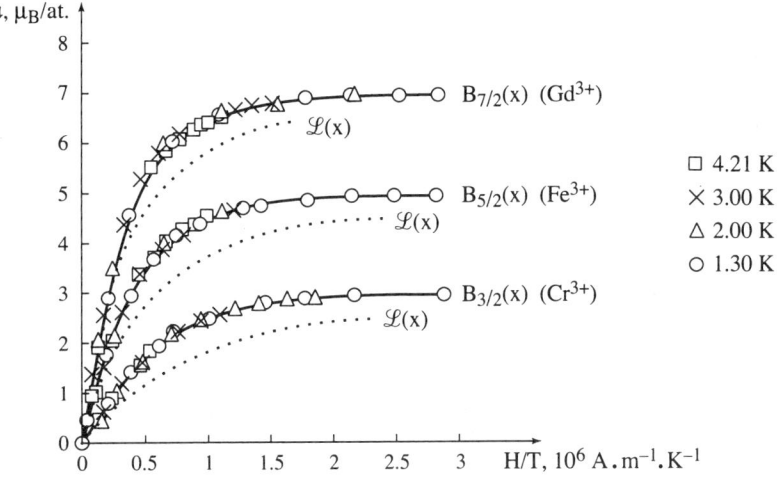

*Figure 4.7 - Magnetization curves, as a function of H/T,
for three salts containing a magnetic ion:*
$CrK(SO_4)_2.12H_2O$, $Fe(NH_4)_2(SO_4)_2.12H_2O$, and $Gd_2(SO_4)_3.8H_2O$, after[1]

The above calculation was based on the classical assumption of moments which can point in any direction. Actually, we will see in chapter 7 that, on account of the quantization of the magnetic moment, its component along the field (which defines the z axis) can only take discrete values: $m_z = -g_J \mu_B M_J$, where the number M_J can have $2J+1$ values spanning the interval from $+J$ to $-J$ by steps of 1 (fig. 4.8), with J having only integer or half integer values (g_J is a constant called *Landé factor* that will be detailed in chapter 7, and M_J is the component of the total angular momentum along the quantization axis). The potential energy in the field H is then: $E = \mu_0 g_J \mu_B M_J H$.

Applying Boltzmann statistics as previously, but now setting: $x = \mu_0 g_J \mu_B J H / k_B T$, the magnetization induced by the field becomes:

$$M = \frac{N}{Z} \sum_{-J}^{+J} -g_J \mu_B M_J \exp\left(-\frac{xM_J}{J}\right) = -\frac{Ng_J \mu_B J}{JZ} \sum_{-J}^{+J} M_J \exp\left(-\frac{xM_J}{J}\right)$$

$$M = M_0 \frac{1}{Z} \frac{dZ}{dx} \qquad (4.25)$$

where the partition function, Z, has the following expression:

$$Z = \sum_{-J}^{+J} \exp(-xM_J/J) \qquad (4.26)$$

$M_0 = N g_J \mu_B J = N m_0$ is the maximum magnetization or saturation magnetization, and m_0 the magnetic moment per atom. $Z = e^{-x}[1 + e^{x/J} + (e^{x/J})^2 + ... + (e^{x/J})^{2J}]$ is the sum of a geometric series which reaches:

$$Z = \sinh\left(\frac{2J+1}{2J}x\right) \Big/ \sinh\left(\frac{x}{2J}\right) \qquad (4.27)$$

Hence: $$M = M_0 \mathcal{B}_J(x) \qquad (4.28)$$

where $\mathcal{B}_J(x)$ is the so called *Brillouin function*:

$$\mathcal{B}_J(x) = \frac{2J+1}{2J}\tanh^{-1}\left(\frac{2J+1}{2J}x\right) - \frac{1}{2J}\tanh^{-1}\left(\frac{1}{2J}x\right) \quad (4.29)$$

This function is shown in figure A6.3 (app. 6) for different values of J. When J reaches large values, one gets closer to the classical case, because $\mathcal{B}_J(x)$ gets closer to $\mathcal{L}(x)$ when J tends toward infinity.

When J = 1/2, which corresponds to the smallest possible magnetic moment, the Brillouin function is reduced to: $M = M_0 \tanh x$.

For small values of x, the following expansion is used:

$$\tanh^{-1} x = 1/x + x/3 + ... \quad (4.30)$$

which leads, to first order, to:

$$\mathcal{B}_J(x) = (J+1)x/3J \quad (4.31)$$

whence:

$$M = M_0(J+1)x/3J \quad (4.32)$$

The susceptibility has the same expression as in (4.15): $\chi = \mathcal{C}/T$, but the Curie constant is now written as:

$$\mathcal{C} = \frac{\mu_0 N g_J^2 J(J+1)\mu_B^2}{3k_B} = \frac{\mu_0 N m_{eff}^2}{3k_B} \quad (4.33)$$

where $m_{eff} = g_J \mu_B \sqrt{J(J+1)}$ is the *effective moment*, which is no more equal to the maximum moment $m_0 = g_J \mu_B J$ measured along the field (as was the case for the Langevin theory), but slightly larger. This results from a quantum mechanical property that we will see in chapter 7: the average value $<J^2>$ of the square of an operator **J** is $J(J+1)$ whereas the maximum value of the component of this operator along the z quantization axis, $<J_z>$, is J. Thus the experimental measurement of the Curie constant allows the determination of the effective moment of the substance.

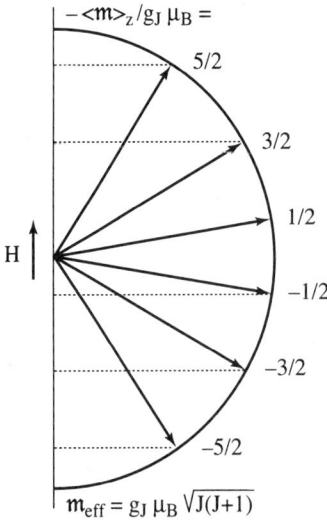

Figure 4.8 - Quantization of the magnetic moment (J = 5/2)

m_{eff} represents the length of the vector describing the moment, and m_z its projection on the vertical axis.

For each of the three compounds presented in figure 4.7, the experimental points plotted as a function of the ratio H/T perfectly fit the three Brillouin functions: $\mathcal{B}_{3/2}$, $\mathcal{B}_{5/2}$, and $\mathcal{B}_{7/2}$, respectively, in agreement with the moment value expected (see chap. 7) for each of the magnetic ions of these materials.

The paramagnetism of free atoms can be observed in substances which contain magnetic atoms, and where the distances between these atoms are large enough so that they do not interact. The substances which satisfy these conditions are the following:
- gases of magnetic atoms or molecules,
- the salts of 3d transition elements, and rare earth elements such as those previously discussed,
- some oxides of rare earth elements.

One calls paramagnetic only those substances that exhibit such a behavior down to very low temperature. We will see hereafter that all materials which magnetically order (ferro-, ferri-, antiferromagnetic...) behave like paramagnets above a critical temperature.

Finally, the paramagnetism of metals (sodium, copper, gold, silver, lead, ...) which have a susceptibility weakly temperature dependent, and weaker than that of Curie paramagnets, will be treated in chapter 8. They feature Pauli paramagnetism, which is due to itinerant electrons.

2.3. SUPERPARAMAGNETISM

We will see farther that very small ferromagnetic or antiferromagnetic particles can behave, below a so called *blocking temperature*, like a *giant magnetic moment* (or macro-spin). Applying a magnetic field will then induce a magnetization which follows a H/T law given by the *Langevin function*, since the situation is that of the classical limit of infinite spins.

However it should be noticed that the blocking temperature varies with the time scale of the measurement, and with the volume of the particle, because the transition of the particle from a blocked state (with a magnetic moment rigidly directed along a given direction) toward the *superparamagnetic state* is a relaxation phenomenon. Thus, at room temperature, fine particles can seem to be superparamagnetic when their magnetization is measured with the classical extraction technique (duration of the experiment of the order of a second) while a neutron diffraction experiment will see them "blocked" as the neutron-spin interaction time is very short.

It is L. Néel who developed in 1949 the superparamagnetism theory (without using the term of superparamagnetism) for ferromagnetic fine particles [2], and extended it in 1961 to antiferromagnetic fine particles [3]. These two basic articles are reproduced in his collected scientific works [4].

3. EXCHANGE INTERACTIONS

3.1. GENERAL

Often, in matter, magnetic moments are not free but interact with each other, and with their surroundings. The result is a collective behavior which manifests itself, below a critical temperature, by the onset of magnetic orders such as ferro-, ferri- or antiferromagnetism. These behaviors are characterised, at low temperature, by a microscopic arrangement of atomic magnetic moments which can be observed by neutron diffraction, a key experimental technique in magnetism. At high temperature, thermal agitation overcomes the ordering, and a paramagnetic behavior is observed. Among the different interactions two are preponderant, and responsible for the microscopic magnetic behavior of the majority of materials: these are the exchange interaction, and the interactions responsible for the magnetocrystalline anisotropy.

The description of the fundamental features of these two types of interactions need a quantum mechanical approach, and will be presented in detail in chapters 7 and 9. Additional interactions are also present but they are only a weak perturbation on the magnetic properties. Some of them will be treated farther.

Obviously, there also exists the classical dipolar interaction between magnetic moments, treated in chapter 2, and responsible for the field created by matter inside and outside itself. However, this interaction is so weak compared to the others that it does not contribute significantly to the microscopic magnetic properties of matter. On the other hand, because it is a long range interaction, it plays a key role in the magnetization processes of ferromagnets (which will be treated in chapters 5 and 6), and in the majority of technological applications.

We will describe hereafter the exchange interaction and its effects, as it is responsible for magnetic orders in a large number of substances.

3.2. EXCHANGE INTERACTION AND MOLECULAR (OR MEAN) FIELD APPROXIMATION

Exchange interaction is an interaction of electrostatic origin which was introduced in 1929 by Heisenberg within the framework of quantum mechanics. It will be shown in chapter 9 that the energy associated with this interaction can be expressed as a function of the magnetic moments of two neighbouring atoms i, and j as:

$$\mathcal{E}_{ij} = -\mu_0 \, n_{ij} \, \mathfrak{m}_i \mathfrak{m}_j \qquad (4.34)$$

Depending on whether the coefficient n_{ij} is positive or negative, the magnetic moments $\mathfrak{m}_i$ and $\mathfrak{m}_j$ tend to align parallel or antiparallel with each other, respectively. The exchange energy density E_{ex} is then written, by summing over the volume unit, as:

$$E_{ex} = -\frac{\mu_0}{2} \sum_{i,j \neq i} n_{ij} \mathfrak{m}_i \mathfrak{m}_j \qquad (4.35)$$

4 - PHENOMENOLOGY OF MAGNETISM AT THE MICROSCOPIC SCALE

This expression can also be written as:

$$E_{ex} = -\frac{\mu_0}{2} \sum_i \mathbf{m}_i \mathbf{H}_i \qquad (4.36)$$

with:

$$\mathbf{H}_i = \sum_{j \neq i} n_{ij} \mathbf{m}_j \qquad (4.37)$$

The 1/2 factor comes from the fact that, in the summation over i and j, the interaction of each pair is counted twice. $\mathbf{H}_i$ can be considered as a local field acting on the moment $\mathbf{m}_i$. This field, in the same way as each of the moment $\mathbf{m}_j$ on which it depends, fluctuates as a function of time. It is therefore impossible to determine exactly the orientation of all magnetic moments for a given applied field and a given temperature. In order to account for ferromagnetism, P. Weiss proposed in 1906 (i.e. 23 years before the Heisenberg model) the molecular field model, which led to significant progress in magnetism. This model consists in neglecting the fluctuating character of $\mathbf{H}_i$, and in considering only its average at a given temperature. The average value of the energy of a sample at the temperature T is written as:

$<\mathcal{E}_{ex}>_T = -\mu_0 \sum_i <\mathbf{m}_i \mathbf{H}_i>_T$, which becomes in the Weiss assumption:

$$<\mathcal{E}_{ex}>_T = -\mu_0 \sum_i <\mathbf{m}_i>_T <\mathbf{H}_i>_T \qquad (4.38)$$

where:

$$<\mathbf{H}_i>_T = \mathbf{H}_m = \sum_{j \neq i} n_{ij} <\mathbf{m}_j>_T \qquad (4.39)$$

$\mathbf{H}_m$ is called the *molecular or exchange field*.

If all considered moments are identical: $<\mathbf{m}_j>_T = <\mathbf{m}>_T$, and:

$$\mathbf{H}_m = w\mathbf{M} \qquad (4.40)$$

where $\mathbf{M} = N<\mathbf{m}>_T$ is the magnetization, N being the number of magnetic atoms per volume unit. $w = \frac{1}{N} \sum_{j \neq i} n_{ij}$ is the molecular field coefficient. $\mathbf{H}_m$ and $\mathbf{M}$ being expressed in the same unit in SI units, w is dimensionless. The molecular field thus has the same effect as an applied field but it is, most of the time, much larger. Therefore the effect that an applied field can realise only at very low temperature, namely the alignment of all moments, is obtained by the molecular field at temperatures sometimes markedly larger than room temperature, in the case of a ferromagnet for instance.

Let us finally notice that, in terms of molecular field, the exchange energy per volume unit (i.e. the energy density) defined at the top of this section becomes, in the above assumption where all moments are identical:

$$E_{ex} = -(1/2)\mu_0 w M^2 \qquad (4.41)$$

3.3. EXPERIMENTAL

We will see farther that it is possible to determine the molecular field coefficient w from the measurement of the critical temperature below which magnetic moments order. It is then easy to deduce the exchange interaction energy between two neighbouring atoms, i and j. One usually considers that figure 4.9 represents the $\mathcal{E}_{ij}$ energy thus obtained as a function of the distance between magnetic shells in metals and alloys of the 3d series.

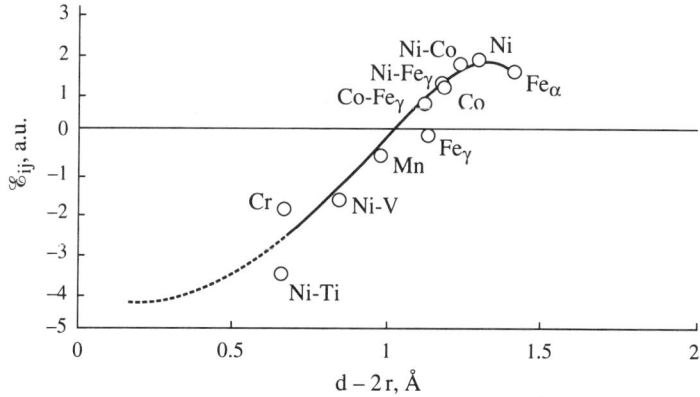

Figure 4.9 - Slater-Néel curve

Exchange interaction ($\mathcal{E}_{ij}$ of equation 4.34 in arbitrary unit) as a function of the distance between magnetic shells (d = distance between two atoms, r = average radius of the magnetic shell) in metals and alloys of 3d elements, after [4].

Actually, this curve also shows the dependence of the exchange interactions as a function of the 3d band filling (see chap. 8). In the itinerant magnetism approach, the interactions are antiferromagnetic at the middle of the band, and more and more ferromagnetic when going towards the end of the band. This explains why the $\mathcal{E}_{ij}$ observed values are negative for Cr and Mn, positive, and stronger and stronger from Co to Ni.

4. FERROMAGNETISM
IN THE MOLECULAR FIELD MODEL

We will qualify here as *ferromagnetic* a substance in which all magnetic moments are, at the microscopic scale, and below a critical temperature T_C, *spontaneously parallel to each other*, i.e. in the absence of any external magnetic field (§ 1.4 and 2 of chap. 3). The result is a spontaneous magnetization $M_s(T)$. Such *magnetic order* can be observed thanks to a neutron diffraction experiment. The case of ferrimagnetic substances, analogous in its macroscopic features, is markedly different at the microscopic scale, and will be treated not here, but in section 6.

Let us come back to the microscopic aspect of ferromagnetism. The simplest model consists in considering identical magnetic moments, coupled through a positive exchange interaction leading to a molecular field $H_m = wM$ ($w > 0$).

This molecular field has the same effect as an external field, and simply adds to the latter in the Brillouin function (quantum approach) or the Langevin function (classical approach) associated with the considered magnetic ion, and in which one thus simply replaces H by $H + wM$.

We do not use vectors but scalars, because in this description we do not take into account the anisotropy, and accordingly magnetization is assumed to be parallel to the applied field.

One therefore has $x = \mu_0 m_0 (H + wM)/k_B T$, which gives in the quantum approach:

$$m = M/M_0 = \mathcal{B}_J(x) = \mathcal{B}_J\left(\frac{\mu_0 m_0 H}{k_B T} + \frac{\mu_0 w M_0^2}{N k_B T} m\right) \quad (4.42)$$

For definite values of H and T, the reduced magnetization m is given by the two following equations:

$$m = \mathcal{B}_J(x) \quad (4.43)$$

$$m = -\frac{H}{wM_0} + \frac{N k_B T}{\mu_0 w M_0^2} x \quad (4.44)$$

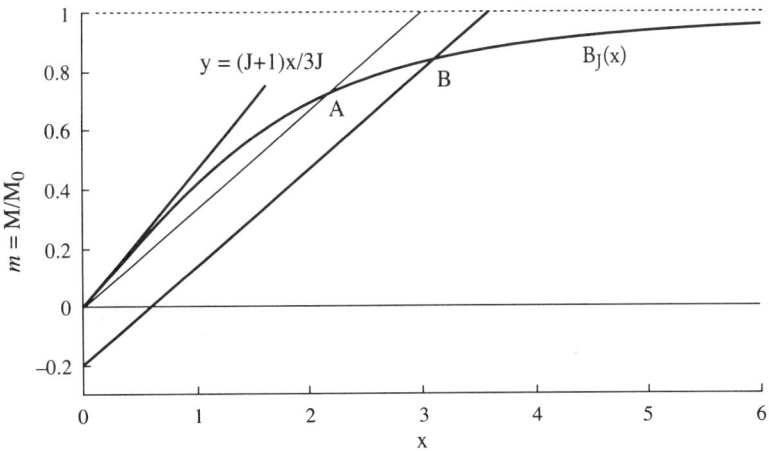

Figure 4.10 - Graphical determination of the magnetization for any field and temperature in a simple ferromagnet, thanks to a Brillouin (or Langevin) function

The resolution of these equations can be achieved either numerically, or graphically as shown in figure 4.10. In this latter approach, in the absence of applied magnetic field (H = 0), the straight line given by equation (4.44) passes through the origin. When temperature is low enough so that its slope is smaller than $(J+1)/3J$, i.e. the initial slope of the Brillouin function, there exists, in addition to the obvious solution $M = 0$, a solution given by point A. The substance then has a spontaneous magnetization $M_s(T) = m\,M_0$. At $T = 0$, this point corresponds to x infinite, and m then reaches 1, which corresponds to $M_s = M_0 = N\,g_J\,\mu_B\,J$.

When temperature is increased, point A moves to the left, $M_s(T)$ decreases, and then vanishes at T_C, the temperature for which the slope of the straight line is $(J + 1)/3J$. One deduces the Curie temperature:

$$T_C = \mu_0 \frac{(J+1)M_0^2 w}{3JNk_B} = \mathscr{C}\,w \qquad (4.45)$$

where $\mathscr{C}$ has been defined previously (eq. 4.33). In a classical description, one also obtains $T_C = \mathscr{C}w$, but $\mathscr{C}$ is given by equation (4.24). We remark that the spontaneous magnetization varies faster as one approaches T_C, as observed experimentally (fig. 3.8).

Now, when an external field is applied at a temperature smaller than T_C, the straight line of equation (4.44) moves towards increasing values of x while keeping the same slope, so that the intersection of the two curves moves in the same direction (point B for $H = H_B$). Magnetization then increases with the applied field. (Let us notice that the field H used in this formalism is always the total or internal field, $H = H_0 + H_d$).

Above T_C, in low field, magnetization is also weak; x being small, one can use expression (4.31) which leads to:

$$y = \frac{J+1}{3J}\mu_0\left(\frac{m_0 H}{k_B T} + \frac{wM_0^2}{Nk_B T}y\right),$$

whence one easily obtains: $\qquad \chi = \mathscr{C}/(T - T_C) \qquad (4.46)$

This is the *Curie-Weiss* law which can also be written as:

$$\chi^{-1} = T/\mathscr{C} - w \qquad (4.47)$$

The thermal variation of the reciprocal susceptibility is linear but, unlike the Curie law which passes through the origin, it cuts the temperature axis at $T = \theta_p$ which is called the *paramagnetic Curie temperature*. Figure 4.11 shows that the exchange interactions simply lead to a shift of the reciprocal susceptibility without interaction, and that the ordering temperature T_C of the compound, i.e. that corresponding to the disappearence of the spontaneous magnetization, has the same value as θ_p. As we noticed in section 3.3, the T_C value given by equation (4.45) is a measure of the exchange interaction. In pure iron, cobalt, and nickel, where the exchange interactions are among the largest, T_C reaches 1,043 K, 1,404 K, and 631 K, respectively.

Qualitatively, the thermal variation of the spontaneous magnetization (in reduced unit) associated with a $\mathscr{B}_J$ function is shown in figure 4.11, together with that of the reciprocal susceptibility: both curves are only valid in the molecular field approximation. We have adopted the usual representation, i.e. M and $1/\chi$ on the same graph (see caption of fig. 3.8).

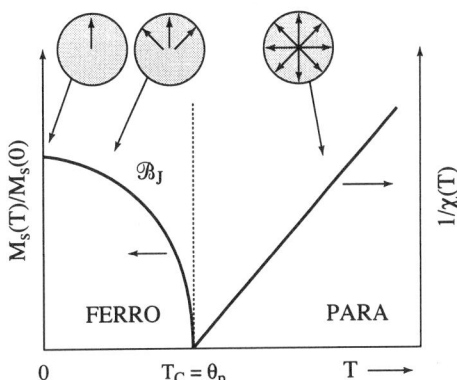

Figure 4.11
Theoretical representation of the thermal variation of M_s, and of paramagnetic reciprocal susceptibility χ^{-1} in a simple ferromagnet

The insets illustrate the thermal agitation effect on moments (molecular field approximation).

Actually, we have seen in figure 3.8 that, experimentally, the Curie-Weiss law is not exactly observed in the vicinity of T_C, a temperature range where the molecular field approximation is less valid. The thermal variation of $1/\chi$ is then written at high temperature:

$$1/\chi = (T - \theta_p)/\mathscr{C} \tag{4.48}$$

θ_p is slightly larger than T_C. The difference $\theta_p - T_C$ depends on the substance, and is associated with the persistence of a short range magnetic order slightly above T_C, as will be seen in section 7 of chapter 10.

Among the best illustrations of the above model, one has to quote the gadolinium based compounds in which moments are well defined, and where no significant additional interaction, in particular magnetocrystalline anisotropy, is present.

5. ANTIFERROMAGNETISM IN THE MOLECULAR FIELD MODEL

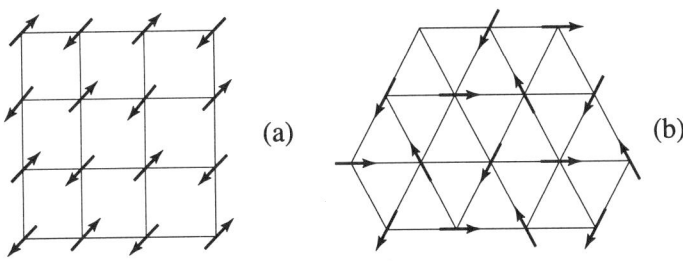

Figure 4.12 - Two examples of antiferromagnetic lattices

(a) Square lattice where the moments of the first nearest neighbours are all antiparallel, and those of the second nearest neighbours all parallel.

(b) Hexagonal lattice in which the moments located on the same vertical lines are parallel, and make an angle of ± 120° with respect to those located on the two neighbouring vertical lines.

One qualifies as antiferromagnetic any substance in which the magnetic moments, below the ordering temperature called *Néel temperature*, can be divided into two sublattices the magnetizations of which are equal and opposite, so that the resulting magnetization is zero (see for instance fig. 3.3-a). More generally, we will classify as antiferromagnetic any substance in which the moments are not necessarily parallel or antiparallel, nor of the same magnitude, but where the resulting magnetization, at the microscopic (atomic) scale, is zero. Figure 4.12 shows two examples of magnetic moment lattices with antiferromagnetic ordering.

Experimentally, antiferromagnetism exhibits the following features: the magnetic susceptibility, sketched in figure 4.13-a, is always small, and presents a maximum at the *Néel temperature* T_N, the ordering temperature of the compound. Above T_N the compound is paramagnetic. Its susceptibility decreases, and follows a Curie-Weiss law as for a ferromagnet (eq. 4.48), but θ_p is always smaller than T_N, and often negative. Below T_N, magnetization increases weakly, and approximately linearly with the field, up to a critical field at which it exhibits a so called *metamagnetic transition* (fig. 4.13-b). Depending on the substance, this transition is more or less sharp, and occurs for more of less large fields. For a given substance, this transition is less marked the closer it is to T_N, and it disappears at this temperature.

In some substances, the critical field is too large to be observed, in others, on the contrary, one can observed several transitions. In a simple antiferromagnet, this field can correspond to the flipping of the magnetization of one of the sublattices, initially opposite to the field, and which becomes parallel to the latter. The molecular field model accounts for the experimental results.

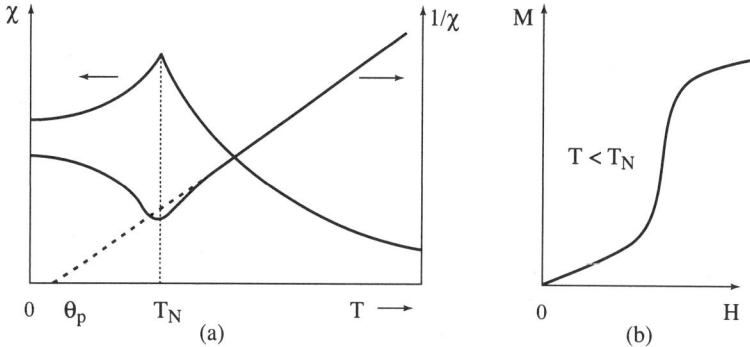

Figure 4.13 - (a) Schematic representation of the experimental variations of the susceptibility, and its reciprocal, for an antiferromagnetic material (b) Shape of the magnetization curve, below the ordering temperature T_N

5.1. NÉEL TEMPERATURE

The simple interpretation of antiferromagnetism, due to L. Néel, consists in considering two identical sublattices A and B; the exchange interactions are assumed

4 - PHENOMENOLOGY OF MAGNETISM AT THE MICROSCOPIC SCALE

to be positive within each sublattice, and negative between the two sublattices. Let $w_{AB} = w_{BA} = -w$, and $w_{AA} = w_{BB} = w'$ (with w and w' > 0) be the molecular field coefficients, between the two sublattices, and within each sublattice, respectively. The total field acting on each sublattice is then, i and j (i ≠ j) representing A or B:

$$\mathbf{H}_i = \mathbf{H} - w\mathbf{M}_j + w'\mathbf{M}_i \qquad (4.49)$$

At any temperature, $\mathbf{M}_i$ is parallel to $\mathbf{H}_i$, and one can write:

$$M_i = M_i{}^0 \mathcal{B}_J\left(\frac{\mu_0 m_0 |H_i|}{k_B T}\right) \qquad (4.50)$$

In the paramagnetic range, the variable of the Brillouin function is small, and all vectors are colinear so that we can write algebraically:

$$M_i = (\mathcal{C}/T)(H - wM_j + w'M_i) \qquad (4.51)$$

with $\mathcal{C} = N'\mu_0 m_{eff}^2/3k_B$, where $N' = N/2$ is the number of magnetic atoms of each sublattice per volume unit. In zero field (H = 0), the system of the two linear equations (4.51) has a solution such that M_A and M_B are non zero only if the determinant of the matrix of coefficients of these equations is zero.

The highest temperature for which such a solution exists is the Néel temperature which then satisfies the equation $(1 - \mathcal{C}w'/T_N)^2 - (\mathcal{C}w/T_N)^2 = 0$, which leads to:

$$T_N = \mathcal{C}(w + w') \qquad (4.52)$$

This formula shows that the Néel temperature is a measure of the sum of the absolute values of the molecular field coefficients.

Comments

♦ *The magnetization of the sublattices being finite at any temperature below T_N, it can seem strange that the equation is satisfied only at T_N, and not at smaller temperatures. Actually, the linear approximation which gives rise to equations (4.51) is valid only down to T_N, but not below, where the magnetization of each sublattice rapidly increases.*

♦ *The theory which has just been presented does not take into account the moment direction of each sublattice with respect to the reference directions of the compound. Indeed the magnetocrystalline anisotropy of the substance has so far not been considered. Actually, the magnetizations of the sublattices tend to be directed along one or several preferential directions called antiferromagnetism directions or easy magnetization directions. This anisotropy can be described with the same formalism as in ferromagnetic substances.*

5.2. PARAMAGNETIC SUSCEPTIBILITY

Above the Néel temperature, when the applied field is small, the system of equations (4.51) leads to: $M = M_A + M_B = 2\mathscr{C} H/(T - \theta_p)$, which gives:

$$\chi^{-1} = (T - \theta_p)/2\mathscr{C} \qquad (4.53)$$

with $\theta_p = \mathscr{C}(w' - w)$. One again obtains a Curie-Weiss law in which, unlike ferromagnetism, the θ_p temperature is smaller than T_N, and can even be negative. At T_N, χ presents a finite and weak value, as can be seen in figure 4.13.

5.3. PERPENDICULAR SUSCEPTIBILITY

Let us consider an antiferromagnetic crystal with uniaxial symmetry. When, below T_N, a small field is applied perpendicular to the antiferromagnetic direction (D), the magnetization of each sublattice rotates by a small angle θ (fig. 4.14) which minimises the total energy density, namely the Zeeman energy E_z due to the applied field, the exchange energy E_{ex}, and the anisotropy energy E_a. Setting: $|M_A| = |M_B| = M'$ for the magnetization of each sublattice, these energy densities (per unit volume) can be written, by limiting the expansion to second order in θ, as:

$$E_z = -2\mu_0 M' H \theta, \quad E_{ex} = \mu_0 [-w'M'^2 - wM'^2(1 - 2\theta^2)], \text{ and } E_a = K_1\theta^2$$

when considering a uniaxial anisotropy.

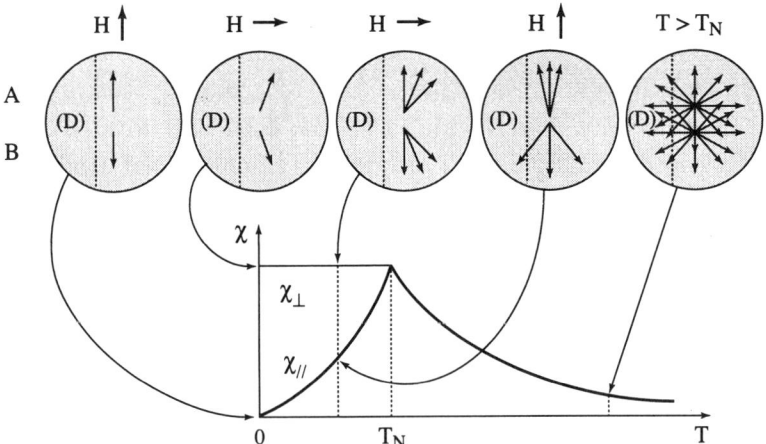

Figure 4.14 - Thermal variation of the susceptibilities parallel and perpendicular to the antiferromagnetism direction in a uniaxial compound in the case where (D) is the high symmetry axis

The insets show the effect of the field and of the thermal agitation on moments.

Minimization of the total energy with respect to θ leads to the equilibrium value:
θ = μ₀M'H/(2μ₀wM'² + K₁), and to the perpendicular susceptibility:

$$\chi_\perp = \frac{2M'\theta}{H} = \left[w\left(1 + \frac{K_1}{2w\mu_0 M'^2}\right)\right]^{-1} \quad (4.54)$$

As the coefficient w is virtually temperature independent, and as $K_1/2w\mu_0 M'^2$ generally remains very small compared to unity and, furthermore, weakly varies with temperature, the perpendicular susceptibility is approximately constant below T_N (fig. 4.14) with a value close to $1/w$.

5.4. PARALLEL SUSCEPTIBILITY

The field is now applied along the antiferromagnetism direction (see fig. 4.14). Under the effect of the field, the magnetizations of the parallel and antiparallel sublattices are markedly modified (if $T \neq 0$) according to the Brillouin law, and are respectively $M' + \Delta M$, and $M' - \Delta M$. In small field, an expansion to first order leads to the parallel susceptibility:

$$\chi_{//} = \frac{2\Delta M}{H} = 2A\mathcal{B}_J'(x_0)/\left[T + A(w-w')\mathcal{B}_J'(x_0)\right] \quad (4.55)$$

where $A = \mu_0 m_0 M'_0/k_B$, $x_0 = (w+w')M'(\mu_0 m_0/k_B T)$, and $\mathcal{B}_J'(x)$ is the derivative of $\mathcal{B}_J(x)$ with respect to x.

At 0 K, the magnetization of each sublattice is maximum, and $\Delta M = 0$. As a result, in the considered model, the parallel susceptibility is zero at this temperature. At T_N, $x_0 = 0$, and $\mathcal{B}_J'(x_0) = (J+1)/3J$. One then shows that $\chi_{//} = 1/w$, and thus has the same value as $\chi_\perp$, and as the paramagnetic susceptibility. Between 0 K and the Néel temperature, $\chi_{//}$ continuously increases. It can be easily shown that both parallel and perpendicular susceptibilities exhibit a change of slope at T_N, and that, above T_N, they are virtually the same (provided the anisotropy is not too large). All these characteristics, sketched in figure 4.14, are in agreement with the experimental results obtained in a large number of antiferromagnetic substances, especially those where the only significant interactions are those used in the model. For other materials, the model qualitatively account for the observed properties.

5.5. METAMAGNETISM

In an antiferromagnet, below T_N, $\chi_\perp$ is larger than $\chi_{//}$. The energy density in a given field being $E = -\chi\mu_0 H^2/2$, the state in which magnetization is perpendicular (or almost) to the magnetic field is energetically favoured compared to that in which magnetization is parallel (or antiparallel). The result is that if the system is initially (i.e. in low field) in the latter configuration, it will tend to make a sudden change toward the former, all the more the higher the field. However this effect works against the magnetocrystalline anisotropy, which favours the configuration stable in zero field. Therefore, in order to study the field effect on an antiferromagnet in the above used simple model, two situations have to be considered:

- If the anisotropy energy is small enough (fig. 4.15), and if an increasing magnetic field is applied along the antiferromagnetism direction (D), the magnetization of each sublattice suddenly switches, for a critical value of the field, into the perpendicular configuration sketched in the figure, so that the total magnetization suddenly increases. In larger fields, the magnetization of each sublattice progressively turns, and tends to align along the field direction. The associated variation of the total magnetization corresponds to the curve (b) of figure 4.15. Such a transition is termed a *spin-flop* metamagnetic transition.

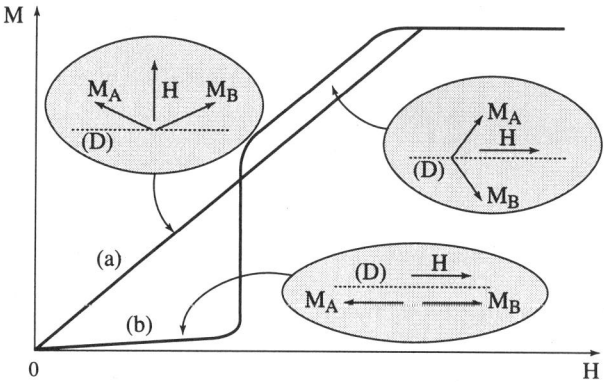

Figure 4.15 - Schematic representation of the low temperature magnetization process of an antiferromagnetic material with a weak uniaxial magnetocrystalline anisotropy
(a) H ⊥ (D) - (b) H // (D): in the latter case, a metamagnetic transition is observed

- If the uniaxial anisotropy is assumed to be large, when a reasonable field is applied parallel to (D), the magnetization of each sublattice remains parallel to this direction (fig. 4.16). In increasing field, the magnetization of the sublattice antiparallel to the field suddenly rotates for a critical value of the field, and becomes parallel to the latter in such a way that the total magnetization suddenly switches from a weak value to a very high value close to saturation (curve (b) of fig. 4.16). Such a transition is called a *spin-flip* metamagnetic transition.

In both cases, when the field is applied perpendicular to (D), magnetization increases approximately linearly up to saturation (curves (a) of fig. 4.15 and 4.16). This variation corresponds to the gradual rotation of moments toward the applied field. Whatever the magnitude of the anisotropy, the model then predicts, in agreement with experimental results, a transition in the magnetization process of a single crystal when the field is applied along the antiferromagnetism direction. This implies that the metamagnetic transition must also be observed on a polycrystalline sample.

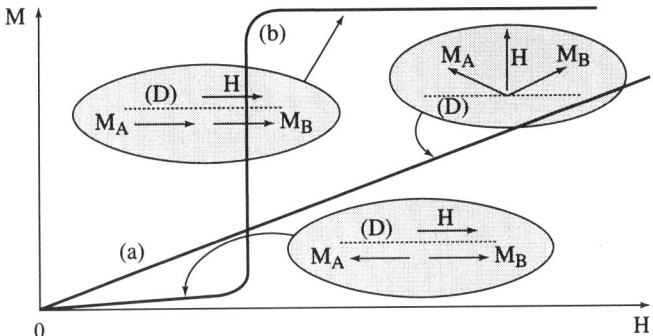

Figure 4.16 - Schematic representation of the low temperature magnetization process of an antiferromagnetic material with strong uniaxial magnetocrystalline anisotropy
(a) H ⊥ (D) - (b) H // (D): in the latter case, a metamagnetic transition is observed

Let us notice that, in a large number of colinear antiferromagnetic compounds with strong uniaxial anisotropy, *multistep metamagnetic processes* are observed, each transition being associated with the partial reversal of the moments of the sublattice initially antiparallel to the field. The magnetic arrangements of the different intermediate states can be determined thanks to neutron diffraction experiments.

6. FERRIMAGNETISM IN THE MOLECULAR FIELD MODEL

One qualifies as ferrimagnetic an antiferromagnetic substance in which the different sublattices do not exactly compensate each other, so that there is a spontaneous magnetization. Such a magnetic arrangement can occur when the sublattices are not identical due to the fact that the number of atoms of each sublattice is different, and / or that they contain different types of atoms (and accordingly different moments). Ferrimagnetism was predicted by Louis Néel before being confirmed by experiment. In the molecular field model, he considered two colinear sublattices with different magnetizations, M_A and M_B. Let $w_{AB} = w_{BA} = -w$ ($w > 0$), $w_{AA} = \alpha_A w$, and $w_{BB} = \alpha_B w$ (w_{AA} and $w_{BB} > 0$) be the various molecular field coefficients. The total field acting on each sublattice is written as:

$$H_i = H - w(M_j - \alpha_i M_i) \tag{4.56}$$

In the paramagnetic range, formula (4.51) becomes:

$$M_i = (\mathscr{C}_i / T) H_i \tag{4.57}$$

where $\mathscr{C}_i$ (i = A or B) is the Curie constant of each sublattice, $\mathscr{C}_i = N_i \mu_0 m_{i(eff)}^2 / 3 k_B$. Solving equations (4.56) allows to deduce the total magnetization:

$$M = M_A + M_B = \chi H \tag{4.58}$$

In this expression, the susceptibility is such that:

$$\frac{1}{\chi} = \frac{T - \theta_p}{\mathscr{C}} - \frac{\gamma}{T - \theta} \qquad (4.59)$$

where $\mathscr{C} = \mathscr{C}_A + \mathscr{C}_B$, with:

$$\theta = w\,(2 + \alpha_A + \alpha_B)\,\mathscr{C}_A\mathscr{C}_B/(\mathscr{C}_A + \mathscr{C}_B) \qquad (4.60)$$

$$\gamma = w^2\,[\mathscr{C}_A(1 + \alpha_A) - \mathscr{C}_B(1 + \alpha_B)]^2\,[\mathscr{C}_A\mathscr{C}_B/(\mathscr{C}_A + \mathscr{C}_B)^3] \qquad (4.61)$$

$$\theta_p = w\,\frac{\alpha_A\,\mathscr{C}_A^2 + \alpha_B\,\mathscr{C}_B^2 - 2\mathscr{C}_A\,\mathscr{C}_B}{\mathscr{C}_A + \mathscr{C}_B} \qquad (4.62)$$

The reciprocal magnetic susceptibility thus follows a hyperbolic law, shown in figure 4.17, with the following asymptotes: at high temperature $1/\chi = (T - \theta_p)/\mathscr{C}$ which cuts the temperature axis at $T = \theta_p$, and, in the vicinity of the ordering temperature T_C, a vertical asymptote $T = \theta$.

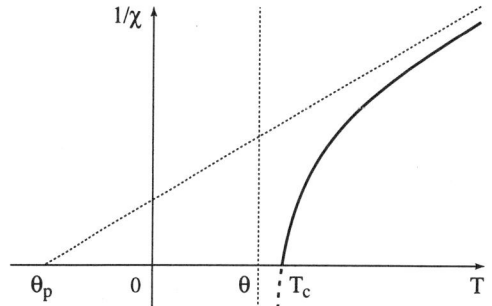

Figure 4.17
Theoretical variation
of the paramagnetic susceptibility
of a ferrimagnetic material

T_C is defined from the vanishing of $1/\chi$ (χ becomes infinite). It is obtained from a second-degree equation, the only realistic solution of which is:

$$T_C = \frac{w}{2}\left[\alpha_A\,\mathscr{C}_A + \alpha_B\,\mathscr{C}_B + \sqrt{4\mathscr{C}_A\,\mathscr{C}_B + (\alpha_A\,\mathscr{C}_A - \alpha_B\,\mathscr{C}_B)^2}\right] \qquad (4.63)$$

Below T_C, and in the absence of an external field, the numerical solution of the two equations (4.57) allows to deduce the thermal dependence of the spontaneous magnetization $|M_s| = ||M_A| - |M_B||$, the possible variations of which are sketched in figure 4.18. Case (N) is the most interesting, because the spontaneous magnetization passes through zero at a temperature, called *compensation temperature* T_{comp}, smaller than the Curie temperature. This behavior occurs when the thermal variation of the magnetization of the sublattice which dominates over that of the other at low temperature is faster than that of the latter.

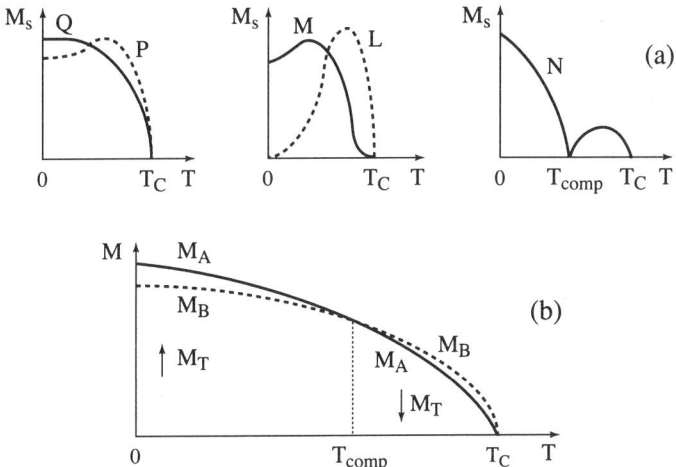

Figure 4.18 - (a) Different types of thermal variations of the spontaneous magnetization in a ferrimagnetic material as proposed by Néel [2] (b) Details of case N: $M_T = M_A - M_B$

7. OTHER TYPES OF MAGNETIC ARRANGEMENTS (HELIMAGNETISM, SINE WAVE MODULATED STRUCTURES...) IN THE MOLECULAR FIELD MODEL

The magnetic arrangements, and the associated magnetic behaviors, that have been described above, are the simplest ones, in particular all their magnetic moments are colinear. Actually, neutron diffraction allowed to observe, in crystallized materials, much more complex magnetic structures, often non colinear, which must be described with a number of sublattices larger than two. We will treat two examples: helimagnetic, and sine wave modulated structures. Finally, in amorphous substances, characterized by the absence of macroscopic anisotropy, spin configurations markedly different from the previous ones, and called speromagnetic, asperomagnetic, and sperimagnetic, are observed.

7.1. HELIMAGNETIC STRUCTURE

Let us consider a uniaxial crystal (with hexagonal or tetragonal symmetry for instance) in which the magnetic atoms, all identical, belong to layers parallel to each other, and perpendicular to the z axis (sixfold or fourfold, respectively). Let also assume that:
♦ within each layer, the moments are parallel to each other, and their magnetization is M_c,
♦ from layer to layer the moments make an angle ϕ (fig. 4.19-a).

For the sake of simplicity, one only considers, which is reasonable, the interactions between moments of the same layer, and between those belonging to first and second nearest neighbour layers. Let w_0, w_1 and w_2 be the molecular field coefficients due to the moments of the same layer, and to those of first and second neighbour layers, respectively. The exchange energy density is then written as:

$$E_{ex} = -\mu_0 M_c^2 (w_0 + 2 w_1 \cos \phi + 2 w_2 \cos 2\phi)/2 \qquad (4.64)$$

Minimizing this energy with respect to ϕ leads to the equation $(w_1 + 4w_2 \cos \phi) \sin \phi = 0$, which is satisfied for three magnetic configurations:
- $\phi = 0$, i.e. ferromagnetism,
- $\phi = 180°$, i.e. antiferromagnetism,
- an angle ϕ given by:

$$\cos \phi = - w_1 / 4w_2 \qquad (4.65)$$

This latter case corresponds to helimagnetism. It is sketched in figure 4.19-a.

Let us notice that this configuration can exist only if $|w_1/4w_2| < 1$. Furthermore, the study of the sign of the second derivative of the energy, and comparison of the energies of these configurations, leads to the phase diagram shown in figure 4.20. The helimagnetic configuration is the most stable when $w_2 < 0$ and $|w_2| > |w_1/4|$. Such a structure is observed in a large number of hexagonal, and to a lesser extent tetragonal rare earth based compounds.

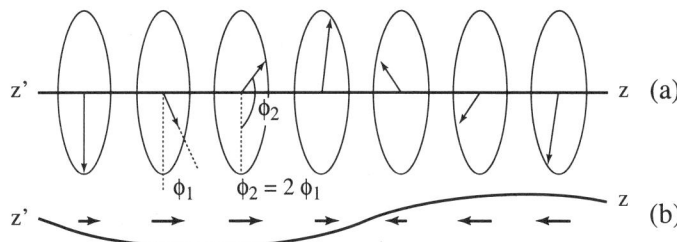

Figure 4.19 - Helimagnetic (a), and sine wave modulated structures (b)

7.2. SINE WAVE MODULATED STRUCTURE

Let us now consider the case where the uniaxial anisotropy forces the moments to be along the z axis, i.e. perpendicular to the layers. The only degree of freedom left to magnetization is then its amplitude: let us assume that it exhibits a sine wave modulation (fig. 4.19-b) with: $M_n = M_{max} \cos (n\phi + \phi_0)$, where $n = \ldots -2, -1, 0, 1, 2, \ldots$ numbers the successive layers.

Considering the same exchange interactions as previously, the energy density is written as:

$$E_{ex} = (\mu_0/2)(M_{max})^2 \cos \phi_0 \{ w_0 \cos \phi_0 + w_1 [\cos (\phi + \phi_0) + \cos (-\phi + \phi_0)]$$
$$+ w_2 [\cos (2\phi + \phi_0) + \cos (-2\phi + \phi_0)] \}$$

which becomes: $E_{ex} = \mu_0 (M_{max})^2 \cos^2 \phi_0 (w_0 + 2w_1 \cos \phi + 2w_2 \cos 2\phi)/2$.

As layer n = 0 can be located in any place of the sine modulation, the average exchange energy is such that $\cos^2 \phi_0 = 1/2$, so that:

$$E_{ex} = \mu_0 (M_{max})^2 (w_0 + 2w_1 \cos \phi + 2w_2 \cos 2\phi)/4 \quad (4.66)$$

To within a factor 2, it is the same energy as in (4.64). It leads to the same phase diagram (fig. 4.20) except that the helimagnetic structure is replaced by a sine wave modulated one with the same period. Thus, for the same exchange interactions, the compound will adopt either a helimagnetic structure or a wave modulated one depending on whether the anisotropy is planar or axial, respectively.

It is worth noticing that, in the modulated structure, a large number of moments have a reduced amplitude. In the description adopted so far, in which the thermal average value of each moment follows the Brillouin function, such a moment reduction can exist only if the temperature is not too low because at 0 K, whatever the field amplitude, magnetic moments have their maximum value.

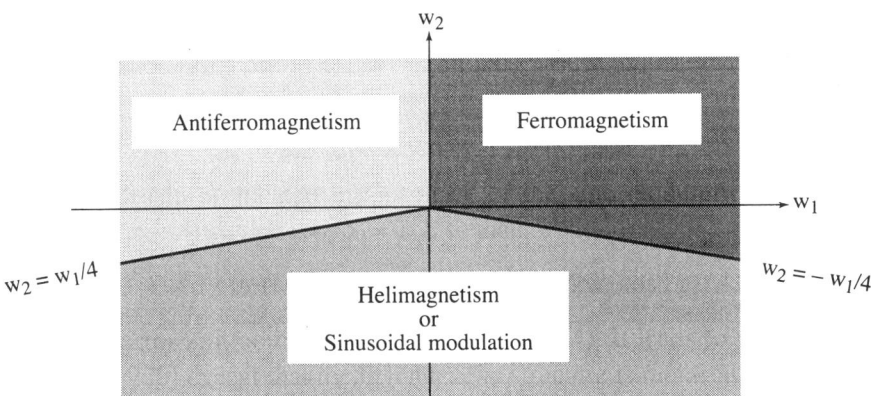

Figure 4.20 - Phase diagram of different structure types in a w_1-w_2 representation

This is the reason why a wave modulated structure is not stable at low temperature. It generally appears just below the Néel temperature, and transforms at low temperature into an equal moment structure. However there exist some particular compounds in which the wave modulated structure remains stable down to very low temperature, the moment reduction being of non-thermal origin.

As the above described structures have zero resulting magnetization, their macrocopic magnetic properties are similar to those of a simple antiferromagnet, and they are hence classified in the same category: the susceptibility is always weak, and exhibits a maximum at the Néel temperature. In the same way, below T_N, a metamagnetic transition occurs, although less sharp than in the case of a simple antiferromagnet.

7.3. MAGNETIC STRUCTURES OBSERVED IN AMORPHOUS SUBSTANCES

In amorphous substances, characterized by a local anisotropy direction which randomly varies from site to site, *ferromagnetic* structures can be observed. It is the case of some metallic glasses which present a very weak local anisotropy: positive exchange interactions are then preponderant, and impose a colinear arrangement of moments.

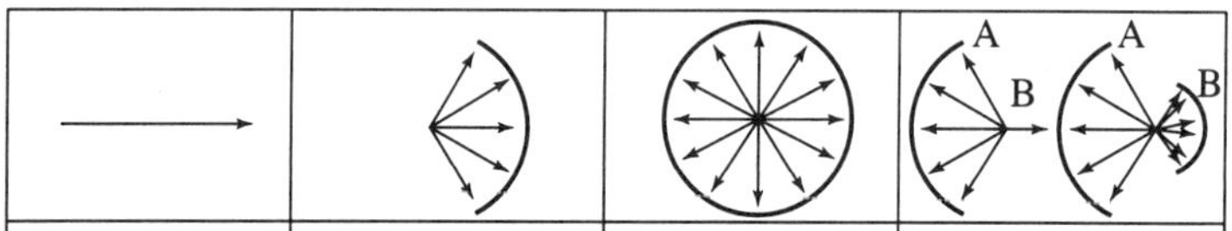

Ferromagnetism	Asperomagnetism	Speromagnetism	Sperimagnetism
Spontaneous magnetisation (a)	Spontaneous magnetisation (b)	No spontaneous magnetisation (c)	Spontaneous magnetisation (d)

Figure 4.21 - The different magnetic structures encountered in amorphous alloys

(a) Colinear ferromagnetism.

(b) Asperomagnetism (ferromagnetism with an easy magnetization direction which randomly varies from site to site: a resulting magnetization is observed at the mesoscopic scale).

(c) Speromagnetism (antiferromagnetism, but the antiferromagnetism direction randomly varies from site to site: no resulting magnetization at the mesoscopic scale).

(d) Sperimagnetism (ferrimagnetism with an easy magnetization direction which randomly varies from site to site: a resulting magnetization is observed at the mesoscopic scale. It results from the unbalance between the two, A and B, sublattice magnetizations; besides, one of them can be ferromagnetic).

A spontaneous magnetization is observed (fig. 4.21-a). When the local anisotropy becomes noticeable, it locally imposes the easy magnetization direction; in the presence of ferromagnetic interactions, magnetic moments locally become more or less parallel to each other. Although, at long range, there is no permanent magnetic moment, locally there does exist a spontaneous magnetization: the magnetic structure is called *asperomagnetic* (fig. 4.21-b). On the contrary, in the presence of antiferromagnetic interactions, with each moment is associated an antiparallel neighbour, and the structure, called *speromagnetic*, has no spontaneous magnetization (fig. 4.21-c). It is the case of the TbAg and $Dy_{21}Ni_{79}$ amorphous alloys.

Finally, in binary amorphous alloys made of two elements A and B, if A-A and B-B ferromagnetic interactions, and A-B antiferromagnetic interactions are present, so called *sperimagnetic* structures can be observed (fig. 4.21-d). They couple an asperomagnet and a ferromagnet (case of $Dy_{23}Co_{77}$ amorphous alloy), or two asperomagnetic "sublattices" (case of $Dy_{21}Fe_{79}$ amorphous alloy). These various behaviors were identified by J.M.D. Coey [5].

8. THE TWO MAIN FAMILIES OF MAGNETIC MATERIALS

The magnetic structures described and discussed above are the simplest ones. Neutron diffraction revealed that actually there exist more complex magnetic structures, all of which mainly result from the interplay between exchange interactions and magnetocrystalline anisotropy. In particular one observes a large number of non colinear complex structures, for instance the triangular structure which is observed in some hexagonal compounds, and which results from the *frustration* of negative exchange interactions between first nearest neighbours (fig. 4.12-b). The word frustration used in this case means that antiparallelism between first nearest neighbours cannot occur on account of the position of these atoms at the vertices of a triangle.

Whatever the magnetic structure, one can classify the substances which order magnetically into two categories with respect to their macroscopic properties:

- On the one hand, the compounds which have a *spontaneous magnetization*, and *ferromagnetic* type behavior with the characteristics presented in section 1.4 and 2.1 of chapter 3, and sketched in figures 3.7 and 3.8. Although the spontaneous magnetization of a ferrimagnet can exhibit in some cases a special thermal variation, theses compounds are generally considered in the category of ferromagnets.
- On the other hand, the compounds *without spontaneous magnetization*, and which present an *antiferromagnetic* type behavior with the characteristics presented in section 1.3 of chapter 3, and sketched in figure 4.13.

9. ARROTT PLOTS

The determination of the macroscopic magnetic characteristics of materials requires the study of low field magnetization. It is the case for the determination of:
- the spontaneous magnetization of a ferromagnet,
- the ordering temperature, and
- the susceptibility in the absence of spontaneous magnetization.

Yet, experimentally it is not always easy to determine these quantities because they are really defined only in zero applied field.

Actually one always applies a finite field, and the measurement is often inaccurate for different reasons such as the weakness of the signal or, more serious, the influence of inhomogeneities or possible impurities (especially if they are ferromagnetic with a Curie temperature larger than the ordering temperature of the material under investigation). However, even in the case of a perfectly accurate experiment on a pure sample, it is necessary –in order to determine the spontaneous magnetization– to make the sample single domain, and to align all moments along the applied field.

This sometimes requires rather intense fields, in order to be out of the linear range of the initial susceptibility: one then measures a magnetization "which approaches saturation" but which is never saturated, and one experimentally defines approach laws allowing to extrapolate "to infinite field" the value of the *saturation magnetization*. The latter can markedly differ from the spontaneous magnetization, especially when approaching the Curie temperature, around which the applied magnetic field is less and less negligible with respect to the molecular field.

It is therefore better to analyse the isothermal variations over a definite field range, over which an expansion involving only low order terms remains valid (actually this field range is often rather large). In order to do that, one uses so called *Arrott plots* [6]. This consists in plotting the isothermal variations of M^2 as a function of H/M or, which is often equivalent, of H_0/M. We will see now what information these plots provide in the various experimental situations that have been encountered previously.

9.1. MATERIAL WITHOUT SPONTANEOUS MAGNETIZATION

This is the case for all materials in the paramagnetic phase, or for antiferromagnetic materials in the magnetically ordered state. M can then always be expanded in increasing powers of H, or H in increasing powers of M. Limiting the expansion to third order, and considering that magnetization is an odd function of the field, the latter can be written:

$$H = M/\chi + AM^3 \qquad (4.67)$$

where χ is the initial susceptibility, and A a constant (which depends on temperature).

One then deduces: $\qquad M^2 = (H/M - 1/\chi)/A \qquad (4.68)$

Arrott plots are thus straight lines which cut the horizontal axis at $1/\chi$.

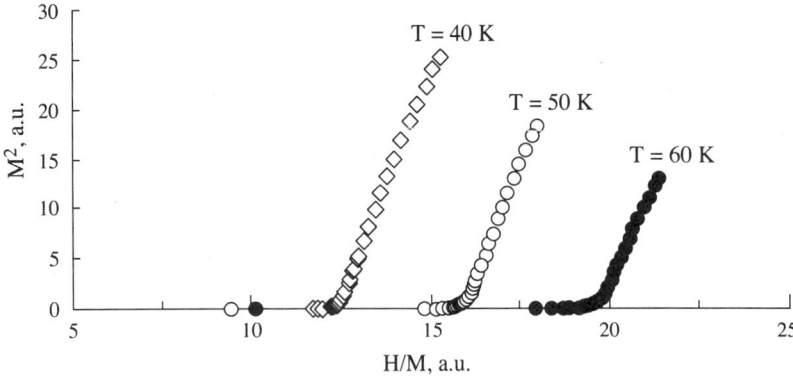

Figure 4.22 - Arrott plots in the paramagnetic range for the DyNi$_2$Si$_2$ antiferromagnetic compound, in which the Néel temperature is $T_N = 6$ K [7]

Figure 4.22 shows an example of such plots for $DyNi_2Si_2$, an antiferromagnetic compound. It is worth noticing that the experimental points of low magnetization, i.e. in low field, diverge from the straight line, which justifies the use of such plots rather than directly measuring magnetization in very low field.

9.2. MATERIAL WITH A WEAK SPONTANEOUS MAGNETIZATION

This is the case of all ferromagnetic (or ferrimagnetic) materials in the vicinity of the Curie temperature T_C, or of some of these materials, in which the magnetization is weak whatever the temperature. Assuming that M is small, the magnetic contribution to the free energy in the presence of an internal field H (applied plus demagnetising field) can be expressed thanks to a so called Landau expansion (see § 6 in chap. 10):

$$F(M) = a M^2/2 + b M^4/4 + \ldots - \mu_0 MH \quad (4.69)$$

where a and b are coefficients which depend on temperature. In the ferromagnetic and paramagnetic phases, b is generally positive. In zero field, F(M) is even, because independent on the sense of magnetization.

In figure 4.23, this function is plotted for two cases. In the case (F), a is negative, and F(M) is minimum for a non zero value of magnetization: the compound is ferromagnetic. In the case (P), a is positive, and F(M) is minimum for M = 0: the compound is paramagnetic. Thus the coefficient a becomes negative at T_C. In the presence of a field H, minimising F(M) with respect to M leads to the equation $aM + bM^3 = \mu_0 H$, hence to:

$$M^2 = (\mu_0/b) H/M - a/b \quad (4.70)$$

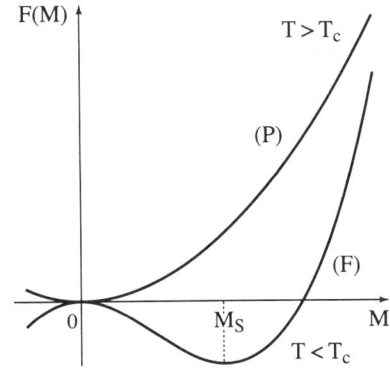
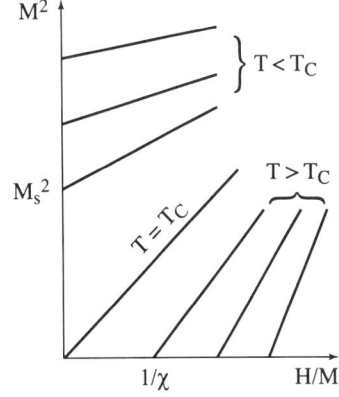
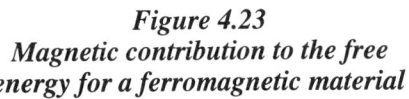

Figure 4.23
Magnetic contribution to the free energy for a ferromagnetic material

Figure 4.24
Schematic representation of the Arrott plots for a ferromagnetic material

One again finds that the Arrott plots are linear (see fig. 4.24). For $T < T_C$ (a < 0), their intersection with the vertical axis – a/b is equal to M_s^2. For $T > T_C$ (a > 0), there is no more spontaneous magnetization, and the intersection with the horizontal axis a/μ_0 is equal to $1/\chi$, i.e. the reciprocal initial susceptibility as shown previously. The Arrott plot which passes through the origin corresponds to the Curie temperature.

The above results are independent of the model used to describe the magnetism of the considered material. It is hence logical to find them in the framework of the localised model of magnetism so far used, and in which magnetization is expressed thanks to the Brillouin (or Langevin) function:

$$m = M/M_0 = \mathcal{B}_J \left[\frac{\mu_0 m_0}{k_B T} (H + wM) \right] \quad (4.71)$$

The expansion of this function up to third order in M leads to: $\mu_0 H = a M + b M^3$, and thus to the relation (4.70) with $a = \mu_0 w (T/T_C - 1)$, and

$$b = \mu_0 [3(2J^2 + 2J + 1) T] / 10 M_0^2 (J+1)^2 \mathcal{C},$$

where $\mathcal{C}$ is the Curie constant. One clearly sees that b is positive, and that a, positive above T_C, vanishes at T_C, and then becomes negative at lower temperature.

In conclusion the Arrott plots are an excellent method to make use of the magnetization measurement data in order to characterize a material in detail.

Comment - *The Arrott plots are often performed using the applied field H_0 instead of the internal field H. In this case, there is a shift of the horizontal axis scale. Assuming that $H_d = -NM$, formula (4.68) becomes:*

$$M^2 = [H_0/M - N - (1/\chi)]/A \quad (4.72)$$

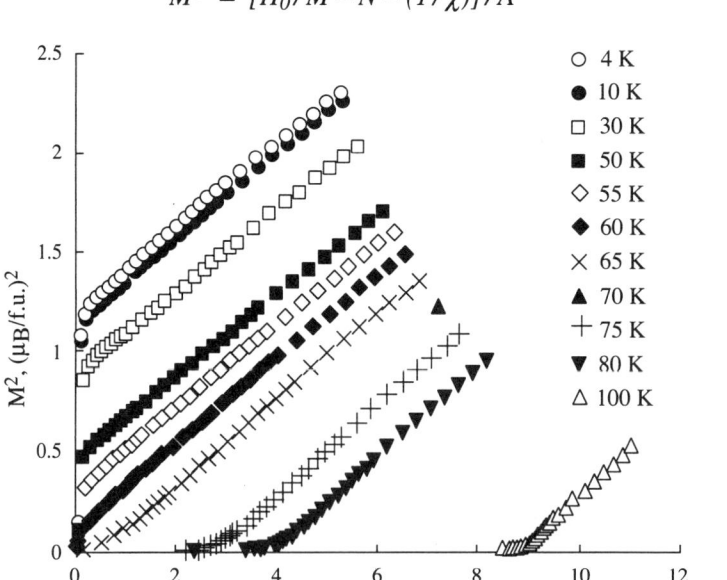

Figure 4.25 - Arrott plots for a very weak ferromagnet, $Lu_3Co_{7.77}Sn_4$ [8]
One will notice that the curve passing through the origin corresponds to the Curie temperature, namely $T_C = 63$ K.

Figure 4.25 illustrates an Arrott plot in the case of a compound showing weak ferromagnetism [8]. Here again, one observes that the straight line passing through the origin corresponds to the Curie temperature, and that the experimental points in low field deviate from the linear variation. It is worth reminding that such plots should be performed only with materials showing very weak ferromagnetism at any temperature, or with strong ferromagnets in their paramagnetic phase, and in their ferromagnetic phase only very close to the Curie temperature.

10. CONCLUSIONS

The phenomenological description of localised magnetism presented in chapter 3 at the macroscopic scale, and in chapter 4 at the microscopic scale, has the advantage of being clear and rather simple. However, it is worth reminding that many substances need another description, namely the itinerant electron model: the electrons responsible for the magnetism of the substance are partially or totally delocalised. This model will be described in chapter 8. However, whether magnetism is localised or itinerant, the basic phenomena will be the same:

♦ at the microscopic scale, the spontaneous magnetic structure will always result from the exchange interaction which forces the moments to be either parallel, or antiparallel to each other, and from anisotropic interactions that will define, on each site, one or several preferential directions for the magnetic moments.

♦ at the mesoscopic scale, the dipolar interaction (which is rather weak compared to the two previous ones, but acts at long range) will play a key role in the domain distribution, as soon as a spontaneous magnetization occurs.

In the two following chapters, we will analyse in more details the mesoscopic and macroscopic physical behaviors of a ferromagnetic substance (Weiss domain distribution, magnetization process, hysteresis phenomena), behaviors with a universal character, and thus independent of the considered substance, and of the localised or itinerant character of the electrons responsible for magnetism.

APPENDIX: DETERMINATION OF MAGNETIC STRUCTURES USING NEUTRON DIFFRACTION

Neutron scattering, in particular neutron diffraction, is an essential tool for the study of magnetism of materials [9, 10].

The neutron, a neutral particle with mass $m_N = 1.675 \times 10^{-27}$ kg, is the world's smallest compass. Its magnetic moment, associated with intrinsic angular momentum described as "spin 1/2", can have a component $\mu_N = \pm 0.966 \times 10^{-26}$ A.m^2 along a given direction. Neutron beams for research on condensed matter are produced either by nuclear reactors, or by spallation sources. In either case their initial kinetic energy is on the order of 10^6 eV, and they have to lose most of it through interaction with a cold, "thermal" or hot source before becoming useful. The De Broglie wavelength associated to neutrons with velocity v is $\lambda = h/(m_N v)$, where h is Planck's constant, and this is on the order of 10^{-10} m (1 Å), hence of interatomic distances, for the "thermal" neutrons most frequently used. Thus, when propagating in a crystal, neutrons sense two spatially periodic contributions to their potential energy, which lead to diffraction. Diffraction is geometrically characterised by the scattering vectors $\mathbf{q}$ such that an incident plane wave with wave-vector $\mathbf{k}_o$ produces an intense diffracted wave with wave-vector $\mathbf{k} = \mathbf{k}_o + \mathbf{q}$.

The first contribution to the potential energy corresponds to the interaction of the neutron with nuclei, which leads to nuclear scattering, characterised by the scattering amplitude $-b$ (b, called the nuclear scattering length, depends on the isotope, can be measured but not calculated, and is of the order of 1 fm). Nuclear scattering produces Bragg peaks characteristic of the crystallographic period, hence essentially the same as those obtained with X-ray scattering.

The second contribution to the potential energy corresponds to the magnetic energy $-\boldsymbol{\mu}_N \mathbf{b}(\mathbf{r})$ of the neutron's magnetic moment with the magnetic field at the microscopic scale $\mathbf{b}(\mathbf{r})$, which varies rapidly as a function of position $\mathbf{r}$. This interaction leads to magnetic scattering. This term has no equivalent (except for a very weak contribution) in X-ray scattering. The magnetic scattering amplitude p for an electronic magnetic moment corresponding to 1 μ_B (one unpaired electron spin), perpendicular to the scattering vector, is $p_1 = 2.7\,f(\mathbf{q})$ fm, with $f(\mathbf{q})$ the magnetic form factor, related to the fact that the magnetic electron is not localised (f = 1 for forward scattering, i.e. q = 0). Because p is of the same order of magnitude as b, the effect on neutrons of the distribution of electronic magnetic moments in atoms is large.

The magnetic contribution to neutron diffraction allows the determination of the magnetic structure of crystallized compounds, i.e. the magnitude and the direction of the magnetic moment of each atom. On the one hand, the position of the Bragg peaks is characteristic of the magnetic periodicity, i.e. it provides the parameters of the magnetic unit cell. In many compounds, in particular in antiferromagnetic materials, the period of the magnetic structure is different from the crystallographic period (this period is often large, and incommensurate with the crystallographic one). In this case, new Bragg reflections appear when the material goes from the disordered (paramagnetic) state to the low temperature ordered (antiferromagnetic) state. This is illustrated in figure 4.26 in the case of the antiferromagnetic tetragonal compound $HoCo_2Si_2$ with Néel temperature $T_N = 13$ K [11]. On the other hand, the intensity of the Bragg reflections is characteristic of the direction and magnitude of the moments of all the atoms in the magnetic unit cell.

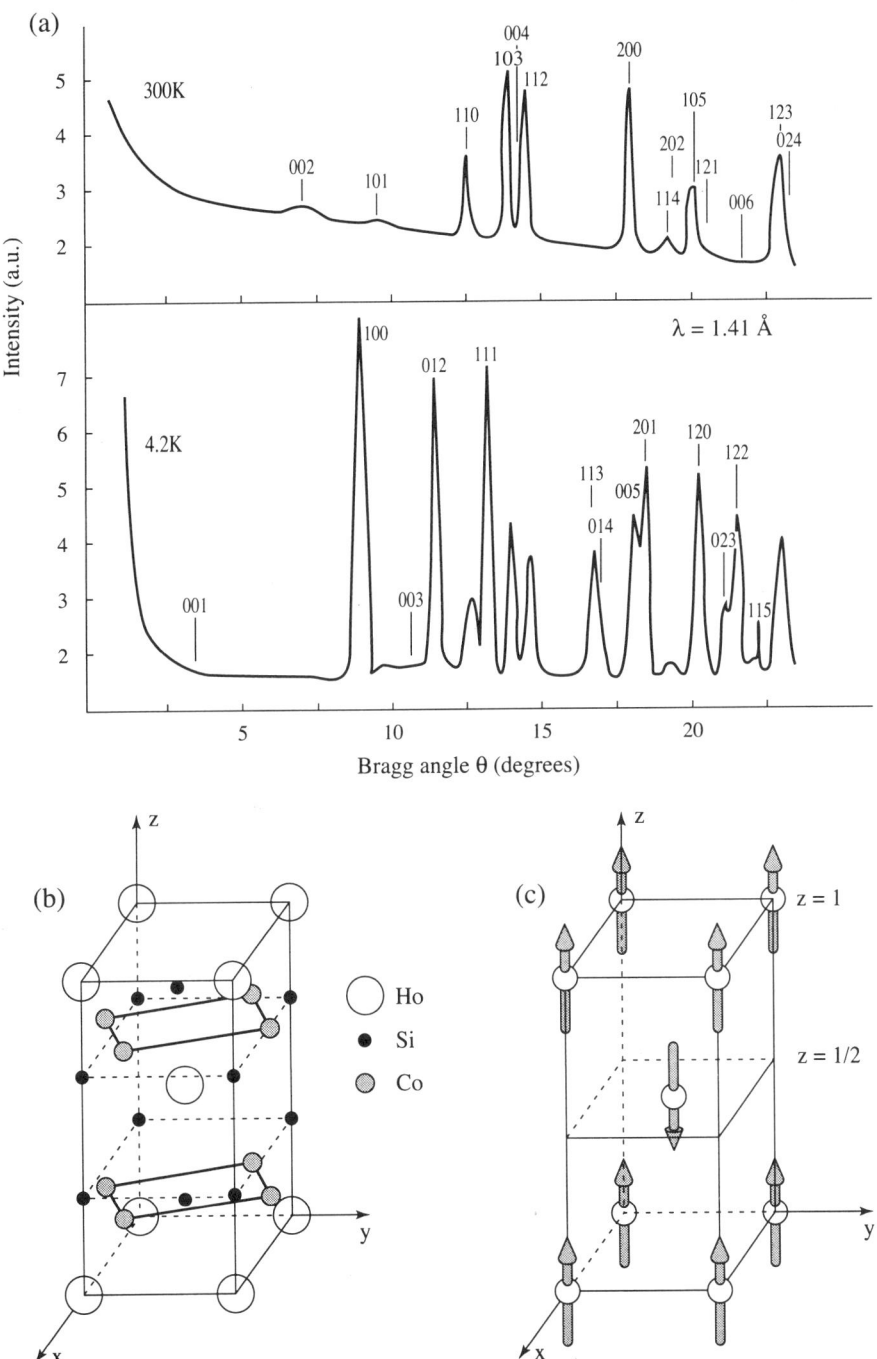

Figure 4.26 - Antiferromagnetic compound HoCo$_2$Si$_2$ with tetragonal structure

(a) Neutron diffraction pattern on a powder sample, above and below the Néel temperature - (b) Crystallographic structure - (c) Magnetic structure. Note that only the Ho ions are magnetic.

140   MAGNETISM - FUNDAMENTALS

The majority of neutron diffraction experiments are performed on powder (or polycrystalline) samples, and with a non polarized incident neutron beam. Neutron diffraction experiments on single crystals, and with a polarized incident beam, lead to more accurate information, in particular it is possible to obtain the magnetic density map in the material.

Finally many other neutron scattering techniques provide complementary information on magnetism. Among these techniques, Bragg diffraction imaging, and its application to domain investigations, is described in chapter 5, section 6.1.4.

EXERCICES

E.1 - The susceptibility of a paramagnetic material, which contains only one type of rare earth ions (number $N = 10^{27}/m^3$), follows a Curie law given by $\chi = 0.162/T$ (in SI units). Determine the effective moment; deduce which rare earth element is present.

E.2 - Pure cobalt crystallises in the hexagonal close-packed structure ($a = 0.251 \times 10^{-9}$ m, and $c = 0.407 \times 10^{-9}$ m) which contains two atoms per unit cell at $(0, 0, 0)$, and $(1/3, 2/3, 1/2)$. It is ferromagnetic with $T_C = 1{,}404$ K, and the magnetic moment of each atom at low temperature is $m = 1.7\ \mu_B$.

E.2.1 - How many first nearest neighbours does each atom have, and at what distance?

E.2.2 - Assuming that the exchange interaction occurs only between first nearest neighbours, and considering that the energy of this interaction is written as $\mathcal{E}_{ex} = -nm^2$, express the molecular field coefficient w as a function of n.

E.2.3 - In the classical case of an effective moment equal to m, give in SI the value of $\mathcal{E}_{ex}$.

E.2.4 - One wants to compare this energy with the dipolar energy. Calculate the latter ($\mathcal{E}_{dip}$) between two cobalt atoms considering that they are at distance a, with their moments parallel to each other, and to the straight line which joins them. Compare $\mathcal{E}_{ex}$ and $\mathcal{E}_{dip}$.

We recall that the dipolar energy between two moments, m_i and m_j, has the following expression:

$$\mathcal{E}_{ij} = -\frac{\mu_0}{4\pi}\left[3\frac{(m_i r)(m_j r)}{r^5} - \frac{m_i m_j}{r^3}\right]$$ where **r** is the vector joining both atoms;

this expression can be deduced from equations (2.24) and (2.80), chapter 2.

SOLUTIONS TO EXERCICES

S.1 $m_{eff} = 7.94\ \mu_B$, the substance contains Gd^{3+} ions.

S.2 *S.2.1* 12, 0.251×10^{-9} m.

S.2.2 w = 12 n/N where N is number of atoms per volume unit.

S.2.3 $\mathcal{E}_{ex} = 4.95 \times 10^{-21}$ J.

S.2.4 $\mathcal{E}_{dip} = 3.14 \times 10^{-24}$ J.

The dipolar energy is more than three orders of magnitude smaller than the exchange energy.

REFERENCES

[1] W.F. HENRY, *Phys. Rev.* **88** (1952) 559.

[2] L. NÉEL, *Ann. Géophys.* **5** (1949) 99-136.

[3] L. NÉEL, *CR Acad. Sci.* **252** (1961) 4075-4080.

[4] L. NÉEL, Œuvres Scientifiques (1978) Editions du CNRS, Paris.
English translation: Selected works of Louis Néel (1988) Gordon & Breach, Sci. Publishers, New York, London.

[5] J.M.D. COEY, P.W. READMAN, *Nature* **246** (1973) 476; J.M.D. COEY, J. CHAPPERT, J.P. REBOUILLAT, T.S. WANG, *Phys. Rev. Lett.* **36** (1976) 1061; R. ARRESE-BOGGIANO, J. CHAPPERT, J.M.D. COEY, A. LIÉNARD, J.P. REBOUILLAT, *J. Phys.* **37** (1976) C6-77.

[6] A. ARROTT, *Phys. Rev.* **108** (1957) 1394.

[7] A. GARNIER, D. GIGNOUX, D. SCHMITT, private communication (1998).

[8] R. SKOLOZDRA, B. GARCIA-LANDA, D. FRUCHART, D. GIGNOUX, J.L. SOUBEYROUX, L. AKSELRUD, *J. Alloy. Compd* **235** (1996) 210.

[9] G.E. BACON, Neutron Diffraction, 3rd ed. (1975) Clarendon Press, Oxford.

[10] G.L. SQUIRES, Thermal Neutron Scattering (1978) Cambridge University Press, (1996) Dover Publications.

[11] V.N. NGUYEN, F. TCHEOU, J. ROSSAT-MIGNOT, *Solid State Commun.* **45** (1983) 209.

CHAPTER 5

FERROMAGNETISM OF AN IDEAL SYSTEM

In an ideal system of magnetic moments, three types of interaction coexist: exchange interactions favouring –as far as this chapter is concerned– ferromagnetic or ferrimagnetic order, interactions with the lattice which give rise to a magneto-crystalline anisotropy, and magnetic dipolar interactions.

The impossibility of simultaneously satisfying the requirements associated with all these interactions frequently leads to the establishment of magnetic domain structures. The configuration of the moments thus obtained results from a compromise, and can, for this reason, be easily perturbed. The size, and relative dimensions of the system under consideration have a radical effect on the equilibrium configuration obtained as well as on the strength of perturbations capable of modifying it.

In a perfect, anisotropic system, large enough to feature a multidomain structure in the demagnetised state, all the magnetization processes starting from this state are perfectly reversible. By contrast, magnetization reversal starting from the fully saturated state implies crossing an energy barrier. In the presence of magneto-crystalline anisotropy, this reversal takes place only for applied reverse fields in excess of the anisotropy field (Brown's theorem). In the absence of magneto-crystalline anisotropy, configurations of non-uniform magnetization allow reversal in weaker fields.

1. INTRODUCTION

When the temperature of a ferromagnetic system is maintained below T_C, the exchange interactions become preponderant, and lead to a parallel ordering of the individual moments. All ferromagnetic materials whose Curie temperature is above room temperature should, therefore, appear as spontaneously magnetised. Experience teaches us that this is, in fact, not the case, and the magnetization of many ferromagnetic systems (iron, nickel, sheets of iron-silicon for example) becomes manifest only in an applied field. However, it is also known that a toroid of any ferromagnetic substance, once saturated by an applied field, retains a remanent magnetization close to M_s even after the inducing field has been brought to zero. This magnetization disappears when a slit or air-gap, however narrow, is introduced into the toroid (fig. 5.1). Remagnetising to saturation a toroid thus sectioned requires the application of a much larger field than that which was necessary to magnetise it without the slit. Furthermore, this magnetization returns to a value close to zero the moment the applied field is suppressed.

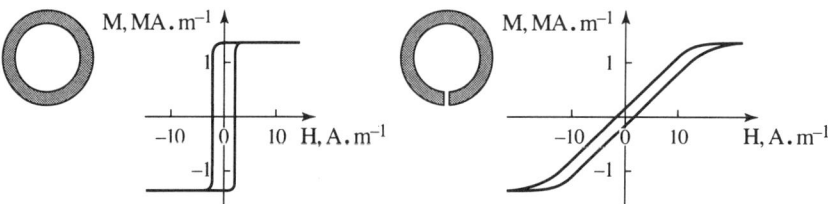

Figure 5.1 - The influence of a slit on the magnetization curve of a toroid of soft iron as a function of the applied external magnetic field

This observation shows that the absence of spontaneous magnetization in most ferromagnetic materials is a response to the need to prevent the formation of poles on the surface of a sample. Exchange interactions, which favour the parallel alignment of the moments, tend to generate such poles while, by contrast, the dipolar interactions tend to make them disappear. We shall recall the essential characteristics of these two interactions, and of a third, which is frequently present: local magnetic anisotropy. We shall see how the competition between them can give rise to extremely varied configurations of moments, first in a zero applied field (§ 2 to 4), then in the presence of a magnetic field (§ 7, and 8). Section 5 deals with the special case of particles of very small size, and section 6 with the techniques for visualising magnetic domains.

2. THE PRINCIPLE OF FERROMAGNETIC CONFIGURATIONS OR THE ART OF COMPROMISE

2.1. EXCHANGE INTERACTION

The exchange interaction is of quantum mechanical origin, and electrostatic in nature. It is very strong but acts between neighbouring moments only, and falls off very rapidly with distance. In ferromagnetic systems where it tends to align moments parallel to each other, it favours a uniform magnetization. However, its influence is weak for large distances (where it acts only through a chain effect), thus permitting the mutual independence of moments sufficiently far apart, for a relatively modest cost in energy. We gave an analytical expression for this interaction energy in a preceding chapter (eq. 4.34). This expression is proportional to the product $m_i m_j$, and hence to $\alpha_{1i}\alpha_{1j} + \alpha_{2i}\alpha_{2j} + \alpha_{3i}\alpha_{3j}$, where the α_n's are the direction cosines of the magnetic moment vectors at sites i, and j, respectively. L. Néel, and C. Kittel have

5 - FERROMAGNETISM OF AN IDEAL SYSTEM

shown [1] that this expression may also be written as a function of the gradients of the direction cosines in the form of an energy density:

$$F_{ex} = A_{ex}\left[(\nabla\alpha_1)^2 + (\nabla\alpha_2)^2 + (\nabla\alpha_3)^2\right] \quad (5.1)$$

For a crystal of cubic symmetry, where the nearest neighbour magnetic atoms are ξ apart, and carry the same magnetic moment, the coefficient A_{ex} is written as:

$$A_{ex} = \mu_0\left(wM_s^2/12\right)\xi^2 \quad (5.2)$$

This exchange coefficient may also be expressed as a function of the Weiss molecular field energy, $E_w = (1/2)\mu_0 w M_s^2$: $A_{ex} = (1/6) \xi^2 E_w$. w is the exchange coefficient in the molecular field model (w = T_C/C, see eq. 4.45). The coefficient A_{ex} (in joules per metre) characterises the strength of the exchange interactions within the material. Expression (5.1) –often referred to as the "exchange term"– expresses the cost in energy density of the shift $\nabla\alpha_i$ between the direction cosines α_i (i = 1, 2, 3) of neighbouring moments.

2.2. MAGNETIC DIPOLAR INTERACTION

Each magnetic moment of the substance is submitted to a magnetic dipolar interaction with other moments. This decreases with the reciprocal of the cube of the distance between the moments, but the number of neighbours to a given moment increases with the cube of this distance (see eq. 2.103). Hence the dipolar interaction, while it is much weaker than the exchange interaction among near neighbour moments, becomes dominant at large distances. To avoid the formation of poles it tends to close the magnetic flux lines inside the magnetised body, thereby opposing the establishment of a resultant uniform macroscopic magnetization. From equation (2.88) one derives the expression for the energy density of the demagnetizing field:

$$F_{dip} = -(1/2)\mu_0 M H_d \quad (5.3)$$

where $H_d = -NM$ is the demagnetising field of the sample under consideration. Like the interaction term with the external applied field, $F_H = -\mu_0 M H_0$ (according to eq. 2.80), F_{dip} shows that the interaction with the demagnetising field would lower the total energy of the system if the demagnetising field was parallel to the magnetization. Since H_d is antiparallel to M (or makes an angle greater than 90° with M), F_{dip} is a positive term, proportional to M^2, which raises the energy of the system. But unlike (2.80) this term contains a factor of (1/2) since it concerns a self-energy: the demagnetising field is not external to the material, rather it is the material itself that produces it.

2.3. COMPETITION BETWEEN EXCHANGE AND DIPOLAR INTERACTIONS

As ferromagnetic exchange interactions and dipolar coupling have conflicting objectives, neither can be wholly satisfied at the same time, and the energy of a ferromagnetic system cannot correspond to an absolute minimum of energy. The coexistence of these interactions is steered by the fact that each of them possesses a privileged zone of influence: the close neighbourhood for exchange, the regions further away for dipolar interactions. A typical distance, known as the exchange length (l_{ex}), is defined as the ratio of the square roots of the energies involved:

$$l_{ex} = \left(A_{ex}/\mu_0 M_s^2\right)^{1/2} \tag{5.4}$$

This characteristic distance separates the zone of influence where exchange prevails ($l < l_{ex}$), from that where the dipolar interaction dominates ($l > l_{ex}$). In the area of influence of exchange, no moment can deviate by much from the direction of its immediate neighbours. However, through a series of tiny movements, moments situated at $l > l_{ex}$ acquire total independence relative to the original moment. Very gradually the magnetic flux may thus close up onto itself inside the material, thus avoiding the creation of poles at the surface. In this way the energy of the demagnetising field is considerably reduced. Such a configuration of moments has been observed by Van den Berg [2] in the classroom case of a flat disc of elliptical shape obtained from an amorphous material which is perfectly isotropic at both the macroscopic and semi-microscopic scales. The moments then arrange themselves spontaneously in such a way that the magnetic flux closes without allowing the magnetic poles to appear (fig. 5.2). At the centre, a domain wall, the width of which is of the order of l_{ex}, separates two zones with opposite orientation. The exchange interactions are severely frustrated inside the wall, and in the neighbourhood of the singular points A, and B, but virtually nowhere else, since the moments are nearly aligned.

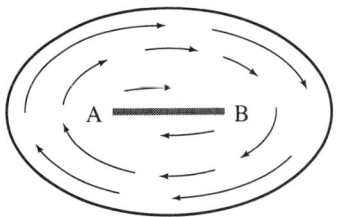

Figure 5.2
Magnetic flux lines observed inside a ferromagnetic sample having no anisotropy, and shaped in the form of an elliptical disc [2]

2.4. THE ROLE OF MAGNETOCRYSTALLINE ANISOTROPY

In most materials, the magnetic moments interact with the electric field of the surrounding ions via their orbital component. A first consequence of this interaction with the lattice is the appearance of magnetocrystalline anisotropy. This acts on each magnetic moment individually, tending to orient it along a lattice direction privileged by its particular symmetry. All directions of the same symmetry are, in theory, equivalent, and the alignment of the moments along each direction is, therefore, equally probable. Thus, the magnetocrystalline interaction leads by itself to the existence of a zero resulting magnetization. The magnetocrystalline anisotropy energy varies considerably from one material to another, and its expression has been given in equations (3.2) to (3.5) for the most common symmetries. These expressions measure the energy density cost for rotation of local magnetization away from one of the easy magnetization directions.

The magnetocrystalline anisotropy and exchange energies can often coexist without frustration. In ferromagnetic systems in particular, together they favour the existence of spontaneous magnetization. In cubic symmetry, magnetocrystalline anisotropy is equally compatible with the dipolar interaction (provided the magnetostrictive effects in closure domains are neglected: see fig. 12.24 further). On the other hand, in uniaxial ferromagnetic systems, the minimization of exchange energy and of magnetocrystalline energy never fully corresponds to the requirements for minimum, i.e. zero, dipolar energy.

2.5. COMPROMISE SOLUTIONS

The notion of compromise between the three types of principal interactions that exist in all ferromagnetic materials forms the basis of the diversity of magnetic behavior. It explains and justifies the variety of materials, and applications of magnetism. The state which results from this compromise is, in the case of large systems, that suggested by Pierre Weiss (1906 -1907): the ferromagnetic material decomposes into elementary domains inside which the magnetization is oriented along one or other of the equivalent easy magnetization directions. In the absence of an applied field and any other constraint there is equipartition of the domain volumes among the various directions: at the macroscopic scale, the resulting magnetization is then zero, in agreement with what experiment usually shows.

Between two domains one can observe a buffer zone, more or less wide, inside which the magnetization rotates: this is a *domain wall* which brings into play a compromise between the above-mentioned three energies. Its width is well defined, and it stores a fixed energy: the width of the wall and its energy are characteristic of the material being considered. In general terms, in an ideal ferromagnetic material, when one passes from a region in which the magnetization is saturated in one particular direction, to another where the magnetization is saturated in a different direction, or when, inside a volume saturated along an x axis a nucleus of saturated magnetization is created along y (y can be – x), a wall separates the saturated regions.

The most general approach in determining the states of magnetic equilibrium in a ferromagnetic object –the configuration of all the moments corresponding to a compromise state of minimum energy– rests on the minimization of the total free energy density, F_T. For a homogeneous ferromagnetic object, this energy density F_T

may be written as the sum of the four principal energy terms (exchange, anisotropy, dipolar and coupling with the applied field):

$$F_T = F_{ex} + F_{an} + F_{dip} + F_H \qquad (5.5)$$

3. THE DOMAIN WALLS

The domain walls are the inevitable interfaces between domains. As with every magnetic moment configuration, they must correspond to the best compromise between the energy terms involved. If two neighbouring domains find themselves in direct contact with each other (with no partition zone between them: fig. 5.3-a), the magnetic moments of the atoms situated at the border make a significant angle with each other. A strong increase in exchange energy results from this, and is higher, the higher the exchange constant A_{ex} (eq. 5.1).

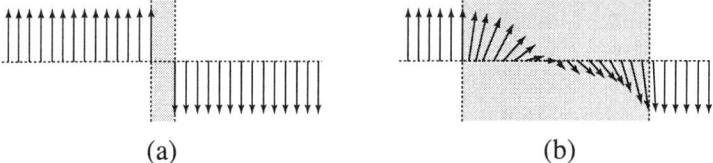

(a) (b)

Figure 5.3 - Narrow, and wide domain walls

A narrow wall (a) without a buffer zone represents a high cost in exchange energy, while a wide wall (b) represents a high cost in anisotropy energy.

Conversely, one can consider reducing the cost in exchange energy almost to zero by spreading the angular transition of the magnetization between the neighbouring domains over a very large number of atoms (fig. 5.3-b). Then the angle between two adjacent moments is very small, but the price paid in anisotropy energy becomes prohibitive. In fact, almost all the moments that participate in the transition deviate from easy magnetization directions, and for every one there is a cost in magnetocrystalline energy, which is greater, the larger the anisotropy constant K is.

The width of a real domain wall is defined by the least demanding compromise. Clearly this depends on the relative values of the constants A_{ex} and K for the material considered. For a given material, the higher the A_{ex}/K ratio, the wider the domain walls, and the smaller A_{ex}/K the narrower the walls. We shall now go into more detail, and quantitatively complement what has been said above.

3.1. DOMAIN WALLS IN ORDINARY SYSTEMS (BLOCH WALLS)

3.1.1. The geometry of Bloch walls

The existence of domain walls allows us to reduce, as far as possible, the energy brought into play in the boundary area between the walls. Inevitably the compromise includes both the exchange energy and the anisotropy energy, but one must also ensure that putting the domain walls into place does not cause a local increase in dipolar energy. This requires that no magnetic mass should appear either on the surface of the domain walls or inside them.

The magnetic mass density at the interface between two regions, 1 and 2, where the magnetizations are M(1) and M(2) respectively, is deduced from the relation (2.42) and, if the unit vector normal to the interface is n, it is written as:

$$\sigma_s = \mathbf{M}(1) \cdot \mathbf{n} - \mathbf{M}(2) \cdot \mathbf{n} \tag{5.6}$$

The desired condition implies that $\sigma_s = 0$, and hence:

$$\mathbf{M}(1) \cdot \mathbf{n} = \mathbf{M}(2) \cdot \mathbf{n} \tag{5.7}$$

The wall being formed from successive planes where the magnetization is uniform, that is to say, where all the moments are parallel, the relation (5.7) is applicable from plane to plane. Thus the normal component of the magnetization should remain unchanged within the entire thickness of the wall. This means that, along the z-axis, perpendicular to the plane of the domain wall, one must have: $dM_z/dz = 0$. The projection of **M** on the xy plane, perpendicular to **n**, describes a circle (fig. 5.4). Such a rotation of **M** about the normal **n** to the plane of the wall, the angle between **M** and **n** remaining constant, forms a *Bloch wall*, so called because it was modelled by Bloch in 1932 [3].

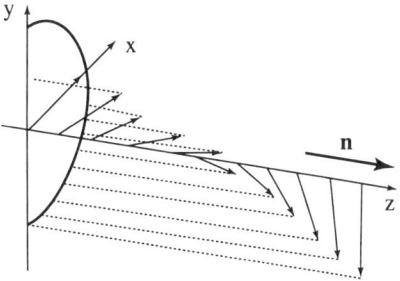

Figure 5.4 - The projection of the magnetization onto the plane of the domain wall describes a circle

When a Bloch wall separates two domains magnetised antiparallel to one another it is called a *180° wall*. In the case of uniaxial systems with an easy magnetization axis, only this type of domain wall exists.

Any direction contained within the plane perpendicular to the easy magnetization axis can serve as normal to the domain wall, which makes the existence of curved or even cylindrical walls possible (see fig. 5.5). 180° walls can also exist in cubic crystals, but in that case it is no longer unimportant whether **n** is parallel to x or to the bisector of xOy. Depending on the case, the magnetization may, or may not, pass through a difficult axis during its rotation. The energy stored in the domain wall is least when the moments are never oriented along the most difficult axis of magnetization.

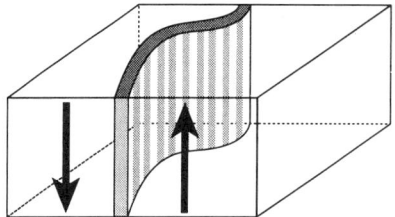

Figure 5.5 - A 180° wall is not necessarily straight

In a crystal of cubic symmetry other types of domain walls may exist: 90° walls which separate domains magnetised at $\pi/2$ from one another (along two fourfold axes in iron for example), 70.53° walls separating domains [111] and [$\bar{1}$11] or at 109.47° between domains [111] and [$\bar{1}$1$\bar{1}$] (in nickel for example). In general, and by misusing the term, all domain walls that separate domains deviating from each other by an angle other than 180° are called 90° walls.

3.1.2. Evaluation of the domain wall energy, and width

Two quantities characterise a Bloch wall:
- its thickness δ, which is also called the domain wall width,
- the local increase in energy, γ, brought about by the configuration of moments it contains. This increase in energy is evaluated relative to the saturation state (the condition that prevails inside the domains in the neighbourhood of the wall). It is normally given per unit area of the domain wall. This is why γ is called the surface energy of the domain wall.

Additional exchange energy due to deviation between the moments within the domain wall

- Consider a system of two adjacent magnetic moments with the same modulus $\mathfrak{m}$. If they do not interact with each other then these moments are independent (this is the zero energy state). In a ferromagnetic system they are coupled through exchange interaction. The lowering of energy corresponds to the ordered state of parallel alignment, and is expressed by the relation (4.34): $\mathscr{E}_{ex}^{sat} = -\mu_0 n \mathfrak{m}^2$, where n designates the local exchange coefficient.

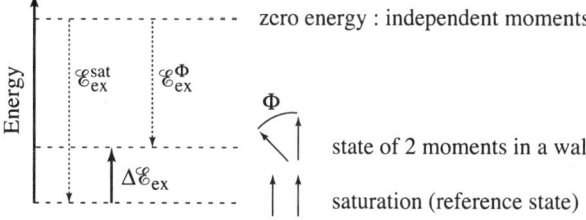

Figure 5.6 - Influence of the local magnetic state on the exchange energy

Let us assume that these two moments find themselves within a domain wall, where they make an acute angle Φ with each other. The exchange interactions taking place between these two moments lower the energy only by

$\mathcal{E}_{ex}^{\Phi} = -\mu_0 n m^2 \cos \Phi$. Consequently, two neighbouring moments in a domain wall deviating by an angle Φ raise the energy of the system by $\Delta\mathcal{E}_{ex} = \mathcal{E}_{ex}^{\Phi} - \mathcal{E}_{ex}^{sat} = \mu_0 n m^2 (1 - \cos \Phi)$ relative to the state of parallel coupling. The diagram presented in figure 5.6 clarifies these diverse energy states.

• We shall now consider a *chain of atomic moments* of lattice spacing a, which goes through a 180° domain wall across its width (fig. 5.7), and we set $\delta = Na$ (N atoms in this chain). Within the domain wall neighbouring moments deviate angularly from one another. Let us assume they make an angle $\Phi = \pi/N$, constant across the wall, with each other.

The energy cost of crossing the domain wall is:

$$\mathcal{E}_{ex}^{chain} = N \mu_0 n m^2 (1 - \cos \Phi) \tag{5.8}$$

Assuming that the wall is sufficiently wide, then the angle Φ is small, and one has: $\mathcal{E}_{ex}^{chain} \approx N \mu_0 n m^2 \Phi^2/2 = \mu_0 \pi^2 n m^2/2N$.

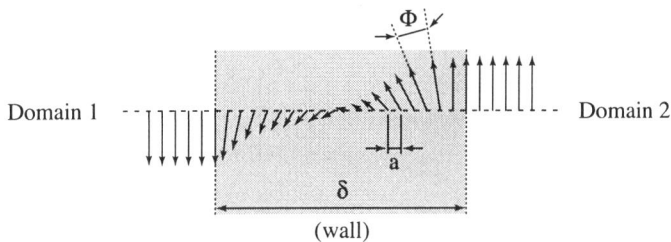

Figure 5.7 - A 180° domain wall

• If *the magnetic atoms form a simple cubic lattice*, there will be $1/a^2$ atomic chains of the above type per unit surface area of the wall. The cost of the wall in exchange energy, per unit area of the wall (surface energy), will then be:

$$\mathcal{E}_{ex}^{wall} = \mu_0 \pi^2 n m^2 / (2N a^2) \tag{5.9}$$

The larger N is, the wider the wall, and the less the cost in exchange energy. The intuitively expected result is thus confirmed.

The cost of the domain wall in anisotropy energy

The anisotropy energy is minimum when the atomic moments are oriented along one of the easy magnetization directions, which is the case inside domains 1 and 2 on both sides of the wall (fig. 5.7). As soon as the moments deviate from these directions there is an energy loss, described by the relations (3.3) to (3.5). For the pth moment, which deviates from the easy direction by an angle θ_p, the anisotropy energy is equal to $K_{at} \sin^2 \theta_p$, where:

$$K_{at} = Ka^3 \tag{5.10}$$

for a simple cubic lattice (in which the anisotropy is nevertheless assumed to be uniaxial), if the expansion is restricted to order 2 ($K = K_1$). For a section of the domain wall corresponding to unit area, the anisotropy energy lost is $Ka \sin^2 \theta_p$. For the entire wall the cost in anisotropy energy per unit surface area is, therefore, approximately NKa $\overline{\sin^2 \theta_p}$, where $\overline{\sin^2 \theta_p}$ represents the average value of $\sin^2 \theta_p$ taken between 0, and π, that is 1/2. One has therefore:

$$\mathscr{E}_{anis}^{wall} = N a K / 2 \tag{5.11}$$

The cost of the domain wall in anisotropy energy is less the narrower the wall (N smaller). Here too one finds the intuitively expected result.

The total energy cost of the domain wall

The extra energy per unit area of the surface, γ, brought into play in a domain wall of width Na, is equal to:

$$\gamma = E_{ex}^{wall} + E_{anis}^{wall} = \frac{\mu_0 \pi^2 n m^2}{2 N a^2} + \frac{NaK}{2} = \frac{\mu_0 \pi^2 n m^2}{2 \delta a} + \frac{\delta K}{2} \tag{5.12}$$

γ is minimized for the value of δ which makes the two contributions (in δ, and $1/\delta$) equal. The width of the domain wall in this simple model is therefore given by the following relation:

$$\delta = \sqrt{\frac{\mu_0 \pi^2 n m^2}{Ka}} = \sqrt{\frac{\mu_0}{a}} \pi m \sqrt{\frac{n}{K}} \tag{5.13}$$

By carrying this expression for δ into that of γ one obtains the total surface energy stored in the domain wall (the total cost of the wall per unit area):

$$\gamma = \sqrt{\frac{\mu_0 \pi^2 n m^2 K}{a}} = \sqrt{\frac{\mu_0}{a}} \pi m \sqrt{nK} \tag{5.14}$$

3.1.3. Exact calculation of γ and δ

In the preceding model, we simplified the calculations by assuming a uniform rotation of the moments within the domain wall. The exact expressions for δ and γ are calculated as shown below.

Figure 5.8 shows the structure of a domain wall. The Oz axis perpendicular to the plane of this wall is the axis along which the wall is crossed widthwise. The origin O is chosen at the middle of the wall. The orthogonal axes Ox and Oy are therefore found in the mid plane of the domain wall, and form with Oz a right-handed cartesian set of axes. To satisfy condition (5.7), the moments must rotate around Oz while remaining in a plane parallel to the xOy plane. The direction of Ox can, therefore, be chosen as that of the moments contained in the median plane xOy, and thus becomes the bisector of the angle enclosed by the moments in the two domains neighbouring the domain wall. Constructed this way, Ox will be taken as the reference direction for the orientation of the moments in the wall. The moments situated on the side $z = z_1$ make, with the Ox axis, an angle θ, and the angle between two successive moments along the z axis will be equal to $\Delta\theta$. We shall consider the variation of θ with z as continuous but not necessarily constant, and let $\Delta\theta = (\partial\theta / \partial z) \Delta z$.

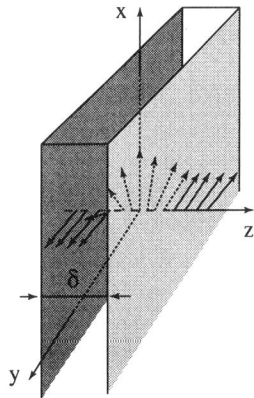

Figure 5.8 - Simplified 180° domain wall

The moments rotate between the two planes shown on the figure, and remain parallel outside these two planes.

The magnetization within the domain wall is not uniform in the direction of Oz. From equation (5.1), the cost in exchange energy per unit volume is:

$$F_{ex} = A_{ex}(d\theta/dz)^2 \qquad (5.15)$$

where A_{ex} is the exchange energy coefficient defined by equation (5.2). For an element of the wall with unit surface area, the exchange energy brought into play is:

$$E_{ex}^{wall} = \int_{-\infty}^{+\infty} F_{ex}\, dz \qquad (5.16)$$

i.e:
$$E_{ex}^{wall} = A_{ex} \int_{-\infty}^{+\infty} \left(\frac{d\theta}{dz}\right)^2 dz \qquad (5.17)$$

The integration between $-\infty$, and $+\infty$ is possible because the function to be integrated is zero within both domains (parallel moments). The above relation is true whatever the geometry considered, provided one uses the value of A_{ex} appropriate to that geometry.

The anisotropy energy is always a function of θ. Let us indicate this by $g(\theta)$, and assume it to be zero along a direction of easy magnetization. The extra anisotropy energy put into play inside the wall with respect to that within the adjoining domain is:

$$E_{anis}^{wall} = \int_{-\infty}^{+\infty} g(\theta)\, dz \qquad (5.18)$$

The total energy per unit surface area of the domain wall created is therefore:

$$\gamma = E_{ex}^{wall} + E_{anis}^{wall} = \int_{-\infty}^{+\infty}\left[A_{ex}\left(\frac{d\theta}{dz}\right)^2 + g(\theta)\right] dz \qquad (5.19)$$

and θ, as a function of z, must take the values which minimise this integral. For this it is necessary that if, at each point z, $\theta(z)$ varies by a small amount $\delta\theta(z)$, the modification resulting from the configuration of the spins does not provoke, to first order, any variation $\delta\gamma$ of the domain wall energy, thus:

$$\delta\gamma = \int_{-\infty}^{+\infty}\left[2A_{ex}\left(\frac{d\theta}{dz}\right)\left(\frac{d\delta\theta}{dz}\right) + \frac{dg(\theta)}{d\theta}\right]dz = 0 \qquad (5.20)$$

hence: $\int_{-\infty}^{+\infty} 2A_{ex}\left(\frac{\partial \theta}{\partial z}\right)\left(\frac{\partial \delta\theta}{\partial z}\right) dz = \left[2A_{ex}\frac{\partial \theta}{\partial z}\delta\theta\right]_{-\infty}^{+\infty} - \int_{-\infty}^{+\infty} 2A_{ex}\frac{\partial^2 \theta}{\partial z^2}\delta\theta\, dz$ (5.21)

since $\partial \theta / \partial z = 0$ when $z \to \pm \infty$, the first term on the right hand side of the above equation is zero, from which:

$$\int_{-\infty}^{+\infty}\{[\partial g(\theta)/\partial \theta] - 2A_{ex}(\partial^2 \theta/\partial z^2)\}\delta\theta\, dz = 0 \quad (5.22)$$

Relation (5.22) must remain true whatever the variation $\delta\theta(z)$ chosen. This implies that: $\partial g(\theta)/\partial \theta - 2A_{ex}\partial^2 \theta/\partial z^2 = 0$. This is Euler's equation of variational calculus. On multiplying by $\partial \theta/\partial z$, and integrating over z, one obtains: $g(\theta) = A_{ex}(\partial \theta/\partial z)^2$ to within an additive constant. But, when $|z| \to \infty$, indicating that one is within the domain, $g(\theta) = 0$, and $(\partial \theta/\partial z) = 0$. The constant is therefore zero, and the equation obtained becomes:

$$g(\theta) = A_{ex}(\partial \theta / \partial z)^2 \quad (5.23)$$

At all points of the domain wall the costs in anisotropy energy and in exchange energy balance each other. In other words, inside the domain wall, where the anisotropy energy is most costly, the angle between adjacent spins is greater, and vice versa. The domain wall energy is $\gamma = \int_{-\infty}^{+\infty} g(\theta) dz$ that is, twice the anisotropy energy. One may extract dz from equation (5.24):

$$dz = \sqrt{A_{ex}/g(\theta)}\, d\theta \quad (5.24)$$

and from there obtain the most general energy expression for a Bloch domain wall, whose spins rotate between the initial angle θ_i, and the final angle θ_f:

$$\gamma = 2\sqrt{A_{ex}}\int_{\theta_i}^{\theta_f}\sqrt{g(\theta)}\, d\theta \quad (5.25)$$

For a 180° wall ($\theta_i = -\pi/2$ and $\theta_f = \pi/2$), and assuming that the local anisotropy energy is of magnetocrystalline origin and can be reduced to its first term, viz $g(\theta) = K_1 \cos^2 \theta$, the energy cost of the domain wall has the expression: $\gamma = 2\sqrt{A_{ex}K_1}\int_{-\pi/2}^{\pi/2}\cos\theta\, d\theta$, and hence:

$$\gamma = 4\sqrt{A_{ex}K_1} \quad (5.26)$$

In the same context of a uniaxial anisotropy, one may calculate z from the relation (5.24):

$$z = \sqrt{\frac{A_{ex}}{K_1}}\int_0^\theta \frac{d\theta}{\cos\theta} = \sqrt{\frac{A_{ex}}{K_1}}\log\left[\tan\left(\frac{\theta}{2}+\frac{\pi}{4}\right)\right]_s^2 \quad (5.27)$$

From this one may deduce the variation of θ with z (fig. 5.9); this is maximum at the centre of the domain wall, at the point where the magnetocrystalline anisotropy energy is maximum. It should also be noted that the domain wall is not clear-cut, which results from the fact that the integral (5.27) cannot be calculated to the boundary $\theta = \pi/2$ or $-\pi/2$. In a first approximation, the thickness δ of the 180° wall

5 - FERROMAGNETISM OF AN IDEAL SYSTEM

is usually defined as what it would be if the angular deviation between the moments which constitute the wall were constant, and equal to its actual value at the centre of the wall, thus:

$$\delta = \pi (\partial z / \partial \theta)_{z=0} = \pi (A_{ex}/K_1)^{1/2} \quad (5.28)$$

where $\theta_f - \theta_i = \pi$ has been assumed, and taking account of the fact that the slope of the tangent at the inflexion point, i.e. at the centre, has the value $(A_{ex}/K_1)^{1/2}$.

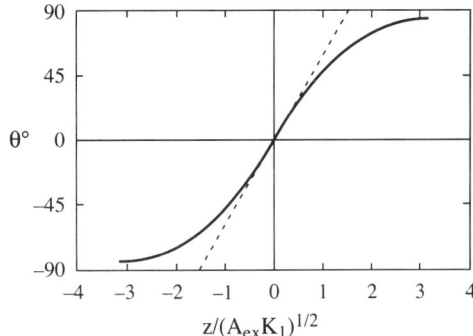

Figure 5.9 - Deviation θ of the magnetic moments inside a 180° wall, as a function of their position in the wall

The classic Bloch domain wall just described is not the only type of domain wall that can exist, just the one that is most common. In the following section we describe the properties specific to Bloch walls when they become very narrow. We then go on to discuss (see § 3.3) the specific types of domain wall that can be found in thin films.

3.2. WALLS IN HIGHLY ANISOTROPIC MATERIALS (NARROW WALLS)

In materials where the anisotropy energy approaches or exceeds the exchange energy, the width of the domain wall $\delta = \pi (A_{ex}/K)^{1/2}$ becomes very small, and the continuous approximation is no longer valid. It is then necessary to take the discrete character of matter into account.

If the deviation of the pth moment in the wall with respect to the anisotropy direction is θ_p, two neighbouring moments, labelled by the indices $p-1$ and p, make an angle $\theta_p - \theta_{p-1}$ between them. The exchange energy stored among first neighbours is given (as we saw above) by: $\mathscr{E}_{ex} = \mu_0 n m^2 [1 - \cos(\theta_p - \theta_{p-1})]$.

Expanded to second order, the uniaxial magnetocrystalline anisotropy energy (expressed in joules) has the value: $\mathscr{E}_{anis} = K_{at} \sin^2 \theta_p$ for the pth atom of the domain wall.

The total energy associated with the pth moment of the wall (which interacts with moments $p-1$ and $p+1$), is thus:

$$\gamma_p = K_{at} \sin^2 \theta_p + aA_{ex} \left[1 - \frac{\cos(\theta_{p+1} - \theta_p) + \cos(\theta_p - \theta_{p-1})}{2} \right] \quad (5.29)$$

For a simple cubic lattice of magnetic ions with parameter a (where the anisotropy is nevertheless assumed to be uniaxial), the total energy cost per unit area of the wall is given by the following relation:

$$\gamma = \frac{1}{a^2} \sum_{p=-\infty}^{+\infty} \gamma_p \qquad (5.30)$$

The extrema in energy correspond to $d\gamma/d\theta_p = 0$. They are obtained by solving an infinite set of non-linear equations of the form:

$$(K_{at}/aA_{ex}) \sin 2\theta_p = \sin(\theta_{p+1} - \theta_p) + \sin(\theta_{p-1} - \theta_p) \qquad (5.31)$$

where p varies from $-\infty$ to $+\infty$ in steps of one.

In general terms one can calculate numerically an equilibrium configuration for the domain wall, a solution of system (5.31), on condition that the directions of two moments are fixed arbitrarily. The minimum energy configuration (fig. 5.10-a), obtained by successive adjustments, is symmetrical, and the angular discontinuity at the centre of the domain wall is a function of the ratio $\eta = (K_{at}/aA_{ex}) = K a^2 / A_{ex}$ for a simple cubic lattice, because of equation (5.10). The maximum energy configuration, also symmetrical, has a moment perpendicular to the easy magnetization axis (fig. 5.10-b).

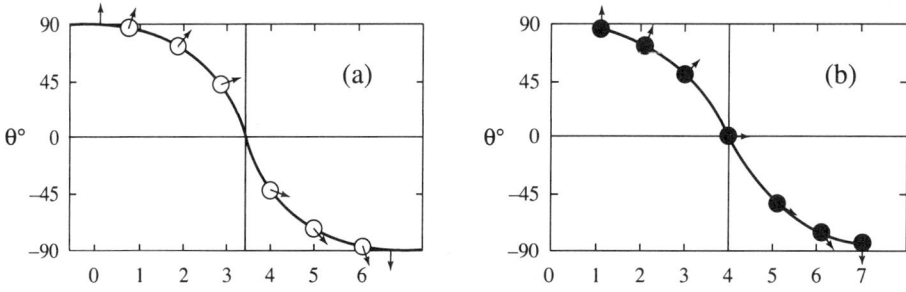

Figure 5.10 - Minimum (a), and maximum (b) energy configurations
The atomic positions inside the domain wall are marked on the abscissa scale.

In the case of narrow domain walls considered here ($\eta \gtrsim 1$), the moments never deviate very much from the easy magnetization direction. On one side of the wall θ_p is close to zero, and on the other, to π. Linearization of equation (5.31) leads to: $\theta_{p+1} + \theta_{p-1} = \rho \theta_p$ with $\rho = 2(\eta + 1)$. The solution to this system of equations is: $\theta_p = \theta_0 e^{-p\psi}$ with: $\psi = \cosh^{-1}(\rho/2)$. When p increases, the angles θ_p decrease as a geometric progression with the common ratio $x = e^{-\psi}$. The domain wall is fully defined by the common ratio of the progression, and by the angle θ_0 at the centre of the wall, which satisfies the equation: $(1 - \eta) \sin 2\theta_0 = \sin[\theta_0(1 - x)]$.

Expanding this relation to third order allows a discussion of the evolution of the wall [4]: $(\theta_0)^2 = [12\eta - 6(1+x)] / [8(\eta - 1) + (1+x)^3]$.

When η increases, θ_0 decreases rapidly. $\theta_0 = 0$ for $2\eta = 4/3$. For $\eta > 2/3$, the width of the wall is no more than one interatomic distance (because $\theta_0 = 0$ is then the only solution to the equation). Figure 5.11 shows the evolution of the reduced thickness $\delta/a = N$ (number of interatomic intervals in the wall) as a function of the ratio η of the anisotropy and exchange constants of the material.

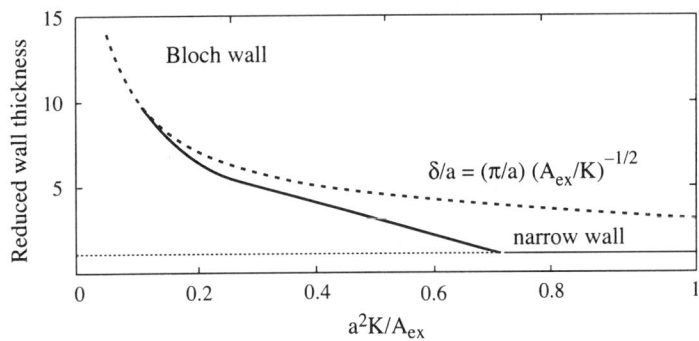

Figure 5.11 - Wall thickness as a function of η

The reduced thickness of the domain wall is expressed in units of interatomic distances.

3.3. DOMAIN WALLS IN VERY THIN FILMS

In thin film systems, the demagnetising fields are very important in the direction corresponding to the small dimension, this is why the magnetization in the domains generally orients itself in the plane of the film. Let us consider two adjacent domains separated by a 180° wall. A Bloch wall, where moments rotate about the normal to the surface of the wall (see fig. 5.12-a), leads to the appearance of magnetic poles which are north on one face of the thin film and south on the other. The associated magnetostatic energy becomes greater as the ratio of the thickness of the film (e) over the width of the domain wall (δ) is reduced. Taking this magnetostatic energy into account leads to a new expression for the the domain wall energy:

$$\gamma = \frac{1}{2}\gamma_0 \left(\frac{\delta}{\delta_0} + \frac{\delta_0}{\delta} + \frac{\mu_0 M_s^2}{2\gamma_0} \frac{\delta^2}{e} \right)$$

where γ_0 and δ_0 are, respectively, the energy, and thickness of the ordinary Bloch wall. This approximation is valid as long as δ remains small compared to e. This domain wall energy is thus the sum of three contributions, respectively magnetocrystalline anisotropy which varies as δ, exchange energy which varies as δ^{-1}, and dipolar energy which varies as δ^2.

γ is not only a function of the properties of the material, but also of the thickness of the film. For small thicknesses other domain wall structures are favoured. Néel considered in particular [5] a domain wall structure in which the moments rotate about the axis perpendicular to the plane of the film (fig. 5.12-b). The contribution of the magnetic poles is eliminated, but the configuration of the moments inside the wall leads to stray field.

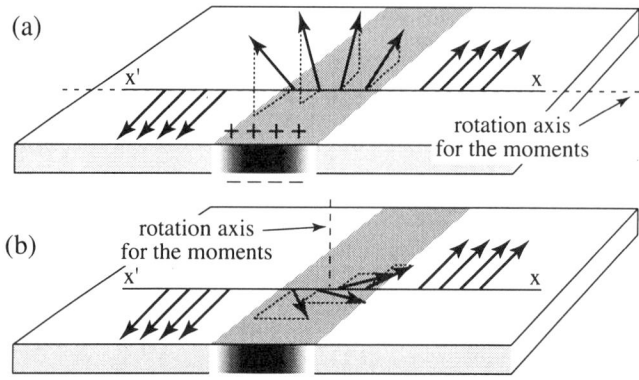

Figure 5.12 - (a) A Bloch wall - (b) A Néel wall

(a) Along a line x'x perpendicular to the domain wall, the magnetic moments turn progressively, leaving the plane of the film, and thus allowing magnetic masses to appear on the surface of the film. In the middle of the wall the moments are perpendicular to the plane of the film.

(b) Here the moments also turn progressively but they remain within the plane of the film, and thus magnetic masses no longer appear at the surface of the film. Here the rotation axis is perpendicular to x'x.

The energy of such a *Néel wall* is: $\gamma_m = \gamma_0 + \mu_0 M_s^2 e$. This expression indicates that the energy associated with the stray field decreases with the thickness of the film. Below a critical thickness the energy of a Néel wall is less than that of a Bloch wall. However, there is no abrupt passage from one type of wall to the other. At intermediate thicknesses complex structures exist; these are mixtures of Bloch and Néel walls. In a film of $Fe_{80}Ni_{20}$ of 100 nm thickness, the separation between domains appears as a dotted line (fig. 5.13). The Bloch wall subdivides into segments with alternating rotation directions, this leads to magnetic masses on each surface of the film which are positive and negative in succession.

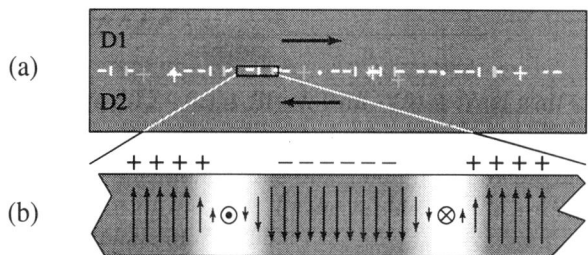

Figure 5.13 - Domain walls observed by the Bitter method on a thin film of Fe-Ni of 100 nm thickness, according to [6]

(a) Thin film seen from above; the domains are marked as D1 and D2.

(b) A sectional view of the film, much enlarged, at the centre of the domain wall, showing the local structure; zones with a strong field gradient (Néel walls) concentrate the magnetic powder, giving the appearance of singular points.

5 - FERROMAGNETISM OF AN IDEAL SYSTEM

Between the Bloch wall segments, short zones appear in which the moments, instead of rotating around the normal to the domain wall, rotate around the normal to the film: these are bits of Néel wall regions. The mixture of diverse structures in the domain wall leads to an alternation of magnetic poles of opposite signs over short distances. The dipolar energy is strongly reduced. If the thickness of the film further diminishes, each Néel wall segment grows at the expense of the Bloch wall segments. When the thickness is around 60 nm, the $Fe_{80}Ni_{20}$ film contains Néel walls only.

A quite regular alternation of two types of Néel walls, characterized by the direction of rotation of the magnetization, is observed. At the centre of the transitions zones from one type to the other, the lines of magnetization are normal to the film as figure 5.14 shows: these are residues of the Bloch wall observed at greater thickness, and they are known as *Bloch lines*.

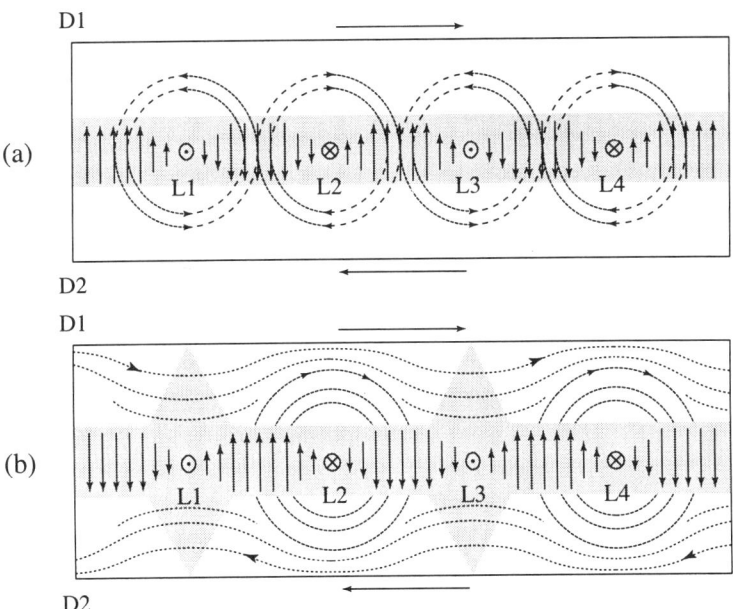

Figure 5.14 - Mechanism by which cross-ties are created

Film seen from above.

(a) At the level of the Bloch lines L1, and L3, the flux lines (dotted) would be oriented opposite to the magnetization inside each domain.

(b) The Néel wall segments, represented by triangles at the level of lines L1, and L3, rectify the flux lines appreciably.

As soon as domain walls of this type appear, they take on the appearance of cross-ties: then, at fairly regular intervals, short segments of Néel wall form perpendicular to the principal domain wall, which crosses them through their centre (fig. 5.15).

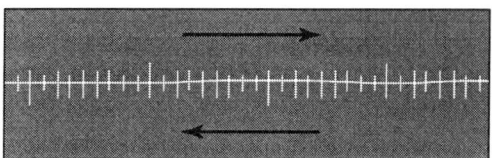

*Figure 5.15 - Néel wall, and Bloch lines
observed on a very thin film of Fe-Ni of 60 nm thickness*

Qualitatively, this can be explained as follows: around each Bloch line, dipolar fields are produced inside the domains by the adjacent Néel wall segments (fig. 5.14-a). Around every other Bloch line, these dipolar fields run opposite to the magnetization in the domains, raising the energy of the system strongly. The Néel wall segments perpendicular to the principal wall appear at the level of these unfavourable Bloch lines; they suppress the unfavourable dipolar fields and ensure the continuity of those dipolar fields aligned parallel to the magnetization of the domains. (fig. 5.14-b).

4. DOMAIN CONFIGURATIONS

We have seen that the distribution of domains in ferromagnets results from a compromise in which the energies involved are not individually minimised. Strong local frustrations linked, for example, to the geometry or the crystallographic structure of the magnetic systems under consideration can exist. These lead to domain subdivision modes suited to each specific situation.

Extremely diverse situations are possible. In the following sections we briefly describe several of the most common and typical of these arrangements. For interested readers, detailed explanations are given in specialised chapters.

4.1. DOMAIN CONFIGURATIONS IN UNIAXIAL CRYSTALS

Let us consider the simplest case of a single crystal in the form of a parallelepiped, cut in a uniaxial ferromagnetic material (with spontaneous magnetization M_s), such that the axis of easy magnetization **c** is perpendicular to one of the faces. Let us assume the moments to be directed along this axis. If the crystal is single-domain it produces a large dipolar field outside the sample, and the associated demagnetising field $\mathbf{H_d}$, prevailing in the material, gives rise to a magnetostatic energy loss:

$$W_{dip} = -(\mu_0/2)\int_{vol} \mathbf{M}_s \mathbf{H_d} \, dv \qquad (5.32)$$

$\mathbf{H_d}$ being oriented antiparallel to $\mathbf{M}_s$, we verify that the magnetostatic energy thus written is positive. This energy can be reduced by a factor of the order of two, of four, of eight, etc. if the crystal subdivides into two, four, eight, etc. domains (fig. 5.16). The demagnetising field at a point inside the crystal can be considered to be the sum of the fields created at this point by the magnetic masses present on the surface.

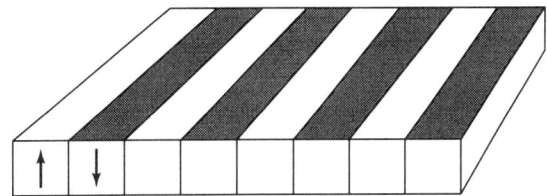

*Figure 5.16 - Decomposition of a parallelepiped
of uniaxial symmetry into a simple slab domain structure*

As the number of surface domains increases, these masses become more and more closely mixed, and, as a consequence, tend to neutralise their effects at ever shorter distances. One might envisage the total disappearance of the magnetostatic energy loss through infinite subdivision of the crystal into domains. Actually the appearance of a growing number of domains is, inevitably, accompanied by the creation of an equivalent number of domain walls.

While the gain in dipolar magnetic energy gets weaker and weaker with successive subdivisions, the loss through domain wall energy is strictly proportional to the surface of the wall, that is to say, to the number of domain walls created.

To illustrate this process of compromise, let us consider the above single crystal, cut to the shape of a small platelet of dimensions $L \times L \times e$ ($e \ll L$), the **c** axis being parallel to e. If the crystal is single-domain, a very strong demagnetising field occurs within the platelet. Assuming the platelet to be a very flat ellipsoid, the magnetostatic energy is:

$$\mathscr{E}_{mag} = -\frac{\mu_0}{2} \int_{vol} \mathbf{M_s H_d} \, dv = -\frac{\mu_0}{2} M_s H_d \int_{vol} dv \qquad (5.33)$$

or $E_{mag} = -(\mu_0/2) M_s H_d = (\mu_0/2) N M_s^2$ per unit volume, since $H_d = -NM_s$, where N is the demagnetising field coefficient. Due to the fact that $e \ll L$, the demagnetization factor is close to unity, and hence: $E_{mag} \approx \mu_0 M_s^2 / 2$.

Per unit area of the platelet, the magnetostatic energy is written as: $E_{mag}^{surf} = E_{tot}^{surf} \approx (\mu_0/2) M_s^2 e$.

Let us now consider the case of a configuration in the form of parallel slabs, such as that shown in figure 5.16. The calculation of the magnetostatic energy makes use of the scalar potential Φ from which H_d is derived and which must satisfy Laplace's equation, $\Delta\Phi = 0$ subject to the condition: $(\partial\Phi/\partial z)_{z>0} = \sigma_{mag}/\mu_0 = M_s$.

The decomposition of the platelet into parallelepiped-like domains corresponds to a magnetostatic energy which can be estimated to: $E_{mag}^{surf} = 0.136 \, \mu_0 M_s^2 d$ per unit surface area of the platelet, where d is the width of the domains, and should remain small compared to the thickness e of the platelet. This expression confirms that E_{mag} becomes smaller the larger the number of domains (proportional to $1/d$).

To this energy, one must add for the whole platelet the energy of the L/d domain walls that it contains, and whose total surface area is L e (L/d), hence the wall energy:

$E_{wall} = \gamma e L^2 / d$, if γ designates the surface energy of the wall. Per unit area of the platelet this gives: $E_{wall}^{surf} = \gamma e / d$.

The total energy is the sum of the magnetostatic and domain wall energies, i.e.:

$$E_{tot}^{surf} = E_{mag}^{surf} + E_{wall}^{surf} = 0.136\, \mu_0\, M_s^2\, d + \gamma(e/d) \tag{5.34}$$

This energy is minimum for $dE_{tot}^{surf}/dd = 0.136\, \mu_0\, M_s^2 - \gamma(e/d^2) = 0$, i.e. for a domain width:

$$d = \sqrt{\gamma e} / \sqrt{0.136 \mu_0 M_s^2} \tag{5.35}$$

The total energy of the system is then equal to:

$$E_{tot}^{surf} = 2\sqrt{0.136 \mu_0 M_s^2 \gamma e} \tag{5.36}$$

Let us apply these results to the case of cobalt. At room temperature, cobalt has a saturation magnetization $M_s = 1.45 \times 10^6$ A.m^{-1} and a domain wall energy $\gamma = 7.6 \times 10^{-3}$ J.m^{-2}.

For a sample of cobalt cut into a platelet of thickness $e = 1$ cm, relations (5.35) and (5.36) give the values:

- Total magnetic energy in the single domain state: $E_{tot}^{surf} = 13.2 \times 10^3$ J.m^{-2}.
- Total magnetic energy in the polydomain state: $E_{tot}^{surf} = 10.5$ J.m^{-2}.
- Width of the domains: $d = 14.5$ μm (compare this with the 12.5 nm wall thickness: the walls are about 1 000 times thinner than the domains which they separate).

The gain in magnetic energy obtained by the division into slab-shaped domains reaches a factor greater than 1 200. The calculation gives 687 domains per centimetre of the platelet. Of course this is just an order of magnitude.

We have just seen that the subdivision of a flat parallelepiped into domains considerably reduces the magnetostatic energy of a system saturated along its small dimension. Are other types of arrangements capable of reducing this energy further? In fact a variety of configurations: checkerboard, polkadot, etc. (fig. 5.17) have been studied and found to lead to a magnetostatic energy slightly smaller than for the simple slabs of figure 5.16. However, it appears that they require a much more significant wall area per domain, and their overall cost balance is not favourable. On the other hand, the formation of highly imbricated domain configurations, whose energy is only very slightly higher than that of the parallel slab configuration, is often observed. The walls, while remaining parallel to the perpendicular easy magnetization axis Oz, curve in the xOy plane without the increase in their surface becoming prohibitive (due to the fact that they are relatively thin). Seen from above, the domains display a characteristic maze appearance (fig. 5.18): the moments are perpendicular to the plane of the figure, and the contrast between black and white domains indicates a reversal of the moments (↑, and ↓).

5 - FERROMAGNETISM OF AN IDEAL SYSTEM 163

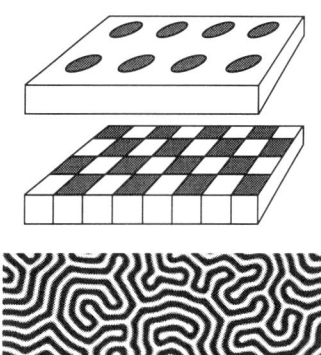

Figure 5.17 - Various domain configurations

Figure 5.18 - Structure of domains visualised by the Faraday effect in a film of magnetic garnet (width of the stripes ≈ 5 μm)

In the thicker systems, the mixture of surface magnetic masses is rather obtained without involving too much extra wall area, by the formation, inside the slab domains, of inverse spike-shaped domains (fig. 5.19), which fan out at the surface (where they compensate the surrounding masses by masses of opposite sign) and do not reach very deep into the surrounding material (in order not to cost too much in wall area).

Figure 5.19 - Cone shaped domains-observed at the surface of cobalt, the moments are parallel to the axis of hexagonal symmetry

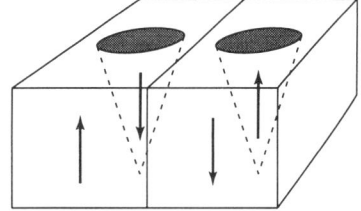

Nevertheless, this conical shape causes magnetic masses to appear on the walls since the component of the magnetization normal to the wall is discontinuous, and the quantity of these masses is bigger, the shorter the cone. As always, therefore, the actual domain structure results from a compromise.

4.2. CLOSURE DOMAINS

In crystals of cubic symmetry, the number of easy directions of magnetization is greater than for uniaxial crystals: 6 <100> type directions, 8 <111> directions. Besides the 180° walls, 90° walls can form in materials whose easy magnetization axes are of the <100> type, e.g. iron, as can 71° or 109° domain walls in materials where the easy directions of magnetization are of the <111> type, nickel for instance.

The existence of domains oriented at 90° to one another makes it possible for the magnetic flux to close within the sample without causing poles to appear either on the surface, or inside the magnetic material. The magnetostatic energy of the system is thus reduced to zero. Figure 5.20 shows the arrangement of such *closure domains*. Contrary to what one might expect, the formation of these domains costs more than the energy of the domain wall alone. The appearance of magnetic order often causes deformation in the crystal lattice inside the domain, in the direction of magnetization; this is magnetostriction (see chap. 12).

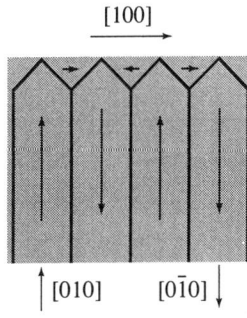

Figure 5.20 - Closure domains observed on the surface of an iron-silicon crystal

Thus, in the case of iron where this magnetostriction is positive in a weak field, a closure domain along [100] should become elongated in the direction of the moments as shown in figure 12.24 (the relative elongation is denoted by λ_{100}) but it is impeded by the adjacent [010] and [0$\bar{1}$0] domains. *Magnetoelastic energy* is thus stored inside the [100] domain, proportional to its volume. To reduce it amounts to reducing this volume, which can be achieved by reducing the width of all the [010] type domains and thus increasing their number. The state of equilibrium results from the compromise established between the magnetoelastic energy stored inside the closure domains, and the number of principal domains brought into play, i.e. the required domain wall area. It should be noted that closure domains may also exist in uniaxial materials when their anisotropy is not very strong.

5. MAGNETIC CONFIGURATIONS IN SMALL PARTICLES. SINGLE DOMAIN PARTICLES

For a single crystal of a uniaxial ferromagnet, cut into the shape of a parallelepiped of thickness e equal to 1 cm, we evaluated in section 4.1 the gain in internal energy associated with its subdivision into slab shaped magnetic domains. More generally the expression for the energy ratio between the two states considered is:

$$\frac{E_{monodom}}{E_{multidom}} = \frac{\mu_0 M_s^2 e}{4\sqrt{0.136 \mu_0 M_s^2 \gamma e}} = \sqrt{\frac{\mu_0}{2.176\gamma}} M_s \sqrt{e} \qquad (5.37)$$

This ratio tends towards zero with e. A critical thickness, e_c, should therefore exist, below which the single domain configuration becomes energetically inferior to the multidomain configuration, and hence represents the stable state of the system.

5 - FERROMAGNETISM OF AN IDEAL SYSTEM

This result expresses the fact that the demagnetising field energy is proportional to $M_s^2 V$, hence to the volume of the sample, whereas the domain wall energy, equal to γS, is proportional to the total surface area of the wall, S. It is well known that for small sizes, the energy of a surface term becomes relatively more important than that of a volume term.

To illustrate this competition between magnetostatic and domain wall energies, Kittel [1] compared the energy of the single domain state for a sphere, with that of a state where the sphere is divided into two domains of equal volume by a plane wall passing through its centre. In the single domain state, the demagnetising field energy is: $E_{mag} = (4\pi/18)\mu_0 M_s^2 R^3$, where R is the radius of the sphere. In the state with a single domain wall, one assumes that the energy of the demagnetising field is divided by two, whence: $E_{tot} = (2\pi/18)\mu_0 M_s^2 R^3 + \gamma 4\pi R^2$. The critical radius, below which the single domain state is lower in energy, is obtained by equating the two energies above, i.e.: $R_K = 9\gamma/\mu_0 M_s^2$.

This relation expresses the functional dependence between the critical dimension and magnetic quantities γ and M_s. If the energy of the domain wall is large, introducing one or several walls is expensive and the critical radius is raised, but this radius is also raised when the spontaneous magnetization is small, leading to a reduced demagnetising field energy. In the case of iron, $\gamma \approx 3 \times 10^{-3}$ J.m^{-2} and $M_s = 1.71 \times 10^6$ A.m^{-1}, from which one deduces $R_K = 7.3$ nm. Note that $\delta \approx 30$ nm.

5.1. THE MOST LIKELY STABLE CONFIGURATIONS

The preceding discussion dealt with the comparison in energy between the uniform and the multidomain magnetization states. Other moment configurations, non-uniform and without domains, may be envisaged (that, for example, described by Van den Berg and mentioned in § 2.3 of this chapter). Such configurations, which generally incur a direct cost in exchange energy, are possible in materials with a very weak magnetocrystalline anisotropy, and depend, above all, on the shape of the sample. The calculation of the internal magnetic energy of these configurations is tricky because it must be based on the theory of micromagnetism, which entails the use of non linear differential equations.

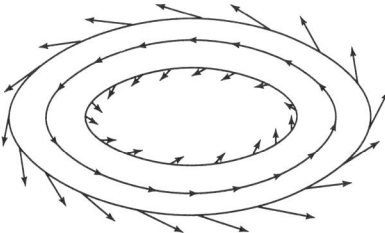

Figure 5.21 - *State of magnetization curling*

For a sphere, a prolate ellipsoid, or a cylinder, the most classic, and the most likely of these non-uniform, non-domain configurations, corresponds to the diagram of figure 5.21 which we shall refer to as "a state of magnetization curling" This is a state close to that described by Van den Berg for a flat disc [2].

Using as a basis the results of the exact calculations by Brown [7] or the approximations by Kittel [1], Zijlstra [8] compared the energies of three states:
* uniform magnetization,
* magnetization curling,
* two domain state,

for a uniaxial material cut to a spherical shape.

By comparing the single domain state with the two other states it can be shown that there exists a radius R_{c0} below which the single domain state definitely has the lowest energy with $R_{c0} = 5.099 (A_{ex}/\mu_0 M_s^2)^{1/2}$, where A_{ex} is the exchange constant. For iron, $A_{ex} \approx 1.95 \times 10^{-11}$ J.m^{-1}, from which $R_{c0} = 11.75$ nm. For R values greater than R_{c0}, it is convenient to single out two cases. The case of a weak magnetocrystalline anisotropy corresponds to $K \equiv K_1 < 0.178 \; \mu_0 M_s^2$ [8]. The uniform magnetization state must, in this case, be compared to the state of magnetization curling. The associated critical radius R_{c1} is thus:

$$R_{c1} = 6.4053 \sqrt{A_{ex}/\mu_0 M_s^2} \left[1 - 5.615(K/\mu_0 M_s^2)\right]^{-1} \quad (5.38)$$

For $K = 0$, $R_{c1} = 1.2562 \, R_{c0}$. For $K > 0.178 \; \mu_0 M_s^2$ (strong magnetocrystalline anisotropy), $R_{c1} \to \infty$, i.e. the state of magnetization curling becomes unstable relative to that of uniform magnetization. The two domain states then, must be compared to the state of uniform magnetization. The critical radius separating them is:

$$R_{c2} = 56.129 \sqrt{A_{ex}/\mu_0 M_s^2} \left[1.5708 + (K/\mu_0 M_s^2)\right] \quad (5.39)$$

For iron $R_{c2} = 160$ nm, greater than the thickness of the wall.

For $R > R_{c2}$, the two-domain state is of lower energy than the uniformly magnetised state, even when the anisotropy is weak (in Brown's sense).

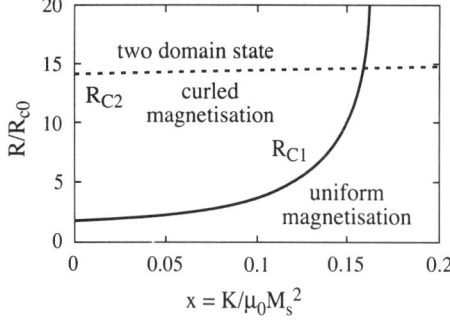

Figure 5.22 - Evolution of the critical radii as a function of the parameter $x = K/\mu_0 M_s^2$

Figure 5.22 shows the evolution of the critical radii R_{c1} and R_{c2}, relative to R_{c0}, as a function of parameter $x = K/\mu_0 M_s^2$, according to Zijlstra-Brown.

5 - FERROMAGNETISM OF AN IDEAL SYSTEM

In particular one has:

for $K = 0$, $R_{c1} = 1.2562\ R_{c0}$ and $R_{c2} = 13.7965\ R_{c0}$

for $K = 0.178\ \mu_0\ M_s^2$, $R_{c1} \to \infty$ and $R_{c2} = 13.7965\ R_{c0}$

for $K = 0.1627\ \mu_0\ M_s^2$, $R_{c1} = R_{c2} = 14.5\ R_{c0}$.

5.2. ELLIPSOIDAL PARTICLES AND SHAPE ANISOTROPY

Let us consider a particle whose shape can be considered as a prolate ellipsoid of revolution, with major axis 2c and minor axes 2a. Let us first assume that the magnetization of the particle is oriented along the major axis of the ellipsoid. The demagnetising field coefficient in this direction, N_c, was given by equation (2.48). In this section we will just return to Kittel's model, and compare the saturated state (1) (uniform magnetization) with a two-domain state (2) with a central wall. Let us recall that this model considers the residual magnetostatic energy in the bi-domain state as being reduced to approximately half of that in the single domain states.

One may then write the magnetic energies of the two states:

$$E(1) = (\mu_0/2)\ N_c\ M_s^2\ V_{ellipsoid} = (\mu_0/2)\ N_c\ M_s^2\ 4\pi a^2 c/3$$
$$E(2) = E(1)/2 + \gamma\ S_{ellipse} = E(1)/2 + \gamma \pi\ a\ c$$

from which $E(1) < E(2)$ for $a < a_c$, such that: $a_c = 3\gamma/\mu_0\ N_c\ M_s^2$.

Consider again the case of an iron particle and assume that it has the shape of a very elongated ellipsoid ($c/a = 10$). One then has $N_c = 2 \times 10^{-2}$, and $a_c = 150\gamma/\mu_0\ M_s^2$.

With the values of γ and M_s of iron (see the preceding section), one obtains as the critical dimension $2\ a_c = 240$ nm for a major axis length $2c$ of 2.4 µm.

Elongated particles thus remain single domain for volumes much greater than the spherical particles. This result is completely consistent with the remarks made at the beginning of this section: with an elongated particle the dipolar effects are less important for comparable size, and the wall is more costly.

In contrast, for an oblate ellipsoid in which the magnetization would be forced to remain oriented along the minor axis, the dipolar effect intervenes more intensely, and the single domain situation exists only for very small particles.

These last two remarks lead to the concept of *shape anisotropy* or *demagnetising field anisotropy*: in every non spherical system, featuring non-zero resultant magnetization **M**, the magnetization tends to align itself along the greatest dimension. For a prolate ellipsoid ($c \gg a$), magnetostatic energy is given by: $E_d = (\mu_0 M^2/2)\ (N_a \sin^2\theta + N_c \cos^2\theta)$, if θ is the angle which **M** makes with its Oz axis of revolution, which leads –on replacing $\cos^2\theta$ by $(1 - \sin^2\theta)$ and ignoring the constant– to a uniaxial anisotropy of the form:

$$E_d = (\mu_0\ M_s^2/2)(N_a - N_c)\sin^2\theta = K_f \sin^2\theta \qquad (5.40)$$

By analogy with the expressions (3.8, 3.9) which define a magnetocrystalline anisotropy field, one can define a *shape anisotropy field*, H_A^f:

$$H_A^{sh} = (N_a - N_c) M_s \tag{5.41}$$

6. OBSERVATION OF DOMAINS AND DOMAIN WALLS

Since the time P. Weiss postulated the existence of domains inside ferromagnetic materials, many experimental techniques have been proposed for displaying these magnetic domains and the domain walls that separate them. The physical approach to the observation of domains differs markedly depending on whether the subject matter is ferromagnetic and ferrimagnetic materials (which feature a spontaneous magnetization M_s) or antiferromagnetic materials ($M_s = 0$).

In the case of the former, the distribution of magnetization acts directly on various probes: *light* through the magneto-optical effects (chap. 13), *electrons* through the Lorentz force, and *neutrons* through the potential interaction energy between the field and the magnetic moment of the neutron, which brings a magnetic contribution to diffraction. Furthermore, magnetization inhomogeneities create a dipolar field, called the *stray or dispersion field*, which can attract magnetic fine particles (*Bitter method*): their increased local concentration can be observed through a microscope.

For antiferromagnets, the only direct method is neutron diffraction imaging.

6.1. FERRO- AND FERRIMAGNETIC DOMAINS

6.1.1. The Bitter method

F. Bitter had the idea of sprinkling a fine ferromagnetic powder on the polished surface of the material to be observed. The particles are attracted by the stray field gradients: the higher the gradient the stronger the attraction. They concentrate on the *domain walls,* whose outcrops at the surface of the sample are detected using an optical microscope, thus revealing the domains.

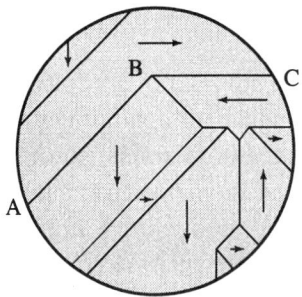

Figure 5.23 - Observation of a Weiss domain structure by the Bitter technique [9]

The current version of this technique uses colloidal suspensions of particles in a liquid, and in particular some of the commercially available ferrofluids. Figure 5.23

(from S. Chikazumi [9]) schematises the view obtained for a crystal of Fe-4%Si. The magnetizations are in the <100> directions which lie close to the surface; thus, the line AB constitutes a 90° wall and the line BC a 180° wall. A derived technique makes use of magnetotactic bacteria (unicellular organisms which carry grains of magnetic material, see fig. 25.2) to visualise the walls.

6.1.2. Observation of magnetic domains by magneto-optical effects

The Faraday effect and the three magneto-optic Kerr effects, discussed in chapter 13 of this book, can be used to observe domains. Observation by the Faraday effect makes use of a rectilinearly polarised light. On crossing the sample a rotation of the polarization plane of the light occurs which is proportional to the projection of the magnetization along the direction of propagation of the beam, and to the length of the path. For two 180° domains, the directions of rotation of the polarization are opposed. The Faraday effect can only be considered for transparent samples; it is easy to use and spectacular, the rotation of the polarization being sizeable. A polarising microscope of medium quality is adequate for this. The observation pleasure is increased by the fact that, because the rotation is wavelength dependent, one sees coloured images. The application of a magnetic field allows one to follow the reorganization of the domains in real time.

For metallic samples, which are non-transparent except in the case of very thin films, the observations are made in reflection mode, and involve the magneto-optical Kerr effects. The polar Kerr effect, analogous to the Faraday effect, reveals the magnetic domains when the magnetization has an appreciable component in the direction normal to the illuminated surface of the sample. The longitudinal Kerr effect and the transverse Kerr effect allow the domains to be revealed when the magnetizations are confined to the surface plane. The longitudinal effect involves rotations of the plane of polarization, which are different for different components of the magnetization direction in the plane of incidence. The transverse effect leads to a difference in reflected intensity caused by the sign of the component of magnetization normal to the plane of incidence. The signals coming from the Kerr effect are weaker than those associated with the Faraday effect. Accordingly, observation is more difficult than for the case of transmission, and requires good instrumentation, as well as very good surface polishing. On the one hand the surface must be optically smooth, in particular in the longitudinal and transverse Kerr effects where the incidence is oblique: this geometry is in fact, particularly sensitive to roughness, and the magnetic contrast is easily drowned by surface undulations. Their influence can be decreased by the use of interference films (see chap. 13) which increase the Kerr rotation and thus the magnetic contrast. On the other hand, as this method of observation only affects a surface layer, typically 20 nm thick, it is vital that the domains should not be complicated close to the surface by the appearance of small parasitic domains, as a result of internal stress *via* magnetostriction. This last effect cannot be compensated for by the use of stratagems. To avoid it one must either resort to mechano-chemical polishing, or carry out electrolytic or chemical polishing after the mechanical polishing stage. With regard to instrumentation, a good polarising microscope may be used for observations using the polar Kerr effect, i.e. for the visualization of domains where the magnetization has a sizeable component normal to the surface.

Longitudinal Kerr effect observations at low magnification (see fig. 5.24) are possible using a simple apparatus made from a small inclined microscope, and a symmetrical illuminator which can be produced by modifying a microscope tube. Only a stripe of the field of observation is then in focus. The contrast is satisfactory provided two precautions are taken.

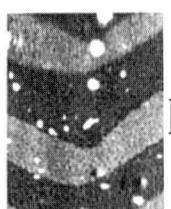

Figure 5.24 - Weiss domains revealed by the longitudinal Kerr effect

One is to suppress the rays that deviate too much from the average plane of incidence, and cannot be extinguished because they acquire an elliptical polarization even apart from any magnetic effect. For this, a diaphragm with a slit shaped opening parallel to the plane of incidence should be placed in the back focal plane of the objective of the illuminator [10]. The other is that the depolarization effects arising from the optical elements of the illuminator and the microscope can be avoided by placing the polariser and the analyser outside the microscope tubes, i.e. immediately before and after the sample. Good dichroic polarisers are very suitable for this purpose.

For high magnification work, this arrangement is not suitable. One can, however, use a very good polarising microscope, with a polarization objective with a large numerical aperture, provided the aperture diaphragm is closed down and off-centred so as to obtain an oblique illumination. It is again preferable to replace the circular aperture diaphragm by a slit. The reduction in the illumination aperture lowers the resolution in comparison with the normal performance of the objective, but it can reach 1 µm approximately.

Scanning Near-field Optical Microscopy (SNOM) has provided very promising preliminary results for the study of ferromagnetic domains, at a resolution of the order of 10 nm, hence well beyond far-field microscopy. Chapter 13 (magneto-optics) refers to this.

6.1.3. *X-ray observation of magnetic domains*

X-ray Bragg diffraction imaging, often called *X ray topography*, makes it possible to obtain images of defects present in a single-crystal sample or in large enough grains. [11]. The principle is that the heterogeneity of diffracted beams, if recorded on a position-sensitive detector (film or CCD camera), provides an image of the sample, generally as a result of deformations in the lattice planes used for the Bragg reflection. This method also reveals domains, or walls between the domains.

In almost all of the works carried out so far, the visibility of magnetic domains or domain walls was indirect. The contrast is actually due to a secondary effect, the difference in distortion, associated with magnetostriction, between domains. Inside a domain, the magnetostriction tends to impose a spontaneous deformation which is dependent on the direction, but not on the sign, of the magnetization. The requirement to match the crystal lattice, if possible without involving any stress, between domains with different magnetization directions, generally also leads to a relative rotation of the lattice in the two domains.

Two domains with different spontaneous magnetization directions are then in slightly different situations with respect to the beams, and in particular Bragg's diffraction conditions. If the effective divergence of the beams is less than their effective orientation difference, the intensities of the beams diffracted by these domains are different, which leads to different grey shades on the corresponding regions of the image. In the opposite case, most frequent using laboratory X-ray sources, where the angular difference between the domains is small compared to the effective divergence of the beam, the domains can no longer be distinguished from one another. At that point, it is the domain walls that are observed, their images being the result of the perturbations which they impose on the propagation of X-rays in the sample.

X-ray topography using X-rays provided by a synchrotron radiation source lends itself to real time observations. The abundance of high energy photons supplied by a machine such as the European Source of Synchrotron Radiation (ESRF) allows the observation of (90°, 71° or 109°) walls or domains inside relatively thick samples (more than one mm for iron). One of the variants of this experimental technique, white beam topography, is instrumentally very simple; it is well adapted to the application of magnetic field and to the study of temperature effects, and tolerates the subdivision of the sample into sub-grains.

Figure 5.25 shows the same domain structure as figure 5.24, now revealed by X-ray transmission topography. Here the walls in the bulk of the sample are visible, in particular as a series of equal thickness fringes.

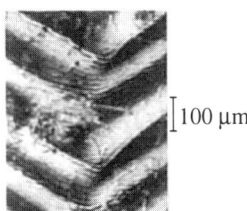

Figure 5.25 - Weiss domains revealed by X-ray topography

Recently, Kawata and Mori [12] have shown the feasibility of *direct* observation of domains by resonant magnetic X-ray scattering, in the neighbourhood of the K edge of iron, in magnetite Fe_3O_4.

X-rays obtained from synchrotron radiation are also used to produce images, but in a radically different fashion, through the effect of dichroism in the vicinity of absorption edges. X-ray magnetic dichroism is the term used to indicate the variation of absorption of the beam as a function of the state of polarization. The relevant factor is in fact the relative orientation of the polarization and of magnetization. Thus, for circular magnetic dichroism, the variation in absorption when going over from right circular to left circular polarization for a given direction of magnetization is obtained just as well, at constant polarization, when the magnetization changes direction.

This effect is also very closely associated with the mechanism of absorption, in which an internal electron is ejected towards a vacant state, the energy difference being that of the absorbed photon. The energy differences are characteristic of the energy or wavelength of the absorption edge of each atom or ion. Dichroism is associated with the fact that the magnetization has a profound effect on the distribution of both the occupied and free states. Stöhr *et al.* [13] make use of the fact that, following absorption, photoelectrons are emitted from different domains with different probabilities and thus different densities, to form the image with the help of an electron-optical setup. Because of the short distance travelled by the photoelectrons, the observation here is concerned only with the part of the sample which is very close to the surface, and must be carried out under very high vacuum. In a slightly different approach based on soft X-ray microscopy, Fischer *et al.* directly use transmitted X-rays, of varying intensity following local absorption, to obtain the same type of images [14]. The noteworthy aspect of this technique is that it involves a single chemical constituent at a time, viz the element the absorption edge of which is being explored. Thus it provides important information about the magnetism of multilayer films, and in particular the coupling between films of different composition.

6.1.4. Neutron observation of magnetic domains

Neutrons interact with matter through two principal mechanisms. On the one hand they are scattered by the nuclei, with which they interact through the very short-range strong force. On the other hand, having a magnetic moment associated with their spin angular moment (spin 1/2), they are affected by the magnetic field produced at microscopic level by the magnetic moments of atoms. This magnetic neutron scattering mechanism is, through neutron diffraction work in the classical approach (the exploration of the reciprocal lattice), the source of most of the available information regarding magnetic structures (see appendix to chap. 4). The same effect, used differently, leads to Bragg diffraction imaging (or neutron topography), analogous in its principle to X-ray topography. This method can provide images of all types of magnetic domains within large single crystals. Unlike the X-ray case, the magnetic information is here associated with the domains directly, and not through a side effect. The most useful technique, in the case of ferro- or ferrimagnetic domains, is based on polarised neutrons. Depending on the relative orientation of the magnetization and of the polarization of the incident neutrons, the nuclear and magnetic contributions to the diffraction add or subtract.

For a given polarization of the beam there is a corresponding contrast between domains (for example across a 180° wall) which reverses when the polarization of the beam is reversed. This method requires long exposure times, and, because of the weak intensity of the beams available, its resolution is mediocre (several tens of µm). Figure 5.26 shows ferromagnetic domains revealed in this way by neutron topography.

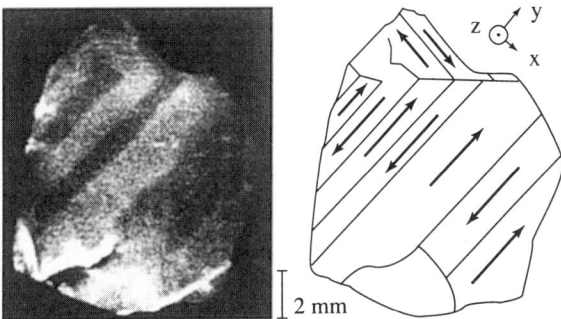

Figure 5.26 - Ferromagnetic domains, revealed by polarised neutron topography
The sample is Fe-3%Si, with surface (001) and thickness 0.12 mm. $\lambda = 0.135$ nm.

Two other approaches use the Larmor precession of the magnetic moment of neutrons around an average magnetic field, or induction **B**, to map domains. The intensity diffracted by an analyser, set to detect one component of the polarization of a fine beam of neutrons transmitted through the sample, varies when the sample is displaced relative to the beam. This intensity depends upon the direction of the magnetic field felt by the beam, hence on the local magnetization [15]. The same principle, usually described in terms of dephasing rather than Larmor precession, provides images of domains thanks to a neutron interferometer [16]. The apparatus is made of three or four strictly parallel and perfect silicon crystals, in which the beam is coherently split, then recombined after the two paths have been subjected to various actions, as in an optical (e.g. Mach-Zehnder) interferometer. The silicon crystals, which actually belong to a single monolithic block, perform the division, deviation, and recombination of the beams by Bragg diffraction. In neutron interferometry too, the use of imaging is less usual than are metrological applications. Successive diffraction, first in one direction then in the other, by two perfect, identical crystals, requires the crystals to be rigorously parallel. The very high resulting angular sensitivity has also been used to detect neutron refraction in a sample placed between the two crystals. An image of domain walls inside very bulky (13 mm diameter) silicon iron samples was thus obtained [17, 18].

6.1.5. Electron observation of magnetic domains

Electron microscopy offers very rich potential for the observation of ferromagnetic domains, and there is abundant literature on the subject.

The most widely used, and most direct technique, is that of transmission electron microscopy (TEM). It is based on dephasing or, in classical physics terms, the deviation due to the Lorentz force which the field **B** exerts on electrons. The reversal of magnetization causes a reversal of the Lorentz force, and hence reversed deviation of the electron beam. The phase object corresponding to the domain structure is not visible in the focused image because the rays stemming from the same point of the object converge at the same point on the image, whatever the deviation. The domain structure becomes visible under sufficient defocusing (*Lorentz microscopy*): the domain walls then appear in the form of dark or light lines. One can also take advantage of the fact that the focal plane of the objective contains the diffraction diagram of the object, each beam direction corresponding to a point on this plane. By using a selective screen (*Foucault's microscopy*), one can remove the beam deflected by one of the domain families. The image then shows black and white domain contrast for 180° domains. This technique requires very thin samples; it also imposes a non-standard operation of the microscope, in which the standard objective lens is not used. In all electron microscopes currently in use, the lenses are magnetic, and the field through which they affect the electron beams is incompatible with the observation of domains and of their evolution under a controlled magnetic field. Accessories allow observations to be made at high or low temperatures. The use of a quadrant detector allows the components of the magnetization in the sample to be determined [19].

The scanning electronic microscope (SEM), operating in reflection mode, can also, though less directly, show the domains lying close to the surface of samples, which in this case may be bulk specimens. The scanning electron microscope with polarization analysis (SEMPA) makes it possible to determine quantitatively, at the surface, the direction of magnetization.

Lastly, the gorgeous experiments by A. Tonomura and his group, using electron interference holography, should be cited. They show quite clearly that a phase change of π of the wave associated with the electrons corresponds to a flux equal to the flux quantum $\Phi_0 = h/2e$, where h is Planck's constant and e the electronic charge. The interferograms obtained, using an extremely fine field emission electron source, show the flux lines, labelled in a quantitative manner, and hence, in particular, the domains [20]. It should be noted that this experiment is particularly spectacular with regard to the observation of vortex lines in superconductors: each vortex line being precisely associated to the flux Φ_0, it constitutes an ideal phase object [21]. Finally, spin polarised scanning tunnelling microscopy (spin-polarised STM) is rich in promise with regard to the direct observation of magnetic structures at very high resolution.

5 - FERROMAGNETISM OF AN IDEAL SYSTEM

6.1.6. Observation of magnetic domains by precipitation anisotropy

Libovicky [22] showed that at room temperature, hence in the frozen-in, or fossil state, it is possible to observe the domains that exist at around 600°C in silicon-iron samples (10 to 15 atomic % Si) due to the anisotropy, associated with the local direction of magnetization, of the precipitation that occurs during tempering. The metallurgical phase segregation is probably complex, and is sensitive to the local direction of magnetization. This technique in no way lends itself to experiments in real time, and it is restricted to materials with complex phase diagrams. Nevertheless, despite its very particular character, it allows observation of domains that exist at high temperatures, without the need for a special microscope stage, and within the volume of the sample.

6.1.7. Magnetic Force Microscopy

Following the advent of Scanning Tunnelling Microscopy (STM) and of Atomic Force Microscopy (AFM) in the 1980's, a large variety of new scanned probe microscopies appeared, giving access to the submicron-scale investigation of various properties of materials. Magnetic Force Microscopy (MFM) was rapidly recognized as being useful both for fundamental and for applied studies of magnetic materials.

In fact, the pioneering work of Y. Martin and H.K. Wickramasinghe [23] occurred with remarkable timing. At the end of the 1980's, both industry and basic research were craving for new tools to investigate domain and field configurations at a sub-micron-scale. As a result, MFM enjoyed an impressive development, with commercial versions on the market at the beginning of the nineties, and widespread use in the second half of the decade. Other techniques, such as Lorentz microscopy, may provide information at a comparable scale, but none offers the same combination of ease, cost and performance.

Basically, MFM relies on a familiar property of magnetism: the attraction or the repulsion between two magnets. Like atomic force microscopy, it is based on scanning a (magnetic) tip over or very near the sample surface, and detecting the small forces induced by the stray fields from the sample (fig. 5.27). The results are impressive. Magnetic resolution in the 30 nm range is easily achieved [24] in a non-destructive measurement implying little, if any, sample preparation. The technique is well adapted to the observation of continuous thin films and nanostructures. Its high sensitivity allows domain configurations to be determined in ultrathin magnetic films (a few monolayers thick), even when working under room conditions.

Because MFM relies on the interaction of the tip with the *stray field* from the sample, and *not directly with the magnetization*, the determination of the domain structure leading to the measured tip to sample interaction is sometimes difficult. Simple calculations also show that different magnetic configurations can induce exactly the same stray field pattern. Thus, the interpretation of the images relies, often implicitly, on some assumptions or previous knowledge of the magnetic properties of the sample.

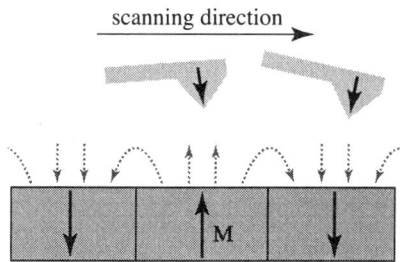

Figure 5.27 - Principle of MFM imaging

As the magnetic tip is scanned close to the sample surface, it is alternatively attracted and repelled by the stray field from the sample (here a thin film with perpendicular magnetization, observed in cross-section). This magnetic force is detected and leads to the contrast observed in MFM images. Usual scanning heights are in the 15 to 100 nm range, usual dimensions of the pyramidal part of the cantilever are a few micrometers, with a radius of curvature of a few tens of nanometers for the tip apex. To prepare the magnetic tips, most groups use commercial AFM silicon cantilevers, with an additional 20-100 nm magnetic coating, such as an CoCr alloy.

Tip to sample interaction

As the tip is brought near the sample surface, it is submitted to a magnetic interaction. The corresponding energy can be written as:

$$E = -\mu_0 \cdot \iiint_{tip} \mathbf{M}_{tip} \cdot \mathbf{H}_{sample} \cdot d\tau_{tip} = -\mu_0 \cdot \iiint_{sample} \mathbf{M}_{sample} \cdot \mathbf{H}_{tip} \cdot d\tau_{sample} \quad (5.42)$$

$\mathbf{M}_{tip\ (sample)}$ is the tip (sample) magnetization, $\mathbf{H}_{sample\ (tip)}$ the stray field created by the sample (tip).

Although it is obvious, the equivalence of these two formulations may help underline an important point: as the tip is moved through the stray field from the sample, the sample is unavoidably submitted to the stray field from the tip. If the term $\mathbf{H}_{tip}$ (within the sample) reaches zero, the interaction and hence the signal vanish. Equation (5.42) is the basis for a fruitful approach of the tip to sample interaction [25] and for many investigations on the perturbation of the sample magnetization by the stray field from the tip [26].

Equation (5.42) also helps understand one of the most serious limitation of MFM. For the investigation of magnetically very soft samples, the field created by the tip on the sample has to be reduced in order to avoid disturbing the sample magnetization. This is usually done by using tips with a smaller magnetic moment (a thinner magnetic coating). If, in addition, the sample magnetization is weak, this leads to a vanishing signal. As a result, one has to maneuver between the two pitfalls of a perturbing measurement and a signal below the detection threshold. In actual fact, however, good MFM can reliably image almost any type of thin film or media.

Removing topography from the images

A significant difficulty arises from the intrusion of topography related contrast in the MFM images. Here, unlike in the above description of Bragg diffraction X-ray or neutron imaging, "topography" refers specifically to surface corrugation. Its spurious effect on the data is easily understood: if the tip reaches contact to the surface, it is very likely that the contact force will dominate over the magnetic interaction. Furthermore, the tip is submitted to other long range interactions, such as Van der Waals forces. These forces may produce undesirable signals, even if the tip does not contact the surface. Thus, interpreting the images obtained on a rough surface may be very difficult.

As magnetic forces decay slowly compared to the others, they will, far enough from the sample, dominate. Hopefully, the magnetic forces will still be large enough to be detected, providing a magnetic image if the tip is scanned at some distance from the sample. Then, the question arises: how can the tip be maintained at a given distance from the sample surface?

To a large extent, on reasonably flat samples, the problem is solved by first detecting the position of the sample, and next, after the height profile has been recorded, scanning at a given height above the sample surface. In the most widespread mode, each line is scanned twice, first with the tip in contact with the sample, and then with the tip flying at a height selected by the operator (fig. 5.28). Unfortunately, this does not always remove all the spurious topography related contrast within the images (fig. 5.28). Other techniques have been proposed, such as the use of an additional electrostatic interaction to stabilise the tip to sample distance, but they are not widely used.

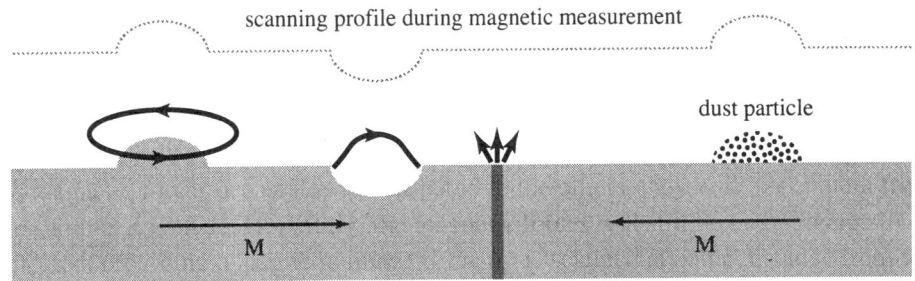

Figure 5.28 - Artefacts due to surface topography

If the sample exhibits in-plane magnetization, the stray field is observed above domain walls. However, protuberances or depressions in the magnetic layer surface may also induce a stray field. To obtain an image, each line is scanned twice, first to detect the sample surface topography, then to record the magnetic information at a given flying height. In this process, non magnetic roughness (dust particles...) will also affect the tip to sample distance during the measurement.

Interpretation of images

In the most crude approximation, the tip was described as a monopole, i.e. a magnetic charge q_m, implying that the opposite charge distribution, which is necessarily present, is at large enough a distance from the sample surface to be neglected. Next, most authors considered it as a magnetic dipole m, either point-like, or extended. The force exerted on the tip may then be written (for a point-like tip, the z axis being perpendicular to the sample surface and the tip magnetised along this z-direction):

$$F_z = \mu_0 q_m \cdot H_{z,sample} \text{ (monopolar tip) or } F_z = \mu_0 m\, z_{tip} \frac{\partial H_{z,sample}}{\partial z} \text{ (dipolar tip)}$$

Very often, the measurement is performed in the so-called AC-mode, where the tip oscillates at its resonant frequency, and where either the amplitude, or the phase of the oscillation is detected. In the simple case of a harmonic oscillator in a force gradient, the shift (Δf) of the resonant frequency (f_0) may be expressed as:

$$\frac{\Delta f}{f_0} \approx \frac{1}{2k} \frac{\partial F_z}{\partial z},$$

k being the force constant of the lever holding the magnetic tip (see fig. 5.27).

As a result, the signal should be proportional to the first or to the second derivative of the stray field from the sample. In fact, neither the monopole nor the dipole description quantitatively explains the experimental data. Retrieving the sample magnetization distribution is a complex task, and it has not often been attempted. Hug *et al.* recently came close to achieving this goal, at the cost of heavy procedures including the calibration of *each* tip on a test sample [27].

Nevertheless, highly useful information such as the geometry of the magnetic domains or the location of the domain walls is easily extracted. As the tip is generally magnetised along its axis (approximately perpendicular to the sample surface); the measurement is sensitive to the perpendicular stray field. Such a geometry leads to the detection of the domain walls for in-plane magnetization, and to the apparent detection of the domains for perpendicular magnetization (fig. 5.29-a). In fact, in the perpendicular geometry, contrast also appears in the vicinity of the domain walls (see fig. 5.29-b), but this is not conspicuous for small enough magnetic domains where the contrasts linked to the presence of neighbouring walls overlap.

Figure 5.29-c corresponds to the observation of the classical closure pattern in a square element. Obviously, magnetic contrast extends far from the domain walls. Here, this can be partly due to the extended tails of the 90° walls (part of the magnetization rotation may extend over large distances). Some authors observed this so-called domain contrast over micrometers, thereby casting suspicion about the possible need to take the perturbation of the sample magnetization by the stray field from the tip into account to explain even basic features of common images [26]. Recent work pointed out the complex dependence of the tip sensitivity to the spatial period of the observed magnetic pattern. Basically, the idea is the following [27]: if λ is the spatial period of the magnetic pattern, and z the distance from the surface, the stray field decays as $e^{-z/\lambda}$. As the tip widens from its apex to its pyramidal part, the active magnetic volume of the tip increases for larger features of the magnetic pattern. This may be the origin of a large part of the observed "domain contrast".

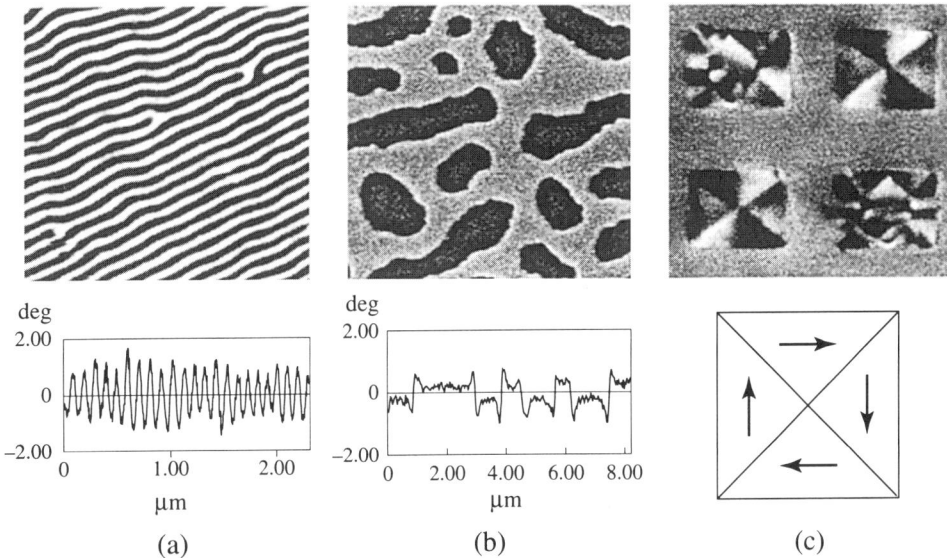

Figure 5.29 - Magnetic force microscopy images
(Images courtesy of (a, b): Y. Samson, A. Marty, R. Hoffmann, C. Beigné,
(c): Y. Samson, P. Warin, A. Marty, C. Fermon)

(a) *2 μm image of a classical stripe configuration within a FePd layer (thickness: 45 nm) with perpendicular anisotropy. The stripe width is 50 nm. Black and white stripes correspond respectively to areas with up or down magnetization. Lower part: signal profile along the grey line.*

(b) *8 μm image of a FePd layer (thickness: 1.6 nm) with perpendicular magnetization. Lower part: signal profile along the grey line. The signal decreases, as does the stray field, far from the domain walls.*

(c) *4 μm image of square dots obtained by e-beam lithography in a 50 nm Co layer. The magnetization lies in the plane of the sample. Lower part: schematic representation of the domain structure of the upper right or lower left dots. The two other dots exhibit more complicated domain configurations. The observed contrasts are mainly induced by the 90° domain walls.*

Perspectives

Instruments have been developed so as to cover a wide range of experimental conditions : strong fields, ultra-high vacuum, low temperatures. As a result, MFM is now an efficient tool to address a large variety of physical problems. In the opinion of the author, the main limitation will remain the lateral resolution (about 30 nm) [24]. This limitation will become serious for the observation of hard disk materials as the bit size approaches 100 nm. Other techniques may then be able to cope with the task, such as spin-polarised STM, but none seems to offer the same ease of use, and the same ability to work under room conditions. MFM is thus likely to retain a significant role in the physics and technology of thin magnetic films in the coming years. In addition, researchers are now considering ways to use the tip as an active probe to induce local magnetization reversal, with possible applications for ultra-high density recording [28].

6.2. OBSERVATION OF ANTIFERROMAGNETIC DOMAINS

Although this chapter is devoted to ferromagnetism, we give some information about the visualization techniques for antiferromagnetic domains. Unlike the case of ferro- and ferrimagnetic domains, for which dipolar (magnetostatic) energy plays an active role, the presence of antiferromagnetic domains does not lead to a lowering of the free energy of a sample. However, their existence is anticipated. Because the transition from the paramagnetic state to antiferromagnetic order is associated with a lowering of symmetry, several individuals are possible. These can be deduced from one another through the symmetry elements lost during the phase transition. Their investigation is more fundamental than that of ferro- or ferrimagnetic domains, and very little research has been devoted to them, principally because antiferromagnetic materials are little used in applications. Ferro-antiferromagnetic coupling, which lies at the root of the giant magneto-resistance devices, is a notable exception. Several types of antiferromagnetic domains exist; and here we shall briefly describe those which we will mention.

6.2.1. Observation of antiferromagnetic domains by optical microscopy

The visualization of antiferromagnetic domains is sometimes possible by optical means. Nickel oxide, which is cubic above its Néel temperature, contains two types of domains. The arrangement of the magnetic moments corresponds to an alternation of planes in which the magnetic moments are aligned, the direction of moments being reversed from one plane to the next. This order, which corresponds to trigonal symmetry, distinguishes four families of domains, depending on whether the ferromagnetic planes are one or other of the four families of {111} planes such as $(\bar{1}11), (1\bar{1}1)$, etc. (fig. 5.30). They are called **q** domains, evoking the propagation vector perpendicular to the alignment planes which characterises this arrangement, or T (as in twin) domains, the name given originally [29].

Furthermore, the magnetic moment orientation corresponds to a further lowering of symmetry, and to other domains called S (as in spin) domains [30]. The birefringence associated with these distortions makes it possible for the domains to be seen under polarised light. The visualization is indirect insofar as the distortion is a side effect of the arrangement (exchange striction), or of the orientation (magnetostriction), of the magnetic moments.

Another family of classical antiferromagnetic materials is that of the transition metal fluorides, MnF_2, CoF_2, FeF_2. The crystallographic structure is that of rutile, with tetragonal primitive Bravais lattice. The magnetic ions (Mn^{2+} for example) occupy the apices and the centre of the cell but their environments (two fluoride ions) are oriented differently. In the absence of an applied field the magnetic moments are parallel, in one or other direction, to the [001] axis. Hence, only 180° antiferromagnetic domains can exist. Domain I, for example, will correspond to the ↑ orientation for the magnetic moment at the apices, and to the ↓ orientation of the magnetic moment at the centre of the cell, and domain II to the opposing orientations (see fig. 5.31).

In the absence of a field these domains are only visible by neutron topography (see § 6.2.3). However, in CoF_2 and FeF_2, for which the induced linear birefringence effect (see chap. 13) is relatively large, they become optically visible in a sufficiently strong field [31].

It is possible to make a simplified physical representation of this effect on the basis of the uncommon piezomagnetism effect, which this family of compounds displays in the antiferromagnetic phase. The application of a magnetic field induces a deformation whose variation with B is linear (magnetostriction, which is always present, is quadratic with respect to the field), and, therefore, changes sign when **B** is reversed. If the fluoride ions get very slightly closer to the magnetic ion whose moment is parallel to the field, the refractive index for light whose polarization is parallel to the direction of their bond will, for example, increase with respect to the refractive index for light whose electric field is normal. The two domains, acquire, therefore, opposing birefringences.

6.2.2. Observation of antiferromagnetic domains by X-rays and electrons

The sensitivity of X-ray Bragg diffraction to crystal distortions enables X-ray topography to reveal domains of the type found in NiO [32]. The mechanism leading to their visibility is, as in the case of the optical birefringence of NiO, associated with the distortions due to the arrangement and to the orientation of magnetic moments (exchange striction and magnetostriction). X-ray topography generally offers, over visible light, the advantage of greater sensitivity to such distortion variations but, here again, the effect is indirect.

Electron microscopy, less sensitive to differences in distortion, but with a much better resolution, also allowed the observation of antiferromagnetic domains on the basis of elastic effects.

6.2.3. Observation of antiferromagnetic domains by neutrons

From the fundamental point of view the most interesting field of application of neutron topography is probably the observation of antiferromagnetic domains. Indeed this is the only technique to show them directly, by means of the magnetic contribution of Bragg diffraction. In the case of NiO referred to in the preceding section, the arrangement of the magnetic moments of the Ni^{2+} ions in layers, within which they lie parallel, leads to a doubling of the spatial period relevant for a probe sensitive to the direction of the magnetic moments of the neutrons (fig. 5.30). Thus purely magnetic (superstructure) Bragg reflections appear.

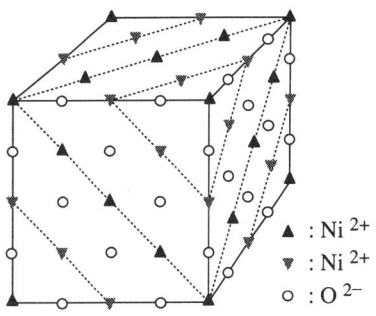

Figure 5.30 - Arrangement of magnetic moments in NiO

In the case shown, the magnetic moments are represented by triangles pointing up or down. The moments contained within (111) planes are aligned ferromagnetically, and their direction is reversed on passing from one plane to the next. This situation defines one of the four T or q domains. The other three correspond to ferromagnetic stacking in the $(1\bar{1}1)$ planes etc.

They are characterised by diffraction vectors of the $\frac{\bar{1}}{2}\frac{1}{2}\frac{1}{2}$ type for the domain whose ferromagnetically aligned planes are $(\bar{1}11)$, or $\frac{1}{2}\frac{\bar{1}}{2}\frac{1}{2}$ for those whose ferromagnetic planes are $(1\bar{1}1)$ etc. When the crystal is oriented in Bragg position for one of these magnetic reflections, $\frac{\bar{1}}{2}\frac{1}{2}\frac{1}{2}$ for example, only the domains corresponding to this propagation vector diffract, and their image will thus be obtained directly. The use of four magnetic reflections of this type in succession will provide both the image and the unambiguous characterization of these **q** or T domains. The S domains can also be visualised directly, by magnetic reflections of the $\frac{3}{2}\frac{\bar{1}}{2}\frac{1}{2}$ type, using the variations of diffracted intensity with the angle between the magnetic moments and the diffraction vector. Experimental studies carried out on NiO have confirmed the conclusions of the indirect techniques [32].

Neutron topography has also made it possible to observe domains that cannot be revealed by other means. Thus, in MnF_2, the 180° domains (fig. 5.31) can be observed directly, and in the absence of magnetic field, using polarised neutrons on a mixed reflection, with nuclear and magnetic contributions. Qualitatively, we can say that the nuclear contribution, sensitive to the fluorine sites, offers a reference by which the apices and centres of the cell may be distinguished through the direction of neighbours, and that the magnetic contribution allows us to distinguish the orientation of the moments.

Quantitatively the 210 reflection displays, at around 20 K, nuclear and magnetic structure factors that are practically equal in modulus: for a given domain they add or subtract depending on the polarization of the incident beam. This leads to diffracted intensity caused exclusively by one of the domain types. The behavior of these 180° domains, during successive coolings through the Néel temperature, shows memory effects which are both surprising and poorly understood [33].

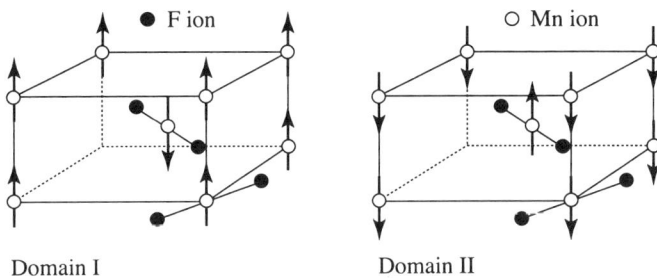

Figure 5.31 - Magnetic structure of the two 180° domains of MnF$_2$

The arrows represent magnetic moments carried by the Mn ions at the apices and the centre of the cell. These sites are distinguished through the orientation of the two neighbouring F ions.

Chirality domains, associated with right or left rotation of the spins in helimagnets, have also been observed in terbium [33]. In this case, the magnetic reflections ("satellite" reflections) are associated with the propagation vector of the helix, which characterises the spatial distribution of the orientation of magnetic moments. This propagation vector is in general incommensurate, i.e. not equal to a simple fraction of reciprocal lattice vectors. Depending on the polarization of the incident neutrons, the left or right helical domains alone are in diffraction position, and hence, once again, one obtains an image in a straightforward and unique fashion.

7. FROM THE MACROSCOPICALLY DEMAGNETISED STATE TO THE SATURATED STATE: MAGNETIZATION PROCESSES UNDER THE EFFECT OF AN EXTERNAL FIELD

Having described as precisely as possible the spontaneous magnetization configuration in ferromagnets, the time has now come to study the effect of an external magnetic field. Our objective in this section is to establish basic models for the behavior of macroscopically demagnetised systems submitted to a magnetic field of increasing strength. The system under consideration is initially demagnetised. The most efficient method of demagnetising a substance is to raise its temperature to above T_C before allowing it to cool down. During the cooling process domains of different orientations nucleate spontaneously, as soon as the exchange interactions impose a short-range ferromagnetic order. The arrangement of the domains is such that the macroscopic magnetization remains zero as a result of compensation by volume.

7.1. EFFECT OF A FIELD APPLIED PARALLEL TO THE EASY AXIS IN A UNIAXIAL SYSTEM: *MAGNETIZATION BY DOMAIN WALL DISPLACEMENT*

Here we consider uniaxial systems in which intrinsic anisotropy forces the moments to orient along the two opposite directions parallel to the preferred **c** axis.

7.1.1. Results of observation: growth of domains parallel to the field, at the expense of others

Let us look at a single-crystal film of yttrium-iron garnet, some micrometers thick, in which the strong uniaxial anisotropy exceeds the shape anisotropy and causes the magnetization within each domain to stay perpendicular to the film. We have already described the case of thin films, and their domains tangled into a maze (see § 4.1). The observation by optical microscopy of such a transparent sample enables the domains to be revealed by the Faraday effect (see § 6, and chap. 13). They appear light or dark respectively, depending on whether their magnetization points up or down.

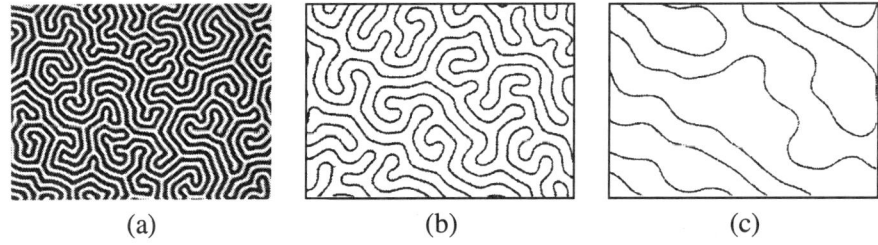

Figure 5.32 - Evolution of a domain structure under the influence of a magnetic field

The field is applied upward, normal to the surface of a ferrimagnetic platelet which shows a uniaxial anisotropy with the easy magnetization direction perpendicular to its plane. The domains appear light or dark according to whether their magnetization is directed up or down; (a) $H_0 = 0$ - (b) $H_0 = 4.2$ kA.m^{-1} - (c) $H_0 = 5.6$ kA.m^{-1}.

By applying a magnetic field normal to the platelet and directed upward, that is to say parallel to one of the directions of magnetization, and antiparallel to the other, the influence of this field may be observed. The domains whose magnetization is parallel to the field grow as the field increases, at the cost of the domains with antiparallel magnetization (fig. 5.32). The latter (the dark zone) disappear completely when the applied field reaches 7 kA.m^{-1} (~ 8.8 mT).

7.1.2. The law of behavior: the demagnetising field straight line

We now consider a uniaxial single crystal of ellipsoidal shape, for example of cobalt (which crystallises in a hexagonal structure with its principal six-fold axis **c** as its easy magnetization axis). In the absence of an applied field, although ferromagnetic at room temperature, the material appears demagnetised. It is subdivided into domains

5 - FERROMAGNETISM OF AN IDEAL SYSTEM

in which the spontaneous magnetization is oriented along the easy axis in one or other direction. This is known as a two-phase system (a phase is defined as the part of the material having its magnetization vector in a specific direction: in other words it consists of all the domains with the same orientation). In the absence of a resultant magnetization each of the two phases occupies the same volume.

In order to parametrise the growth of one phase at the expense of the other, in a magnetic field, we shall designate by u the proportion of the volume of the sample occupied by the domains oriented in the direction of the applied field (phase 1). The proportion by volume of reversed domains (phase 2) will, of course, be $1 - u$, and in zero field ($H_0 = 0$): $u = 1 - u = 1/2$ (fig. 5.33-a).

The effect of a very weak applied field is to cause the volume fraction of phase 1 to grow at the expense of that of phase 2 (fig. 5.33-b). A resulting magnetization M appears, which has the value:

$$M = [u - (1 - u)] M_s = (2u - 1) M_s \qquad (5.43)$$

A demagnetising field H_d is associated with this magnetization value M. Assuming that this field is equal to what it would be if the magnetization M of the sample were uniform, and assuming that the sample is ellipsoidal in shape, one obtains: $\mathbf{H_d} = -N\mathbf{M}$; $\mathbf{H_d}$ is antiparallel to $\mathbf{M}$.

The magnetostatic energy density of the crystal associated with the magnetization M is then written as:

$$E_{mag} = (\mu_0/2) N M^2 = (\mu_0/2) N (2u - 1)^2 M_s^2 \qquad (5.44)$$

whereas the interaction energy with the external field H_0, $E_H = -\mu_0 \mathbf{M} \cdot \mathbf{H_0}$ is:

$$E_H = -\mu_0 (2u - 1) M_s H_0 \qquad (5.45)$$

The variation of the macroscopic magnetization M in the field is determined by the variations in the above two energy terms. One can thus write the total energy of the system:

$$E_{tot} = E_0 + E_{mag} + E_H \qquad (5.46)$$

where E_0 contains all the terms independent of u (one neglects in particular the energy stored in the domain walls). For each value of H_0, the equilibrium state of the system is defined by $\partial E_{tot}/\partial u = 0$, which leads to $(2u - 1) M_s = H_0/N$, hence: $M = H_0/N$. It is to be noted that the internal field ($H = H_0 - NM$) remains zero throughout this process.

Thus, as long as the domains with magnetization anti-parallel to the applied field H_0 have not disappeared entirely (that is to say for $1/2 \leq u \leq 1$, fig. 5.33-b), the macroscopic magnetization M of the sample, measured along the direction of the field, varies linearly with the applied field, and the slope of the straight line $M(H_0)$ is equal to the reciprocal of the demagnetising field coefficient: for example, the slope of the straight line $M(H_0)$, $1/N$, is equal to 3 for a single crystalline sample of spherical shape ($N = 1/3$ in all directions). This straight line is referred to as the *"demagnetising field line"*. Its equation was already obtained in chapter 2 (§ 1.3.5 and fig. 2.18) without the mechanism for the appearance of magnetization being considered.

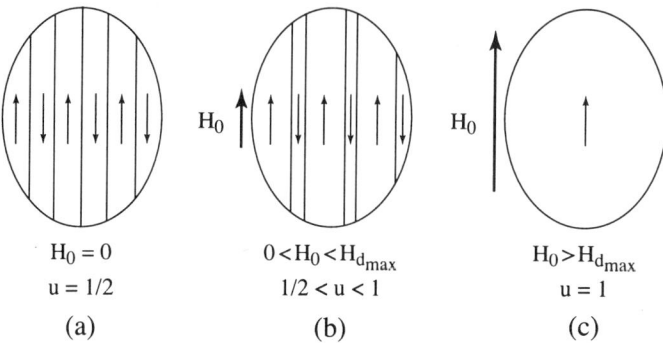

Figure 5.33 - Magnetization of a uniaxial crystal through the displacement of domain walls

This part of the magnetization curve does not depend on M_s, i.e. on the nature of the magnetic material being considered. Only the shape of the sample is relevant. It is only when saturation is reached ($u = 1$, fig. 5.33-c) that the nature of the ferromagnetic material shows up through the characteristic value of its spontaneous magnetization M_s. One has then $M = (2u-1) M_s = M_s$, and the applied field which corresponds to saturation magnetization has the value $N M_s$. As long as $H_0 < N M_s$ the variation in magnetization induced by the application of the field H_0 is such that it allows the internal field to be cancelled at any time. The field $N M_s$ is exactly equal (to within the sign) to the demagnetising field, H_{dmax}, in the saturated single-domain sample. When the applied field grows beyond this value, the magnetization M remains equal to M_s: this is the saturation phenomenon. In a case where a susceptibility, superimposed on the saturation, shows up, it reveals an effect of the field on the modulus of the individual moments.

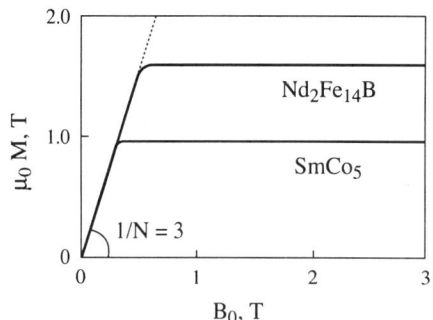

Figure 5.34 - Magnetization curves of $SmCo_5$ and $Nd_2Fe_{14}B$ single crystal spheres

The dotted line has the slope $1/N = 3$.

Figure 5.34 shows the magnetization curves measured along the easy magnetization axis of the uniaxial compounds $SmCo_5$ and $Nd_2Fe_{14}B$. The saturation values differentiate the two compounds, but the straight demagnetising field lines are identical because the measurements were performed on two single crystal samples both of which were spherical in shape.

7.1.3. The mechanism of wall displacement in perfect systems where exchange is dominant

So, when a ferromagnetic substance is subjected to a field oriented along one of its easy magnetization axes, the change in its macroscopic magnetization arises from the increase in size of the domains whose magnetization is parallel to the applied field, at the expense of the other domains. At the microscopic level this growth results from the displacement of walls between contiguous domains whose magnetization directions remain fixed.

The displacement of domain walls results from the fact that they are the preferred place for the existence of a torque due to the field and acting on the local magnetization. Subject to this torque and seeking to minimise their internal energy (width more or less constant), the walls between antiparallel domains react in three stages (fig. 5.35-a, -b and -c), which correspond to:
- slight displacement, and loss of symmetry causing an increase in wall energy γ,
- displacement by half an interatomic distance, and passing through the symmetrical configuration of maximum energy $\gamma + \Delta\gamma$: configuration with moments normal to the direction of anisotropy at the centre of the wall (see fig. 5.10-b, § 3.2),
- return to the initial symmetrical configuration after displacement of the wall by one interatomic distance a.

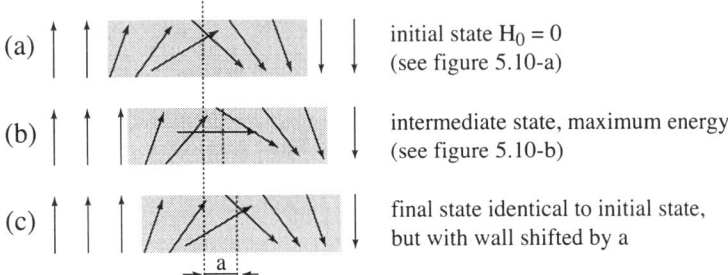

Figure 5.35 - Displacement mechanism for a 180° wall

If the system considered contains no defect, all stable equilibrium positions of the wall, distant by one lattice period in the propagation direction of the wall, correspond to the same value of its internal energy γ. In order to pass from one position of stable equilibrium to the next, it is necessary to cross the energy barrier $\Delta\gamma$ (fig. 5.36), with the help of the energy brought in by the applied field. This barrier is weaker the larger the width of the wall. One can show that $\Delta\gamma$ is proportional to $e^{-\delta/a}$ [34, 35], where a is the lattice parameter in the propagation direction. In magnetic systems at room temperature, exchange is dominant ($A_{ex} \gg K$), and the walls between domains are Bloch walls, many tens of interatomic distances wide (see fig. 5.11); the energy barrier which tends to oppose the displacement is extremely small; one can consider the walls to be completely free to react to the slightest pressure exerted on them by the field.

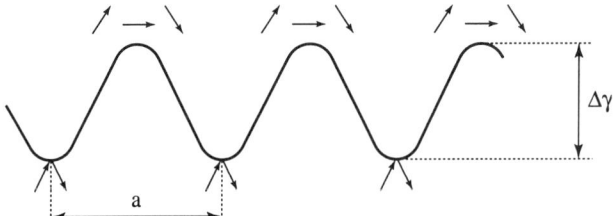

Figure 5.36 - Intrinsic wall energy barrier

Equivalent pressure exerted by the field on a domain wall

The local torque, applied to the moments within the wall, can be considered as the equivalent of a pressure acting on a larger scale over the entire wall.

When an induction field $B_0 = \mu_0 H_0$ is applied along one of the easy magnetization directions of a ferromagnetic system subdivided into domains, leading to the displacement Δx of a given wall (fig. 5.37), the variation of moment, Δm, following the displacement of this wall: $\Delta m = (\mathbf{M}_i - \mathbf{M}_j) \Delta x$ (per unit area of the wall), is limited to the volume swept by the wall. This corresponds to an energy variation: $\Delta E = -\Delta m \cdot \mathbf{B}_0 = -[(\mathbf{M}_i - \mathbf{M}_j) \cdot \mathbf{B}_0] \Delta x$ (per unit wall area).

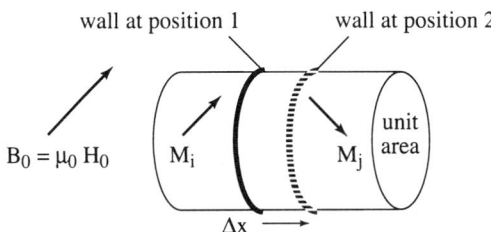

Figure 5.37 - Pressure exerted on a wall, after application of $B_0 // M_i$

Everything occurs as if the wall was submitted, by the field, to an equivalent pressure Π given by:

$$\Pi = (\mathbf{M}_i - \mathbf{M}_j) \cdot \mathbf{B}_0 \qquad (5.47)$$

Since the displacement of the wall causes no appreciable friction, it can continue for as long as a field exists inside the magnetic substance, i.e. until the internal field vanishes.

7.1.4. Wall displacement mechanism in perfect systems with very strong intrinsic anisotropy

In systems where the magnetocrystalline anisotropy energy is of the order of magnitude of the exchange energy, the walls are very narrow, and the energy stored within them satisfies relations (5.31) and (5.32) (see § 3.2 of this chapter). If the anisotropy is larger than the exchange, the wall does not extend over more than one interatomic distance. While the value of the intrinsic propagation field H_p is negligible for wide Bloch walls, as we saw in the preceding section, it grows progressively when the wall becomes narrower, i.e. as the ratio K/A_{ex} increases (fig. 5.38).

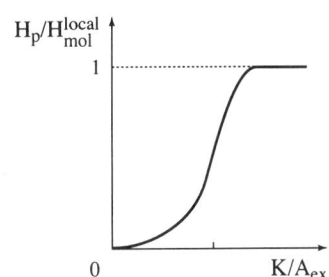

Figure 5.38 - Influence of parameters of the material (K, A_{ex}) on the propagation field [35]

The theoretical limiting value of the propagation field obtained for very large K/A_{ex} is that of the local molecular field, H_{mol}^{local}, acting at the centre of the wall. The magnetic moments oriented antiparallel to the field reverse one after the other; the intrinsic propagation field, H_p, does not depend on the anisotropy. The compound Dy_3Al_2, in which the ratio K/A_{ex} varies strongly with temperature, made it possible to demonstrate this behavior experimentally. At 30 K, the magnetization process for this compound is the classical one: the approach to saturation follows a simple demagnetising field straight line of slope equal to $1/N$. At 4.2 K, the field B_0 –still applied along the easy magnetization axis– produces almost no variation in the magnetization unless it reaches 2.1 T. Above this threshold value, which corresponds to an intrinsic propagation field, the magnetization increases abruptly until it reaches saturation (fig. 5.39).

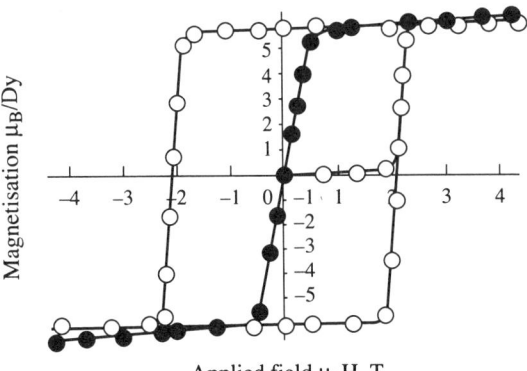

*Figure 5.39
Hysteresis loops
of the compound Dy_3Al_2
at 4.2 K (○) and 30 K (●) [4]*

7.2. EFFECT OF THE FIELD APPLIED NORMAL TO THE EASY AXIS IN A UNIAXIAL SYSTEM: *MAGNETIZATION THROUGH MOMENT ROTATION*

In this section, we continue to consider a uniaxial system, but now the field is applied normal to the easy magnetization axis **c**. The individual moments, belonging to one or the other type of domain, are subjected, by the externally applied field B_0, to a torque which tends to align them in the direction of the field (fig. 5.40). The field is applied in a symmetrical fashion to the two types of domains. Consequently there is no reason to differentiate between them.

At equilibrium, under the influence of $B_0 = \mu_0 H_0$, the magnetization of each of the two domains has rotated by the same angle θ with respect to the easy magnetization axis along which they were aligned before the application of the field (fig. 5.41).

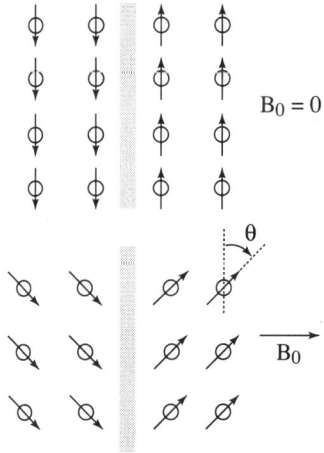
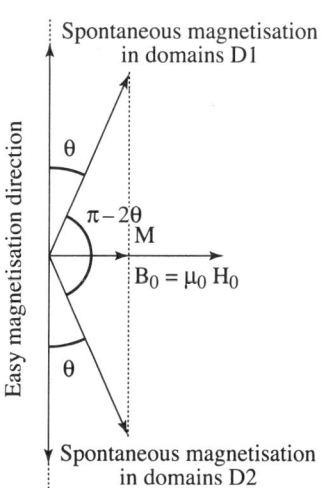

Figure 5.40 - The field applied perpendicular to the easy axis rotates the moments in each domain

Figure 5.41 - The magnetizations of the two types of domains evolve symmetrically

A macroscopic magnetization M, such that $M = M_S \sin \theta$, results in the direction of the applied field. The components along $\pm c$ of the magnetization in each of the domains compensate macroscopically because the two domains –neither of which is preferred– each occupy one half of the volume of the sample. In the presence of the external field, the total energy F_t of the system contains three terms in which the angle θ appears (the others are grouped into a single term F_0): the anisotropy energy term $F_A = K_1 \sin^2 \theta$, the magnetostatic energy term $F_m = (\mu_0/2) N M_s^2 \sin^2 \theta$ (where N is the demagnetising field coefficient along H_0), and the term representing interaction with the field $F_B = -M_s B_0 \sin \theta$. The equilibrium state of the system is determined by $dF_t/d\theta = 0$, giving:

$$(2 K_1 + \mu_0 N M_s^2) \sin \theta \cos \theta - M_s B_0 \cos \theta = 0,$$

which leads to the equilibrium value: $\theta = \sin^{-1}[M_s B_0/(2K_1 + \mu_0 N M_s^2)]$ which corresponds to the energy minimum, and since $M = M_s \sin\theta$, this means that:

$$M = \frac{M_s H_0}{(2K_1/\mu_0 M_s) + NM_s} \qquad (5.48)$$

The graph $M(H_0)$ representing this equation is displayed in figure 5.42. In weak fields the variation of M with H_0 is linear with a slope:

$$dM/dH_0 = (N + 2K_1/\mu_0 M_s^2)^{-1} \qquad (5.49)$$

The magnetization reaches its saturation value, M_s, when the field reaches the critical value H_s defined by the equation:

$$H_s = NM_s + 2K_1/\mu_0 M_s \qquad (5.50)$$

For $H \geq H_s$, the equilibrium state of the system corresponds to $\cos\theta = 0$, i.e. $\theta = \pi/2$, the magnetization then remains saturated: $M = M_s$.

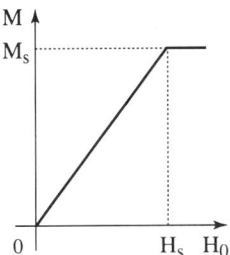

Figure 5.42 - The threshold field H_s marks the end of the rotation

Thus the threshold field, H_s, corresponds to the value that the external field must reach in order to align all magnetic moments in the material in the proper direction. This field must, at the same time, compensate for the demagnetising field arising from the appearance of macroscopic magnetization, and create a torque equal to the restoring anisotropy torque. It is often useful to consider the anisotropy through a fictitious equivalent field H_A, which corresponds to the internal field along the hard direction of magnetization, and which counterbalances the torque resulting from the anisotropy. This fictitious field is called the *anisotropy field*, and its value is (see chap. 3, § 2.4.1.1):

$$H_A = 2K_1/\mu_0 M_s \qquad (5.51)$$

The mechanism through which the macroscopic magnetization appears and develops in this applied field configuration is the simple rotation of individual magnetic moments. It is a reversible mechanism.

7.3. FIELD APPLIED AT AN ANGLE TO THE EASY AXIS

When the field applied to a demagnetised uniaxial single crystal makes an angle φ other than 0° or 90° with the easy axis (see fig. 5.43), the mechanisms of wall displacement and of magnetization rotation both intervene in the magnetization process. The field component along the easy axis Oz gives rise to the displacement of walls, resulting in an increase in the volume fraction (u) of the domains whose magnetization is approximately along this component of the field.

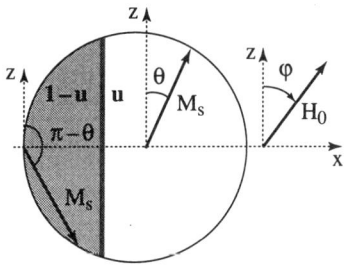

Figure 5.43 - Field applied in an arbitrary direction

The displacement continues until the internal field along the easy axis is cancelled. One then has: $H_0 \cos \varphi = (2u-1) N M_s \cos \theta$. The internal field is thus directed along Ox, and the magnetic moments in one domain and the other domain are oriented symmetrically with respect to Ox. From this one deduces:

$M_z = [u - (1-u)] M_s \cos \theta = (2u-1) M_s \cos \theta$ and $M_x = [u + (1-u)] M_s \sin \theta = M_s \sin \theta$.

For a given value of $\mathbf{H}_0$ ($H_0 \cos \varphi$, $H_0 \sin \varphi$), the energy density of the system is written:

$$E = K_1 \sin^2 \theta - \mu_0 M_s H_0 \sin \theta \sin \varphi - \mu_0 (2u-1) M_s H_0 \cos \theta \cos \varphi \\ + (\mu_0/2) N M_s^2 \sin^2 \theta + (\mu_0/2) N (2u-1)^2 M_s^2 \cos^2 \theta$$

and equilibrium corresponds to the value of θ for which $dE/d\theta = 0$. For $\theta \neq \pi/2$, the combination of equations $H_z = 0$ and $dE/d\theta = 0$ leads to:

$$\sin \theta = H_0 \sin \varphi / [N M_s + (2 K_1 / \mu_0 M_s)].$$

The macroscopic magnetization measured in the direction of the applied field, M_H, is obtained from: $M_H = \mathbf{M} \cdot \mathbf{H}_0 / H_0 = M_s [\sin \theta \sin \varphi + (2u-1) \cos \theta \cos \varphi]$.

By replacing $\sin \theta$, and u by their expressions found above, one has (as long as $u \leq 1$):

$$M_H = H_0 \frac{\sin^2 \varphi}{N + (2K_1/\mu_0 M_s^2)} + \frac{\cos^2 \varphi}{N} \quad (5.52)$$

Up to the field H_s for which $u = 1$, the characteristic $M(H_0)$ is a straight line whose slope α has a value given by:

$$\alpha = \frac{\sin^2 \varphi}{N + (2K_1/\mu_0 M_s^2)} + \frac{\cos^2 \varphi}{N} \quad (5.53)$$

♦ for $\varphi = 0$ (magnetic field parallel to the easy axis), one finds again the slope $1/N$,

♦ for $\varphi = \pi/2$ (field normal to the direction of easy magnetization), one again finds the result obtained earlier: $\alpha_{(\pi/2)} = [N + (2 K_1 / \mu_0 M_s^2)]^{-1}$,

5 - FERROMAGNETISM OF AN IDEAL SYSTEM

• in the general case, where $0 < \varphi < \pi/2$, the slope α lies between these two values (see fig. 5.44). The threshold field, H_s, corresponds to the point where the sample becomes single domain, the process of magnetization by domain wall displacement having ended (u = 1). Substituting u = 1 into the relation given above one obtains the component of H_s along Oz, $H_s \cos \varphi = N M_s \cos \theta$,

from which: $H_s = \dfrac{NM_s}{\cos \varphi} \cos \theta = \dfrac{NM_s}{\cos \varphi} \sqrt{1 - \dfrac{H_0^2 \sin^2 \varphi}{[NM_s + (2K_1/\mu_0 M_s)]^2}}$,

where $H_0 = H_s$,

thus: $H_s = M_s \Big/ \sqrt{\dfrac{\cos^2 \varphi}{N^2} + \dfrac{\sin^2 \varphi}{[N + (2K_1/\mu_0 M_s^2)]^2}}$ (5.54)

For $H_0 = H_s$, the magnetization inside the sample is not co-linear with the applied field.

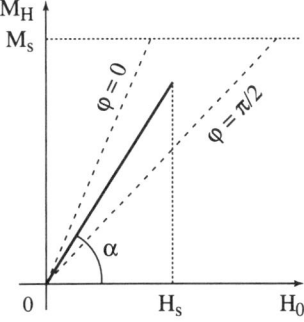

Figure 5.44 - Magnetization process $(H_0 < H_s)$

When the applied field H_0 grows beyond H_s, the magnetization, now equal to M_s, rotates until it is aligned along $\mathbf{H}_0$. The sole magnetization mechanism which then intervenes is the rotation of individual moments. The total energy of the system contains two terms in which the orientation θ of M_s occurs: the anisotropy term and the interaction with the applied field (if the sample is not spherical, the interaction with the demagnetising field gives rise to an anisotropy term, which is also a function of θ: this effect is ignored here). This energy is written as:

$$F_{tot} = F_0' + (\mu_0/2) N M_s^2 + K_1 \sin^2 \theta - \mu_0 M_s H_0 \cos(\varphi - \theta)$$

or, by grouping together all the terms independent of θ into F_0:

$$F_{tot} = F_0 + K_1 \sin^2 \theta - \mu_0 M_s H_0 \cos(\varphi - \theta).$$

Equilibrium requires $\partial F_t/\partial \theta = 0$, giving $K_1 \sin 2\theta - \mu_0 M_s H_0 \sin(\varphi - \theta) = 0$, from which:

$$H_0 = K_1 \sin 2\theta / \mu_0 M_s \sin(\varphi - \theta) \quad (5.55)$$

This means that $\theta \to \varphi$ when $H_0 \to \infty$. The law of approach to saturation is an asymptotic law (fig. 5.45) contrary to the cases already considered (field applied parallel or normal to the preferred axis).

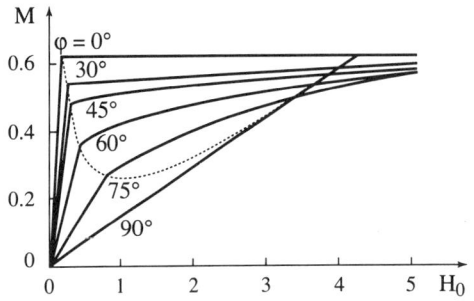

Figure 5.45
The magnetization process for different orientations of the magnetic field

The dotted curve represents the threshold points for which the sample becomes single domain ($H_0 = H_s$).

7.4. CUBIC SINGLE CRYSTAL SYSTEMS (NÉEL'S PHASE RULE)

Under the effect of the applied field the process of macroscopic magnetization of cubic single crystals brings into play the two mechanisms described above: domain wall displacement and rotation of magnetization in the domains. In general the two mechanisms intervene one after the other, because the domain wall displacement mechanisms in weak fields allow the strict cancellation of the internal field.

In a very weak field the wall displacement mechanism leads to the growth of those domains whose magnetization direction is nearest to the field, to the expense of the others. The mechanism of magnetization rotation intervenes when the displacement of the wall no longer allows cancellation of the internal field in all directions. This process is summarised in figure 5.46.

In a cubic system in zero field, the number of phases (domains oriented differently) present depends on the symmetry of the easy magnetization axes: 6 phases correspond to three four-fold axes, 8 phases to the 4 three-fold axes and 12 phases to the 6 two-fold axes. The volumes occupied by these phases are equivalent in the demagnetised state.

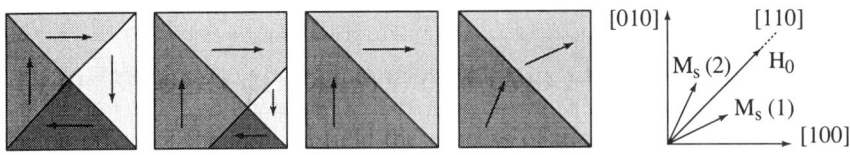

Figure 5.46 - The magnetization process for a single crystal with cubic symmetry in which the four-fold axes are the easy magnetization directions

After displacement of 90° domain walls, only two domains remain. As the field continues to grow, the moments rotate toward the [110] direction along which the field is applied.

5 - FERROMAGNETISM OF AN IDEAL SYSTEM

In the most general case, where the field direction is not a symmetry direction for the crystal, only one of the phases present is favoured. If the field $\mathbf{H}_0$ is applied along a symmetry direction of the crystal, several phases may be favoured. For example, consider a single crystal of a magnetic material belonging to the cubic system, and let us assume that the <100> type axes are easy magnetization directions (as in the case of iron). In such a system, Néel [36] expressed the above remarks in the following way:

- When the internal field is zero, the six phases can coexist in equilibrium with one another.
- If the internal field is not zero, let its direction be defined by the direction cosines p, q and r. For $p > q \geq r$, equilibrium between phases is impossible: the crystal will have only one phase.
- In order for equilibrium to occur between two phases, the direction of the internal field must be such that two of its direction cosines be equal, the third being smaller than the others.
- In order for equilibrium to be established between three phases, the internal field must be parallel to a threefold axis.

In cubic systems, one can distinguish four magnetization modes according to the number of phases present and the orientation of the internal field (Néel's phase rules). Table 5.1 summarises the characteristics of these modes (numbered from I to IV). In general, the magnetization curve of a single crystal subdivides into several parts which join at angular points. Each part corresponds to a different magnetization mode (fig. 5.47).

Table 5.1 - Phase rule for a crystal with cubic symmetry whose easy axes are along the fourfold axes

Magnetization mode	Internal field Intensity	Internal field Direction cosines	Number of phases
I	$H_i = 0$		6
II	$H_i > 0$	$p = q = r$	3
III	$H_i > 0$	$p = q \geq r$	2
IV	$H_i > 0$	$p > q \geq r$	1

When all six phases are present the initial magnetization state of the crystal belongs to mode I. The internal field is zero, and the demagnetising field just compensates the applied field. This condition determines the magnetization law: $M = H_0 / N$ (in the case of an ellipsoid), which recalls to mind the fact that in weak fields the magnetization is both parallel and proportional to the field. Its variation is independent of the orientation of the external field. It depends only on the external shape of the sample. This magnetization mode, during which the relative volumes of different phases are modified by wall displacements, lasts as long as there are more than three phases present. The magnetization M* reached at the end of mode I is given by:

$$M^* = M_s/(l+m+n) \tag{5.56}$$

if l, m and n are the direction cosines of the applied field H_0 [37].

One then goes on to mode II, or directly to III, or even to mode IV. If the field is applied along a threefold axis, mode II starts for $M^* = M_s/\sqrt{3}$, while if the field is applied along a twofold axis, mode III begins at $M^* = M_s/\sqrt{2}$ (fig. 5.47). In all cases, the magnetization obtained after the field has been brought to zero, from the saturation state, without any wall being created, is called the *remanent magnetization*, M_r.

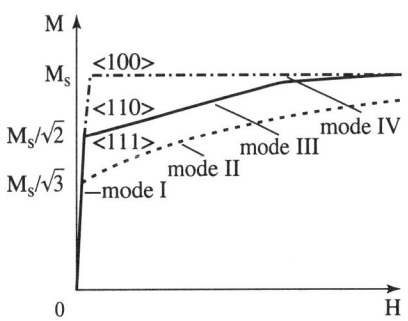

Figure 5.47 - Magnetization curves of a crystal with cubic symmetry

7.5. MAGNETIZATION PROCESS IN POLYCRYSTALLINE SYSTEMS

A polycrystalline material is made up of a collection of small, randomly oriented single crystals. If one assumes the crystallites to be independent, the expected properties are directly deduced from those of single crystals through statistical combination. Thus the remanent magnetization M_r can, theoretically, be linked to the spontaneous magnetization M_s. The values obtained depend on the crystal structure and on the type of easy magnetization axes. In the case of cubic systems [38], $M_r/M_s = 0.83$ if the fourfold axes are the easy axes, and $M_r/M_s = 0.86$ if they are the threefold axes. In the case of uniaxial systems, where the preferential axis is the easy magnetization axis, $M_r/M_s = 0.5$.

The magnetization law of a cubic polycrystal (fig. 5.48) merges with the demagnetising field straight line as long as the wall displacement mechanism alone is brought into operation, i.e. until $M = M_r$. Following this the theoretical magnetization law is the average of the magnetization curves of the crystal for different orientations of the field with respect to its crystallographic axes. The magnetization curve of a uniaxial polycrystal is also the sum of the magnetization curves of the constituent crystals, oriented randomly with respect to the field.

In contrast to the case of cubic polycrystals, the magnetization law, in this case, does not result in the appearance of the demagnetising field straight line, which is only a function of the shape of the sample (fig. 5.49).

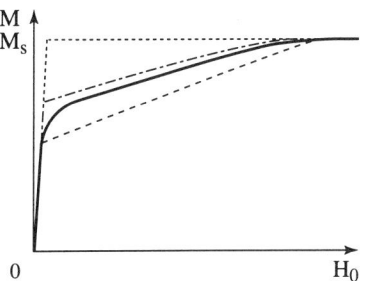
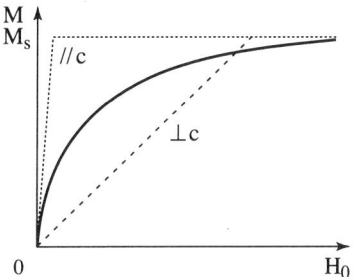

Figure 5.48 - Magnetization curve of a polycrystalline material with cubic symmetry

Figure 5.49 - Magnetization curve of a polycrystalline material with uniaxial symmetry

The curves referring to the single crystal are shown by dotted lines.

For all the crystallites whose easy axis is not parallel to the field (see § 7.2 and 7.3) the mechanism of magnetization rotation inside the domains comes into play from the start. In very high fields, the law of approach to saturation in polycrystals is, in principle, determined by the reaction of the anisotropy to the torque exerted by the applied field. This torque is equal to $M_s H \sin \psi$, if ψ is the angle between the field and the magnetization M_s of a single-domain crystallite (fig. 5.50).

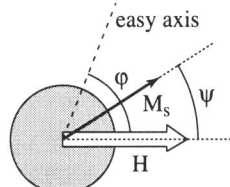

Figure 5.50 - Single-domain crystallite subjected to an internal field H

Near saturation, ψ is small. Hence one can write: $M_s H \psi = \Gamma_A$, so that $\psi = \Gamma_A / M_s H$ where Γ_A, the restoring torque due to the anisotropy, is given by $\Gamma_A = -\left(\dfrac{\partial E_A}{\partial \psi}\right)_{\psi \approx 0}$.

For a single crystallite, per unit volume, $E_A = K_1 \sin^2(\varphi - \psi)$ where φ is the angle between the field and the easy axis of the crystallite (fig. 5.50), and

$$\dfrac{\partial E_A}{\partial \psi} = -2K_1 \sin(\varphi - \psi) \cos(\varphi - \psi)$$

$$= -2K_1 (\sin \varphi \cos \psi - \sin \psi \cos \varphi)(\cos \varphi \cos \psi - \sin \varphi \sin \psi).$$

When ψ is small: $\left(\dfrac{\partial E_A}{\partial \psi}\right)_{\psi \approx 0} \approx -K_1 \sin 2\varphi \cos^2 \psi$. Thus for a polycrystal $\Gamma_A \sim \cos^2 \psi \approx 1$ and can be considered as constant for ψ small. On the other hand $M_H = M_s \cos \psi = M_s (1 - \psi^2/2 + ...)$, whence: $M_H = M_s (1 - b/H^2 + ...)$ with $b = \Gamma_A^2 / 2 M_s^2$. The existence of such a term has been verified, and the values of Γ_A deduced are of the expected order of magnitude [39].

Theoretically, the expansion of M_H cannot contain terms involving a/H, because the presence of this term leads to an infinite magnetization energy, as we can see by writing this energy as:

$$W = \mu_0 \int_M^{M_s} H dM = \mu_0 \int_H^{\infty} H \frac{dM}{dH} dH = \mu_0 \int_H^{\infty} M_s \left(\frac{a}{H} + 2\frac{b}{H^2} + \cdots \right) dH,$$

or:
$$W = M_s \left(\left| a \log H \right|_H^{\infty} - \left| \frac{2b}{H} \right|_H^{\infty} + \cdots \right) \to \infty.$$

However, experiment shows that the law of approach to saturation is often described by a function of the field of the form:

$$M_H = M_s[1 - (a/H) - b/H^2] + \chi H \tag{5.57}$$

The term χH is often said to correspond to "superimposed susceptibility", arising from the increase in the modulus of M_s, for reasons related to the quantum mechanical definition of moments (chap. 8). The term involving (a/H) has an effect only in intermediate fields. Néel attributes it to the effects of defects [40] inside the crystallites and even more to grain boundaries.

8. MAGNETIZATION REVERSAL FROM THE SATURATED STATE AND COERCIVITY

Here we analyse the behavior of a ferromagnetic system which, after saturation in a very high field, is subjected to a reverse field that makes, with the direction of remanent magnetization, an angle φ that lies between $\pi/2$, and π (fig. 5.51), so that the field tends to cause the reversal of magnetization.

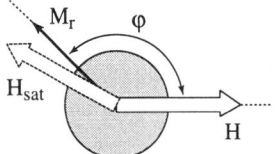

Figure 5.51 - Application of a field in the half-space opposite to the initial field H_{sat} on a magnetised ferromagnet ($\pi/2 \leq \varphi \leq \pi$)

The study of the mechanism of magnetization reversal allows one to understand the properties of many types of magnetic materials, notably the permanent magnets and magnetic memory devices. As a prelude to this study, let us recall that when the initial saturation field is progressively reduced, the local magnetization reorients in a reversible manner towards the nearest easy magnetization direction. For a single crystal magnetised along an easy magnetization direction, the remanent magnetization M_r obtained in zero internal field is equal to the spontaneous magnetization, M_s. More generally, the ratio M_r/M_s is given by the relations established in section 7.5. As soon as the external field becomes less than the demagnetising field, the internal field in the material is in the direction opposite to that of the magnetization. The magnetization reversal mechanisms then set in.

8.1. REVERSIBILITY OR IRREVERSIBILITY?

Are the phenomena described earlier reversible or not? One might expect the first effect of the appearance of an internal field opposite to the magnetization to be the reversible displacement of the walls in the opposite direction, so as to lead to a magnetization configuration with lower energy. But this requires domain walls to exist, which is not the case at saturation. Thus, the first step consists of creating a small *nucleus* of reverse magnetization, and naturally the domain wall associated with it. In section 8.3.1 we will show that this *nucleation* brings, in the case of magnetic systems without defects, a rise in overall energy higher than that involved in the processes envisaged below (§ 8.2 and 8.3). Thus the behavior of a ferromagnetic system brought to saturation becomes irreversible when the internal field changes sign.

8.2. REVERSAL BY UNIFORM COLLECTIVE ROTATION OF MOMENTS (ROTATION OF SATURATION MAGNETIZATION): THE STONER-WOHLFARTH MODEL [41]

Because of its simplicity, the Stoner-Wohlfarth model serves, in general, as a reference for other possible theoretical descriptions of magnetization reversal. It is assumed that any variation in magnetization can result only from a rotation, in unison, of magnetic moments. Note that, in this section, the remanent magnetization (considered as the starting point) is defined as the magnetization in zero *applied* field (that is to say in a slight reverse *internal* field).

8.2.1. Presentation and formulation of the model system

Consider a ferromagnetic single crystal of uniaxial symmetry, whose **c** axis is the easy magnetization axis. Let us assume that a single crystal is cut from it in the shape of an prolate ellipsoïd of revolution, whose long axis Oz is co-linear with **c**: magneto-crystalline anisotropy and shape anisotropy thus act in the same direction. The orientation of the Oz axis is chosen in such a way that (Oz, $\mathbf{H}_{sat}$) ≤ $\pi/2$ (fig. 5.52).

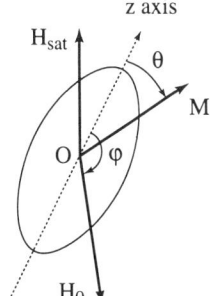

Figure 5.52 - Ellipsoid-shaped uniaxial single crystal

After it has been saturated, the magnetization returns along Oz: it is positive and its value is equal to that of the spontaneous magnetization, M_s. The study of magnetization reversal is reduced to defining the behavior of magnetization under the action of an increasing external magnetic field, H_0, making with it an initially obtuse angle. The conventions used to denote the angles are those of figure 5.52.

The equilibrium of the system brings into play three energy densities:

- the magnetocrystalline anisotropy energy (here limited to second order):

$$E_A = K_1 \sin^2 \theta \qquad (5.58)$$

- the magnetostatic (or demagnetising field) energy which is at the origin of the shape anisotropy (see eq. 5.40):

$$E_{dip} = \frac{1}{2}\mu_0 \left[N_{//} M_s^2 + (N_\perp - N_{//}) M_s^2 \sin^2 \theta \right] \qquad (5.59)$$

where $N_\perp$ and $N_{//}$ are the demagnetising field coefficients relative to the directions perpendicular and parallel to the Oz axis, respectively. When $\theta \neq 0$, these two anisotropy terms raise the energy of the system: they oppose the rotation of magnetization by forming an energy barrier.

- the interaction energy with the applied field:

$$E_H = -\mu_0 M_s H_0 \cos(\varphi - \theta) \qquad (5.60)$$

The closer the moments to the direction of H_0, the more this term lowers the energy of the system.

The stable equilibrium conditions, corresponding to a given configuration $[\varphi, \theta]$, is obtained by minimising the total energy $E = E_A + E_D + E_H$. They are therefore solutions of the following system of equations:

$$dE/d\theta = \mu_0 M_s H_0 \sin(\theta - \varphi) + K'\sin 2\theta = 0 \qquad (5.61)$$

$$d^2E/d\theta^2 = \mu_0 M_s H_0 \cos(\theta - \varphi) + 2K'\cos 2\theta > 0 \qquad (5.62)$$

where $K' = K_1 + K_{sh} = K_1 + \mu_0 (N_\perp - N_{//}) M_s^2/2$ defines an equivalent total anisotropy, the sum of intrinsic anisotropy (parameter K_1) and of shape anisotropy (parameter $K_{sh} = \mu_0 (N_\perp - N_{//}) M_s^2/2$).

8.2.2. Field applied antiparallel to M_r

The magnetization law

When H_0 is applied antiparallel to M_r, that is $\varphi = \pi$, with H_0 assumed to be positive, equations (5.61) and (5.62) become: $dE/d\theta = 2K'\sin\theta\cos\theta - \mu_0 M_s H_0 \sin\theta = 0$ and $d^2E/d\theta^2 = 2K'\cos 2\theta - \mu_0 M_s H_0 \cos\theta > 0$. $dE/d\theta$ is zero for $\sin\theta = 0$ or $\cos\theta = \mu_0 M_s H_0 / 2K'$. θ being confined between zero and π, three solutions are possible $\theta = 0$, $\theta = \pi$ and $\theta = \cos^{-1}(\mu_0 M_s H_0 / 2K')$.

For $\theta = \pi$, $d^2E/d\theta^2$ is positive. Alignment of magnetization along the field, naturally, always corresponds to a minimum of energy.

As long as $H_0 < 2K'/\mu_0 M_s$, $d^2E/d\theta^2 > 0$ for the solution $\theta = 0$, and $d^2E/d\theta^2 < 0$ for the solution $\theta = \cos^{-1}(\mu_0 M_s H_0/2K')$ which, therefore, corresponds to a maximum of energy. Thus the initial configuration, with magnetization antiparallel to the field, is a metastable state of minimum energy. The magnetization can only flip over and align along the field when H_0 reaches a critical value $2K'/\mu_0 M_s$ above which the configuration defined by $\theta = 0$ is no longer of minimum energy.

These results, plotted in the form of a classical $M(H_0)$ curve, by considering H_0 as positive or negative according to whether it is applied parallel or antiparallel to the initial M_s, give the portion of the hysteresis loop shown in figure 5.54. The critical value, $2K'/\mu_0 M_s$, of H_0 for which the magnetization flips over is that of the anisotropy field defined by relation (5.51), in which K is replaced by K', representing the total equivalent anisotropy defined in the previous section.

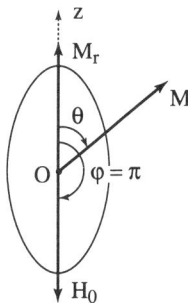
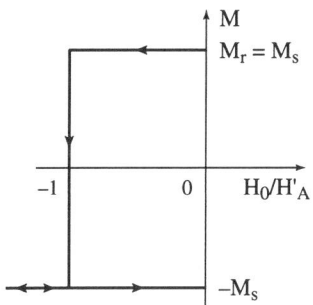

Figure 5.53 - Field opposing remanent magnetization

Figure 5.54 - Portion of resultant hysteresis loop

Analysis and remarks: effect of the field on the anisotropy energy barrier

In the present configuration, the field is always either parallel, or antiparallel to the magnetization; the torque due to the field is zero; it is normal for the magnetization not to rotate at all before tipping over.

A representation of the energy of the system as a function of the orientation of M_s, for various values of the field applied antiparallel to M_r, clarifies the behavior of this system (fig. 5.55).

In zero field the two directions parallel to the Oz ($\theta = 0$, and $\theta = \pi$) axis correspond to equally stable states: minima with the same energy E_0, defining the two directions of easy magnetization. Between these two minima there exists a state of maximum energy E_1, which defines the energy barrier between the stable states. The position of this barrier is given by $dE/d\theta = 0$ and by:

$$-\mu_0 M_s H_0 \cos\theta + 2K'\cos 2\theta < 0 \tag{5.63}$$

With $H_0 = 0$, one obtains $\theta = \theta_{B0} = \pi/2$. The position of the energy barrier between the stable states corresponds here to an orientation of magnetization M_s normal to Oz, that is to say to the maximum of anisotropy energies.

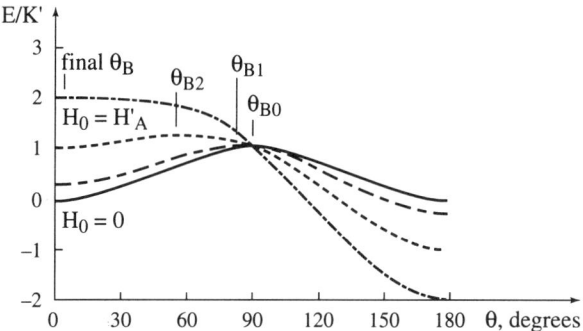

Figure 5.55 - Evolution of the energy barrier with the applied field

The height of the energy barrier corresponds to the difference $\Delta E = E_1 - E_0$ between the energy of the maximum and the energy of the stable state occupied: $E_1 = (\mu_0/2) N_{//} M_s^2 + K'$ and $E_0 = (\mu_0/2) N_{//} M_s^2$, giving: $\Delta E = K'$; this indeed refers to an energy barrier provided by the anisotropy.

When H_0 increases in magnitude from 0 to $2K'/\mu_0 M_s$, the position of the maximum of the barrier, marked by an arrow on each curve of figure 5.55, evolves from $\theta_{B0} = \pi/2$ for $H_0 = 0$ to $\theta_B = 0$ for $H_0 = H'_A = 2K'/\mu_0 M_s$. At the same time, the height of the energy barrier, ΔE, decreases from its original value to zero (see fig. 5.55). The state defined by $\theta = \pi$ also becomes more and more stable since its energy is $E_0 - \mu_0 M_s H_0$, so that the energy of the occupied stable state increases as $E_0 + \mu_0 M_s H_0$.

Magnetization reversal arises from the disappearance of the energy barrier between the occupied state, which becomes unstable (maximum energy), and the stable state corresponding to $\theta = \pi$. This occurs when $H_0 = H'_A$. In the present process, where magnetization reversal occurs by uniform collective rotation of the individual magnetic moments, each moment as well as the resulting magnetization must go through the direction of difficult magnetization. We thus understand why the reversal field (the coercive field H_C) was found equal to the anisotropy field.

8.2.3. Inverse field applied at an angle to the direction of M_r

In this case, $\pi/2 \leq \varphi < \pi$ (fig. 5.56). Solving equation (5.61) is a little more complicated than when φ is equal to π (or when it is exactly equal to $\pi/2$). This calculation was carried out for several values of φ by Stoner and Wohlfarth [41].

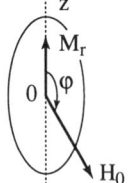

Figure 5.56 - Oblique inverse field

It leads to the M (H₀) behaviors shown in figure 5.57. This time, the field H_0 exerts a torque on the magnetization M_s of the system. This torque gets stronger –for a given value of H_0– the closer the angle φ, through which the field is applied, gets to $\pi/2$. The progressive decrease in the measured magnetization, M, results from the rotation of M_s under the effect of this torque. The flipping over of the magnetization is always a sudden process. It occurs for values H_R of H_0 which depend on the direction of the applied field (fig. 5.58), but which remain less than the values H'_A obtained for $\varphi = \pi$ or $\pi/2$. For $\varphi = 3\pi/4$, the value of the reversal field is minimum and is equal to $H'_A/2$.

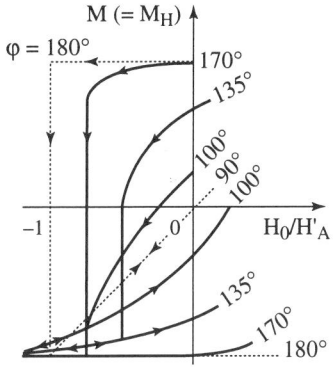

Figure 5.57 - Behavior of magnetization vs field

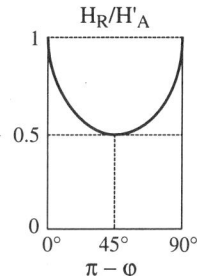

Figure 5.58
Variation of H_R/H'_A
as a function of $\pi - \varphi$

If one sets $\psi = \pi - \varphi$, the value of the reversal field H_R as a function of the angle of the applied field H_0 is given by:

$$H_R = H'_A (\sin^{2/3} \psi + \cos^{2/3} \psi)^{-3/2} \quad (5.64)$$

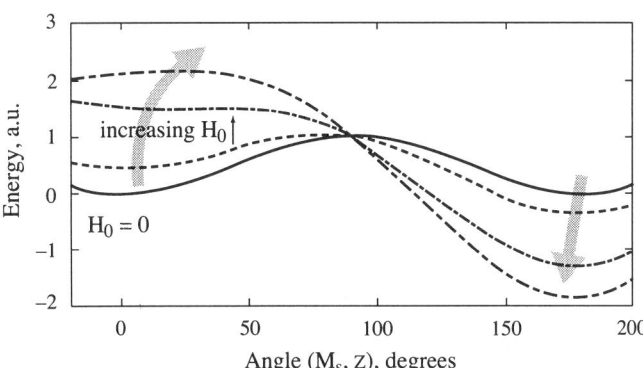

Figure 5.59 - Evolution of the energy of the system
with the orientation of M_s with respect to Oz

The different curves correspond to several values of the applied field between 0 and H_R for the same angle $\varphi \approx 170°$. The evolution of the energy minima is shown in grey.

The variation of the total magnetic energy of the system as a function of the possible orientation of M_s, for several values of H_0, is shown in figure 5.59. As in the case where $\varphi = \pi$, the positions of the occupied stable state and of the peak of the barrier shift towards each other as H_0 increases.

Remark - *Assume one applies to a collection of crystallites of the preceding type, whose c axes are distributed over all directions of space in a totally random manner, firstly a saturation field which orients all the moments of the crystallites in the same half-space, and then a field increasing in the direction opposite to the saturation field (hysteresis loop procedure). The theoretical result expected according to the Stoner-Wohlfarth model is shown in figure 5.60: the resulting magnetization cancels out in a field equal to 0.49 H'_A, under the sole effect of reversible rotations of the moments of the grains, while the inflection point corresponding to the maximum of irreversible susceptibility occurs for 0.5 H'_A.*

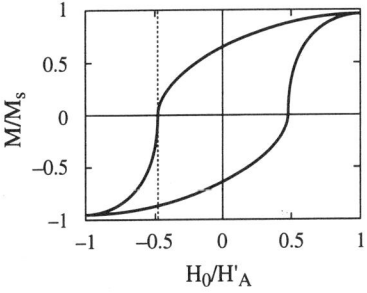

Figure 5.60 - Hysteresis loop predicted in the Stoner-Wohlfarth model

Note that the loop intercepts the horizontal axis for $H_0 = -0.49 H'_A$.

8.3. ARE OTHER MODES OF COLLECTIVE REVERSAL POSSIBLE?

The Stoner-Wohlfarth model defines a reference for the energy barrier that corresponds to the *collective and uniform* rotation of the moments of the system. It remains to explore whether collective but *non-uniform*, processes might allow an easier reversal. For that we again consider the model case of an isolated single crystal grain, originating from a perfect ferromagnetic material, in the two limiting cases of a very strong, and then vanishing magnetocrystalline anisotropy.

8.3.1. Systems with strong magnetocrystalline anisotropy: Brown's inequality

Whether the rotation process of moments in the system takes place uniformly or *via* more complex configurations, it is always the same anisotropy energy that comes into play. It acts on every individual moment which, in the course of magnetization reversal, is led to deviate by the same angle, whatever process is taking place. However, exchange energy associated with the non-parallel state of the moments, will be involved in a process of non-uniform rotation, but not if the rotation occurs in a uniform manner.

5 - FERROMAGNETISM OF AN IDEAL SYSTEM

Thus, in the case of systems with strong magnetocrystalline anisotropy, any collective, non-uniform mode of magnetization reversal will cost more than coherent rotation. This observation led Brown to state that the theoretical reversal field is:

$$H_R \geq 2 K_1 / \mu_0 M_s \qquad (5.65)$$

in an ideal homogeneous system governed by a uniaxial anisotropy of magnetocrystalline origin [8]. In the following chapter we will see that this result is not observed in real systems where defects play a fundamental role.

8.3.2. Systems without magnetocrystalline anisotropy

Let us assume that the single crystal grain considered is a prolate ellipsoid, having no magnetocrystalline anisotropy ($K_1 = 0$). It has a shape anisotropy defined by equation (5.59). In zero field the magnetization is oriented along the major axis **c** of the ellipsoid.

When this grain is placed in an increasing reverse field, the coercive field associated with uniform moment rotation is equal to the anisotropy field, of the form $H_A^{sh} = (N_\perp - N_{//}) M_s$, where $N_{//}$ and $N_\perp$ are the demagnetising field coefficients defined above (eq. 2.48 and 5.43). Within this limit, the earlier discussion of the Stoner-Wohlfarth model applies in full.

But one can imagine processes in which the variation of magnetization in an applied reverse field results from the formation of non-uniform configurations of moments. For every configuration in which the total magnetization decreases while remaining parallel to Oz, the demagnetising field energy steadily decreases until the magnetization vanishes. Thus it cannot form a barrier. In contrast, an increase in exchange energy is associated with the formation of non-uniform configurations of the moments. Therefore it is the exchange which tends to oppose the reversal.

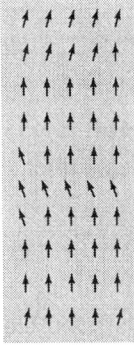

Figure 5.61 - Magnetization mode through buckling

Different modes of non-uniform reversal have been considered: *curling* of the local magnetization (fig. 5.21), *buckling* (fig. 5.61), etc. The reversal mode actually brought about will be that which corresponds to the lowest energy barrier. In energy terms the non-uniform modes have generally been found to be more economical than the collective uniform mode. The coercive field is thus smaller than the shape anisotropy field H_a^{sh}, and hence Brown's inequality is not valid in this case.

For the case of a prolate ellipsoid of revolution, several authors [42] have compared uniform rotation with non-uniform processes such as those we cited above. The result depends on the length of the minor axis of the ellipsoid. There exists a minimum critical radius R_c (the semi-minor axis of the ellipsoid), below which magnetization reversal takes place through uniform rotation (then $H_c = H_A^{sh}$). R_c is defined by the intrinsic properties of the material. In the case of iron, $R_c = 8$ nm [7]. For $R > R_c$, the reversal mode via local magnetization curling is, in general, favoured. For an infinitely long cylinder of iron with radius of the order of $10\,R_c$, H_C is typically of the order of $H_A^{sh}/10$. Figure 5.62 compares the reversal fields associated with the uniform and non-uniform rotation modes, as a function of the diameter of an elongated grain [42].

For other grain shapes, other modes could possibly be involved in the magnetization reversal. But *in all cases, uniform rotation tends to occur for grains with small volume and non-uniform rotation for larger grains.*

When an elongated grain ($R \sim 10\,R_c$) is placed in a field that makes an angle φ varying between $\pi/2$ and π with the **c** axis of the ellipsoid (inset of fig. 5.63), magnetization reversal takes place for values of the field which increase considerably with $\pi - \varphi$ [43]. Note that the variation of the coercive field roughly follows a $1/\cos\varphi$ (fig. 5.63) law. This behavior is explained by considering that the component of the applied field normal to **c**, the source of reversible rotation of moments, remains too small (up to $H_c \sim H_A^{sh}/10$) to allow the moments to rotate appreciably; the reversal (irreversible rotation) is due entirely to the antiparallel component of the applied field which is equal to $H\cos\varphi$.

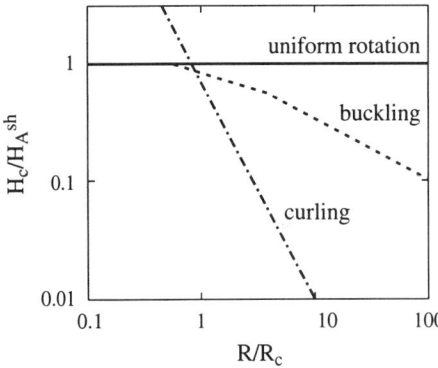
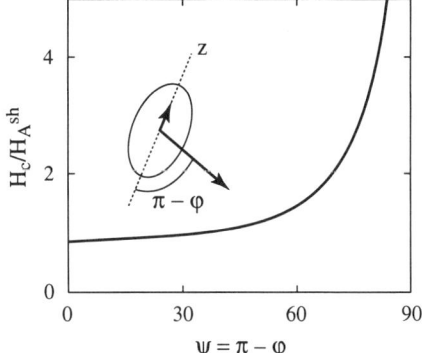

Figure 5.62 - Comparison of reversal fields according to different processes as a function of grain diameter in the case of a material without magnetocrystalline anisotropy

Figure 5.63 - Computed angular dependence of the irreversible reversal field H_c, when the reversal is not preceded by any reversible rotation of moments ($1/\cos\theta_H$ behavior)

This behavior therefore characterises the fact that the coercive field is very much weaker than the anisotropy field. It is not observed for reversal by uniform rotation (see fig. 5.58).

Remark - *The reader wishing to delve deeper into the physics of magnetic domains is invited to consult the excellent work of A. Hubert and R. Schäfer [44].*

REFERENCES

[1] L. NÉEL, *Cah. Phys.* **25** (1944) 1; C. KITTEL, *Rev. Mod. Phys.* **21** (1949) 541.

[2] H.A.M. VAN DEN BERG, *Thesis* (1984) Delft University, the Netherlands.

[3] F. BLOCH, *Z. Phys.* **74** (1932) 295.

[4] B. BARBARA, *J. Phys.* **34** (1973) 1039.

[5] L. NÉEL, *C.R. Acad. Sci. Paris* **241** (1955) 533.

[6] S. METHFESSEL, S. MIDDELHOEK, H. THOMAS, *I.B.M. J. Res. Dev.* **4** (1960) 96.

[7] W.F. BROWN JR., Micromagnetism (1963) Interscience Publ., New York.

[8] H. ZIJLSTRA, *in* Ferromagnetic Materials, Chap. 1 (1982) **III**, E.P. Wohlfarth Ed., North Holland, Amsterdam.

[9] S. CHIKAZUMI, Physics of Ferromagnetism (1997) Clarendon Press, Oxford.

[10] J. KRANZ, A. HUBERT, *Z. Angew. Phys.* **15** (1963) 220.

[11] M. SCHLENKER, J. BARUCHEL, X-ray diffraction topography, *in* Handbook of Microscopy Vol. I (1997) S. Amelinckx, D. van Dyck, J. van Landuyt, G. van Tendeloo Eds, VCH, Weinheim.

[12] H. KAWATA, K. MORI, *Rev. Sci. Instrum.* **66** (1995) 1407.

[13] J. STÖHR, Y. WU, B.D. HERMSMEIER, M.G. SAMANT, G.R. HARP, S. KORANDA, D. DUNHAM, B.P. TONNER, *Science* **259** (1993) 658; W. SWIECH, G.H. FECHER, CH. ZIETHEN, O. SCHMIDT, G. SCHÖNHENSE, K. GRZELAKOWSKI, C.M. SCHNEIDER, R. FRÖMTER, H.P. OEPEN, J. KIRSCHNER, *J. Electron Spectrosc. and Related Phenomena* **84** (1997) 171.

[14] P. FISCHER, T. EIMÜLLER, G. SCHÜTZ, P. GUTTMANN, G. SCHMAHL, K. PRUEGL, G. BAYREUTHER, *J. Phys. D. Appl. Phys.* **31** (1998) 649.

[15] M. SCHLENKER, C.G. SHULL, *J. Appl. Phys.* **44** (1973) 4181-4184.

[16] M. SCHLENKER, W. BAUSPIESS, W. GRAEFF, U. BONSE, H. RAUCH, *J. Magn. Magn. Mater.* **15-18** (1980) 1507-1509.

[17] K.M. PODURETS, V.A. SOMENKOV, R.R. CHISTYAKOV, S.SH. SHILSTEIN, *Physica B* **156-157** (1989) 694-697.

[18] K.M. PODURETS, D.V. SOKOL'SKII, R.R. CHISTYAKOV, S.SH. SHIL'SHTEIN, *Sov. Phys. Solid State* **33** (1991) 1668-1672.

[19] J.N. CHAPMAN, *J. Phys. D. Appl. Phys.* **17** (1984) 623.

[20] A. TONOMURA, *Adv. Phys.* **41** (1992) 59.

[21] S. HASEGAWA, T. MATSUDA, J. ENDO, N. OSAKABE, M. IGARASHI, T. KOBAYASHI, M. NAITO, A. TONOMURA, *Phys. Rev. B* **43** (1991) 7631.

[22] S. LIBOVICKÝ, *Phys. Status Solidi A* **12** (1972) 539.

[23] Y. MARTIN, H.K. WICKRAMASINGHE, *Appl. Phys. Lett.* **50** (1987) 1455. In this work, the authors investigated the field pattern produced by a thin film recording head. From the very beginning, MFM was then closely connected to the applications in the magnetic recording industry.

[24] L. ABELMANN, S. PORTHUN, M. HAAST, C. LODDER, A. MOSER, M.E. BEST, P.J.A. VAN SCHENDEL, B. STIEFEL, J.J. HUG, G.P. HEYDON, A. FARLEY, S.R. HOON, T. PFAFFELHUBER, R. PROKSCH, K. BABCOCK, *J. Magn. Magn. Mater.* **190** (1998) 135.

[25] C.D. WRIGHT, E.W. HILL, *Appl. Phys. Lett.* **67** (1995) 433.

[26] L. BELLIARD, A. THIAVILLE, S. LEMERLE, A. LAGRANGE, J. FERRÉ, J. MILTAT, *J. Appl. Phys.* **81** (1997) 3849.

[27] P.J.A. VAN SCHENDEL, H.J. HUG, B. STIEFEL, S. MARTIN, H.J. GUNTHERODT, *J. Appl. Phys.* **88** (2000) 435.

[28] M. KLEIBER, F. KUMMERLEN, M. LOHNDORF, A. WADAS, D. WEISS, R. WIESENDANGER, *Phys. Rev. B* **58** (1998) 5563.

[29] W.L. ROTH, *J. Appl. Phys.* **31** (1960) 2000.

[30] H. KONDOH, T. TAKEDA, *J. Phys. Soc. Jpn* **19** (1964) 2041.

[31] N.F. KHARCHENKO, *Ferroelectrics* **162** (1994) 173.

[32] S. SAITO, *J. Phys. Soc. Jpn* **17** (1962) 1287.

[33] M. SCHLENKER, J. BARUCHEL, *IEEE Trans. Magn.* **MAG 17** (1981) 3085.

[34] H.R. HILZINGER, H. KRONMÜLLER, *Phys. Status Solidi B* **54** (1972) 593.

[35] B. BARBARA, D. GIGNOUX, C. VETTIER, Lectures on Modern Magnetism, Science Press Beijing and Springer Verlag, Berlin- Heidelberg (1988) 166; B. BARBARA, *J. Magn. Magn. Mater.* **129** (1994) 79.

[36] L. NÉEL, *J. Phys. Rad.* **5** (1944) 241.

[37] J.C. GORTER, *Nature* **132** (1933) 517.

[38] R. GANS, *Ann. Phys.* **15** (1932) 28.

[39] E. CZERLINSKY, *Ann. Phys.* **13** (1932) 80.

[40] L. NÉEL, *J. Phys. Rad.* **9** (1948) 184.

[41] E.C. STONER, E.P. WOHLFARTH, *Phil. Trans. Roy. Soc. London A* **240** (1948) 599.

[42] E.H. FREI, S. SHTRIKMAN, D. TREVES, *Phys. Rev.* **106** (1957) 446; A. AHARONI, S. SHTRIKMAN, *Phys. Rev.* **109** (1958) 1522.

[43] A. AHARONI, Introduction to the Theory of Ferromagnetism (1996) Oxford University Press, Oxford.

[44] A. HUBERT, R. SCHÄFER, Magnetic domains (1998) Springer Verlag, Berlin.

CHAPTER 6

IRREVERSIBILITY OF MAGNETIZATION PROCESSES AND HYSTERESIS IN REAL FERROMAGNETIC MATERIALS: THE ROLE OF DEFECTS

Here we consider the effect of defects on the magnetization processes of real ferro-, and ferrimagnetic materials. Defects induce irreversibility in the initial magnetization process, whereas, in the ideal system, there is none (wall displacement mechanism). They can, on the other hand, markedly reduce the hysteresis related (in an ideal system) to magnetization reversal starting from the saturated state. The value of the anisotropy, the influence of various types of defects, as well as the effects of particle size, and shape, give rise to many ranges of magnetic materials with specific functions, notably –at the extremes– soft, and hard magnetic materials.

Thermal activation –capable of concentrating energy at a given point in the substance, in a volume of variable but often small size– produces remarkable effects, which can affect the intrinsic properties of the material: we consider these effects at the end of this chapter.

1. INTRODUCTION: THE PERFECT MATERIAL MODEL IS INADEQUATE

The theoretical mechanisms of the evolution of magnetization presented in the preceding chapter lead to predictions which do not fully agree with experiment. Consider, for example, a ferromagnetic substance whose strong uniaxial anisotropy is characterised by the single second order constant K_1. A spherical sample of this substance, large enough to contain many domains after thermal demagnetization, should, according to theory, become magnetised in a reversible manner along a straight line of slope M/H equal to 3 until its magnetization M, measured in the direction of the field $\mathbf{H}$//Oz, reaches its saturation value M_s.

If it is then subjected to a field oscillating between $+H_{sat}$, and $-H_{sat}$, the magnetization M of the sample should describe a rectangular loop with width equal to twice the anisotropy field H_A $(= 2K_1/\mu_0 M_s)$ (see fig. 6.1).

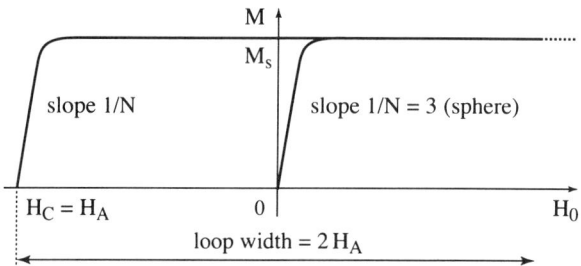

Figure 6.1 - Theoretical loop for a material without defects

Reality is not quite like this. The most obvious disagreement concerns the width of the loops obtained after saturation, which is much smaller than anticipated (fig. 6.2). It does not even correspond to a constant fraction of H_A, or to any other intrinsic magnetic parameter. It depends on the uniaxial substance considered and its history, and can vary from a few per cent of H_A to values of around $2 H_A/3$.

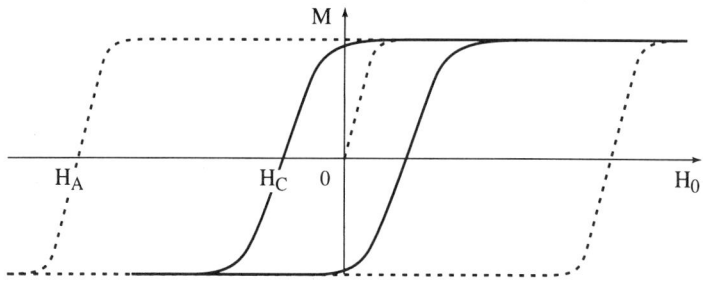

Figure 6.2 - Real (solid line), and theoretical loops (dots)

However, the initial magnetization curve of a ferromagnetic substance appears to vary linearly with the field (slope 1/N), as predicted by theory, only because it is not being observed closely enough. Actually, if it is freed from macroscopic demagnetising effects by the choice of samples that are very elongated along the easy axis (N << 1) or, better, of toroidal samples (N = 0), the irreversible character of the initial magnetization curve becomes clear. This characteristic appears to be a general law affecting all materials (fig. 6.3).

All these disagreements between observations and modelled behavior stem from the fact that until now we have considered only ideal, that is to say perfectly homogeneous, magnetic materials, at all points of which the various magnetic parameters are absolutely identical. So far we have not taken into account the role played by defects (local deviations from homogeneity).

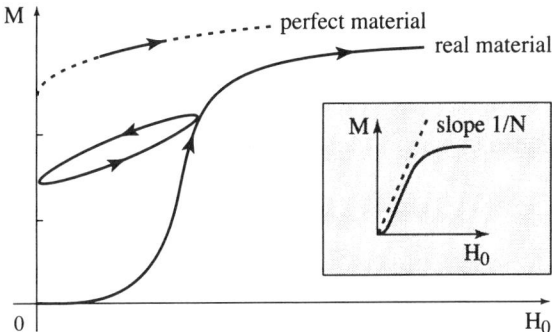

Figure 6.3 - Theoretical (dotted), and actual (solid line) initial magnetization curves as a function of internal field; inset: as a function of external field

2. THE ROLE OF DEFECTS IN THE IRREVERSIBILITY OF THE MAGNETIZATION PROCESS BY DOMAIN WALL DISPLACEMENT

We have seen that in a homogeneous material, where the domain walls are quite wide Bloch walls (material dominated by exchange interactions), the process of magnetization by wall displacement can be considered as reversible, the only effect being the competition between the demagnetising (or dipolar) field and the applied external field. The imbalance, in one or other direction, between these two fields gives rise to a field internal to the material (positive or negative) which is the source of the pressure exerted on the walls. This pressure disappears the moment field compensation is reached as a result of wall displacement. The assumption underlying this analysis is that, in a perfect material, the domain walls move around freely because they cost exactly the same energy whatever their position inside the sample.

Clearly, such a hypothesis is invalid in the case of a real material, which is never homogeneous. Its magnetic energy varies as a function of the position of the walls that it contains. Kersten [1] was the first to take this fact into account. He considered the behavior of an element of a magnetic system containing a single domain wall whose internal energy depends on the position of this wall in a sinusoidal periodical way. This simple, though hardly realistic case, will be dealt with later on in connection with wall-stress interactions: while it reveals the phenomenon of irreversibility, it does not explain the observed behaviors. The formalism best adapted to describe a non-free displacement of the walls within real ferro- or ferrimagnetic materials was proposed by Néel [2]; its principle is given below and it will be discussed in greater detail in section 4.2.

2.1. PRINCIPLE OF THE MODEL: INTRODUCING AN IRREGULAR OSCILLATING POTENTIAL

Like Kersten, Néel considered a simple magnetic system with a single domain wall, assumed to be rigid and with vanishing thickness (fig. 6.4-a).

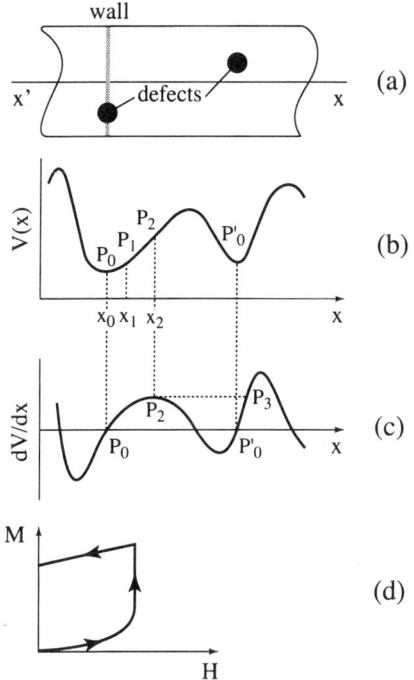

Figure 6.4 - Néel's model

The energies are expressed per unit area of the wall, which can, in the small region under study, be taken as a plane allowed to move along the direction of its normal: a single co-ordinate x is then sufficient to describe its position. The effect of defects, which act as obstacles to the wall displacement, is expressed simply by the evolution of the potential energy V of the system as a function of the position x of this wall. Thus V(x) oscillates irregularly along Ox (fig. 6.4-b), depending on the defects encountered and on their capacity to pin the wall more or less efficiently. V(x), therefore, presents a succession of alternating minima and maxima of varying amplitudes. The minima are possible stable equilibrium positions of the wall. The maxima constitute energy barriers between these equilibrium positions. At each point x of the material, one can write the spontaneous restoring force (per unit surface) that acts on the wall to bring it to a point of equilibrium:

$$F_r = dV/dx \qquad (6.1)$$

Thus, in zero field, the wall is pinned in one of the minima of V(x), for example at x_0 corresponding to point P_0 of figure 6.4-b, where $F_r = 0$. Because the magnetizations M_1, and M_2 on either side of the wall are different, the application of a magnetic field **H** produces on this wall a pressure given by (see eq. 5.47):

$$F_H = \mu_0 (M_1 - M_2) H \tag{6.2}$$

or:
$$F_H = 2 \mu_0 M_S H \cos \theta \tag{6.3}$$

if **H** makes the angle θ ($M_S = |M_i|$) with the nearest magnetization direction, M_i.

When the field is applied parallel to M_1, and when $M_2 = -M_1$ (180° wall), this force per unit area becomes:

$$F_H = 2 \mu_0 M_S H \tag{6.4}$$

As soon as a field H appears –however small it might be– the domain wall of figure 6.4-a is submitted to a non zero force ($F_H + F_r$), and it moves. It stops at the point of abscissa x_1 nearest to x_0 where the force $F_r(x_1)$ compensates F_H exactly. At this point, denoted by P_1 in figure 6.4, one has:

$$F_r(x_1) + F_H = 0 \tag{6.5}$$

giving:
$$2\mu_0 M_S H = \left(\frac{dV}{dx}\right)_{x_1} \tag{6.6}$$

If H then returns to zero, the wall returns to its initial position: its movement between x_0, and x_1 is, therefore, reversible. By contrast, if H is large enough to push the wall beyond the inflection point P_2 (with abscissa x_2), where the restoring force towards x_0 –denoted by $F_r(x_2)$– is maximum, the wall crosses the potential barrier close to the minimum at P_0. At this point its displacement becomes irreversible. The crossing of barriers corresponds to forces, measured by the slopes at the inflection points, on either side of the wall position; it is, therefore, more convenient to discuss the problem using the graph of the derivative of V(x), where the inflection points appear as maxima for one or the other direction of the applied field. In this new representation (fig. 6.4-c) the wall, having reached the peak P_2, jumps abruptly to point P_3, causing a sudden and irreversible variation in the magnetization of the system. If the field is brought back to zero, the system finds its equilibrium at point P_0', not at the initial point P_0. The volume of the material located between points P_0, and P'_0 sees its magnetization reverse during the cycling of the field, producing an irreversible variation of magnetization, ΔM (fig. 6.4-d).

The field H that is relevant in the above discussion is that which is felt by the domain wall. It is the internal field, the resultant of the applied external field, H_0, and the demagnetising field, H_d, associated with the macroscopic magnetization of the system. With the help of relation (6.6), one can associate a value H to each value of dV/dx:

$$H = (2\mu_0 M_S)^{-1} (dV/dx) \tag{6.7}$$

The displacement of the domain wall is associated with a variation of the magnetization described by the loop in figure 6.5-b, if H varies from H_1 to H_2, and returns to H_1.

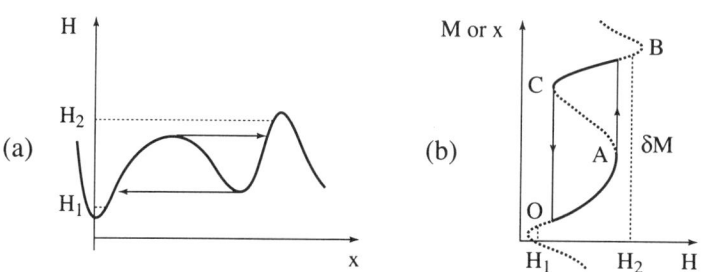

Figure 6.5 - *Variations of the internal field with the position of the domain wall (a), and the associated variation of magnetization (b)*

2.2. MECHANISMS OF DOMAIN WALL PINNING TO SPECIFIC DEFECTS

A variety of interaction mechanisms exist between the domain walls and lattice inhomogeneities or defects. The most classical of these are based on:
- the local modification of the intrinsic magnetic parameters of the material (A_{ex}, K, etc.),
- the appearance, in the vicinity of discontinuities in magnetization, of localised demagnetising fields,
- and a purely geometric effect of local reduction of the wall area.

  We will illustrate these mechanisms through simple examples in the case of planar walls, and then estimate the value of the maximum critical field capable of unpinning the walls from the defects on which they are pinned.

2.2.1 Inhomogeneous stress field: induced uniaxial anisotropy that locally modifies the domain wall energy

The mechanism on which the interaction between a domain wall and a stress inhomogeneity rests is magnetoelastic coupling, dealt with in chapter 12. This phenomenon is the origin of magnetostriction, characterised by a constant λ_S that measures the deformation relative to saturation in isotropic materials (see chap. 12, sections 1.4, 1.5, and 6.2.2). For the moment we will assume that, in an *isotropic* magnetic material, a uniaxial stress σ results in a magnetic anisotropy of the form:

$$E_\sigma = K_u \sin^2 \theta + \text{Constant} \qquad (6.8)$$

where $K_u = 3/2\, \lambda_S\, \sigma$ (see eq. 12.39).

This anisotropy must be taken into account, particularly for the calculations of the energy γ and of the width δ of walls in the material subjected to stresses. The anisotropy energy comes in through a function g(θ) (see eq. 5.18, and following), and it must now involve E_σ. This makes its integration more complicated. Following

Kersten [1], Chikazumi [3] makes the approximation that the functional form of the anisotropy energy, $g(\theta)$, is common to the terms in K_1, and K_u.

When an inhomogeneous stress field prevails inside the material, which means that σ varies from one point to another, γ, and δ have different values depending upon the position of the wall in the material. This variation in the energy of each wall with its position results in the walls locking on to favourable zones and hinders their displacement when a field is applied.

If, within this simplified model, and over and above the approximation stated, we adopt Kersten's hypothesis of a sinusoidal variation of σ only in the direction Ox, viz:

$$\sigma = \text{Constant} + \sigma_0 \sin(2\pi x / \ell),$$

ℓ being the wavelength of the variation (see fig. 6.6), and if we assume δ to be small compared to ℓ, we may consider that σ remains constant inside the wall.

Figure 6.6 - The stress varies as the sine of position (Kersten's model)

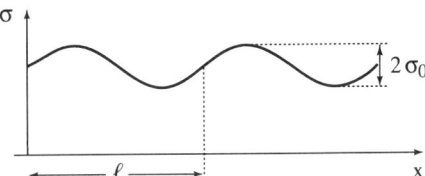

The surface energy of the domain wall is then written as [3]:

$$\gamma = 2\{A_{ex}[K_1 - (3/2)\lambda_s \sigma(x)]\}^{1/2} \tag{6.9}$$

and assuming that $K_1 \gg \lambda_s \sigma$, it can be expanded in the form:

$$\gamma = 2(A_{ex} K_1)^{1/2}[1 - (3/4)(\lambda_s \sigma / K_1) + \ldots] \tag{6.10}$$

The restoring force per unit area which *pins* a wall in the vicinity of the values of x that minimise the above function (equilibrium positions in zero field), is given by $d\gamma/dx = F_r = -(3/2)(A_{ex}/K_1)^{1/2} \lambda_s (d\sigma/dx)$, where $3(A_{ex}/K_1)^{1/2}$ measures approximately the wall width δ. This gives the following expression for the restoring force:

$$F_r = -(1/2)\lambda_s \delta (d\sigma/dx) \tag{6.11}$$

and, in the case of a sinusoidal variation of σ:

$$F_r = -(\pi \sigma_0 \lambda_s \delta / \ell)\cos(2\pi x/\ell) \tag{6.12}$$

The existence of an interaction between the walls and the inhomogeneous stresses in the material was suggested by Becker as early as 1932 [4]. The experimental proof was given by Kersten [1] who studied the effect of internal stresses on the value of the coercive field –defined by equation (6.16)– for nickel rods (fig. 6.7): it is clear that sample (a) is strongly work-hardened following the mechanical treatments involved in its preparation. Sample (b), which was annealed to release internal stress, presents a much smaller coercive field. The interpretation of these effects was given by Becker and Döring [5], Kondorsky [6], and Kersten [1]. A good review of this subject can be found in the already old work of K. Hoselitz [7].

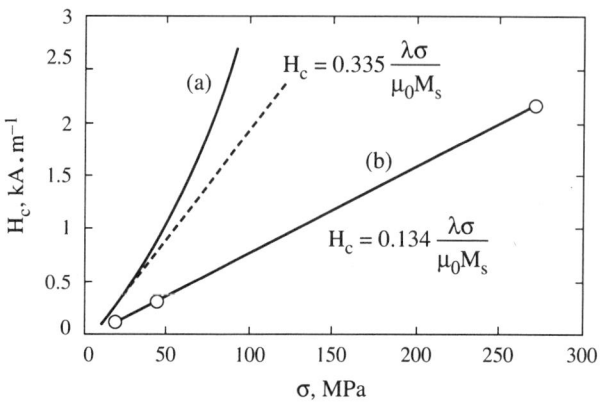

Figure 6.7 - Variation of the coercive field with stress

Work-hardened (a) and annealed (b) nickel samples, according to Kersten [1]. In the formulas linking H_C to σ, H_C is expressed in $A \cdot m^{-1}$, σ in Pa, $\lambda = 36 \times 10^{-6}$, and $\mu_0 M_s = 0.603$ T. The numerical coefficients are then dimensionless.

2.2.2. Large non-magnetic voids or inclusions: local reductions of the wall area and of magnetostatic energy

Geometric effect

The principle of this model is very simple and is based on Kersten's observation [8] that, because *a domain wall costs an energy proportional to its area, it is in equilibrium when it occupies a position for which its area is minimum.*

Let us consider non-magnetic cavities or inclusions, assumed to be spherical with radius r, that are distributed inside a ferromagnetic material and centred on the nodes of a simple cubic lattice with parameter a (see fig. 6.8-a).

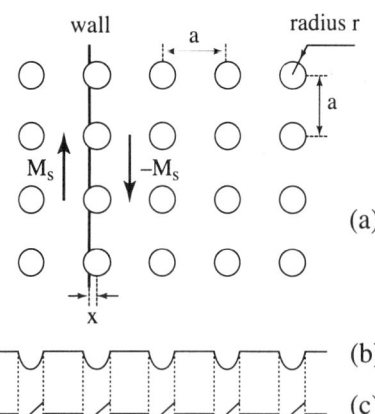

Figure 6.8
Kersten's inclusion model

(a) Lattice of spherical inclusions.
(b) Energy profile of this lattice.
(c) Restoring force acting on the wall.

A 180° wall, assumed to be plane, rigid and parallel to one of the faces of the cube of defects, has a surface energy γ. When the position of the wall coincides with any one of the lattice planes of defects, its surface area is minimal because the surface of the holes produced by the defects inside the wall is maximal. Per unit area of the wall, there are $1/a^2$ holes, each with an area $\mathcal{A} = \pi r^2$. Thus these lattice planes are energy minima where, in the absence of a field, the wall is in equilibrium. When displaced from one of these equilibrium positions by an amount x less than r, the wall area is now decreased by the defects by only an area $\mathcal{A}'$. The relative variation of the area is $\delta\mathcal{A}/a^2 = (\mathcal{A} - \mathcal{A}')/a^2 = \pi(r^2 - x^2)/a^2$.

The change in energy of a wall in the neighbourhood of the lattice planes is then given by:

$$\Delta E_p(x) = \pi x^2 \gamma / a^2 \tag{6.13}$$

which, for a magnetic system of unit surface area containing only one wall (fig. 6.8-a), leads to an energy profile schematised in figure 6.8-b. A restoring force in the direction of the equilibrium position exists only within the zone of width r on either side of each of the minima (fig. 6.8-c). Its value is:

$$F_r = -dE_p/dx = -2\pi x \gamma \tag{6.14}$$

Decrease of magnetostatic energy associated with the presence of a defect

Although it is not negligible, the role of the demagnetising field in the neighbourhood of non-magnetic defects was completely ignored in the preceding model. The magnetostatic energy associated with a non-magnetic spherical void or inclusion with radius r, completely immersed in a magnetic domain (fig. 6.9-a), is given by:

$$E_{d_0} = \frac{2}{9}\mu_0 \pi M_s^2 r^3 \tag{6.15}$$

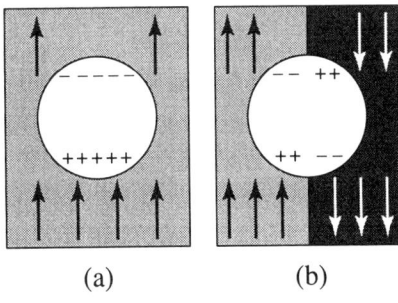
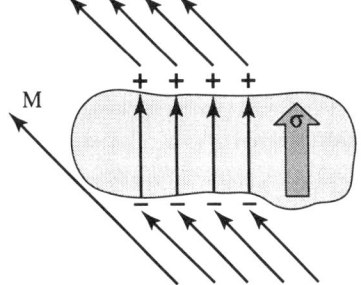

Figure 6.9 - The strong magnetostatic energy linked to the state (a) of a spherical inclusion placed in a medium with uniform magnetization is reduced if the wall crosses this spherical inclusion (b)

Figure 6.10 - The stress σ causes the moments to deviate locally, and thus creates regions where magnetic masses appear

This energy is reduced when a domain wall crosses the defect (fig. 6.9-b), and it passes through a minimum when the wall cuts the sphere into two equal parts [9, 10]. It is then equal to $0.46\,E_{d_0}$. According to this scheme the reduction in energy of the magnetic system is, in general, much more significant than that obtained by considering only the decrease in the surface of the wall. Néel [3] was the first to take into account these dipolar contributions, which also exist in the vicinity of stress inhomogeneities, as shown in figure 6.10 [10].

2.3. THE CRITICAL "UNPINNING" FIELD ASSOCIATED WITH HINDERED DOMAIN WALL DISPLACEMENT

The irreversibility of the magnetization process by domain wall displacement is due to the energy barriers arising from a variety of defects inside the magnetic material, and it is the origin of hysteresis. The largest of the hysteresis loops M(H), obtained by cycling the magnetic field between two values large enough to reach saturation, makes it possible to define the remanence and the coercive field (see fig. 3.5) associated with the process of non-free (hindered) domain wall displacement.

2.3.1. Coercive field associated with the oscillating potential model

In the one dimensional models described above, where a potential energy V(x) expressed per unit area of the wall represents the effect of defects, the pinning force that the structure of the material exerts on the wall at a point x_1 is measured by $(dV/dx)_{x=x_1}$.

The coercive field H_C, by which the acquired magnetization may be cancelled, corresponds to the maximum unpinning field. This is the field which allows the highest barriers to be crossed, by creating a driving force (see eq. 6.6) of the form $2\,\mu_0\,M_S\,H$ at least equal to the greatest of the pinning forces that the material can provide, i.e. that which corresponds to $(dV/dx)_{max}$. By taking account of the deviation θ between the direction of the field and that of the magnetization of the prefered phase:

$$H_c = (1/2\,\mu_0\,M_s)\,(dV/dx)_{max} = F_{max}/2\,\mu_0\,M_s\,\cos\theta \qquad (6.16)$$

With the mechanisms described in section 2.2, this analysis gives the results that follow.

Large non-magnetic defects (Kersten's model)

The restoring force given by equation (6.14) increases up to a value F_{max} for which $x = r$. The magnetic field which allows the wall to be moved to that position is H_C. As soon as $H > H_C$, the wall crosses the whole of the material, the restoring force nowhere taking on values higher than that which has just been overcome. Figure 6.11 shows the corresponding magnetization curve.

Figure 6.11 - Magnetization jump

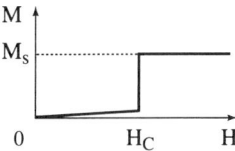

The value of H_C is then:

$$H_C = -F_{max}/2\mu_0 M_s = \pi r \gamma/\mu_0 M_s \qquad (6.17)$$

Periodic inhomogeneous stress field

In Kersten's model the stress varies sinusoidally with distance, $\sigma = \text{Constant} + \sigma_0 \sin(2\pi x/\ell)$. The pinning force is then given by relation (6.12). Its maximum value is $(F_r)_{max} = 2c\lambda_S \sigma_0 \delta/\ell$, which gives a coercive field: $H_C = \pi \sigma_0 \lambda_S \delta/2\mu_0 M_s \ell$.

2.3.2. The critical field in the case of a deformable domain wall

If a domain wall is pinned at its boundaries, and under pressure from a field H, it swells as if it were an elastic membrane (fig. 6.12). The energy involved in this distortion is equal to the surface energy of the wall, γ, multiplied by the increase in area resulting from its distortion $S(H) - S(0)$, thus:

$$E = \gamma [S(H) - S(0)] \qquad (6.18)$$

We shall assume that the distortion is cylindrical. As a function of the radius of curvature r reached, the restoring force per unit surface area of the wall is then written as:

$$F = \gamma/r \qquad (6.19)$$

by analogy with the expression obtained when an elastic membrane is subjected to a difference in pressure [10, 11].

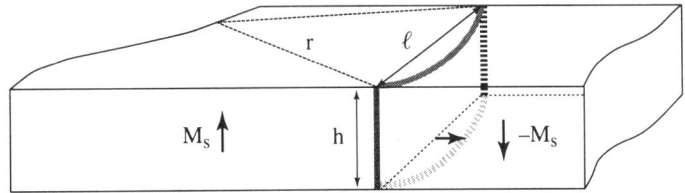

Figure 6.12 - Cylindrically deformable wall

The interaction of a deformable wall with lattice defects may be envisaged in different ways depending on the intensity of the pinning force compared to the energy involved in the deformation of the wall.

♦ If the wall is only slightly deformable, that is to say, if the restoring force caused by its deformation is greater than the pinning force (weak pinning), one can consider that the wall remains flat, and the coercive field is given directly by equation (6.16). In the case where the pinning sites are regularly distributed in the material, giving rise to a potential energy V(x) of the form: $V(x) = V_0 \sin(2\pi x/\ell)$, then $\left(\dfrac{dV}{dx}\right)_{max} = \dfrac{2\pi V_0}{\ell}$, and:

$$H_C = \dfrac{\pi V_0}{\ell \mu_0 M_s \cos\theta} \tag{6.20}$$

The number of pinning sites enters this expression through their distance ℓ.

♦ If, on the contrary, the wall is highly deformable, that is to say that its surface energy γ is small compared to the energy involved in pinning on sites a distance ℓ apart (strong coupling), then the wall swells under the effect of the applied field. The restoring force associated with this deformation is given by equation (6.19). It balances, at equilibrium, the force exerted by the field.

If one considers a 180° wall, equilibrium corresponds to:

$$2\mu_0 M_s H \cos\theta = \gamma/r \tag{6.21}$$

The critical situation is reached when the radius of curvature reaches half the distance between sites, $\ell/2$ (fig. 6.13). Then the wall deforms irreversibly in jumps. This situation corresponds to a critical field H_C equal to:

$$H_C = \gamma/\ell \mu_0 M_s \cos\theta \tag{6.22}$$

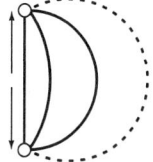

Figure 6.13 - Swelling of a highly deformable cylindrical domain wall

3. THE ROLE OF DEFECTS IN MAGNETIZATION REVERSAL STARTING FROM THE SATURATED STATE

3.1. BROWN'S PARADOX

Now let us return to an ideal material, that is assumed to be homogeneous and free of defects. We are now seeking to reverse its magnetization, starting from the saturated condition, by applying a magnetic field antiparallel to this magnetization. The application of this field provides a uniformly distributed coupling energy with the magnetization, which produces a reversal involving all the moments at the same time. If this process corresponds to a coherent and uniform rotation of the moments, the field capable of producing it is necessarily equal to the total anisotropy field of the system (see chap. 5, § 8.2.2), the sum of the intrinsic anisotropy field (magnetocrystalline anisotropy) and of a possible anisotropy field of macroscopic origin (shape anisotropy).

Although one could imagine other (non-uniform) processes which would allow the substitution of a lower energy barrier for the barrier due to shape anisotropy (a field of lower value then replaces the shape anisotropy field), in no case it is possible –theoretically– to reverse the magnetization of a homogeneous ferromagnetic system with an applied field less than the magnetocrystalline anisotropy field.

This is the expression of Brown's theorem [12].

However, reality contradicts this theoretical principle, because whatever the ferro- or ferrimagnetic materials with large magnetocrystalline anisotropy being considered, the reversal fields measured are always found to be considerably smaller than the magnetocrystalline anisotropy field (see fig. 6.2). This disagreement between theory and experiment is known as *Brown's paradox* [13].

No explanation could be offered for this disagreement within the framework of an ideal material. Thus the measured value of the field for magnetization reversal does not represent an intrinsic property of an ideal material. Rather, it characterises the presence within the material of defects: inclusions, voids, or simply inhomogeneities or lattice irregularities. One is led to associate with these faults the formation of *nuclei* at the point where local magnetization reversal begins. Inevitably, domain walls appear between these nuclei, where the magnetization reverses, and the remainder of the material. Complete reversal of the magnetization of the material occurs when these walls, pushed by the applied field, cross the entire sample more or less easily (see fig. 6.14).

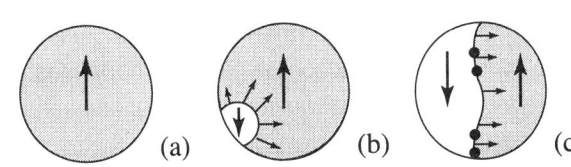

Figure 6.14
Magnetization reversal
(a) Saturated material
(b) Nucleation - (c)
Pinning

According to this outline, magnetization reversal must be considered as being linked to an essentially non-collective process that proceeds in several stages. Among these, we note two essential steps:

♦ the formation of reverse magnetization nuclei, with the emergence of associated domain walls: this is called *nucleation*,

♦ and the sweeping of the material by the walls under pressure from the field, with pinning, and unpinning of these walls. This is the phenomenon known as *pinning* which was mentioned in section 2 of this chapter.

Each of these mechanisms is associated with a critical field, which sets off the effect. The coercive field (identified with the magnetization reversal field) is, in theory, the largest of these critical fields.

3.2. DEFECTS AND NUCLEATION

Starting from the saturated state, nucleation is always at the origin of magnetization reversal if the sample, or the particles of which it is composed, are sufficiently large. This phenomenon involves the magnetic decoupling of one or several zones of the material with respect to their immediate environment. Such a decoupling costs energy (wall energy) and always affects the local situation of the zone involved. However, it can imply it in two very different ways:

- either the decoupled zone is subjected to the same pressures as its environment and is simply less fit to resist these pressures;
- or the pressures that act on the reversal zone are specific and are not suffered by the environnment in the same way or to the same degree.

We shall consider these two types of "defects", and will see that they seem to lead very different behaviors.

3.2.1. Defects giving rise to reversed domain nucleation in zero field

Now let us return, for instance, to the case of the isolated non-magnetic inclusion, immersed in a magnetic domain (fig. 6.9-a). It entails a large loss of demagnetising energy (approximate expression given in 6.15), which is not accepted, as it is, by the system. As Néel has shown [9, 10], a profound alteration of the magnetic environment of the defect inevitably occurs to minimise the losses associated with its presence: it was seen above (§ 2.2.2.), for example, that the dipolar energy losses could be reduced by a factor of more than 2 by introducing a wall that passes through the centre of the inclusion.

Let us reconsider this case by following Néel's approach. For convenience he considered a non-magnetic inclusion (or cavity), in the shape of a cube of side d, which he assumed to have two faces AB, and CD perpendicular to the spontaneous magnetization M_s in the magnetic domain containing the defect (fig. 6.15-a). A total positive charge $M_s d^2$ appears on face AB, while on face CD the opposite charge appears. The dipolar energy involved by these charge distributions is approximately of the order of the demagnetising field energy in a spherical cavity of the same volume (E_{d_0}, see expression 6.15). This situation cannot remain, because it produces high magnetic fields in the neighbourhood of the faces AB and CD (fields close to one tesla, for the spontaneous magnetization of iron), which rotate the magnetization in these regions to the point of bringing it parallel to these faces. Qualitatively it appears clear that, in order to reduce the magnetostatic energy, it would be good for the charges to spread over a larger area and to get farther apart from one another. This can take place only through the creation of new domains and associated walls.

The structure that Néel proposed as suited to reducing the magnetostatic energy comprises (fig. 6.15-b) four small closure domains whose cost is reduced if the lattice is cubic with a four-fold easy axis, or if the anisotropy is weak, and two long spike shaped domains, magnetised antiparallel to the principal domain in which the inclusion is immersed. The total length of this structure is αd (α: coefficient characterising the elongation of the spike-shaped domains).

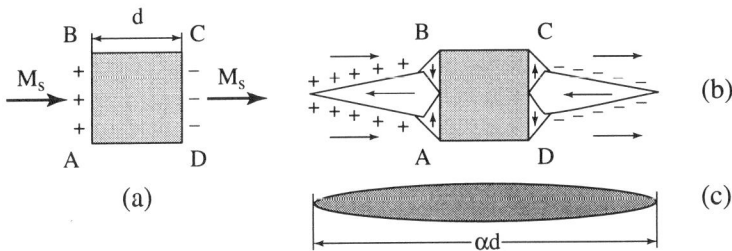

Figure 6.15 - Usefulness of the closure domains

The cubic inclusion (a) immersed in a uniformly magnetised domain is very costly in energy. The domains shown in (b) considerably reduce the magnetostatic energy of the assembly which may be modelled by a prolate ellipsoid with the same aspect ratio α.

A simple approximate calculation can be made by replacing this domain arrangement by a single reversed domain shaped as a prolate ellipsoid of revolution (a×a×c: fig. 6.15-c), with length $2c = \alpha d$ and with equatorial cross-section: $\pi a^2 = d^2/2$. The total energy it involves, E_{tot}, is a compromise between wall energy (around the ellipsoid) and magnetostatic energy. It would give an idea of the saving brought by the appearance of inverse domains in response to the presence of the void.

The volume V and the surface area S of the ellipsoid considered, are given by the following relations:

$$V = \frac{4}{3}\pi a^2 c = \frac{1}{3}\alpha d^3 \quad ; \quad S = \pi^2 a c = \sqrt{\frac{\pi^3}{8}}\alpha d^2 \qquad (6.23)$$

For a very elongated ellipsoid (k = c/a >> 1), the demagnetising field coefficient in the direction of the major axis is given by:

$N_c = [\ln(2k) - 1]/k^2$, with $k = \sqrt{\frac{\pi}{2}}\alpha$ ($\alpha >> 1$), thus:

$$N_c = \frac{2}{\pi\alpha^2}\left[\ln\left(\alpha\sqrt{2\pi}\right) - 1\right] \qquad (6.24)$$

The corresponding magnetostatic energy: $E_{mag} = \mu_0 N_c M^2 V / 2$ is computed by noting that the spike, with uniform magnetization M_s, is not isolated in a non-magnetic environment, but is immersed in a phase with magnetization $-M_s$, which doubles the surface charges (in the preceding relation, $M = 2M_s$) and quadruples the energy. This gives:

$$E_{mag} = \frac{4}{3}\mu_0 \frac{M_s^2}{\pi\alpha} d^3 \left[\ln\left(\alpha\sqrt{2\pi}\right) - 1\right] \qquad (6.25)$$

As for the wall energy, $E_{wall} = \gamma S$, it has a value:

$$E_{par} = \gamma \sqrt{\frac{\pi^3}{8}} \alpha d^2 \tag{6.26}$$

The value of α which minimises $E_{tot} = E_{mag} + E_{wall}$ is the solution of:

$$\frac{3\pi^2}{8} \sqrt{\frac{\pi}{2}} \frac{\gamma}{\mu_0 M_s^2 d} = \frac{1}{\alpha^2} \left[\ln\left(\alpha\sqrt{2\pi}\right) - 2 \right] \tag{6.27}$$

This result demonstrates the influence of the size of the defect. Thus, for inclusions a few tenths of a microns in diameter immersed in iron (parameters used: $M_s = 1.71 \times 10^6$ A.m^{-1}, and $\gamma = 1.4 \times 10^{-3}$ J.m^{-2}), the energy difference between an isolated simple inclusion and an inclusion with spike-shaped domains is minimal: E_{tot}/E_{d_0} is of the order of 0.8 to 0.9 (with $\alpha \sim 10$). But, when the diameter of the inclusion reaches a few microns ($\alpha \sim 100$ to 200), the configuration with inverse spike-shaped domains reduces the energy by a factor of 5 to 10.

The numerical result of this calculation does not reflect reality exactly. However, it confirms that the wider the non-magnetic inclusion, the more it favours the spontaneous appearance (i.e. in zero field) of reverse magnetization domains. Nucleation, the first step of magnetization reversal, has already occurred when the saturation field returns to zero. For a material that contains such defects, the value of the inverse field which must be applied to reverse its magnetization after saturation (coercive field of the material) depends only on the pinning force of the walls by defects (similarly to what happens when one applies a reverse field before saturation).

In a general manner, defects capable of inducing the formation of nuclei of inverse magnetization in a zero (or slightly positive) field are those which produce significant local dipolar losses: non-magnetic inclusions, voids, acute angles at the particle edges, etc.

3.2.2. Defects giving rise to nucleation only in a strong inverse field

The second category of defects does not induce significant local dipolar losses, since the magnetization discontinuities at the junction between the defect and its immediate environment are weak. The defects considered here are simply zones with inferior magnetic properties, notably with regard to the anisotropy. They therefore resist less well and, above all, for a shorter time, the effect of an increasing reverse applied field. In this case, the energy required for the creation of domain walls associated with reversal is provided by the coupling of the magnetization in the defect with the reverse applied field.

If the reverse field that produces this nucleation has a magnitude greater than the critical fields associated with the growth and with the displacement of the domain wall produced across the whole sample, the nucleation field can be taken as the coercive field of the material.

4. HYSTERESIS AND IRREVERSIBILITY: EXPERIMENT AND SIMPLE MODELS

The magnetization of a ferromagnetic substance responds in a complex fashion to an applied external field that alternately increases and decreases.

This response depends on many factors: the initial magnetic state of the substance, the shape of the sample studied, the amplitude of the variation in field considered, the direction of this variation compared to that which preceded it etc. It is therefore important to specify very clearly the experimental conditions used.

4.1. OBSERVED BEHAVIORS

4.1.1. The initial magnetization curve, starting from a demagnetized state

Let us consider a ferro- or ferrimagnetic sample of toroidal shape –this eliminates the effects of the demagnetising field– *which has been demagnetised in an alternating evanescent magnetic field* by the process that will be described in section 4.1.3. When it is submitted to an increasing magnetic field, the magnetization of the sample increases along the curve OABC (fig. 6.16), called the initial magnetization curve. This curve is traditionally divided into three regions:

- that of *weak fields* where magnetization evolves in conformity with Rayleigh's laws (stated in the following section),
- the *intermediate region*, where the magnetization increases very rapidly in a highly irreversible fashion (a significant hysteresis appears each time the field is reversed), and where a magnification of the figure would reveal that it is in fact made up of a series of magnetization jumps separated by brief segments of steady increase: the *Barkhausen jumps* (see inset in fig. 6.16),
- and the region of *strong fields,* where the variation in magnetization changes curvature and tends asymptotically towards a limit, denoted M_s, the *saturation magnetization*; the rotation processes are predominant in this third region.

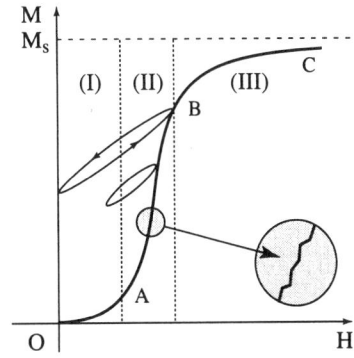

Figure 6.16 - Initial magnetization curve of a ferromagnetic substance

For small variations in the field around a given value, the reversible susceptibility, χ_{rev}, can be obtained for every point of the initial magnetization curve (fig. 6.17).

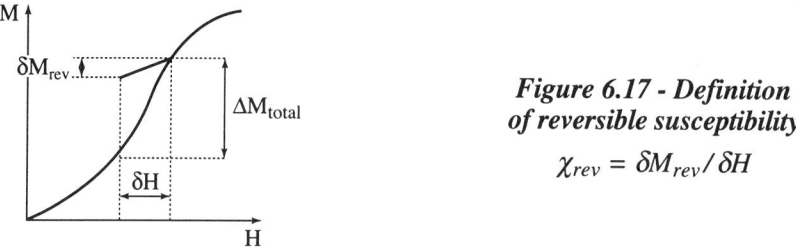

Figure 6.17 - Definition of reversible susceptibility

$$\chi_{rev} = \delta M_{rev} / \delta H$$

From this value, compared to that of the total differential susceptibility $\Delta M_{total}/\delta H$, one deduces the value of the irreversible susceptibility χ_{irr}.

Rayleigh's experimental laws

Study of the initial part of the initial magnetization curve, that which corresponds to weak fields, reveals the following facts which Rayleigh [14] observed as early as 1887:

♦ The curve takes the form of a simple parabola satisfying the following equation:

$$M = \chi H + R H^2 \qquad (6.28)$$

♦ If, starting from a point with coordinates $[H_m, M_m]$ inside this zone, one goes backwards, the tangent of the recoil curve is parallel to the initial tangent, and the recoil curve is again a parabola (see fig. 6.18) with equation:

$$M_\downarrow - M_m = \chi (H - H_m) - (R/2)(H - H_m)^2,$$

where $M_\downarrow$ indicates the magnetization obtained by reducing the value of the field ($H < H_m$). This holds true if one increases the field again after a period of decreasing field. The constants χ, and R are called, respectively, the initial susceptibility and the Rayleigh constant.

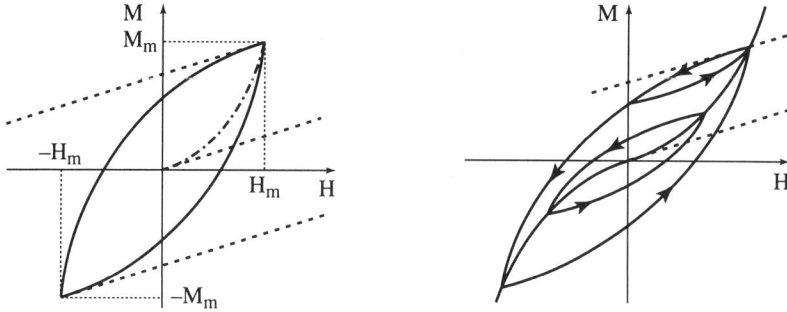

Figure 6.18 - Rayleigh loop   *Figure 6.19 - Collection of Rayleigh loops*

Brown stated these findings in the form of a law (known as *Rayleigh's law*): a variation ΔH of the magnetic field, in the direction opposite to the variation which immediately preceded it, leads to a variation in magnetization with the same sign as ΔH, and with the absolute value:

$$|\Delta M| = \chi |\Delta H| + (R/2) |\Delta H|^2$$

This law is valid only if the field remains in the interval $(-H_m, +H_m)$, H_m being the maximum absolute value reached by the field since the last demagnetization. Figure 6.19 shows a series of possible loops according to Rayleigh's law. The initial magnetization curve is the locus of the peaks of the loops that are symmetric with respect to the origin (field varying between $-H_m$, and $+H_m$). The equations of the rising and falling branches of the symmetrical loop obtained between $-M_m$ and $+M_m$ are respectively:

$$M\uparrow = (\chi + R H_m) H + (R/2) (H^2 - H_m^2) \qquad (6.29)$$

$$M\downarrow = (\chi + R H_m) H - (R/2) (H^2 - H_m^2) \qquad (6.30)$$

From which one infers (on setting $H = 0$ in the equation of the descending branch):

- that the remanence has the value:

$$M_r = (R/2) H_m^2 \qquad (6.31)$$

- that the energy per unit volume dissipated on describing this symmetrical loop is:

$$W = \mu_0 \int H dM = \frac{4}{3} R H_m^3 \qquad (6.32)$$

- and that, when H passes from $-H_m$ to $+H_m$, the variation of magnetization ΔM, written as: $\Delta M = 2\chi H_m + 2R H_m^2$ contains a reversible part (the first term) and an irreversible part (the second term, where H_m appears squared). As a consequence:

$$(\Delta M)_{irr} = 2R H_m^2 \qquad (6.33)$$

and:
$$M_r = (\Delta M)_{irr} / 4 \qquad (6.34)$$

The domain of validity of Rayleigh's laws differs from one material to another. It is, typically, of the order of a few tenths of the coercive field. As we shall see, it corresponds to a zone where the barriers crossed by the walls, thanks to the field, are more or less uniformly distributed.

4.1.2. *The magnetization of a given substance can take all the values contained in its major hysteresis loop*

Loops symmetrical with respect to the origin, and plotted from any point located before saturation on the initial magnetization curve are called *minor loops*. They are all different and their areas are greater the higher the field corresponding to the starting point (loops 1 and 2: fig. 6.20). When saturation is reached, a limiting loop referred to as the *major loop* is obtained; it no longer varies according to the starting point (loop 3: fig. 6.20). Unless otherwise indicated, when reference is made to the *hysteresis loop* of a substance, this refers to the major magnetization loop.

Figure 6.21 shows a typical major loop, with two returns to zero in the negative field range (recoil loops, the average slope of which gives the values of the reversible susceptibility). The intersections of the major loop with the axes define the remanence and the coercivity, quantities used to characterise the material. The remanent magnetization M_r is the residual magnetization in zero field after saturation; the coercive field H_C corresponds to the inverse field required to obtain zero magnetization after the material has been saturated.

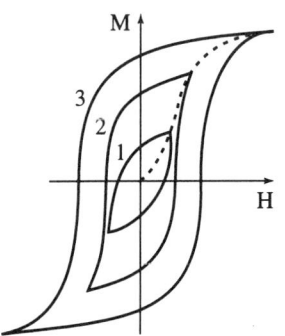

Figure 6.20 - Major and minor loops

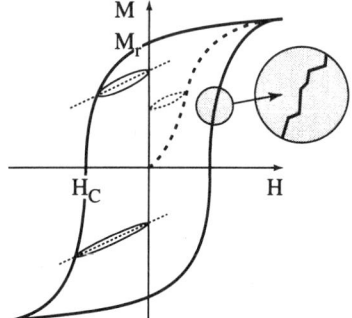

Figure 6.21 - Major loop and recoil loops

An enlargement of the regions of rapid rise and fall of magnetization shows –notably in the case of soft materials– the existence of Barkhausen jumps (inset of fig. 6.21). These jumps can be revealed, and their amplitude measured, through the electromotive force they induce across a coil surrounding the sample (*Barkhausen noise*). Through their amplitude, they reveal that in these zones –as in the central part of the initial magnetization curve (see fig. 6.16)– the irreversible wall displacements are accompanied by magnetization reversal of significant volumes of the material. In a given sample, the amplitude of Barkhausen noise passes through a maximum twice per cycle when the walls perform their largest irreversible jumps, generally in the vicinity of the coercive field $\pm H_C$.

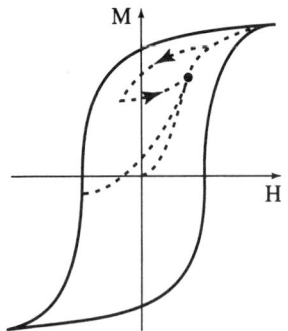

Figure 6.22 - An infinity of paths can lead to the same point P within the major loop

It is also possible to define a remanent magnetization and a coercive field on the minor loops. However, the values obtained are not characteristic of the material, because they are also dependent on the chosen starting point and, possibly, on the magnetic history of the material. Any point of the [H, M] plane located within the major loop may be reached in very many ways: it is the end point of various paths plotted on this plane. This is the consequence of the fact that the system does not retrace a path through its earlier states when the direction of the field reverses (fig. 6.22).

4.1.3. The anhysteretic magnetization curve

We have just seen that for a given value of the field, the magnetization of a substance depends on its magnetic history. In fact, all points of the segment $M_{min}^{H_1}$ $M_{max}^{H_1}$ (fig. 6.23) are accessible, each one by a number of different paths. Are there no reproducible states that can serve as references? So far we have encountered two: the thermally demagnetised state, and the saturation state. A third one may exist, accessible only in cases where the coercitivity is due to domain wall pinning by defects. This is a state of zero magnetization, obtained by applying an alternating field whose amplitude decreases to zero from a high value, of the order of several times H_C. This procedure (fig. 6.24) produces a demagnetised state at the macroscopic scale, which is often used as a state of reference; but the microscopic configurations which it covers up are numerous, and non-reproducible. Only on starting from this state is it possible to observe Rayleigh's laws [15].

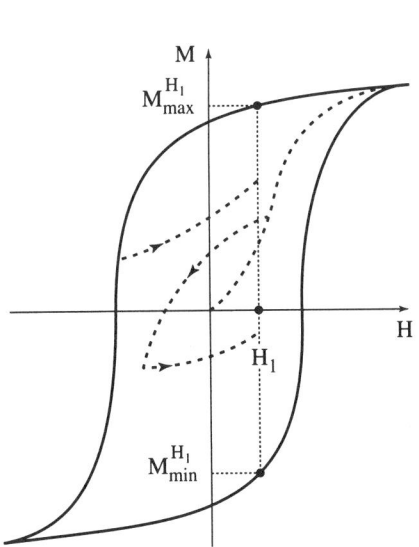

Figure 6.23 - All the values of M lying between $M_{min}^{H_1}$ and $M_{max}^{H_1}$ are accessible by various paths

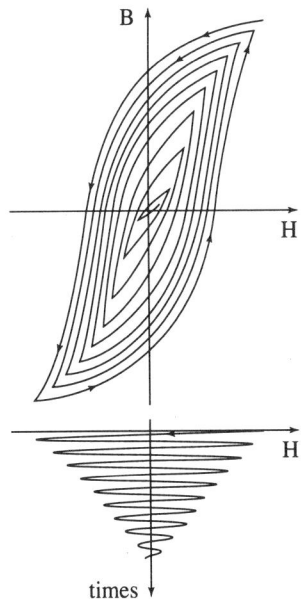

Figure 6.24 - Demagnetization in an evanescent AC field

This evanescent alternating field procedure does not lead to a reference state when the coercivity is largely dominated by nucleation, as is the case with many hard materials.

Let us return to soft materials. The $M_a(0) = 0$ state, obtained by demagnetization in an alternating evanescent field, is only the first point of a possible curve known as the anhysteretic magnetization curve (see § 2.5.1, chap. 3). Point $M_a(H_1)$ of the anhysteretic curve corresponding to a given field H_1 is obtained in the same way that $M_a(0)$ was obtained, by submitting the sample to a field made up of H_1 –which remains fixed–, and of an alternating field with zero average value, whose amplitude is initially sufficient to overcome the saturation field, and which, little by little, is brought back to zero (these two fields can be produced by different coils or by the same one). Only at the end of this operation, shown schematically in figure 6.25, is the value $M_a(H_1)$ measured. The anhysteretic curve, thus constructed point by point, is independent of the prior magnetic history of the sample, and it is reproducible. It is always located above the initial magnetization curve and does not feature an inflection point. The procedure for obtaining the *ideal* anhysteric curve (devoid of hysteresis) comes down to "shaking up the domain structure in every possible way around a given field" [16]. It will, accordingly, supply the lowest energy configuration in this field.

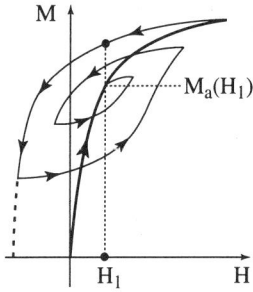

Figure 6.25
Anhysteretic magnetization curve

4.2. MODELLING OF THE BEHAVIOR OBSERVED IN RAYLEIGH'S DOMAIN

The irregularly oscillating potential energy model offers a qualitative representation of the hysteresis of a non-saturated ferro- or ferrimagnetic system (see fig. 6.5). But the passage from the qualitative to the quantitative in actual cases is still beyond the reach of models since it is difficult to determine a realistic V(x) function. Moreover, when field variations reach an amplitude sufficient for saturation to be reached, the nucleation of reverse domains, as well as other –so far poorly defined (see chap. 15)– mechanisms, the proper treatment of which is as yet unknown, also contribute to hysteresis. However, the results obtained in very weak fields, in Rayleigh's domain, can be derived, starting from Néel's theory and using a very simple construction developed by Preisach.

4.2.1. Preisach's representation

Preisach [17], taking up an idea of Weiss, and Freudenreich [18], imagined a collection of small elements each possessing a rectangular, and symmetric magnetization loop (fig. 6.26), close in shape to that obtained earlier (see fig. 6.5-b). These loops, with half-width e, are centred on an arbitrary field value a distance ℓ from $H = 0$ (e, and ℓ thus measure fields); e is considered in absolute value (always > 0), while ℓ may take either positive or negative values.

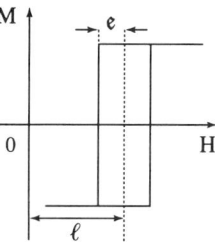

Figure 6.26 - Preisach's elementary loop

In zero magnetic field, all elements for which $\ell > e$ are negatively magnetised, while all those for which $\ell < -e$ are positively magnetised. The others have a positive or negative magnetization depending on the magnetic history of the sample containing the set of elements under consideration.

If, in zero field, the sample is completely demagnetised, this implies that as many elements are magnetised positively as negatively. The points representing the elements of the sample are then uniformly distributed in a single half-plane of the $[\ell, e]$ diagram by which Preisach represented the state of the system: see figure 6.27-a, where the zones of positive magnetization (dark grey), negative (light grey) or other (medium grey) are differentiated.

When a positive field H is applied, all the elements for which the loop corresponds to the field ℓ such that $\ell < H - e$, are positively magnetised. The others retain the magnetization state in which they were before the field was applied. Figure 6.27-b accounts for this situation: the elements for which the representative points are inside triangle OHP had their magnetization reversed; the same goes for those elements of the A'OPB' strip whose magnetization was previously negative.

The return to zero field brings us back to a diagram of type a (fig. 6.27-c), but, while the elements represented by points within the triangle OHP again reversed, and recovered a negative magnetization, this does not occur for those in strip A'OPB', the magnetization of which remains unchanged.

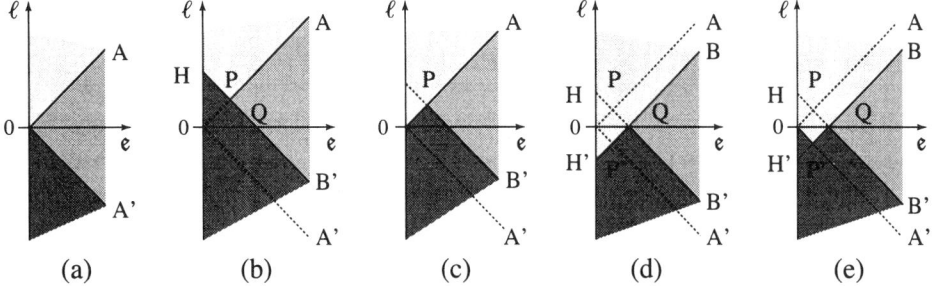

Figure 6.27 - Preisach diagrams (see text)

When, to obtain a symmetrical loop, a negative field, H' = –H, of the same magnitude as previously (in the case of fig. 6.27-b) is applied, the situation is as shown in figure 6.27-d. All the elements represented by points for which $\ell > (-H + e)$ are magnetised negatively. Suppressing this field completes the loop, and leads to the diagram of figure 6.27-e. As before, the magnetization of the elements represented by points inside triangle OH'P' flip towards positive values. All the others remain unchanged.

Between the two extreme values of the field (symmetrical values +H and –H) corresponding to the passage from diagram b to diagram d in figure 6.27, the elements whose magnetization reversed are those whose representative points occupy the triangle QHH'. Let us assume that the density n of these points in the half-plane $[\ell, e]$ is constant. Then the number of elements affected is proportional to H^2, the area of the triangle. The irreversible part of the magnetization variation ΔM_{irr} of the sample when the field varies from +H to –H is thus proportional to H^2. Taking into account the fact that the magnetization varies by 2 M_s in the course of a reversal, we can write that:

$$|\Delta M_{irr}| = 2 n M_s H^2 \qquad (6.35)$$

On the other hand, having made a complete symmetrical loop from H = 0 to H = 0, i.e. from diagram (a) to diagram (e) through stages b, c, and d, the sample retains a remanent magnetization proportional to the area of the square OPQP'. The magnitude of the remanent magnetization M_r may thus be written as: $M_r = n M_s (H/\sqrt{2})^2$, because $(H/\sqrt{2})^2$ is the area of the square OPQP', giving:

$$M_r = n M_s H^2 / 2 \qquad (6.36)$$

or:
$$M_r = \Delta M_{irr}/4 \qquad (6.37)$$

The results given by relations (6.35), and (6.37), when compared with relations (6.33) and (6.34) obtained by direct observation, show that Preisach's representation, with the assumption of a uniform distribution of the elementary loops), is a good picture of the behavior of magnetic materials in Rayleigh's domain, based on the elementary loops that Preisach thought of as real. In his model, presented below, Néel demonstrated the fictitious nature of these loops for a bulk material; in contrast, such loops are quite realistic for fine particles.

4.2.2. Theory of hysteresis in Rayleigh's domain (Néel's random potential)

To better represent the irregularity of the defects in a material, Néel [2] constructed the V(x) potential corresponding to any ferromagnetic substance from a broken line, each segment of which has the same projection Δx on the x axis, the ordinates at the changes in slope being spread randomly around a zero average value. This broken line is then replaced by a succession of pieces of parabolas, to suppress the sharp angles which do not correspond to any physical reality. This results in a function

6 - IRREVERSIBILITY OF MAGNETIZATION PROCESSES AND HYSTERESIS

(dV/dx) that also takes the form of a broken line $A_1A_2A_3$ ... all the segments of which have a projection equal to Δx, and which resembles that shown in figure 6.28-a. On this figure x is drawn vertically because the displacement of the wall along x produces a proportional variation of the magnetization M; the axis bearing (dV/dx) is graduated in magnetic field units, through equation (6.7).

Let us assume that when the magnetic field H takes the value H_1, the representative point of the system is at M on our diagram.

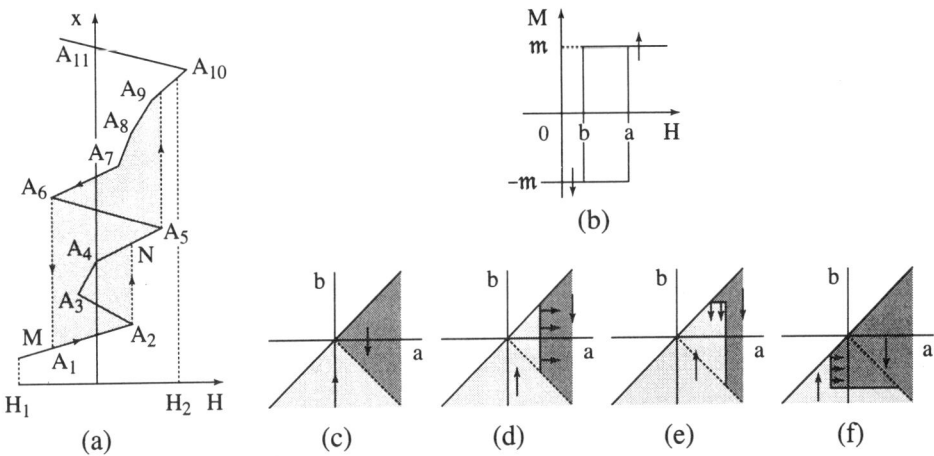

Figure 6.28 - Néel's random potential model

When the field increases, this representative point comes reversibly from M to A_2 (equality 6.7 remaining satisfied), then jumps from A_2 to N (an irreversible jump of the domain wall), and so on. Finally, if the field describes a small loop between the two values H_1 and H_2, the representative point describes a loop shown in grey on the diagram (fig. 6.28-a). The hysteresis loop of the substance thus modelled results from the superposition of very many such partial loops.

Néel decomposes these partial hysteresis loops into simpler elementary, fictitious loops. Those to which the irreversible terms in the expression of magnetization (see eq. 6.28, 6.29, 6.30 or 6.33) correspond are, to second order, rectangular loops with fixed height 2m whose vertical branches correspond to H = a and H = b (with a > b). For the whole sample, a and b are spread with uniform density around the value 0. This was the starting point for Weiss and Freudenreich's theory, and then for Preisach and Kondorsky, but in these loops they saw real hysteresis loops of elementary domains with fixed outline.

Taking up Preisach's model (slightly modified), Néel plots on the semi-plane [a, b] (a > b) the points representing each of the fictitious loops thus obtained (fig. 6.28-c to 6.28-f). After demagnetization in an evanescent alternating magnetic field, half the cycles (those for which a + b > 0) are on one side of the wall, and those for which a + b < 0 are on the other side, as shown in figure 6.28-c.

This actually reproduces the form of the irreversible term in Rayleigh's laws: a term varying as H^2 for the initial magnetization curve (fig. 6.27) according to equation (6.28); an $H^2/2$ term for the descending branch of a Rayleigh loop (fig. 6.27-e) according to equation (6.29); also a term in $H^2/2$ for the ascending branch (fig. 6.27-f) according to equation (6.30). Subject to a hypothesis about the distribution of points A_1, A_2, A_3, ... around Ox (a gaussian for example), this theory also provides a relation between the Rayleigh constants, and the coercive field. Furthermore it shows that Preisach's diagram depends on the position of the walls in the demagnetised state.

4.3. FROM SOFT TO HARD MATERIALS

The hysteresis loops of existing ferro- or ferrimagnetic substances or materials all conform more or less to the scheme presented. However, the scale of field H varies across a very large range from one type of material to another. At the extremes are materials described as soft, with very narrow loops (iron-nickel alloy for example: $10^{-1}\,A.m^{-1} \leq H_C \leq 10^2\,A.m^{-1}$, see fig. 6.29-a), and hard materials, whose loops are, by contrast, very large (SmCo$_5$ magnet for example: $10^4\,A.m^{-1} \leq H_C \leq 2\times 10^6\,A.m^{-1}$, see fig. 6.29-b), which gives a ratio greater than 10^7 between the extremes of the coercivities.

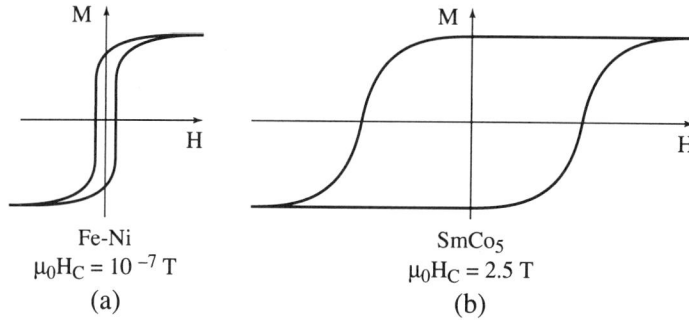

Fe-Ni
$\mu_0 H_C = 10^{-7}\,T$
(a)

SmCo$_5$
$\mu_0 H_C = 2.5\,T$
(b)

Figure 6.29 - Hysteresis loops of Fe-Ni (soft) and of SmCo$_5$ (hard)

The soft materials are used principally to guide and concentrate the flux, and also as signal amplifiers. The prime qualities expected from them are a maximal differential permeability and a hysteresis loop with minimal width. The material from which they are made must, therefore, have a very weak intrinsic anisotropy, which leaves the moments free to rotate easily.

Ideally, these materials should also always contain a large number of extremely mobile domain walls, allowing significant changes in magnetization under the effect of the smallest field. As the hysteresis loops show, nucleation has already occurred when the applied field reverses (fig. 6.29-a).

As for the hard materials, which are used in the manufacture of magnets, the larger their hysteresis loop is, the more efficient they are (fig. 6.29-b). That can be obtained only if the base material has a strong uniaxial anisotropy.

The coercitivity defines the width of the major loop; to maximise it, two methods are available:

♦ The most common method: to prevent for as long as possible, in a growing inverse field, the nucleation of reverse domains and of the walls that appear with them. One thus tries to avoid any nucleation, be it spontaneous or in a weak inverse field. For that one suppresses all defects whose magnetization will be very different from that of the matrix and which might thus constitute an efficient source of local dipolar effects. The presence of defects that can be activated by the inverse field is very difficult to avoid, if only at the edge of the particles. One must try to restrict their effectiveness by bringing together, as far as possible, the magnetic characteristics (in particular the anisotropy) of the defects with those of the matrix. One then speaks of nucleation coercivity (fig. 6.14-b).

♦ Alternative method: to block the propagation of domain walls which have already formed, with the help of a network of defects with enough pinning power. This capacity is generally based on a strong local modification of the wall energy, rather than on local demagnetising effects which would result, through the existence of too many walls, in a softening of the material. This is referred to as *pinning coercivity* (fig. 6.14-c).

A comparison of the field ranges over which the initial magnetization, from the demagnetised state to saturation, and the passage from $-M_r$ to $+M_s$ on the major loop, occur can provide information on the principal process of coercivity. If the range of fields involved in producing both variations is the same, the coercivity must result from pinning, or from lasting immobilization, of the walls (see soft materials: fig. 6.30-a, and some hard materials: fig. 6.30-b). If, starting from the demagnetised state, saturation is reached in a much weaker field than that which is necessary for magnetization reversal from the remanent state, from $-M_r$ to $+M_s$ (most modern hard materials: fig. 6.31), it is probable that the wall pinning mechanism does not play a leading role in coercitivity and it becomes necessary to consider the different stages of nucleation (see chap. 15).

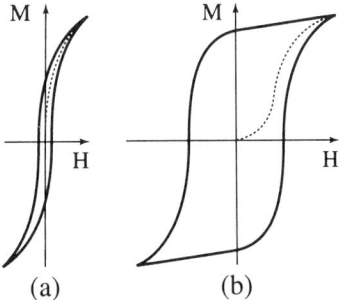

Figure 6.30 - Dominant process: wall pinning

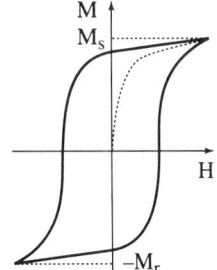

Figure 6.31 - Dominant process: nucleation

5. MAGNETIC AFTER EFFECTS:
EFFECTS DELAYED AND AMPLIFIED AT THE SAME TIME

The existence of time-dependent magnetic effects in ferromagnetic materials clearly stems, above all, from their domain configurations. The delicate balance of a configuration results from the competition between various energy terms, any of which may be greatly altered locally by defects of various origins within the lattice.

Three very different types of time evolution of the magnetic properties must be distinguished:
- A*ging*, related to irreversible modifications of the structure of the material (for example, the formation of precipitates). These structural modifications can accompany mechanical treatments, or be due to chemical effects (oxidation or others), or simply result from an atomic diffusion phenomenon.
- In the case of conductors, the (purely electromagnetic) effect of eddy currents which, for a few seconds or fractions of a second, prevent the penetration into the metallic material of a magnetic field which has just been applied. The displacement velocity of domain walls, measured by Sixtus and Tonks inside metallic filaments [19], agrees with computations which take into account the delay imposed by the eddy currents.
- The *magnetic after-effect* which characterises the fact that, when at time t_0 a magnetic field is suddenly applied to a ferromagnetic substance (whether insulator or conductor), it induces a change in its magnetization that begins at a time t_i very close to t_0, but which may continue for a very long time beyond t_i.

Here we are only interested in the third type of time-dependent phenomena: the magnetic after effect. The aim of the present section is to go into more detail with regard to the characteristic manifestations of magnetic after-effect, the conditions for its existence and the mechanisms that can explain it (already mentioned in chap. 5, § 2.5). These mechanisms distinguish between the diffusion after-effect and the fluctuation after-effect.

5.1. THERMAL ACTIVATION AS THE ORIGIN OF BOTH TYPES OF MAGNETIC AFTER EFFECT

Generally speaking, magnetic after-effects involve not only a time delay between the moment the application of a field step to a material ends and the instant the evolution of its magnetization begins, but also –and above all– the duration of this evolution, which can be significant for minutes or hours, or even days. Two different mechanisms are put forward to explain this behavior. Both involve temperature in a way we did not take into account so far: thermal activation.

The statistical thermal effect consists of the contribution of an energy density distributed uniformly throughout the volume of the sample: it gives rise to a uniform lowering of the barriers and to a, generally reversible, modification of the intrinsic magnetic properties of the substance (K, M, A_{ex}, etc.).

The other temperature effect is not uniform; by accumulating energy first at one point, and then at another via random agitation, it allows irreversible processes, capable of radically modifying the state of a system, to become activated. The two known mechanisms by which the microscopic configuration of a magnetic substance can be modified by thermal activation are:

- the diffusion of foreign species (small atoms, vacancies, …) from one interstitial site to another. This induces a supplementary uniaxial anisotropy in the vicinity of the domain walls, and can alter their position;
- the appearance of local fluctuations resulting in the (irreversible) crossing of energy barriers which, until then, had maintained a metastable state.

In the latter case, the smaller the volume concerned, the more important are the effects of thermal activation. They give rise to the magnetic *fluctuation* after-effect only when the system considered is close to its conditions of irreversibility, that is to say when the applied field, close to a critical value, has already considerably reduced the height of the energy barriers: the remaining barriers are then low, and thermal activation allows them to be crossed within the duration of the experiment. According to this description, thermal activation causes local fluctuations with regard to the statistical state, which show little sensitivity to the absolute value of the temperature.

By contrast, in the case of atomic diffusion, the energy barriers brought into play are characteristic of the crystal structure, and the field contributes only through magnetostriction. If the temperature is too low, the diffusion times are too long to give rise to appreciable effects in the course of accessible experiments. If the temperature is too high, diffusion is quasi-instantaneous, and no retardation effect is observed.

The diffusion after-effect, which is very sensitive to temperature, therefore only appears within definite temperature ranges called *after-effect bands*.

5.2. THE FLUCTUATION MAGNETIC AFTER-EFFECT. EXAMPLE OF HARD MATERIALS NEAR THE COERCIVE FIELD

This type of magnetic after-effect appears as a spontaneous and continuous variation of magnetization with time, after the application of a magnetic field step (fig. 6.32-a).

This variation may occur each time the material is in the immediate neighbourhood of critical conditions: either, for example, in the zones where Barkhausen jumps exist (fig. 6.32-b), or during magnetization reversal, particle by particle, of a permanent magnet near its coercive field (fig. 6.32-c).

The magnetization varies according to a law of behavior, generally logarithmic, featuring a variation that decreases with time. The magnetic relaxation coefficient (or the magnetic viscosity parameter), denoted as S, characterises this behavior; it is defined by the rate of change of M as a function of the logarithm of time, i.e:

$$S = \partial M / \partial \ln t \tag{6.38}$$

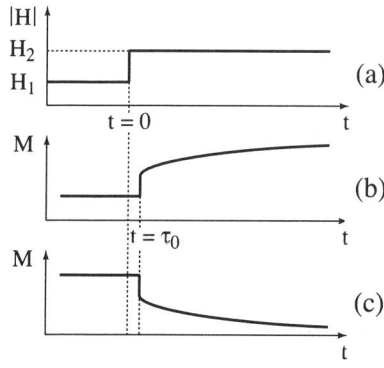

Figure 6.32 - Response to a step in magnetic field

5.2.1. Foundation of the analysis: application to magnetization reversal in permanent magnets

The irreversible after-effect behavior of magnetization in ferromagnetic materials is generally interpreted as [20, 21] being fundamentally connected to a thermally activated process, the crossing of energy barriers, in two-level metastable systems (fig. 6.33). Their mathematical treatment rests on two classical hypotheses:

♦ The first hypothesis concerns the variation of magnetization with time. It stipulates that the crossing of a given barrier Δ is perfectly random and thus does not depend on time that has already passed. Put in another way: during a small time interval dt, the probability of crossing Δ is the same, whatever the moment when the interval dt is considered. This probability is proportional to the duration of the interval dt and thus is written dt/τ where τ has the dimensions of time (because a probability is a dimensionless number). That leads to a classic Poisson probability law, where the time τ appears as the average crossing time for a barrier of height Δ: this is the *relaxation time* associated with Δ.

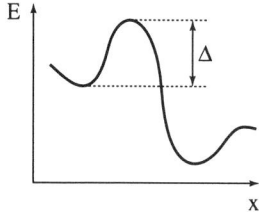

Figure 6.33 - A two level system

♦ According to the second hypothesis, Δ and τ are related by the classic Boltzmann-Arrhenius law:

$$\tau = \tau_0 \exp(\Delta/k_B T) \tag{6.39}$$

where τ_0 is the intrinsic relaxation time and corresponds to the crossing time of a barrier of zero height. Due to the complexity of some irreversible magnetization processes in real systems (magnetization reversal for example), τ_0 may not correspond to a simple intrinsic property of the material [22, 23].

Starting from these two hypotheses, the logarithmic variation of the observed magnetization as a function of time may be understood and interpreted on condition that only the time span normally available to the experiment is taken into account. The shape of the distribution $f(\Delta)$ of the heights of the energy barriers (see fig. 6.34) has no influence on the logarithmic character of the variation of M provided the range in experiment time scales is reasonable (from fractions of a second to several hours).

By way of example let us consider the major irreversible phenomenon of ferromagnetic materials, magnetization reversal, and study it in a sample made up of a collection of independent ferromagnetic particles (a model for a magnet). We consider this sample as placed in an inverse field slightly lower (in modulus) than the coercive field. Starting from the first assumption, simple algebra shows that, for a given particle, the probability P(t) that magnetization reversal occur between times 0 and t can be written as:

$$P(t) = 1 - \exp(-t/\tau) \qquad (6.40)$$

For a collection of independent particles identical to the above, with only one type of energy barriers with height Δ, the above relation also gives the proportion, $\delta N/N$, of particles whose magnetization has reversed during time t. It therefore provides the relative variation of the magnetization, $\delta M(t)/2M_0$, where M_0 is the remanent magnetization of the assembly.

For a system containing particles that involve a distribution $f(\Delta)$ of the energy barriers (fig. 6.34), the relative irreversible variation of magnetization with time is given by:

$$\frac{M(t)}{M_0} = m(t,T) = \int_{\Delta=0}^{\infty} \left[2\exp\left(-\frac{t}{\tau}\right) - 1 \right] f(\Delta) d\Delta \qquad (6.41)$$

with:

$$\int_{\Delta=0}^{\infty} f(\Delta) d\Delta = 1.$$

The expression of the magnetic relaxation coefficient, S, is then:

$$S = \frac{\partial M}{\partial \ln t} = t\frac{\partial M}{\partial t} = 2M_0 \int_{\Delta=0}^{\infty} \lambda \exp(-\lambda) f(\Delta) d\Delta \qquad (6.42)$$

where λ is the reduced time, i.e: $\lambda = t/\tau(\Delta)$.

In this expression of S, the distribution function $f(\Delta)$ is weighted by the function $\lambda e^{-\lambda}$, which defines what can be called an *energy window*.

Owing to the existence of this window (of this weighting), only a small section of $f(\Delta)$, around $\tau(\Delta) = t$ (that is to say characterised by $\lambda = 1$), contributes to S (fig. 6.34).

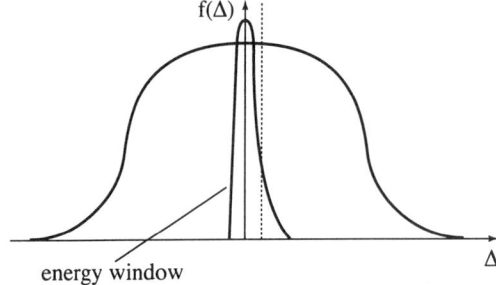

Figure 6.34 - Distribution function of energy barriers f(Δ), and window selecting active barriers at time t = 1 s at 300 K

The vertical dotted line indicates the position that the centre of the window will occupy at time t = 1 000 s.

The position and width of the window are determined with the help of the second assumption. The position Δ_c of the window, at a given time t and a fixed temperature T, corresponds to:

$$\Delta_c(t, T) = k_B T \ln(t/\tau_0) \qquad (6.43)$$

The effects of time and of temperature are thus to shift the energy window towards higher energy values, logarithmically with time, linearly with temperature. One can show that [24] the width of the window is approximately equal to $k_B T$.

The part of $f(\Delta)$ that contributes effectively to S (see eq. 6.42), is defined by the width of the window displaced during the time of observation of the after-effect. At low temperature the width of the window is small. The part of the energy barrier distribution which contributes to S is essentially defined by the displacement of Δ_c between the initial time t_i and the final time t_f of the measuring period. At 300 K, for experiments between $t_i = 1$ s and $t_f = 1,000$ s, the width and the displacement of the window contribute about equally (fig. 6.34).

Bearing in mind the preceding remarks, after-effect experiments may be analysed on the assumption that $f(\Delta)$ is constant and equal to $f(<\Delta_c>)$ ($<\Delta_c>$ being the value Δ_c taken at the average moment t_m, between t_i and t_f). The variation of M is then a logarithmic function of time.

Taking into account the fact that magnetization reversal corresponds to a decrease in M, that is to say ∂M is negative in equation (6.42), one thus obtains:

$$S = 2 M_0 f(<\Delta_c>) \int_{\Delta=0}^{\infty} \lambda \exp(-\lambda) d\Delta \qquad (6.44)$$

with:

$$\lambda = \frac{t}{\tau(\Delta)} = \frac{t}{\tau_0} \exp\left(-\frac{\Delta}{k_B T}\right)$$

hence:

$$S = 2 M_0 k_B T f(<\Delta_c>) \qquad (6.45)$$

If one now considers the most general case where the variation of magnetization will be observed over an arbitrarily long period of time, starting from an initial time close to zero, the time variation of the magnetization will no longer be logarithmic.

Combining the second hypothesis with equation (6.41), one obtains the expression:

$$m(t, T) = 1 - 2 \int_0^{\Delta_c(t)} f(\Delta) d\Delta \qquad (6.46)$$

This relation means that the variation of magnetization with time is then directly linked to the form of the distribution function $f(\Delta)$. As $m(t, T)$ is a function of Δ, a scaling law in $T \ln (t/\tau_0)$ may be inferred. Labarta et al. have plotted, on a single curve, the values of $m(t, T)$, obtained at different temperatures, as a function of the variable $T \ln (t/\tau_0) = \Delta_c / k_B$, by assigning suitable values to τ_0, that is to say, by considering it to be an adjustable parameter [24]. Values varying between 10^{-12} s, and 5×10^{-7} s have thus been obtained for τ_0.

5.2.2. Variation of magnetization with time or with field: the fluctuation field

The mechanisms involved in magnetization reversal under the effect of thermal activation, particularly in a hard material, are not precisely known. They certainly differ from those that operate when reversal occurs under the influence of an increasing reverse applied field. However, a comparison of the results produced by these two processes allows the effects of thermal activation to be quantified by introducing a *coefficient of magnetic viscosity* S_v and a *fluctuation field*, first proposed by Néel in 1950 [20].

The coefficient of magnetic viscosity is defined as: $S_v = S / \chi_{irr}$, where χ_{irr} represents the irreversible susceptibility in the absence of effects of thermal activation. Around a given value of $M(t=0, T)$, a small variation of M is proportional to the corresponding value of $f(\Delta)$, and χ_{irr} may be written as:

$$\chi_{irr} = 2 M_0 f(\Delta) \partial \Delta / \partial H \qquad (6.47)$$

Thus, the expression of S_v is:

$$S_v = \frac{2 M_0 f(\Delta) k_B T}{2 M_0 f(\Delta) \partial \Delta / \partial H} = \frac{k_B T}{\partial \Delta / \partial H} \qquad (6.48)$$

This expression is only valid if the same value of $f(\Delta)$ is involved in the calculations of S and of χ_{irr}, which requires that precautions be taken when carrying out the experiments [25, 26].

The combination of equations (6.45) and (6.47) offers access to the expression of the variation in field $\delta H_{i,j}$ which will lead to the same variation in magnetization as the magnetic after-effect between times t_i and t_j:

$$\chi_{irr} \delta H_{i,j} = S \ln (t_i / t_j) \qquad (6.49)$$

This relation implies that the magnetization curves measured in the vicinity of the coercive field, with different delays t_i and t_j after the field steps cease, must be displaced with respect to each other by a field $\delta H_{i,j}$ which has the value:

$$\delta H_{i,j} = S_v \ln (t_i / t_j) \qquad (6.50)$$

By extension, the curve obtained at a time t after the field steps have ceased (time origin: t = 0), is displaced with respect to the hypothetical curve which would be obtained at time t = 0, by a field H_{eff}:

$$H_{eff} = S_v \ln(t/\tau_0) \qquad (6.51)$$

H_{eff} is known as the *fluctuation field* and, rather than the coefficient of magnetic viscosity (its usual denomination), S_v appears as the fluctuation field coefficient. Portions of the hysteresis loop drawn point by point for a number of times t in the vicinity of $H = H_C$ (fig. 6.35), have shown relation (6.50) to be true. The curve that an experiment, impossible to carry out, would have yielded at the time origin is deduced from equation (6.51).

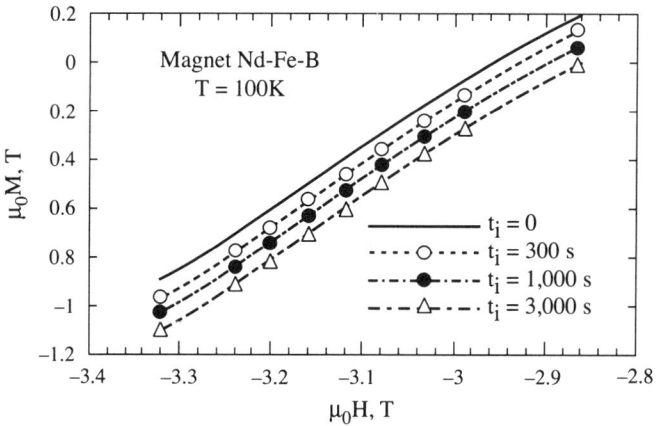

Figure 6.35 - Portions of the same area of the hysteresis loop close to the coercive field of a Nd-Fe-B magnet, obtained point by point at 100 K, after waiting for a variable time t_i

The full line, inaccessible to experiment, is deduced from the others with the help of equation (6.51).

The existence of fluctuation after-effects dictates caution and the need to take the effects of time into account, notably during measurements of the coercive field with the aid of field sweeping instruments (fig. 6.36).

By examining several series of materials with coercive fields spread over more than four orders of magnitude, Barbier has shown a possible link between H_C and S_v, which translates into the empirical relation: $S_v = H_C^{1.59}$ [28]. This relation still holds good today when one adds to it the points corresponding to modern magnets (NdFeB). An explanation was given (for the relationship $S_v \approx H_C^{1.5}$) [26], based on the concept of an activation volume, introduced by Wohlfarth in 1985 [29]. This activation volume, V, is the volume over which the effective fluctuation field H_{eff} could produce the effects of an actual field (see § 7.3.2. of chap. 15). In the area of hard materials and magnetic memory materials, these notions of field fluctuation and activation volume are proving very important for the analysis of observed behaviors.

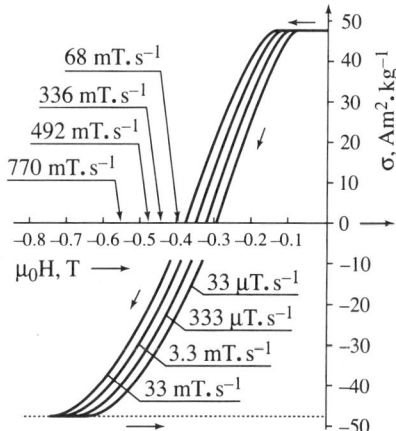

Figure 6.36
Portions of hysteresis loops of $SmCo_{3.5}Cu_{1.5}$ [27]

The loops are very sensitive to the velocity of field sweep. The higher it is, the higher is the measured coercive field.

For the highest sweep velocities, only the coercive field is indicated.

5.3. DIFFUSION AFTER-EFFECT

This phenomenon, related to reversible diffusion of atoms and/or of vacancies between interstitial sites of the lattice of a magnetic material, was first observed at the end of the 19th century, and has since been widely studied, by Richter [30] and many others [5]. It exists only in substances that can contain small size impurities in interstitial sites (carbon, nitrogen…), and it is visible only in the temperature range where diffusion of these impurities is high enough.

5.3.1. A localised, thermally activated uniaxial anisotropy

The material that best displays this phenomenon is iron (α-Fe, body centred cubic lattice), where carbon atoms can ocupy the sites, denoted as sites x, y, and z, respectively (fig. 6.37), located mid-way between two iron atoms along the three reference directions, Ox, Oy, and Oz. Snoek's [31] explanation is that when a carbon atom occupies a site, it increases the distance between the iron atoms surrounding it, thereby reducing the expenditure in magnetoelastic energy that accompanies the orientation of moments along the associated direction (in the case of positive magnetostriction).

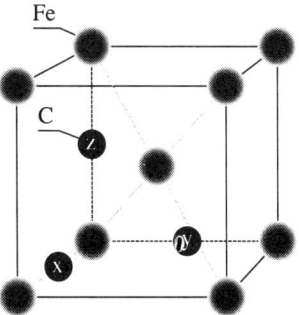

Figure 6.37 - Interstitial sites for carbon in α-iron

There exists, therefore, a relationship between the preferential occupancy of a family of sites and the direction of magnetization in the immediate vicinity of these sites. In other words, the total magnetic energy of a system at equilibrium, corresponding to a certain configuration of domains, is reduced progressively (constituting a potential well that becomes deeper and deeper) as the carbon atoms diffuse towards the sites associated with the direction of magnetization in the various domains. Snoek illustrates this process by comparing the magnetic system to a billiard ball lying on a support that has been overlaid with a putty-like substance that will give way under the weight, provided the ball stays in the same place for long enough (fig. 6.38).

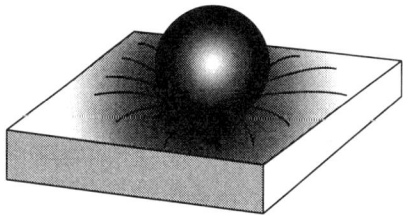

Figure 6.38 - Snoek's billiard ball model

Thus, considering that the magnetostrictive stress due to the presence of an atom in an interstitial position favours the alignment of the magnetic moments in the corresponding direction, Snoek describes this phenomenon through a local uniaxial anisotropy, induced by the stress. While retaining the basic principles of the mechanism imagined by Snoek, Néel assigns the anisotropy induced by atomic diffusion mainly to a dipolar effect (anisotropy of exchange) rather than to the magnetoelastic effect. This was confirmed by experimental work on α-Fe [32]. Furthermore a magnetic after-effect is also observed in substances showing no elastic deformation after-effect, in some ferrites for example.

Whatever the origin of the local uniaxial anisotropy produced by the diffusion of atoms, the importance of the model lies in the existence of this anisotropy induced by thermal activation, *via* diffusion. In the temperature range where diffusion is significant but not instantaneous, the application of a magnetic field cannot immediately result in an equilibrium configuration. The macroscopic magnetization evolves over a period of time, and accompanies the diffusion of impurities in the vicinity of domain walls, particularly 90° walls.

5.3.2. Disaccommodation

This phenomenon is a spectacular manifestation of the diffusion after-effect. It is shown in figure 6.39 and consists of a decrease, with time, of the AC permeability, after demagnetization through an evanescent AC field.

As a function the time t elapsed between the demagnetization and the measurement, the permeability varies as:

$$\mu(t) = \mu(\infty) + [\mu(0) - \mu(\infty)] \, \psi(t) \tag{6.52}$$

where $\psi(t)$ is a function with the value 1 when $t = 0$ and which falls exponentially towards zero when t tends towards infinity. The difference $\mu(0) - \mu(\infty)$ can go up to about 50% of $\mu(0)$: the phenomenon is thus considerable.

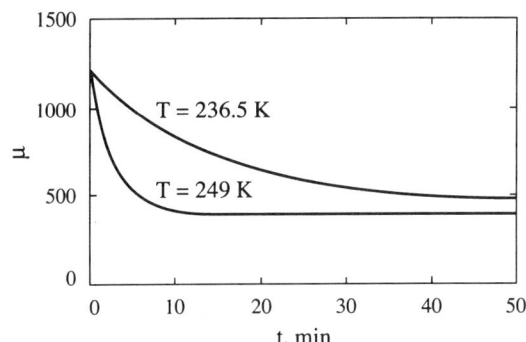

**Figure 6.39
Time evolution
of magnetic permeability**

The relative permeability of carbon steel changes the more quickly the higher the temperature.

Its causes are well represented by the model proposed by Snoek, as schematised in figure 6.38. When the billiard ball, which represents the magnetic system, rolls from one position to another, it is very mobile, and can be displaced easily even by weak forces. If, however, if it stops at some point and remains there for a while, it progressively loses mobility as it sinks deeper and deeper into the soft putty-like bed on which it lies (fig. 6.40).

Figure 6.40 - Like a billiard ball (with weight) resting on soft ground, the wall "digs its hole" after some period of time: a greater effort is then required to displace it

The permeability in weak fields measures the mobility of the walls. Its strong decrease as a function of the time elapsed since the last significant alteration of the field indicates that the walls have stabilised in their equilibrium position. This effect arises from the appearance of local uniaxial anisotropy resulting from diffusion in both the domains and the walls at the same time, since the direction of this anisotropy corresponds to the local direction of magnetization. The walls, by remaining in the same position for a long time, see their energy lowered. Every wall displacement, therefore, raises their energy by pulling them out of the potential well dug out at the time they were at rest. The force which tends to hold them in this well can be regarded as a restoring force, to be added to those already considered in section 2 of this chapter, which were associated with the lattice defects. Thus the equilibrium relation (6.5) must now be written as:

$$F_r + F_d + F_H = 0 \tag{6.53}$$

where F_d, which takes the anisotropy induced by thermal activation *via* diffusion into account, will take the form:

$$F_d(u) = K_a f'(u) \tag{6.54}$$

or

$$F_d(u) = K_a f''(u) \tag{6.55}$$

respectively, for a 90° or a 180° wall moved over a distance u [33]; K_a is a constant, characteristic of the material, and is related to the uniaxial anisotropy energy by:

$$E_K = -K_a \left(\alpha_1^2 \beta_1^2 + \alpha_2^2 \beta_2^2 + \alpha_3^2 \beta_3^2 \right) \tag{6.56}$$

The α_i and β_i (i = 1, 2, 3, ...) are the direction cosines of magnetization and of the induced anisotropy direction, respectively. The forms of the functions f'(u) and f''(u) given by Néel are shown in figures 6.41, and 6.42. In the second case, the function goes to zero if u goes to infinity, since no long-range magnetostrictive stress is induced in the presence of a 180° wall.

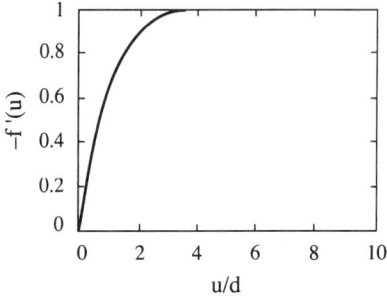
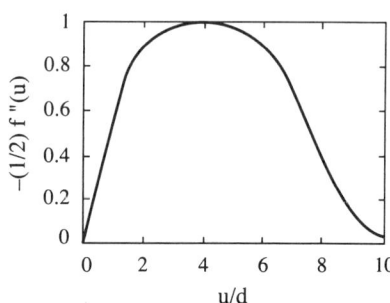

Figure 6.41 - Variation of $-f'(u)$ as a function of u/d

Figure 6.42 - Variation of $-f''(u)/2$ as a function of u/d

The distance d is defined by relation: $d = d_0(E/6K)^{1/2}$, where d_0 is the distance between magnetically interacting atoms, E is the molecular field energy and K the magnetocrystalline anisotropy energy.

REFERENCES

[1]   M. KERSTEN, *Phys. Zeits.* **39** (1938) 860; M. KERSTEN, *in* Probleme der Technischen Magnetisierungskurve (1938) 42, R. Becker Ed., Springer, Berlin.

[2]   L. NÉEL, Complete works *A* **45** (1978) 269 CNRS Publications, Paris; *Cahiers de Physique* **12** (1942) 1; *Cahiers de Physique* **13** (1943) 1.

[3]   S. CHIKAZUMI, Physics of Magnetism (1964) Wiley & Sons, New York.

[4]   R. BECKER, *Phys. Zeits.* **33** (1932) 905.

[5]   R. BECKER, W. DÖRING, Ferromagnetismus (1939) J. Springer, Berlin.

[6]   E. KONDORSKY, *Phys. Z. Sowjet.* **11** (1937) 597.

[7]   K. HOSELITZ, *in* Ferromagnetic Properties of Metals and Alloys (1952) Oxford University Press, Oxford.

[8]   M. KERSTEN, *Phys. Zeits.* **44** (1943) 63.

[9]   L. NÉEL, *Cahiers de Physique* **25** (1944) 21.

[10] L. NÉEL, *Annales Univ. Grenoble* **22** (1946) 299.

[11] M. KERSTEN, *Z. Angew. Phys.* **7** (1956) 313.

[12] W.F. BROWN JR., *Micromagnetics* (1963) 68 Interscience Publishers, New York.

[13] W.F. BROWN JR., *Rev. Mod. Phys.* **17** (1945) 15.

[14] LORD RAYLEIGH, *Phil. Mag.* **23** (1887) 225.

[15] P. MOLHO, *Thesis* (1986) University of Grenoble.

[16] J.L. PORTESEIL, Techniques de l'Ingénieur, Electronique E **1730** (1989) 1.
[17] F. PREISACH, Z. Phys. **94** (1935) 277.
[18] P. WEISS, J. DE FREUDENREICH, Arch. Sci. (Genève) **42** (1916) 449.
[19] K.J. SIXTUS, L. TONKS, Phys. Rev. **37** (1931) 930; Phys. Rev. **42** (1932) 419.
[20] L. NÉEL, J. Phys. Rad. **11** (1950) 49; J. Phys. Rad. **12** (1951) 339.
[21] R. STREET, J.C. WOOLLEY, Proc. Phys. Soc. A **62** (1949) 562.
[22] J.M. GONZALEZ, R. RAMIREZ, R. SMIRNOV-RUEDA, J. GONZALEZ, Phys. Rev. B **52**, n°22 (1995) 16034.
[23] J.C. TOUSSAINT, B. KEVORKIAN, D. GIVORD, M.F. ROSSIGNOL, Proc. 9th Intern. Symposium on Magnetic Anisotropy and Coercivity in Rare-Earth Transition-Metals Alloys, Vol. 2 (1996) F.P. Missel, V. Villas Boas, H.R. Rechenberg Eds, World Scientific, Singapore.
[24] A. LABARTA, O. IGLESIAS, L. BALCELLS, F. BADIA, Phys. Rev. B **48** (1993) 10240.
[25] D. GIVORD, M.F. ROSSIGNOL, V. VILLAS-BOAS, F. CEBOLLADA, J.M. GONZALEZ, Proc. 9th Intern. Symposium on Magnetic Anisotropy and Coercivity in Rare-Earth Transition-Metals Alloys, Vol. 2 (1996) 21, F.P. Missel, V. Villas Boas, H.R. Rechenberg Eds, World Scientific, Singapore.
[26] Y. ESTRIN, P.G. MCCORMICK, R. STREET, J. Phys. Condens. Mat. **1** (1989) 4845.
[27] M. UEHARA, J. Appl. Phys. **49** (1978) 4155.
[28] J.C. BARBIER, Ann. Phys. **9** (1954) 84.
[29] E.P. WOHLFARTH, J. Phys. F **14** (1985) L155.
[30] G. RICHTER, Ann. Phys. **29** (1937) 605; Ann. Phys. **32** (1938) 683.
[31] J.L. SNOEK, Physica **5** (1938) 663; Ann. Phys. **6** (1939) 161.
[32] G. DE VRIES, D.W. VAN GEEST, R. GERSDORF, G.W. RATHENAU, Physica **25** (1959) 113.
[33] L. NÉEL, J. Phys. Rad. **13** (1952) 249.

THEORETICAL APPROACH TO MAGNETISM

CHAPTER 7

MAGNETISM IN THE LOCALISED ELECTRON MODEL

(This chapter can be skipped on first reading; it assumes that the reader is aware of the basics of quantum mechanics).

In this chapter the two contributions (spin and orbital) to the angular momentum (and thus to the magnetic moment) of an electron are introduced. The way in which each electron in a free atom or ion contributes to the total magnetic moment is described using quantum mechanics. While most elements have a magnetic moment when isolated, only a few of these retain a moment when they form part of an arrangement of atoms such as a molecule, liquid or solid. Such elements are those which have unfilled 3d, 4f or 5f shells of electrons, and thus appear in the iron, lanthanide (or rare earth), and actinide series of elements. Their properties are briefly described.

1. MAGNETISM OF A FREE ATOM OR ION

1.1. A SINGLE ELECTRON

1.1.1. Orbital magnetic moment

In chapter 2 the magnetic moment m associated with a current density j occupying a volume V is given as:

$$m = \frac{1}{2}\int_V \mathbf{r} \times \mathbf{j}(r)dV \qquad (7.1)$$

Now consider an electron within an atom. Let $\mathbf{v}$ be its velocity, and $\mathbf{r}$ its position at a given time, thus:

$$\mathbf{j}(\mathbf{r'}) = -e\,\mathbf{v}\,\delta(\mathbf{r'}-\mathbf{r}) \qquad (7.2)$$

where $-e$ is the charge of the electron ($e = 1.6 \times 10^{-19}$ C). The distribution $\delta(\mathbf{r})$ has dimensions of inverse volume due to its integral over space being unity.

Putting this expression into equation (7.1), one obtains the orbital magnetic moment (i.e. that corresponding to the movement of the electron in its orbit):

$$m_o = -(e/2)\, r \times v = -(e/2m_e)\, \pounds_o \qquad (7.3)$$

where $\pounds_o = r \times m_e v$ is the orbital angular momentum of the electron, and m_e is its mass. This general result shows that *the orbital magnetic moment of a charged particle is proportional to its angular momentum.*

It is straightforward to arrive at equation (7.3) using the simple minded representation given in figure 7.1 of an electron travelling with velocity v on a circular orbit of radius r.

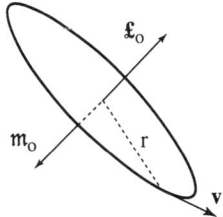

Figure 7.1 - Schematic representation of the orbital moment of an electron

In order to develop this idea further, one needs to make use of quantum mechanics. The stationary states of an electron experiencing the potential of the nucleus and the average potential from all of the other electrons are characterised by 4 quantum numbers n, ℓ, m, and σ.

Remember that:
- the *principal* quantum number n takes the values 1, 2, 3, 4, ...
- for a given n, the *orbital angular momentum* quantum number ℓ can take the integer values such that $0 \le \ell \le n-1$. For $\ell = 0, 1, 2, 3, ...$ the states are known as s, p, d, f, ...
- for a given ℓ, the *magnetic* quantum number m can take the integer values such that $-\ell \le m \le \ell$,
- the *spin* quantum number σ can take the values $\pm 1/2$.

The orbital angular momentum associated with these states is written as:

$$\pounds_o = \hbar \ell \qquad (7.4)$$

where $\hbar$ is Planck's constant divided by 2π, $\hbar = h/2\pi = 1.054 \times 10^{-34}$ J.s, and ℓ is a dimensionless vector operator often called the orbital angular momentum. The values of ℓ^2 and ℓ_z are characterised by two integer quantum numbers ℓ and m such that, for a state where these are good quantum numbers:

$$<\ell^2> = \ell(\ell+1) \quad \text{and} \quad <\ell_z> = m, \qquad (7.5)$$

where the z axis is the quantization axis (e.g. the direction of a magnetic field). For a state having a given ℓ, m can take the $2\ell + 1$ integer values: $m = \ell, (\ell-1), ..., 0, ..., -(\ell-1), -\ell$.

7 - MAGNETISM IN THE LOCALISED ELECTRON MODEL

The fact that the angular momentum and orbital magnetic moment (eq. 7.3) are proportional leads to:

$$m_o = -(\hbar e/2m_e)\,\ell = -\mu_B\,\ell \quad (7.6)$$

where $\mu_B = (\hbar e/2m_e) = 0.92732 \times 10^{-23}$ A.m² is the "*Bohr magneton*".

This is the smallest possible value in this situation of an electronic magnetic moment. This is the reason why it is often used as the unit of measurement of magnetism at the atomic level.

Even though a spatial representation cannot take into account all of the subtleties of quantum mechanics, it does allow one to visualise the properties of magnetic moments and orbital angular momentum. Figure 7.2 is an example of such a representation for $\ell = 2$.

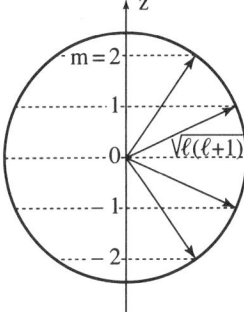

Figure 7.2 - Spatial representation of orbital angular momenta

While in classical mechanics the angular momentum can take any direction and any length, quantum mechanics shows that its length, and projection onto a given axis, can only take discrete, well defined values.

1.1.2. Spin magnetic moment

Orbital magnetism is not the only contribution that the electrons make to the overall magnetism. Stern and Gerlach showed experimentally that the electron also has a magnetic moment deriving from an intrinsic angular momentum which has come to be known as *spin* due to the practical, but inaccurate, way of imagining this angular momentum as due to the electron spinning around its axis. The electron can only have two spin states characterised by $\sigma = \pm 1/2$. The associated angular momentum is written as:

$$\mathcal{L}_s = \hbar s \quad (7.7)$$

where **s** is a dimensionless vector operator called the "*spin*". The eigenvalues of s^2 and s_z are characterised by the quantum numbers $s = 1/2$ and $\sigma = \pm 1/2$. For a given state σ:

$$\langle s^2 \rangle = s(s+1) = 3/4 \quad \text{and} \quad \langle s_z \rangle = \sigma \quad (7.8)$$

In a similar way to the orbital magnetic moment, the spin magnetic moment is proportional to the angular momentum. In this case however the constant of proportionality is twice as large as that for the orbital moment:

$$\mathfrak{m}_s = -2\mu_B \mathbf{s} \qquad (7.9)$$

It turns out that $<(m_s)_z> = \pm 1\,\mu_B$. The Bohr magneton is again apparent, and this confirms its role as the fundamental quantity of atomic magnetism.

The properties of spin magnetic moments are shown in a schematic way in figure 7.3.

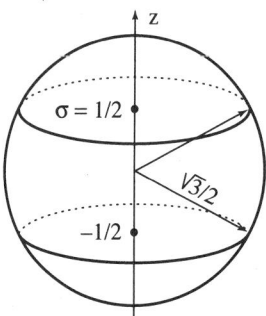

Figure 7.3 - Space representation of the spin angular momenta

Equations (7.6) and (7.9) show that the total magnetic moment $\mathfrak{m}_t = \mathfrak{m}_0 + \mathfrak{m}_s$ is not in general collinear with the total angular momentum $\mathbf{\pounds}_t = \hbar(\boldsymbol{\ell} + \mathbf{s})$, apart from special cases which will be returned to later on.

Note - *Every particle has a magnetic moment, and an intrinsic angular momentum. As an example consider the proton:* $\mathfrak{m} = g(\hbar e/2m_p)\,\pounds$ *where e and m_p are the charge and mass of the proton respectively. The factor g is not equal to 2 as in the case of the electron, but 2.793. The neutron does not carry an electric charge, but it does have a magnetic moment, and an intrinsic angular momentum linked by the same formula, with g = 1.913. It is worthy of special note that, because the masses of these particles are approximately 2,000 times that of the electron, their magnetic moments are three orders of magnitude smaller than that of the electron. Their moments are thus negligible on a macroscopic scale.*

1.1.3. States of individual electrons or hydrogen like atoms

The main steps leading to the description of the atomic states of a single electron are reviewed. For further details the reader should consult one of the many books on the subject [1].

The stationary states of an electron experiencing a central potential V(**r**) are obtained from Schrödinger's equation:

$$\mathcal{H}\phi = E\phi \qquad (7.10)$$

where the hamiltonian $\mathcal{H}$ is given by $p^2/2m_e + V(r)$*. For the hydrogen atom, the potential is simply $V(r) = -e^2/4\pi\varepsilon_0 r$ where r is the distance between the electron and the nucleus. For other atoms, it will become apparent that this is the central potential of the nucleus ($-Ze^2/4\pi\varepsilon_0 r$), plus an average spherical distribution coming from the other electrons.

The solution of equation (7.10) leads to the eigenenergies $E_{n,\ell}$, associated with wavefunctions $\phi_{n,\ell,m}(\mathbf{r}) = R_{n,\ell}(r) Y_\ell^m(\theta,\varphi)$ which are the product of radial, $R_{n,\ell}(r)$, and angular parts (the $Y_\ell^m(\theta,\varphi)$ are *spherical harmonics*, defined in appendix 6, § A6.1). The probability density of an electron being present at a point $\mathbf{r}$ is equal to the product $\phi\phi^*$.

For hydrogen, the energy levels only depend on n, and can be written as $E_n = -E_R/n^2$ where the "Rydberg" $E_R = 13.6$ eV corresponds to the energy of the ground state (the lowest possible energy level).

For other atoms (or ions) each $E_{n,\ell}$ depends on the atomic number of the element, and on the number of electrons considered in the central potential. Figure 7.4 shows schematically the energy levels (or rather their square root) for an electron when the lower levels are filled up with other electrons. This scheme allows one to understand the way in which each shell of electrons fills up.

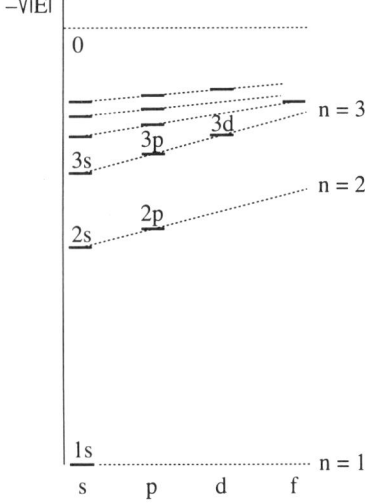

Figure 7.4 - Energy levels of an electron in a atom

The total wave function ϕ of an electron is the product of the orbital wave function above, and a spin wavefunction: $\psi_{n,\ell,m,s} = \phi_{n,\ell,m}(\mathbf{r}) \cdot \chi_\sigma$, where χ_σ corresponds to the spin state characterised by σ.

Each energy level, or each *shell*, is thus $2(2\ell + 1)$ times degenerate, i.e. there are $4\ell + 2$ different electronic states with the same energy.

---

* **p** is the momentum operator.

1.2. MANY ELECTRON ATOMS

1.2.1. Hartree's method - The central field approximation: configurations

The electronic hamiltonian of an atom or ion with many electrons is given by:

$$\mathcal{H} = \mathcal{H}_e + \mathcal{H}_{s.o.} \tag{7.11}$$

If Ze is the charge of the nucleus, and N is the number of electrons, $\mathcal{H}_e$ is written:

$$\mathcal{H}_e = \sum_{i=1}^{N}\left[\frac{p_i^2}{2m_e} - \frac{Ze^2}{4\pi\varepsilon_0 r_i}\right] + \sum_{i>k}\frac{e^2}{4\pi\varepsilon_0 r_{ik}} \tag{7.12}$$

$\mathcal{H}_{s.o.}$ is the spin-orbit coupling hamiltonian. Its origin will become clear later on. Its contribution is small compared to the other terms. Schrödinger's equation for such a system is impossible to solve directly. However, this problem can be circumvented by re-writing $\mathcal{H}_e$ in the following way:

$$\mathcal{H}_e = \mathcal{H}_0 + \mathcal{H}_1 \tag{7.13}$$

where:

$$\mathcal{H}_0 = \sum_i\left[\frac{p_i^2}{2m_e} - \frac{Ze^2}{4\pi\varepsilon_0 r_i} + U(r_i)\right] \tag{7.14}$$

and:

$$\mathcal{H}_1 = \sum_{i>k}\frac{e^2}{4\pi\varepsilon_0 r_{ik}} - \sum_i U(r_i) \tag{7.15}$$

U(r) is a fictitious potential such that $\mathcal{H}_1$ is sufficiently small compared to $\mathcal{H}_0$ so that it can be treated as a perturbation. U(r) is found by an iterative procedure (self consistent Hartree field), and represents reasonably well the average repulsive spherical potential acting on an electron as a result of all of the other electrons. In a first approximation we only consider $\mathcal{H}_0$, thus considering a collection of independent electrons subjected to a central potential. The description of the state of an atom is then very simple: one just has to state that amongst the N electrons there is one in a $\psi_{n1,\ell1,m1,s1}$ state, another in a $\psi_{n2,\ell2,m2,s2}$ state, and so on. The energy of the atomic state is thus:

$$E = E_{n1,\ell1} + E_{n2,\ell2} + \ldots + E_{nN,\ell N} \tag{7.16}$$

Such a state is called a *configuration*. The configuration of lowest energy is found by successively filling the individual states of lowest energy conforming to the level series in figure 7.4. An easy way to find it is to use the filling rule shown in figure 7.5. For an iron atom the atomic number is $Z = 26$, the ground state configuration is written as: $1s^2\,2s^2\,2p^6\,3s^2\,3p^6\,4s^2\,3d^6$, which is the configuration of Ar + $3d^6$ where Ar refers to argon.

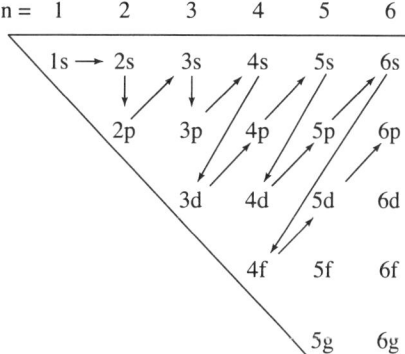

Figure 7.5 - Filling rule of the electronic energy levels

The ground state configurations of neutral atoms are generally shown in the periodic table of the elements: see appendix 3 at the end of this book, where the configurations are only given by the occupation of the outermost shells, for example $3p^3$ for phosphorus, but $3d^8 4s^2$ for nickel because the 4s shell is populated before the 3d shell is fully filled. This shows that there are small irregularities to the filling rule when the shells have energies which are very close. This is the case for the 4s and 3d states, as well as the 4f and 5d states.

Note - *A configuration where all of the shells are full is non degenerate, i.e. there is only one wavefunction associated with each energy level. On the other hand, a configuration where the outer shell is not full is degenerate.*

Consider for example the carbon atom, which has two p electrons: if one looks at all of the possible values of m and σ for each electron, and considers these to be indistinguishable, one obtains a degeneracy of 15 as shown in figure 7.6: the 15 electronic states have the same energy.

Figure 7.6 - Example of different states associated with the configuration $1s^2$, $2s^2$, $2p^2$ of the carbon atom (Z = 6): the multiplicity (the number of different states) is 15

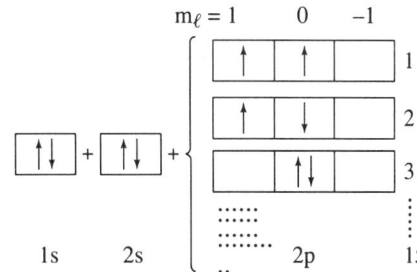

Total orbital and spin angular momenta: The individual orbital and angular momenta add up to give resultant orbital and spin angular momenta:

$$\mathbf{L} = \Sigma \, \boldsymbol{\ell} \quad \text{and} \quad \mathbf{S} = \Sigma \, \mathbf{s}_i \qquad (7.17)$$

For a full shell:

$$\langle L_z \rangle = \Sigma \langle \ell_{iz} \rangle = 0 \quad \text{and} \quad \langle S_z \rangle = \Sigma \langle s_{iz} \rangle = 0 \qquad (7.18)$$

The same result applies for the x and y components. *A full shell is not magnetic*, and thus does not carry an intrinsic magnetic moment.

1.2.2. Terms

The interesting cases to consider from the point of view of magnetism are the cases where there are unfilled shells. The term $\mathcal{H}_1$, which describes *intra-atomic correlations* between the electrons, partially lifts the degeneracy of each configuration, and leads to energy levels known as *"terms"*. Each term is characterised by the quantum numbers L and S, which in turn characterise the eigen states and eigen values of $\mathbf{L}^2$ and $\mathbf{S}^2$. States of a given term are of the type $|L, S, M_L, M_S\rangle$ where M_L and M_S can take any integer value from $-L$ to $+L$, and $-S$ to $+S$ respectively. Each term is thus $(2L + 1)(2S + 1)$ times degenerate.

The shift in energy of the terms is typically of the order of 10 eV, (i.e. about 10^5 K: physicists often measure atomic energies in units of temperature using $E \leftrightarrow k_B T$, i.e. 1 kelvin is equivalent to an energy equal to $k_B = 1.38 \times 10^{-23}$ J (app. 2)).

The values of L and S of the term with the lowest energy are given by the relationship:

$$L = |\Sigma m_i| \quad \text{and} \quad S = |\Sigma \sigma_i| \tag{7.19}$$

where the individual m_i and σ_i are determined by *Hund's rules*, which state that (see chap. 8, § 6.2):
- the values of the individual spins σ_i are those which maximise S, and are compatible with the Pauli exclusion principle. Thus for carbon, $S = 1/2 + 1/2 = 1$.
- the values of the individual orbital angular momenta m_i are those which maximise L, and are compatible with the first rule and the Pauli exclusion principle. For carbon, $L = 1 + 0 = 1$.

Hund's rules are easy to apply using the classic representation of figure 7.6 for the case of carbon. The first rule takes into account on the one hand the fact that the electrons prefer to have the same component of spin, which reduces the probability of passing from one orbit to another that is already occupied (Pauli's principle), and on the other, the desire to be in different orbits as they have less chance to find themselves near one another.

1.2.3. Spin-orbit coupling

Before embarking on a detailed description of electronic energy levels, the fundamental concept of *spin-orbit* coupling must be introduced. This is the interaction between the individual orbital and spin angular momenta. It has the form $-\lambda_{ij} \ell_i s_j$, and is also known as *Russell-Saunders* coupling. It turns out that this coupling is negligible for terms where $i \neq j$, compared to those where $i = j$. The origin of this coupling is the following: in the referential of the electron, the motion of the nucleus produces a magnetic field which interacts with the spin magnetic moment. As a consequence of the coupling between the spin and orbital magnetic moments, leading to different *terms*, the individual couplings reduce to the following interaction energy:

$$\mathcal{H}_{so} = -\lambda \, \mathbf{L}.\mathbf{S} \tag{7.20}$$

where λ is negative for a shell less than half full, and positive for the opposite case, so that the spin and orbital moments prefer to be anti-parallel and parallel respectively. In particular λ is negative in the case of a single electron [2].

1.2.4. Multiplets

The degeneracy of each term is partially lifted by the spin-orbit interaction outlined above. Each new energy level, known as a "*multiplet*", is characterised by an integer or half integer quantum number J which in turn characterises the total angular momentum $\hbar \mathbf{J}$ such that:

$$\mathbf{J} = \mathbf{L} + \mathbf{S} \tag{7.21}$$

Within a multiplet, there exists a basis of 2J + 1 states | L, S, J, M_J > such that M_J can take the values + J to – J in steps of one. They are the eigen states of $\mathbf{J}^2$ and J_z such that:

$$\mathbf{J}^2 | L, S, J, M_J \rangle = J(J + 1) | L, S, J, M_J \rangle \tag{7.22}$$

and

$$J_z | L, S, J, M_J \rangle = M_J | L, S, J, M_J \rangle \tag{7.23}$$

Each multiplet is thus 2J + 1 times degenerate. The values of J of the different multiplets arising from the same term vary from L + S to |L – S| in steps of one. By squaring equation (7.21) it can be deduced that $\mathcal{H}_{so} = -(\lambda/2)(\mathbf{J}^2 - \mathbf{L}^2 - \mathbf{S}^2)$. The energy of a multiplet is thus given by:

$$E_J = \langle \mathcal{H}_{so} \rangle = -(\lambda/2)[J(J+1) - L(L+1) - S(S+1)].$$

The ground state multiplet is such that J = L + S if the shell is more than half full, and J = |L – S| when it is less than half full. If the shell is exactly half full, J = S as Hund's rules give L = 0. The spin-orbit coupling increases with the atomic number of the element under consideration. The difference between two multiplets is typically of the order of 10^{-2} to 10^{-1} eV (10^2 to 10^3 K) for elements in the iron group, and 1 eV (10^4 K) for the rare earths. The schematic diagram in figure 7.7 shows, for the case of the Cr^{3+} ion, the different stages leading to the ground state energy of a free atom (or ion).

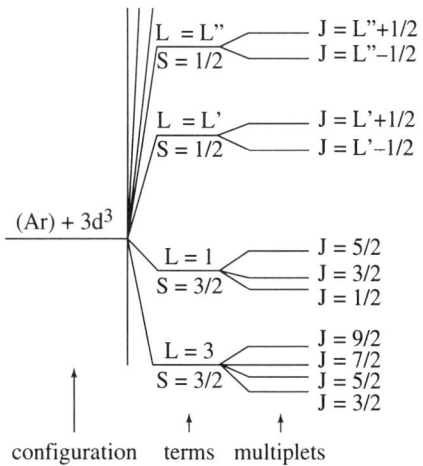

Figure 7.7 - *Splitting of the ground state configuration of the Cr^{3+}ion into its terms, and multiplets*

At this stage, the further liftings of the degeneracy can only take place as a result of external perturbations such as a magnetic field (Zeeman effect), or the effects of neighbouring atoms when the atom is no longer free, and is part of a solid. The multiplets are generally referred to by symbols such as $^2S_{1/2}$, $^3P_0$, $^4F_{9/2}$, etc. This is the *spectroscopic notation* where the letters S, P, D, F, G, H, I, ... refer to the total orbital angular momentum L = 0, 1, 2, 3, 4, 5, 6, ... respectively. The numbers at the top left, and bottom right are 2S + 1, and J respectively.

As a result of the energy differences between the different multiplets, one can neglect the effect of all but the lowest lying multiplet when considering the rare earth elements (4f series). At normal temperatures the occupation of the higher multiplets is negligible.

In addition to the electronic structure of the free atom, the periodic table in appendix 3 also gives the spectroscopic ground states. It can be seen from table 3.1 that all non magnetic free atoms in their ground state are characterised by J = 0 (spectroscopic levels $^1S_0$, $^3P_0$, $^5D_0$, etc.). It will be shown in what follows that the magnetic moment of an atom or ion is strictly linked to its quantum number J.

Consider the magnetic moment associated with the orbital angular momentum of a multiplet. Using equation (7.6), the orbital magnetic moment is written as:

$$m_o = -\mu_B L \tag{7.24}$$

Similarly, using equation (7.9), the spin magnetic moment is written as:

$$m_S = -2\mu_B S \tag{7.25}$$

As has been already mentioned, equations (7.24), and (7.25) show that the total magnetic moment $m = m_o + m_S$ has no constraint that forces it to be collinear with the total angular momentum $\hbar J = \hbar(L + S)$. In fact, quantum mechanics reveals that *within each multiplet* the total magnetic moment and the total angular momentum can be considered to be collinear, and linked by the formula:

$$m = -g_J \mu_B J \tag{7.26}$$

where g_J, the *Landé g factor*, characterises the multiplet under consideration. It is expressed as a function of J, L, and S such that:

$$g_J = 1 + \frac{J(J+1) + S(S+1) - L(L+1)}{2J(J+1)} \tag{7.27}$$

g_J is 1 or 2 when only the orbital or spin contributions are present respectively, but is not forced to lie within these values. As an example one can see from table 7.1 that g_J is less than one when L > S, and when J = L − S.

The main magnetic properties of R^{3+} ions in the rare earth series are given in table 7.1. It is in this valence state that these elements are found in most materials. Note however that cerium can also be tetravalent, and samarium, europium, and ytterbium can be divalent.

7 - MAGNETISM IN THE LOCALISED ELECTRON MODEL

Note - *The above result has been established for the almost ideal case of isolated atoms. It will become apparent in what follows that it holds for substances where the magnetic atom is not too perturbed by interatomic interactions. Irrespective of the substance under consideration, it is always possible to write that the magnetic moment* m *and the angular momentum* $\pounds$ *of the atoms are proportional:*

$$m = -g \frac{e}{2m_e} \pounds = \gamma \pounds \tag{7.28}$$

γ is the gyromagnetic factor. Its value allows the estimation of the relative size of the orbital and spin contributions ($\gamma = -(e/2m_e)$ and $\gamma = -2(e/2m_e)$ respectively) to the magnetic moment in a given substance.

Points to remember:
- Filled electronic shells are not magnetic.
- Only non saturated shells have a magnetic moment.
- The magnetic moment of free atoms or ions is given by: $m = -g_J \mu_B \mathbf{J}$ where $\hbar \mathbf{J}$ is the total angular momentum. For a given configuration, the quantum number J, and thus g_J, are determined by Hund's rules.

Table 7.1 - The main properties of free R^{3+} ions of the elements in the rare earth or lanthanide group R

Ion 4f	$^{2S+1}L_J$	L	S	J	gJ	m_0 (μ_B)	m_{eff} (μ_B)
Ce^{3+} ($4f^1$)	$^2F_{5/2}$	3	1/2	5/2	6/7	2.14	2.54
Pr^{3+} ($4f^2$)	$^3H_4$	5	1	4	4/5	3.20	3.58
Nd^{3+} ($4f^3$)	$^4I_{9/2}$	6	3/2	9/2	8/11	3.27	3.62
Pm^{3+} ($4f^4$)	$^5I_4$	6	2	4	3/5	2.40	2.68
Sm^{3+} ($4f^5$)	$^6H_{5/2}$	5	5/2	5/2	2/7	0.71	0.85
Eu^{3+} ($4f^6$)	$^7F_0$	3	3	0	–	0	0
Gd^{3+} ($4f^7$)	$^8S_{7/2}$	0	7/2	7/2	2	7.00	7.94
Tb^{3+} ($4f^8$)	$^7F_6$	3	3	6	3/2	9.00	9.72
Dy^{3+} ($4f^9$)	$^6H_{15/2}$	5	5/2	15/2	4/3	10.00	10.65
Ho^{3+} ($4f^{10}$)	$^5I_8$	6	2	8	5/4	10.00	10.61
Er^{3+} ($4f^{11}$)	$^4I_{15/2}$	6	3/2	15/2	6/5	9.00	9.58
Tm^{3+} ($4f^{12}$)	$^3H_6$	5	1	6	7/6	7.00	7.56
Yb^{3+} ($4f^{13}$)	$^2F_{7/2}$	3	1/2	7/2	8/7	4.00	4.53

The concepts of the modulus of the moment (m_0) and the effective moment (m_{eff}) were introduced in chapter 4, section 2.2. La^{3+} and Lu^{3+}, with their 4f shell being respectively empty and full, are non magnetic.

2. MAGNETISM OF BOUND ATOMS

2.1. LOCALISED AND ITINERANT MAGNETISM

The description of the magnetism of atoms making up different substances such as molecules, liquids, solids, ... is more complicated than that of free atoms as this becomes a many body problem that is impossible to solve exactly.

Two models have been developed, both of which are only approximations of an often more complicated reality. The first assumes that the magnetic electrons are localised at the atomic sites, and can be found in states that are similar to those of the free atom or ion. This is the model of *magnetism of localised electrons*. On the other hand *the model of itinerant electrons* considers that the magnetic electrons are the conduction electrons which are totally delocalised, and free to travel anywhere in the sample.

In reality, in a solid, the localised electrons –if these stay linked to the atomic site– find their orbits more or less perturbed by the electric field created by the neighbouring atoms or ions: the model of an isolated ion is thus not always a good starting point to work from. The free electrons in a metal feel the periodic electric potential created by the ionised atoms (due to the atoms losing their conduction electrons): these electrons are thus not really free, and the periodic potential of the lattice (in the case of a metal in the crystalline state) will modify the predictions of the itinerant electron model.

The *itinerant electron model* is a reasonable approximation for some transition metals and their alloys, but it remains rather qualitative. It will be treated in chapter 8.

The *localised electron model* applies essentially to insulators, and to the majority of materials with rare earth elements. It allows a more quantitative treatment. The remainder of this chapter focuses on this approach.

2.2. NON MAGNETIC MATERIALS

In order to understand what is happening when atoms become part of a substance, one can make use of certain results obtained in section 1. Thus, all filled atomic shells are not magnetic. In atomic systems, the electrons are often, in the ground state, in orbits that can only hold two electrons: if such orbits are occupied by two electrons, the resultant orbital and spin angular momenta, and magnetic moments are zero. A permanent magnetic moment thus comes from *unpaired electrons*.

It has already been stated in chapter 3 that most elements that are magnetic as a free atom or ion do not remain so once they form part of a solid substance because the bonds (molecular, metallic...) are obtained from bringing together on the same orbit electrons from different elements. Thus orbitals are formed, which are generally non magnetic. This result is illustrated by comparing tables 3.1 and 3.2.

The simplest case is hydrogen, which is magnetic in the atomic state with one electron. The molecule H_2 is not magnetic because, in the ground state, the two electrons are in the same orbital which has zero angular momentum, with opposite spins. This is exactly what happens in numerous organic molecules such as CH_4, where the four bonds are made by grouping two electrons on each of the four orbits linking the carbon to the hydrogen.

In the case of ionic compounds, the ionic bond is often due to the loss and gain of electrons on different elements, in such a way as to only have filled, i.e. non magnetic, shells. This is the case in common salt NaCl: the free atoms of Na and Cl are magnetic, but not when they become Na^+ and Cl^- ions.

The situation for metals, which will be developed in the next chapter, is more complicated. In a very simplified picture, the electrons forming the *conduction band* occupy orbits in pairs, their spins cancelling each other out in pairs. This is the case for transition metals having filled 4d (Y, Zr…) or 5d shells (Hf, Ta, …): these shells are relatively delocalised, and the electrons that occupy them can take part in bonding, at least in the case of non magnetic atoms.

In this way one labels as being *non magnetic* all substances which do not have a permanent magnetic moment. But even so they can still display weak magnetism: diamagnetism, and sometimes Pauli paramagnetism.

2.3. WHICH SUBSTANCES DISPLAY SIGNIFICANT MAGNETISM?

Only non saturated internal electronic shells (i.e. those protected by shells further out from the nucleus) can remain unfilled when an atom is incorporated into a multi-atomic system: these thus retain their magnetism.

This is the case for transition elements in the iron group (3d) and the rare earths (4f). Another series, the actinides characterised by occupation of the 5f shell, should be mentioned despite the fact that it has been less well studied due to the radioactivity of most of its members. Interest in these is thus from an academic rather than an applied perspective.

It is important to bear in mind that the magnetism of an element strongly depends upon its environment, and the value of its magnetic moment can display large variations depending upon the material it is contained in: the case of iron is particularly significant as is apparent from table 3.3.

Finally, the number of substances containing moment-carrying atoms (under no applied field) is relatively limited, and the number of elements that make a significant contribution to the magnetism of materials is about 15. These are the 3d elements Cr, Mn, Fe, Co, and Ni, as well as most of the rare earths, also known as the *lanthanides* (see tab. 3.2).

2.4. THE TWO FUNDAMENTAL SERIES OF MAGNETIC ELEMENTS

Two series of elements play an essential role in magnetism, as much at a fundamental level as in technological applications: on the one hand there are the iron group elements, characterised by a progressive filling of the 3d shell, and on the other the rare earth elements, characterised by a progressive filling of the 4f shell. The importance of these two series comes from their partially filled shells (3d, and 4f) which can hold a relatively large number of electrons, and which are deep enough to remain incomplete, even in a multi-atomic system: they thus carry a permanent magnetic moment.

2.4.1. The spatial distribution of electronic orbitals

The spatial extents of the outermost electrons of the 3d and 4f series are rather different. For example, the radial distribution fonctions ($|rR_{n,\ell}|^2$, i.e. the probability density of finding an electron at a distance r from the nucleus) of the shells have been calculated from their wavefunctions, and are shown in figure 7.8 for cobalt (3d) and gadolinium (4f) metals. One notes that the 4f shell is more localised than the 3d one. As a result, the magnetism of the 4f elements is less affected by bonding than that of the 3d elements.

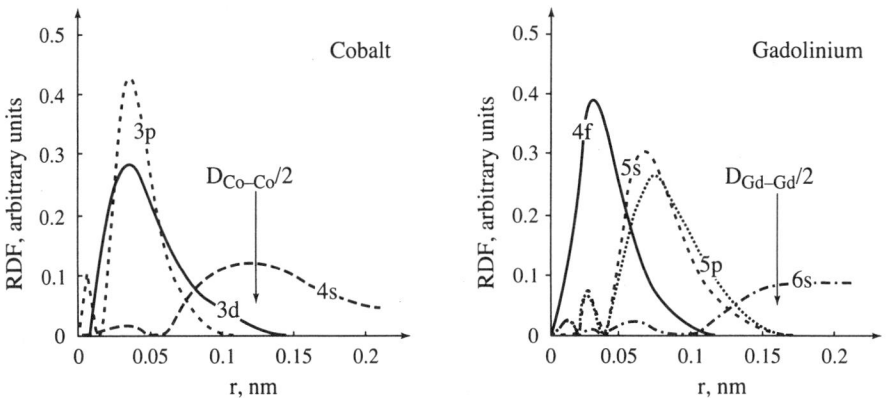

*Figure 7.8 - Radial distribution functions (RDF)
of the outer electronic shells of cobalt, and gadolinium*

RDF's are given in arbitrary units. The 3d electrons for Co, and the 4f electrons for Gd, are responsible for their magnetism. $D_{Co\text{-}Co}$, and $D_{Gd\text{-}Gd}$ represent the smallest Co-Co, and Gd-Gd distances, respectively in pure cobalt and gadolinium.

2.4.2. The influence of neighbouring atoms: crystal field effects

One frequently observes that metallic ions in multi-atomic structure have a magnetic moment different from that of the free ion $g_J J \mu_B$ (or $g_J [J(J+1)]^{1/2} \mu_B$ for the effective moment). This effect is particularly pronounced with 3d ions, where the magnetic moment can get close to the saturation value $2S\mu_B$ (or $2[S(S+1)]^{1/2} \mu_B$ for the effective moment) that one would observe if only the spin participated in the

7 - MAGNETISM IN THE LOCALISED ELECTRON MODEL

magnetic properties of the substance. It is as if the *orbital magnetic moment was practically zero*, and thus insensitive to the application of a magnetic field. Only the spin magnetic moment is left to "follow" the applied magnetic field.

This quenching of orbital moment is due to the crystalline electric field which will be discussed in section 4 of this chapter. It does not apply to the ions (Mn^{2+}, Fe^{3+}) having zero orbital moment (S states): their magnetic moment stays the same as that of the free ion, even in a multi-atomic structure. This effect will be discussed in section 4.3.1.

Table 7.2 shows the same data as table 7.1, but for the 3d ions. In addition the magnetic moment $2S\mu_B$ and effective moment $2[S(S+1)]^{1/2}\mu_B$, that are expected in case of the quenching of the orbital moment, are reported.

Table 7.2 - Magnetic properties of 3d ions

Ion 3d	$3d^n$	$^{2S+1}L_J$	g_J	m_0 (m_B)	2S	m_{eff} (m_B)	m^* (m_B)
Ti^{3+}, V^{4+}	$3d^1$	$^2D_{3/2}$	4/5	6/5	1	1.549	1.732
Ti^{2+}, V^{3+}	$3d^2$	$^3F_2$	2/3	4/3	2	1.633	2.828
V^{2+}, Cr^{3+}	$3d^3$	$^4F_{3/2}$	2/5	3/5	3	0.775	3.872
Cr^{2+}, Mn^{3+}	$3d^4$	$^5D_0$	–	–	4	–	4.899
Mn^{2+}, Fe^{3+}	$3d^5$	$^6S_{5/2}$	2	5	5	5.916	5.916
Fe^{2+}, Co^{3+}	$3d^6$	$^5D_4$	3/2	6	4	6.708	4.899
Co^{2+}, Ni^{3+}	$3d^7$	$^4F_{9/2}$	4/3	6	3	6.633	3.872
Ni^{2+}	$3d^8$	$^3F_4$	5/4	5	2	5.590	2.828
Cu^{2+}	$3d^9$	$^2D_{5/2}$	6/5	3	1	3.550	1.732

Expressed in units of μ_B/ion, the magnetic moment expected in the presence of chemical bonds will lie between that of the free ion, $m_0 = g_J J$ (and $m_{eff} = g_J\sqrt{J(J+1)}$ for the effective moment), and that for an ion with complete quenching of the orbital moment (grey columns): $m'_0 = 2S$ (and $m^* = 2\sqrt{S(S+1)}$ for the effective moment).

Even if the crystal field perturbs the 4f ions, J will still remain a good first approximation to the magnetic moment. J is however no longer a "good quantum number" to describe the magnetic properties of 3d ions.

Table 7.2 allows one to interpret the magnetization curve of the chromium salt $CrK(SO_4)_2 \cdot 12H_2O$ shown in figure 4.7: its saturation moment is 3 μ_B per molecule, whereas the free ion model predicts only 0.6. The Cr^{3+} ion, the only magnetic component in this substance, thus has $\langle L \rangle = 0$. It is no longer J, but S, in this case, that must enter in the summation (4.25) which describes the magnetization as a function of H/T.

3. SOME EXAMPLES OF LOCALISED MAGNETIC MOMENTS

3.1. IRON OXIDES

- Iron monoxide, FeO, has per molecule one Fe^{2+} ion, whose effective magnetic moment, as deduced from the variation of the paramagnetic susceptibility, is 5.33 μ_B [3]. This value lies between the free ion value of 6.71 μ_B, and that expected for quenched orbital angular momentum, viz 4.89 μ_B. This suggests a *partial quenching of the orbital moment*. The value observed here is in good agreement with those encountered in other compounds containing the Fe^{2+} ion: FeF_2 (5.59 μ_B), $FeCl_2$ (5.38 μ_B), FeS (5.24 μ_B) [4].

- Magnetite, Fe_3O_4, or $Fe^{2+}O^{2-}$ ($2Fe^{3+}3O^{2-}$), is a ferrimagnet: half of the Fe^{3+} ions, and the Fe^{2+} ions form the first sublattice, with the remaining Fe^{3+} ions forming the second one. In the free ion model, one expects to have the structure: $\uparrow(6\mu_B)\uparrow(5\mu_B)\downarrow(5\mu_B)$ which would give a resultant moment of $g_JJ = 6\ \mu_B$ per formula unit. However the spontaneous magnetization is found to be 4.1 μ_B per formula unit, which is very close to the value of 4 μ_B expected for complete quenching of the orbital moment of the Fe^{2+} ions.

- In ferrimagnetic maghemite (γ-Fe_2O_3) there are five Fe^{3+} ions making up one sublattice, and three on another. This gives a resultant moment of: $[(5-3)/(5+3)] \times 5\ \mu_B = 1.25\ \mu_B$ per iron atom, which is in excellent agreement with experiment as these are S state ions (§ 2.4.2). This oxide is used in powder form for magnetic memories (tapes, disks).

- Hematite, (α-Fe_2O_3), is more complicated as it is an antiferromagnet with a weak ferromagnetic component which is interpreted as arising from the anti-symmetric exchange interaction that will be discussed in section 2.2 of chapter 9.

- Yttrium-iron garnet $Y_3Fe_5O_{12}$ is a mixed oxide of iron and yttrium. With Y^{3+} being non magnetic, only the iron ions contribute to the magnetism. This oxide is ferrimagnetic with 3 Fe^{3+} ions on one sublattice, and two on the other: the expected resultant moment is thus that associated with Fe^{3+}, i.e. 5 μ_B. The experimental value (5 μ_B per formula unit at 0 K) is in perfect agreement with theory: these ions are again in the S state. This garnet is used in microwave electronics.

3.2. OTHER IONIC COMPOUNDS OF 3d METALS

Manganese monoxide, MnO, is an antiferromagnet. The effective moment deduced from the temperature dependence of the magnetic susceptibility above the Néel temperature (5.9 μ_B), is in excellent agreement with the theoretical value of $\sqrt{35}\ \mu_B$. The agreement does not come about by chance: as for the Fe^{3+} ion, Mn^{2+} is in an S state with zero orbital moment.

On the other hand, the effective moment deduced from paramagnetic susceptibility measurements is no more than 3.5 μ_B for NiF_2, and 3.3 μ_B for $NiCl_2$. These values lie between the values of 5.59 μ_B predicted for a free Ni^{2+} ion, and 2.83 μ_B expected for complete quenching of the orbital moment.

A similar situation exists for the Co^{2+} ion, with an effective moment of 6.63 μ_B predicted for the free ion, and 3.87 μ_B expected for complete quenching of the orbital moment. The experimental results of 4.9 μ_B for CoO, and 5.12 μ_B for CoF_2 are again between the two expected values. For the Ni^{2+}, and Co^{2+} ions there is thus a *partial quenching* of the orbital moment.

It has already been shown that there is complete quenching of the orbital moment of Cr^{3+} ions in an alum of chromium and potassium.

Finally, in certain situations, the influence of the crystal field is such that not only is J no longer a good quantum number, but Hund's rules are no longer valid. This is the case for Yttrium-iron garnet (YIG) when some of the Fe^{3+} ions are changed for Ru^{3+} ions. This should not change the magnetic moment as the electronic structures of ruthenium ($4d^6\ 5s^2$), and iron ($3d^6\ 4s^2$) are similar. It turns out that the Ru^{3+} ions are in a *low spin* quantum state such that L = 6, and S = 1/2 (↑↓↑↓↑) instead of L = 0 and S = 5/2 (↑↑↑↑↑).

The large orbital moment is thus completely quenched (L = 6 but <L> = 0) by the very strong crystal field, and only the weak moment from the spin is able to contribute to the magnetization of this substituted garnet YIG:Ru. A detailed analysis of these phenomena can be found for example in the work of Zeiger and Pratt [1].

In conclusion, the localised electron model in the approximation of free ions only gives good results for 3d ions when these are in S states (Fe^{3+}, Mn^{2+}). Otherwise the orbital moment is partly quenched, and there is a reduction of the magnetic moment. What has been left out here is a discussion of 3d, 4d, and 5d atoms in the metallic state as the localised electron model does not apply at all in these cases. The itinerant model has to be invoked to explain this, and is covered in the next chapter.

3.3. IONIC COMPOUNDS OF THE RARE EARTH ELEMENTS

Unlike the 3d ions, the effective magnetic moment observed in paramagnetic susceptibility measurements on a large number of ionic compounds with trivalent rare earth ions are very close to the theoretical values given in the last column of table 7.1. A few examples of this are the following [4]: $g_J [J(J+1)]^{1/2}$ = 2.51 (th.: 2.54) for CeF_3, and $CeCl_3$; 3.58 (th.: 3.62) for Nd_2Se_3; 7.94 (th.: 7.94) for Gd_2O_3; 9.67 (th.: 9.70) for Tb_2O_3; 7.28 (th.: 7.60) for Tm_2O_3. Note that it is again the ions with zero orbital angular momentum (Gd^{3+} is in an S state) for which the agreement is best.

On the other hand, the magnetic moment deduced from magnetization curves at low temperatures for rare earth compounds which are magnetically ordered are often different from the theoretical values given in the second from last column of table 7.1. In the rare earth garnets $R_3Fe_5O_{12}$ (RIG, where rare earth R^{3+} ions replace the Y^{3+} ion in YIG § 3.2), Chikazumi gives the following experimental values: $m_0 = 0.14\ \mu_B$ (th. 0.71) for Sm^{3+}; 0.74 μ_B (th. 0) for Eu^{3+}; 7 μ_B (th. 7) for Gd^{3+}; 7.7 μ_B (th. 9) pour Tb^{3+}; 7.3 μ_B (th. 10) pour Dy^{3+}; 6.7 μ_B (th. 10) for Ho^{3+}; 5.1 μ_B (th. 9) for Er^{3+}; 1.3 μ_B (th. 7) for Tm^{3+}, and 1.7 μ_B (th. 4) for Yb^{3+}. Excellent agreement is observed only for the Gd^{3+} ion; the general disagreement observed for the other ions is mainly due to the non collinear ferrimagnetic structures in these garnets, and to the mixing of levels having different M_J that can be attributed to the strong crystal field.

3.4. INTERMETALLIC RARE EARTH COMPOUNDS

Here we are dealing with metals, but the conduction and binding are due to electrons which only have a small contribution to the magnetic moment of the substance: most of the moment comes from the strongly localised 4f electrons on the rare earth atoms (R). Table 7.3 gives the saturation and effective moments, expressed in units of μ_B, for three intermetallic series RAl_2, RCu_2 [5], and RNi_2 [6].

Table 7.3 - Saturation and effective magnetic moments per formula unit, measured on three rare earth intermetallic series

R	m (RAl_2)	m_{eff} (RAl_2)	m (RCu_2)	m_{eff} (RCu_2)	m (RNi_2)	m_{eff} (RNi_2)
Ce	–	2.53-2.64	0.8	2.59	–	–
Pr	2.6	3.46-3.5	2.3	3.51	0.86	3.55
Nd	2.27-2.47	3.1-3.59	1.9	3.56	1.78	3.67
Sm	0.2	–	0.1	–	0.22	–
Eu	0.79-1.21	7.84-8.05	5.8	7.4	–	–
Gd	6.9-7.1	7.92-7.94	6	8.4	7.1	7.9
Tb	8.1-8.9	9.81-9.82	7.4	9.8	7.7	9.76
Dy	9.62-9.89	9.12-10.7	8.7	10.75	8.8	10.5
Ho	9.16-9.18	10.7	9.2	10.5	8.8	10.6
Er	7.05-7.92	9.2-9.56	5.6	9.35	6.9	9.55
Tm	4.8	–	4.2	7.49	3.27	7.42

RAl_2 and RNi_2 alloys have cubic symmetry, RCu_2 has orthorhombic symmetry.

The table shows that the effective moments deduced from magnetic susceptibility measurements are in general in good agreement with the theoretical values in table 7.1, except in the case of europium; as far as samarium is concerned, it does not follow a Curie-Weiss law. Van-Vleck and Frank explained these disagreements by an overlap of the ground state multiplet with the first excited state for the Sm^{3+} and Eu^{3+} ions.

The values of the moments found from low temperature magnetization measurements show a lot of scatter, and are in general less than the theoretical values (except for europium where the magnetic moment is not zero, and often quite sizeable).

3.5. RARE EARTH METALS

Pure rare earth metals in the paramagnetic state show excellent agreement between the measured and theoretical effective moments, with three exceptions (measured/theory): samarium 1.74 μ_B/0.85 μ_B (1.55 μ_B with the Van Vleck-Frank correction, see § 3.4), europium 8.3 μ_B/0 μ_B (3.40 μ_B with the Van Vleck-Frank correction), and ytterbium (0 μ_B/4.53 μ_B). The profound disagreement seen for europium and ytterbium comes from these ions being divalent rather than trivalent, giving a half filled 4f shell ($4f^7$) for europium and a full one ($4f^{14}$) for ytterbium.

In the ordered state, the light rare earths display magnetic structures which make the determination of the saturation moments difficult. For the heavy rare earths, the agreement between the saturation moments measured and expected (measured/theory) is rather better than in those compounds previously mentioned: Gd (7.55 μ_B/7 μ_B), Tb (9.34 μ_B/9 μ_B), Dy (10.2 μ_B/10 μ_B), Ho (10.34 μ_B/10 μ_B), Er (8.0 μ_B/9.0 μ_B), Tm (3.4 μ_B/7.0 μ_B). Note however that gadolinium has a moment which differs appreciably from the theory for Gd^{3+}, the difference being ascribed to the contribution from the conduction electrons.

We now realize how difficult it is to give exact predictions on what the magnetic moment of a given compound will be. The reason is that the ground state of the ion is always perturbed to a certain extent by the electrostatic environment within the material. This leads us to consider in more detail the electric field which acts locally on the electronic orbitals in the material.

4. MAGNETOCRYSTALLINE ANISOTROPY: THE CRYSTALLINE ELECTRIC FIELD

Amongst the interactions giving rise to the various magnetic structures found in different materials we introduced (see § 2.3 of chap. 3) the idea of the magnetocrystalline anisotropy, and its phenomenological description as an energy depending only on the angle between the magnetic moments and the directions of high symmetry of the crystallographic structure. This anisotropy originates from two microscopic mechanisms: the anisotropy of the exchange interactions (which will be discussed in chapter 9), and the *crystalline electric field* (or *crystal field*), which is generally the more important mechanism, and will be treated in this chapter.

Before going deeper into the formalism used to describe this effect, one can say in simple terms that, as a result of the electrostatic potential due to the environment, the electrons responsible for the magnetism can no longer occupy with equal ease any 3d (4f) orbital as is the case for the free ion.

An anisotropy in the orbital angular momentum results, and hence the same occurs to the associated magnetic moment. The anisotropy of the total magnetic moment thus arises through the spin-orbit coupling. Remember that the effect of the crystal field is not only to favour certain directions for the magnetic moments, as implied by the phenomenological description, but also to change the amplitude of the magnetic moment.

To start out, we will examine in section 4.1 the simple case of a single d electron. Section 4.2 will then make the distinction between the 3d and 4f elements, and highlight the relative importance of the crystal field with respect to the other interactions. In sections 4.3 and 4.4, the effects of the crystal field will successively be examined for these two types of element. Finally, in the last section, the description of the magnetic anisotropy of uniaxial compounds of the rare earths will be developed.

4.1. A SINGLE d ELECTRON IN A UNIAXIAL ELECTROSTATIC POTENTIAL DUE TO ITS SURROUNDINGS

The orbital contributions to the five wavefunctions that allow one to describe the state of a d electron are the following spherical harmonics:

$$Y_2^0 = \sqrt{5/16\pi}\,(3\cos^2\theta - 1)$$
$$Y_2^1 = -Y_2^{-1*} = -\sqrt{15/8\pi}\,\sin\theta\cos\theta\,e^{i\varphi} \quad (7.29)$$
$$Y_2^2 = Y_2^{-2*} = \sqrt{15/32\pi}\,\sin^2\theta\,e^{2i\varphi}$$

The complete wavefunction also contains a radial component which is common to these five orbitals. These five orbitals lead to the three spatial distributions shown in figure 7.9. Spherical harmonics are described in more detail in appendix 6, section A6.1.

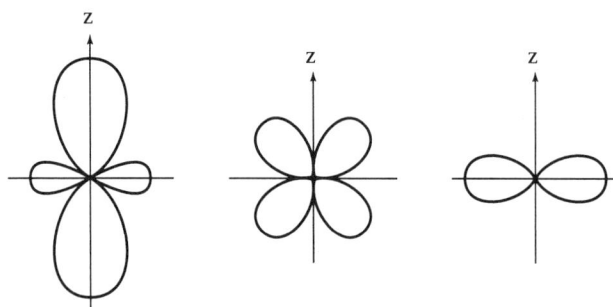

Figure 7.9 - The spatial distribution in a polar representation of the modulus of the wavefunction for $|Y_2^0|$ (left), $|Y_2^1|$ (centre), and $|Y_2^2|$ (right)

7 - MAGNETISM IN THE LOCALISED ELECTRON MODEL

In the absence of a magnetic field or surrounding atoms (a free atom), the probability that each of the five orbitals is occupied is the same, and one can show that the spatial distribution of the d electron is spherical: $\left|\sum_m Y_2^m\right|^2 = \text{constant}$.

One can also show that the orbital magnetic moment induced by a field is *isotropic* with magnitude 2 μ_B in the direction of the field.

Now imagine that the atom is surrounded by two charges (ions for example) $-e$ located symmetrically either side of the atom on the z axis. As a result of the Coulomb repulsion between the d electron and the two neighbouring charges, it is clear, from figure 7.9, that the five orbitals are no longer energetically equivalent, and that (at 0 K) only the Y_2^2 and Y_2^{-2} orbitals are occupied, with the same probability. The ground state is no longer fivefold, but twofold degenerate, and in Dirac notation the states are written as $|m_\ell\rangle = |2\rangle$, and $|-2\rangle$. Taking into account the two possible spin states, the ground state is fourfold degenerate, showing that the four basis states: $|m_\ell, \sigma\rangle = |2, 1/2\rangle, |2, -1/2\rangle, |-2, 1/2\rangle$, and $|-2, -1/2\rangle$ correspond to the same energy, and hence occupation.

Let us now switch on the spin-orbit coupling $\mathcal{H}_{so} = -\lambda \boldsymbol{\ell} \cdot \mathbf{s} = |\lambda| \boldsymbol{\ell} \cdot \mathbf{s}$ (since $\lambda < 0$ in this case of a single electron), which is much smaller than the crystal field. A simple first order perturbation calculation lifts the degeneracy, and leads to two doublets separated by an energy of $\Delta E = 2|\lambda|$, and so the ground state is made up of the basis states: $|2, -1/2\rangle$, and $|-2, 1/2\rangle$. (This result is obtained by diagonalising the 4 by 4 matrix made up of the elements $\langle a|\mathcal{H}_{so}|b\rangle$, where $|a\rangle$ and $|b\rangle$ designate one of the four states).

Finally, let us look at the effect of a perturbing field **H**, first along the z direction, and then perpendicular to this direction (along x for example).

The associated perturbing Zeeman hamiltonians are written respectively as:

$$\mathcal{H}_Z = -\mu_0 m_z H = \mu_0 \mu_B (\ell_z + 2s_z) H \quad (7.30)$$

and

$$\mathcal{H}_Z = -\mu_0 m_x H = \mu_0 \mu_B (\ell_x + 2s_x) H \quad (7.31)$$

When the field **H** is along z, the perturbation calculation for the ground state doublet leads to two singlets of energies $\pm \mu_0 \mu_B H$ (with respect to the doublet energy). The new singlet ground state is $|-2, 1/2\rangle$ the magnetic moment of which is $\langle -2, 1/2|m_z|-2, 1/2\rangle = 1 \mu_B$. When **H** is along x, there is no lifting of the degeneracy to first order, and thus no magnetic moment (the matrix elements $\langle \mp 2, \pm 1/2|\ell_x + 2s_x|\mp 2, \pm 1/2\rangle$ are all zero). One finds that the energy is at a minimum, and the moment is at a maximum, when the field is applied along the z direction, which is thus the *easy axis of magnetization*.

Let us now examine the opposite situation, where the two charges on the z axis, either side of the ion, are now positive. At 0 K, only the Y_2^0 orbital with zero orbital angular momentum, and thus zero orbital magnetic moment, is occupied as a result of the Coulomb attraction ($<0|\ell_i|0> = 0$, irrespective of whether i = x, y or z). The ground state is thus the doublet $|0, \pm 1/2>$. It can be shown that to first order the spin-orbit coupling does not lift the degeneracy ($<0, \pm 1/2|\mathcal{H}_{so}|0, \pm 1/2> = 0$). To first order, there is thus no orbital moment, and hence no spin-orbit coupling. The degeneracy is lifted by an applied field independently of whether this field is applied parallel or perpendicular to z. In either case, the ground state singlet has the same energy $-\mu_0 \mu_B H$, and the same magnetic moment $<m_z> = <m_x> = 1\ \mu_B$. As a result the magnetic moment is purely spin, and isotropic. Note that second order calculations show that z direction is actually a more difficult magnetization direction.

4.2. ORDERS OF MAGNITUDE OF THE CRYSTAL FIELD

The importance of the crystal field compared to other interactions is different for 3d and 4f elements. Figure 7.10 shows for both cases the orders of magnitude of the energies acting within the materials. One notes that the crystal field interaction is much weaker in the 4f elements than is the case for the 3d elements. This comes from the fact that the 4f shell is much better shielded by the outer shells than the 3d shell (see fig. 7.8).

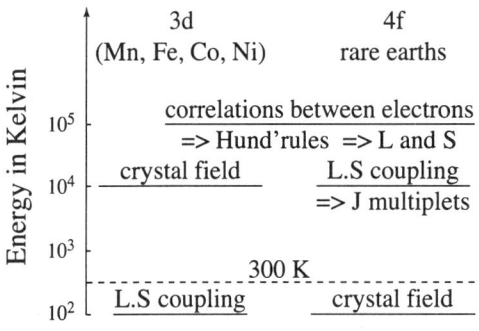

Figure 7.10
Orders of magnitude
of the different interactions acting
on 3d and 4f atoms in materials

4.3. THE EFFECT OF THE CRYSTAL FIELD ON THE MAGNETIC ANISOTROPY OF THE 3d ELEMENTS

As is shown in figure 7.10, the case of the 3d elements corresponds in general to the situation treated in section 4.1, where the spin-orbit coupling is a perturbation with respect to the crystal field. As a result of the rather strong delocalization of the magnetic electrons, the situation of 3d ions is often complicated, in particular in the case of metals and alloys. Only a qualitative description of the main effects of the crystal field for these materials will be given.

4.3.1. Quenching of the orbital moment

We have already seen that, in insulators, the crystalline electric field often gives rise to a near complete quenching of the orbital angular momentum, which means that, in the ground state, $<L_i> = 0$ due to the crystal field (irrespective of whether i = x, y or z).

The magnetic moment thus derives almost completely from spin: this is true in the previously mentioned case of the Cr^{2+} ion in a chromium and potassium alum. *The magnetocrystalline anisotropy is thus small*, and is due to the residual orbital moment that can take part in the spin-orbit coupling. Amongst others, this effect is observed in the case of certain oxides (e.g. ferrites), when the environment of the magnetic ion is cubic or tetrahedral.

4.3.2. L = 2 states

Still considering insulators, when the environment is octahedral or tetrahedral, and $L = 2$ (as is the case for an ion with a single, or nine, d electrons), the five orbital states are split by the crystal field into two levels: a doublet e_g of zero orbital moment, and a triplet t_{2g} of non zero orbital moment. If the doublet is the ground level, which is often the case, the magnetocrystalline anisotropy is very weak. On the other hand, if the triplet is the ground level, then this anisotropy can be pronounced.

4.3.3. Fe^{3+} and Mn^{2+} ions in insulators *(e.g. ferrites)*

These ions have a $3d^5$ configuration. The ground state *term* is such that $S = 5/2$, and $L = 0$ (orbital S state). There is thus no crystal field effect, and thus no magnetocrystalline anisotropy up to high orders of approximation, irrespective of the symmetry. This explains the extremely weak intrinsic anisotropy of γ-Fe_2O_3 (the anisotropy of the particles of this material, used for magnetic recording, is a dipolar one due to the elongated shape of the grains).

4.3.4. Low spin states

Very intense crystal fields can sometimes counteract the electrostatic intra-atomic repulsion that gives rise to Hund's rules. The population of the electronic 3d states is done in such a way as to firstly maximise the orbital angular momentum, and then that of the spin. The states: $|2+>, |2->, |1+>, |1->, |0+>, |0->, |-1+>, |-1->, |-2+>, |-2->$ are successively populated, and not the reverse as predicted by Hund's rules: $|2+>, |1+>, |0+>, |-1+>, |-2+>, |2->, |1->, |0->, |-1->, |-2->$ (here $|m_\ell \pm >$ signifies that the spin is $+1/2$ or $-1/2$). Instead of reaching 5 μ_B, the moment oscillates between 0 and 1 μ_B. This is a low spin state. The magnetocrystalline anisotropy can be strongly perturbed: for example, in the case of the Ru^{3+} ion (§ 3.2), the orbital angular momentum is $L = 6$ instead of $L = 0$, and an anisotropy much larger than that expected from an S state is observed.

4.3.5. The influence of symmetry

Let us finally note that the magnetocrystalline anisotropy is one or several orders of magnitude larger in uniaxial symmetry, in particular hexagonal or quadratic, than in cubic symmetry. This is the reason why soft magnetic materials are generally cubic (e.g. spinel ferrites, see chap. 17), while hard materials have lower symmetries such as hexagonal (e.g. barium hexaferrite, see chap. 15).

4.4. THE EFFECTS OF THE CRYSTAL FIELD UPON RARE EARTH IONS

As shown in figure 7.10, the crystal field is a perturbation acting on each J multiplet, whose degeneracy is hence lifted. Taking into account the large energy differences between the multiplets, only the effect of the crystal field on the ground state multiplet is considered. The hamiltonian of the perturbation is: $\mathcal{H}_{cf} = -|e|\sum_i V(\mathbf{r}_i)$ where $V(\mathbf{r}_i)$ is the electrostatic potential due to the environment of the 4f shell.

The summation takes in all of the 4f electrons of the element under consideration. This hamiltonian can always be written in the form [7]:

$$\mathcal{H}_{cf} = \sum_i \sum_\ell r_i^\ell \sum_{m=-\ell}^{+\ell} A_\ell^m Y_\ell^m(\theta_i, \varphi_i) \qquad (7.32)$$

where $Y_\ell^m(\theta, \varphi)$ are the spherical harmonics, and A_ℓ^m are the *crystal field parameters* which depend on the environment. The number of terms, i.e. the values that ℓ and m can take, is limited by the symmetry, and is lower the higher the symmetry. In addition, ℓ is always even, and terms with $\ell > 6$ do not take part.

In cubic symmetry, the lowest order term is of order 4, whereas for other symmetries the lowest order terms are of order 2, and generally dominant. Thus, for hexagonal symmetry the perturbing hamiltonian is:

$$\mathcal{H}_{cf} = \sum_i \left[r_i^2 A_2^0 Y_2^0 + r_i^4 A_4^0 Y_4^0 + r_i^6 A_6^0 Y_6^0 + r_i^6 A_6^6 \left(Y_6^6 + Y_6^{-6} \right) \right] \qquad (7.33)$$

The perturbation calculation requires matrix elements of the type $<J, M_J|\mathcal{H}_{cf}|J, M_J>$ to be determined, where the states $|J, M_J>$ are relatively complicated functions of the individual wave functions of each f electron. An elegant method has been proposed in order to simplify the determination of these matrix elements, and consists of replacing each operator of the type $\sum_i r_i^\ell A_\ell^m Y_\ell^m(\theta_i, \varphi_i)$ by an operator equivalent that depends upon the components J_x, J_y, and J_z of the total angular momentum $\mathbf{J}$, and which acts directly, and simply, on the $|J, M_J>$ states. For a term of order 2, which is most frequently the case if the symmetry is lower than cubic, one replaces $\sum_i r_i^2 A_2^0 Y_2^0(\theta_i, \varphi_i)$ by the operator:

$$\alpha_J <r^2> A_2^0 O_2^0 = B_2^0 O_2^0 = B_2^0 \left[3J_z^2 - J(J+1) \right] \qquad (7.34)$$

The operators O_ℓ^m are the *Stevens operator equivalents*, and α_J (β_J and γ_J for terms of order 4 and 6 respectively) are *multiplying factors* characteristic of the distribution of the 4f electrons of each of the rare earth ions, sometimes known as the *Stevens coefficients*. These dimensionless coefficients are tabulated, and vary in sign and amplitude from one rare earth to another [8].

In what follows, only the second order term described above (eq. 7.34) will be considered. With the exception of cubic symmetry, it plays the most important role in most cases, and allows the crystal field effects in a large number of compounds, particularly hexagonal or tetragonal ones, to be understood. It is also this second order term which is behind the magnetoelastic coupling for all symmetries (even cubic). Three situations will be treated. For each of them we will discuss first the effect at 0 K of a crystal field, and then a Zeeman hamiltonian $\mathcal{H}_Z = -\mu_0 g_J \mu_B \mathbf{J}\cdot\mathbf{H}$. Actually for a more accurate (but less pedagogical) treatment these two effects (the crystal field, and the Zeeman effect) have to be considered at the same order of perturbation, so that one can directly diagonalise $\mathcal{H}_{cf} + \mathcal{H}_Z$.

4.4.1. A J = 4 multiplet with $B_2^0 < 0$

An example of this is the Pr^{3+} ion. A simple calculation shows that the splitting of the ground state multiplet (which is taken as having zero energy) leads to a ground state doublet $|\pm 4\rangle$ with energy $-28|B_2^0|$, and a first excited state doublet $|\pm 3\rangle$ with energy $-7|B_2^0|$ (fig. 7.11) (see the appendix of this chapter for a more detailed description of this type of calculation). A field **H** along z will lead to a singlet $|-4\rangle$ as the ground state with energy $-28|B_2^0| - 4\mu_0 g_J \mu_B H$, and a magnetic moment of 3.2 μ_B, corresponding to that of the free ion.

On the other hand, a field along x does not lift the degeneracy of the crystal field ground state doublet, so that the associated moment is then zero.

The easy axis is thus the z axis.

4.4.2. A J = 4 multiplet with $B_2^0 > 0$

The ground state due to the crystal field is in this case (fig. 7.11) the singlet $|0\rangle$ with energy $-20 B_2^0$, and the first excited state is the $|\pm 1\rangle$ doublet with energy $-17 B_2^0$. Such a singlet is non magnetic, since $\langle 0|m_z|0\rangle = \langle 0|m_x|0\rangle = 0$. The effect of a field is second order via the matrix elements of the Zeeman hamiltonian between the ground state singlet, and the first excited state. One can show that one progressively induces a moment when the field is perpendicular to z, but a field along z itself has practically no effect. The z axis is the hard axis of magnetization.

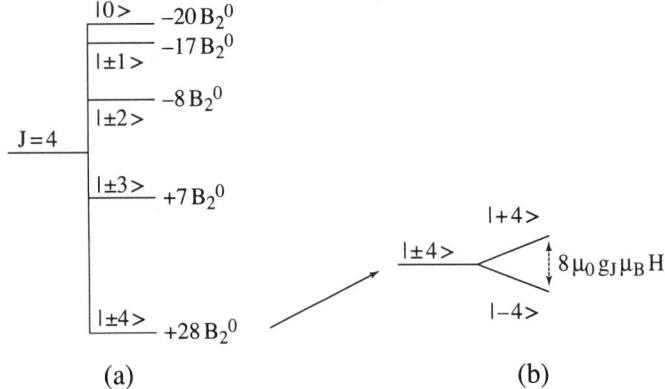

Figure 7.11 - (a) *Decomposition of the J = 4 multiplet by the second order term of the crystal field $B_2^0 O_2^0$. The ground state is the $|M_J = \pm 4\rangle$ doublet if $B_2^0 < 0$, and the $|M_J = 0\rangle$ singlet if $B_2^0 > 0$* (b) *Decomposition of the $|M_J = \pm 4\rangle$ doublet by a field H // Oz*

4.4.3. A J = 5/2 multiplet with $B_2^0 > 0$

This corresponds to the case of the Ce^{3+} ion. The multiplet splits into three doublets $|\pm 5/2\rangle$, $|\pm 3/2\rangle$, and $|\pm 1/2\rangle$ having the energies $10 B_2^0$, $-2 B_2^0$, and $-8 B_2^0$, respectively (fig. 7.12).

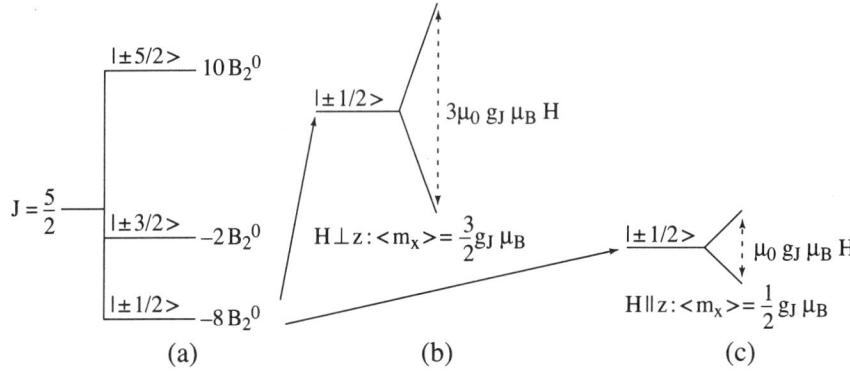

Figure 7.12 - (a) Decomposition of the J = 5/2 multiplet by the second order term of the crystal field $B_2^0 O_2^0$ - (b) Decomposition of the $|\pm 1/2\rangle$ doublet by a magnetic field $H \perp Oz$ - (c) Decomposition of the same doublet by a magnetic field $H // Oz$

The action of a field along Oz on the $|\pm 1/2\rangle$ state gives rise to a singlet ground state $|-1/2\rangle$, of energy $-8 B_2^0 - \mu_0 g_J \mu_B H/2$, and magnetic moment $\langle m_z \rangle = g_J \mu_B/2$. On the other hand a field perpendicular to Oz leads to a singlet $[|1/2\rangle - |-1/2\rangle]/\sqrt{2}$, with energy $-8 B_2^0 - 3 \mu_0 g_J \mu_B H/2$, and magnetic moment $\langle m_x \rangle = 3 g_J \mu_B/2$ (see the appendix of this chapter for a detailed presentation of this calculation). For the same value of the field, the energy is larger along the z axis which is thus the hard axis of magnetization. *The plane perpendicular to z is thus the easy plane of magnetization,* because one finds the same energy independent of the direction of **H** in this plane.

These three examples give a feeling for the large diversity of situations that can arise when one considers all of the crystal field terms. In ferromagnetic systems where the Curie temperature is high enough, the molecular field is such that the magnetic moment is equal, or very close, to the maximum value. If the compound has hexagonal or tetragonal symmetry, the cases just discussed lead to the following conclusion: *the z axis is the easy or hard axis depending upon whether B_2^0 is negative or positive respectively.*

4.4.4. Kramers, non-Kramers ions

In 1933, Kramers derived the following theorem: Let us consider a system with n electrons (contributing to magnetism) which are not submitted to a magnetic field (internal (i.e. applied plus demagnetizing) and/or molecular field). When n is even, there can exist non degenerate levels; whereas when n is odd all the levels are degenerate (see tab. 7.1). Taking into account this property, rare earth ions have been classified as Kramers and non-Kramers ions depending on whether the number of 4f electrons is odd or even, respectively. Thus in the presence of the crystal field all the levels of a Kramers ion are at least doublets, whereas for a non-Kramers ion the degeneracy of the free-ion ground state can be totally lifted. The importance of the Kramers theorem in magnetism is due to the following theorem: a permanent magnetic moment can be associated (in the absence of a magnetic field) only with a degenerate quantum level. As a result for a non-Kramers ion (Pr^{3+}, Tb^{3+}, Ho^{3+}, Tm^{3+}), if the ground state due to crystal field effects is a singlet, it is non magnetic and the magnetic moment at 0 K is only induced by the field. Among the most dramatic consequences of such properties, one can quote the existence of sinc wave modulated magnetic structure down to 0 K, as observed and quantatively analyzed in the compound $PrNi_2Si_2$ [9]. The modulation of the magnetic moments follows that of the molecular field. This situation is opposite to that usually observed, viz a magnetic ground state, where at 0 K, whatever the molecular field, all atoms have the maximum moment. As a result sine modulated structures can normally be observed only just below the Néel temperature and transform into structures with equal moments at low temperature.

4.5. THE ANISOTROPY OF UNIAXIAL 4f COMPOUNDS WITH HEXAGONAL OR TETRAGONAL SYMMETRY

As was previously mentioned, the second order term $B_2^0 O_2^0$ is dominant in most cases. Limiting the discussion to this term, the crystal field effects, in particular the easy axis, depend jointly on the parameters A_2^0 and α_J, and in particular their sign. A_2^0 characterises the potential due to the environment. Imagine two negative charges on the z axis either side of the ion of interest. This leads to a positive value of A_2^0. If the charges are positive, A_2^0 is negative. α_J characterises the form of the electronic distribution of the 4f shell, in particular its quadrupole moment.

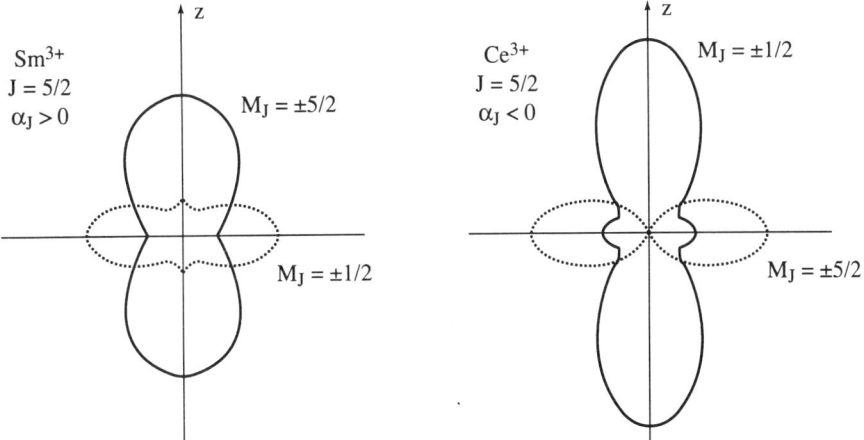

Figure 7.13 - Angular distribution of the 4f electrons for J = 5/2 ions: Sm^{3+} with $\alpha_J > 0$, and Ce^{3+} with $\alpha_J < 0$ [9]

As is shown in figure 7.13, for an ion where α_J is positive (Sm^{3+}), the 4f distribution is elongated along z (prolate) for the $|M_J = \pm J\rangle$ states, and squashed (oblate) for the $|M_J = \pm 1/2\rangle$ or $|M_J = 0\rangle$ states depending upon whether J is half integer or integer respectively. In an ion for which α_J is negative (Ce^{3+}) the 4f distribution is oblate for the $|M_J = \pm J\rangle$ states, and prolate for the $|M_J = \pm 1/2\rangle$ or $|M_J = 0\rangle$ states depending upon whether J is half integer or integer. The absolute value of α_J is larger, the larger the deviation from a spherical distribution. The values of this parameter for the main rare earths are given in table 7.4. Note that the Gd^{3+} ion is not shown: due to its orbital moment being zero, the 4f shell has a spherical distribution, and thus α_J is zero.

Table 7.4 - The Stevens coefficient α_J for the rare earth ions

Ion	Ce^{3+}	Pr^{3+}	Nd^{3+}	Sm^{3+}	Tb^{3+}	Dy^{3+}	Ho^{3+}	Er^{3+}	Tm^{3+}	Yb^{3+}
$10^2 \alpha_J$	–5.71	–2.10	–0.64	+4.13	–1.01	–0.63	–0.22	+0.25	+1.01	+3.17

Within a series of rare earth compounds (i.e. having the same stoichiometry, and the same crystallographic structure, which is the case most of the time), A_2^0, which characterises the environment, retains the same sign, and only changes its absolute value slightly when going from one rare earth to another. $<r^2>$ is obviously always positive, and regularly decreases by a factor of about two in going from cerium to thulium. The B_2^0 term, and in particular its sign, thus follows, more or less, the variation of α_J. In order to discuss the direction of easy magnetization, one can use the approach outlined above, or more physically, take into account the Coulomb interaction between the 4f electrons on the one hand, and the neighbouring charges on the other. One can deduce the easy axis following the respective signs of A_2^0 and of α_J, as is shown schematically in table 7.5.

Table 7.5 - Influence of the signs of α_J and A_2^0 on the form of the orbitals and the magnetocrystalline anisotropy

Environment		α_J is positive	α_J is negative
$A_2^0 > 0$		oblate orbital occupied such that $|M_J| = 1/2$ or 0 $\Rightarrow$ the z axis is the hard axis, in agreement with $B_2^0 > 0$	oblate orbital occupied such that $|M_J| = J$ $\Rightarrow$ the z axis is the easy axis, in agreement with $B_2^0 < 0$
$A_2^0 < 0$		prolate orbital occupied such that $|M_J| = J$ $\Rightarrow$ the z axis is the easy axis, in agreement with $B_2^0 < 0$	prolate orbital occupied such that $|M_J| = 1/2$ or 0 $\Rightarrow$ the z axis is the hard axis, in agreement with $B_2^0 > 0$

In this way, if, in a series of compounds, the z axis is the easy axis for rare earth elements where α_J is positive, it is the hard axis for those for which this parameter is negative, and conversely. This approach is particularly well verified in compounds based on rare earths R which are used as permanent magnets, such as the hexagonal RCo_5, and the tetragonal $R_2Fe_{14}B$. As we will see later on, in order to have a good magnet it is necessary that the z axis be the easy axis. In the RCo_5 series, A_2^0 is negative, so that a magnet can only be obtained with a rare earth where α_J is positive. This is why the best magnet of the series is based on $SmCo_5$. On the other hand, in the $R_2Fe_{14}B$ series, A_2^0 is positive, and the best magnet in the series is based on Nd. These two types of magnet are actually the best performing ones that are available at the moment.

APPENDIX: DETAILED TREATMENT OF A MAGNETIC MOMENT CALCULATION

We derive here the result given in section 4.4.3. We deal in detail with the effect of a magnetic field along x, in the case of the $|\pm 1/2\rangle$ doublet, corresponding, in $|J, M_J\rangle$ notation, to $|5/2, \pm 1/2\rangle$. This is the ground level of energy $-8 B_2^0$ resulting from the effect, on the $J = 5/2$ multiplet, of the crystalline electric field term of order 2, with uniaxial symmetry along z.

We use a first order perturbation treatment, as described for example in *Quantum Mechanics*, by Cohen-Tannoudji, Diu, and Laloë, Wiley & Sons (1992), for the case of a degenerate level.

The perturbing hamiltonian here is the Zeeman term. Using the basic relation (eq. 7.26 in this chapter) $m = -g_J \mu_B J$, it can be written in the form $\mathcal{H}_Z = \mu_0 H_x g_J \mu_B J_x$.

It is convenient to use the operators $J_+ = (J_x + iJ_y)/2$, and $J_- = (J_x - iJ_y)/2$, the main property of which is, as shown in textbooks on Quantum Physics, that the only non zero matrix elements are such that:

$$\langle J, M_J \pm 1 | J_\pm | J, M\rangle = \sqrt{J(J+1) - M_J(M_J \pm 1)}$$

In the subspace spanned by the doublet we are interested in, we immediately see that the only non zero matrix elements are

$$\langle 5/2, -1/2 | J_- | 5/2, 1/2\rangle = \langle 5/2, 1/2 | J_+ | 5/2, -1/2\rangle = 3$$

Going over to $J_x = 1/2(J_+ + J_-)$, we directly see that the eigenstates of J_x are

$$|a\rangle = \frac{1}{\sqrt{2}}(|5/2, 1/2\rangle + |5/2, -1/2\rangle) \quad \text{and}$$

$$|b\rangle = \frac{1}{\sqrt{2}}(|5/2, 1/2\rangle - |5/2, -1/2\rangle).$$

This is immediately confirmed by noting that

$$J_x |a\rangle = (3/2) |a\rangle \quad \text{and} \quad J_x |b\rangle = -(3/2) |b\rangle,$$

which furthermore indicates that the eigenvalues of J_x are respectively $\langle J_x \rangle_a = 3/2$, and $\langle J_x \rangle_b = -3/2$.

Therefore the energies under the influence of the perturbation, i.e. in the magnetic field H_x, are

$$\langle E \rangle_a = -8B_2^0 + \langle a | \mu_o H_x g_J \mu_B J_x | a \rangle = -8B_2^0 + (3/2) \mu_o H_x g_J \mu_B, \text{ and}$$
$$\langle E \rangle_b = -8B_2^0 + \langle b | \mu_o H_x g_J \mu_B J_x | b \rangle = -8B_2^0 - (3/2) \mu_o H_x g_J \mu_B.$$

We conclude that the ground state, in the presence of the magnetic field along x, is $|b\rangle$, and that its magnetic moment is $\langle m_x \rangle_b = (3/2) g_J \mu_B$. Note that for a more accurate (but less pedagogical) treatment, it is necessary to consider the crystalline electric field, and Zeeman hamiltonians at the same order of perturbation. This involves diagonalising directly $\mathcal{H}_{cf} + \mathcal{H}_Z$ within the basis of the $2J + 1$ states of the ground state multiplet.

EXERCISES

E.1 - Determine for the Sm^{2+} and Eu^{2+} ions, which have six and seven 4f electrons respectively, the values of S, L, J, g_J, and the maximum magnetic moment of the ground state.

E.2 - Iron, in its divalent and trivalent states, has $3d^6$ and $3d^5$ external configurations, respectively. Furthermore, these ions behave in many insulating materials as if their orbital moment was zero (by virtue of the crystal field). In this framework, determine, for Fe^{2+} and Fe^{3+}, the values of S, L, J, g_J, and the maximum magnetic moment of the ground state.

E.3 - The spin-orbit coupling coefficient, λ, of the Sm^{3+} and Tb^{3+} ions is -630 K and 580 K respectively. Give the values of S, L, J, g_J, and the maximum magnetic moment of the first excited multiplet for these ions. Also give, in kelvin, the energy difference between this, and the ground state.

E.4 - Let two identical charges q be located at $z = \pm a$ on the z axis. Write the expansion to second order in r/a of the potential $V(\mathbf{r})$ at a point M near the origin of a spherical co-ordinate system r, θ, φ. Deduce from this the second order term A_2^0 of the crystal field.

SOLUTIONS TO THE EXERCICES

S.1 Sm^{2+}: $S = 3$, $L = 3$, $J = |L - S| = 0$, $\mathfrak{m} = 0$.

Eu^{2+}: $S = 7/2$, $L = 0$, $J = 7/2$, $g_J = 2$, $\mathfrak{m} = 7\,\mu_B$.

S.2 Fe^{2+}: $S = J = 3$, $g_J = 2$, $\mathfrak{m} = 6\,\mu_B$. Fe^{3+}: $S = J = 5/2$, $g_J = 2$, $\mathfrak{m} = 5\,\mu_B$.

S.3 Sm^{3+}: $S = 5/2$, $L = 5$, $J = |L - S| + 1 = 7/2$, $g_J = 0.825$, $\mathfrak{m} = 2.89\,\mu_B$, $\Delta E = 2,205$ K.

Tb^{3+}: $S = 3$, $L = 3$, $J = L + S - 1 = 5$, $g_J = 3/2$, $\mathfrak{m} = 7.5\,\mu_B$, $\Delta E = 3,480$ K.

S.4 $A_2^0 = -|e|q/\varepsilon_0 a^3 \sqrt{5\pi}$ where $|e|$ is the charge of an electron. Note that $A_2^0 < 0$ for $q > 0$, and vice versa.

REFERENCES

[1] C. COHEN-TANNOUDJI, B. DIU, F. LALOË, Quantum Mechanics (1992) Wiley & Sons; R.P. FEYNMAN, R.B. LEIGHTON, M. SANDS, The Feynman Lectures on Physics, Vol. 3 Quantum Mechanics (1964) Addison-Wesley; L.D. LANDAU, Quantum Mechanics (1997) Butterworth-Heinemann; H.J. ZEIGER, G.W. PRATT, Magnetic Interactions in Solids (1973) Clarendon Press, Oxford; J.C. SLATER, Quantum Theory of Atomic Structure, Vol. 1 (1960) McGraw-Hill Book Company.

[2] A. MESSIAH, Quantum Mechanics (2000) Dover.

[3] F.B. KOCH, M.E. FINE, *J. Appl. Phys.* **38** (1967) 1470.

[4] G. FOËX, C.J. GORTER, L.J. SMITS, Constantes sélectionnées diamagnétisme et paramagnétisme (1957) Masson & Cie dépositaires, Paris.

[5] K.H.J. BUSCHOW, *Rep. Prog. Phys.* **42** (1979) 1373.

[6] K.H.J. BUSCHOW, *Rep. Prog. Phys.* **40** (1977) 1179.

[7] M.T. HUTCHINGS, in *Solid State Physics* **16** (1964) 227, F. Seitz & D.T. Turnbull Eds, Acad. Press, New York.

[8] K.H.J. STEVENS, *Proc. Phys. Soc. London A* **65** (1952) 209.

[9] J.A. BLANCO, D. GIGNOUX, D. SCHMITT, *Phys. Rev. B* **45** (1992) 2529.

[10] D. SCHMITT, *J. Phys.* **47** (1986) 677.

CHAPTER 8

MAGNETISM IN THE ITINERANT ELECTRON MODEL

(this chapter deals with fundamentals, and may be skimmed over during a first reading)

In chapter 7, we showed under what circumstances a magnetic moment can exist on an isolated atom, and then in a solid made up of these atoms (or ions) under the assumption that the electrons remain localized on the atoms. In some metals, however, electrons can propagate throughout the solid. How can such a system be magnetic? This is the basic question we address in this chapter.

1. GENERALITIES

We have seen in chapter 3 that many elements, in the free atom state, possess a magnetic moment (tab. 3.1). However, in the solid state, the majority of them are diamagnetic: only the metals of the third row, from *chromium to nickel,* and most of the *rare earth* elements and *actinides* carry magnetic moments (tab. 3.2). This means that the formation of a chemical bond significantly alters the electronic orbits that are responsible for magnetism. The transition metals (3d series) are in fact far more sensitive to these effects, and the magnetic moments observed in these elements, their compounds, and alloys, generally differ markedly from those of the free atoms or isolated ions. We have already treated, in the preceding chapter, the materials in which the electrons responsible for the magnetism are localized. In other situations, most importantly in the metals and alloys of the 3d transition series, the magnetic moment carried by a magnetic atom is rarely the same as that of the isolated atom because the valence electrons now propagate throughout the material, and their wavefunctions are very different from those in the isolated atom. In this case we cannot say anything *a priori* about the magnetism of a given material. Understanding the magnetism of metals and alloys is therefore more subtle than for the compounds and oxides described in the previous chapter. The rare earth elements are a special case, as their magnet moment arises from the f electrons, while the valence electrons responsible for conduction are of s and d character. Localized magnetic moments are then observed in metallic environments, and we will return to this situation later.

In this chapter, we will show how the electrons responsible for the electronic conductivity can at the same time give rise to magnetism. We will establish that this magnetism arises from the interplay of Coulomb repulsion between electrons with the Pauli exclusion principle, and that in a metallic system the existence of the magnetic moment on an atom depends very strongly on its local environment.

Before we explain why some materials are magnetic, we must remember why the majority are not. According to the Pauli principle, in a system of atoms the electronic states which extend over all the atoms can be occupied by just two electrons with opposed spins. In most solids, all these states are doubly occupied, and correspondingly the substance remains diamagnetic.

2. SPECIAL PROPERTIES OF MAGNETIC METALS

Some experimental properties of magnetic metals cannot be explained by the simple models of the preceding chapter, e.g. the saturation moments observed at low temperatures are equal to 0.61, 1.72, and 2.22 μ_B per atom respectively for nickel, cobalt, and iron. As the Landé factor in these metals is very close to two, the orbital contribution to magnetism is below 10%, even for cobalt ($m_L = 0.17\ \mu_B$). We would therefore expect a saturation moment close to an integral value in terms of Bohr magnetons, which is far from true here. Also, the effective magnetic moment, obtained from the Curie-Weiss law above the critical temperature, is systematically too high, e.g. for nickel $m_{eff} \approx 1\ \mu_B$. For the ferromagnet $ZrZn_2$, the moment derived from the Curie-Weiss law is ten times larger than the saturation moment.

The rules for localized magnetism do not apply any better to chromium which presents for $T < T_N = 310$ K an antiferromagnetic sinusoidal structure that is incommensurate with the period of the lattice. As the temperature is changed between 0 and T_N, the wavelength of the magnetization varies between 21 and 27 times the interatomic distance. Its saturation moment is of the order of 0.6 μ_B per atom, while the effective magnetic moment is practically zero.

All these results, and many others, show that the theory of magnetism developed within the localized electron model is not at all valid when the electrons responsible for magnetism are *itinerant* electrons. Since *band theory* explains many experimental observations well, we will start from this approach, and introduce correlations between electrons, which we previously assumed to be negligible. But firstly, we will introduce the various types of metallic magnetic materials, because the interpretations of their properties differ in each case.

2.1. VERY WEAK FERROMAGNETS

Some materials with a very weak magnetization are ferromagnets even though their constituent atoms are not. One of the best examples is $ZrZn_2$, which has a Curie temperature of 35 K, and a saturation moment of 0.18 μ_B per formula unit. Neither of the two elements is magnetic on its own, and the moment density is not localized on a particular atom. Another example is the compound $Ca_{0.995}La_{0.005}B_6$ which has a moment of 0.07 μ_B per formula unit, and a Curie temperature of 600 K [1]. Stoner's theory, discussed below, applies well to this type of material, which at present has no industrial use.

2.2. TRANSITION METALS, AND THEIR ALLOYS

In the crystallised state, these often have transition temperatures significantly higher than room temperature. Addition of carbon, phosphorus, or other metalloids allows some of them to be obtained in the amorphous state. The magnetic state of an amorphous material can be very different from that of the crystalline state, often with a lower Curie temperature. The theory that we will present here applies only qualitatively, because no complete theory has yet been developed for these materials, which are currently of technological interest.

2.3. RARE EARTH METALS

The magnetic moments of the rare earth elements are associated both with the spin and the orbital angular momentum of the f electrons. Their small spatial extension makes them weakly sensitive to their local environment. The *s* or *d* electrons delocalise to some extent to become conduction electrons, and it is through them that the localized magnetic moments of the rare earths are coupled. In general, the rare earth elements carry a large magnetic moment, as we have seen in chapter 7, but the transition temperatures are less than those of the 3d elements. Most of them also have a very large magnetocrystalline anisotropy, an important feature for some technological applications, as well as an anisotropy of the magnetic moment.

2.4. RARE EARTH - TRANSITION METAL COMPOUNDS

These compounds are of great industrial interest because they combine the large magnetic moment of the rare earth metals, with their often huge magnetic anisotropy, with the high critical temperatures of the transition metals. It is therefore in this class of materials that the best permanent magnets, e.g. $Nd_2Fe_{14}B$ or $SmCo_5$, are encountered.

3. MAGNETISM OF COMPLETELY FREE ELECTRONS

3.1. SIMPLE DESCRIPTION OF A METAL

Before explaining ferromagnetism in metals, it is useful to recall the simplest model for describing the properties of metals such as copper or aluminium: they are considered as a box in which the valence electrons of the atoms can freely move, while those of the internal shells remain localized. To a first approximation the potential inside the box can be considered to be constant, i.e. the periodic potential due to the ions can be ignored. In quantum mechanics, each electron is described by a wavefunction, which in the case of a constant potential is a plane wave. The wave vector **k** of this wave is related by the de Broglie equation to the momentum of the particle **p**: **p** = $\hbar$**k**, where k = $2\pi/\lambda$.

As the electrons are in a box, their wavefunctions must fulfill the boundary conditions, and only certain waves can exist. The possible values of k are therefore quantized. The energy of the state characterized by a wave vector **k** is given by:

$$\varepsilon_k = \frac{\hbar^2 k^2}{2m} \tag{8.1}$$

where m is the electron mass. According to the Pauli principle each state can accommodate two electrons with opposite spins. At zero temperature, the total energy of the system should be minimized. This involves filling all the states, from that with the lowest energy up to an energy ε_F, called the Fermi energy, which has an associated wave vector k_F. k_F is clearly a function of the density N_e of valence electrons in a metal, and is given by the relation:

$$k_F^3 = 3\pi^2 N_e \tag{8.2}$$

The state we have just constructed is nonmagnetic because it contains the same number of electrons with spin up and spin down. Is this state paramagnetic or diamagnetic?

If we apply a magnetic induction **B** to the metal, the electrons which have their magnetic moments parallel to **B** will have their energy reduced by the quantity $\mu_B B$, while those with the opposite spin will have their energy increased by the same quantity. The state with equal numbers of up and down spins is no longer the lowest energy state because electrons can be transferred from a state with spin up into the lower energy down states by reversing the spin. Thus the ground state in the presence of the magnetic induction will not have the same number of electrons with the two spin directions. Rather, a majority of the electrons will have their magnetic moments in the direction of the field, and the metal will correspondingly possess a magnetic moment in this direction. The metal is therefore paramagnetic.

3.2. DENSITY OF STATES

We now introduce a fundamental quantity for metals, namely the density of states. It is characteristic of a given metal, and determines many of its physical properties. The number of states between energies ε and $\varepsilon + d\varepsilon$ is by definition given by $N(\varepsilon)d\varepsilon$, where $N(\varepsilon)$ is the density of states for a particular direction of spin, per unit volume, and per unit energy. For the model of free electrons with no interaction, this density is given by the relation:

$$N(\varepsilon) = \frac{mk}{2\pi^2 \hbar^2} = \frac{\sqrt{m^3 \varepsilon}}{\sqrt{2\pi^2 \hbar^3}} \sqrt{\varepsilon} \qquad (8.3)$$

This density of states can also be calculated in more complicated models. It then depends on the nature and the crystallographic structure of the metal, as we will see later. We will assume it to be known in the general case.

3.3. PAULI PARAMAGNETISM

In order to evaluate the magnetic susceptibility, we search for the ground state, i.e. the state with minimum energy, in the presence of a magnetic induction **B**. To achieve this state, it is necessary to transfer the electrons which are in a slice of thickness (in terms of energy) $\delta\varepsilon = \mu_B B$ from one spin direction to the other (fig. 8.1). There are $N(\varepsilon_F)\mu_B B$ electrons in this slice, and so, after the transfer, twice this number of electrons will contribute to a magnetic moment oriented in the direction of the field. This creates a total magnetic moment $m = 2(\mu_B)^2 N(\varepsilon_F) B$.

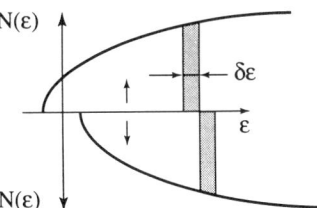

Figure 8.1
Density of states of a simple metal in the presence of a magnetic induction

The susceptibility of free noninteracting electrons is therefore:

$$\chi^0 = 2\mu_0 (\mu_B)^2 N(\varepsilon_F) \qquad (8.4)$$

This susceptibility is due to the valence electrons of a metal, and is called the *Pauli susceptibility*. It is largely independent of temperature, which can be demonstrated by generalizing this calculation to finite temperatures. This gives:

$$\chi^0(T) = \chi^0(0) \left\{ 1 + \frac{\pi^2}{6} k_B^2 \left[\frac{N''}{N} - \left(\frac{N'}{N}\right)^2 \right] T^2 \right\} \qquad (8.5)$$

where $N = N(\varepsilon_F)$, $N' = dN/d\varepsilon$, and $N'' = d^2N/d\varepsilon^2$.

The first correction term comes from the change in the effective density of states (see eq. 8.20) that arises from the broadening over $k_B T$ of the Fermi distribution. The second term accounts for the displacement of the Fermi level that is necessary to keep the number of electrons constant. As the density of states for d electrons varies appreciably over an energy range of the order of 0.1 eV, which corresponds to ~1,000 K, the temperature dependent term is therefore very small, and negligible to a first approximation.

3.4. LANDAU DIAMAGNETISM

We have said that diamagnetism is a universal phenomenon that is always present even though, as is often the case, it is totally hidden by phenomena such as paramagnetism, ferromagnetism, ... Free electrons do not escape this rule, and Landau calculated the susceptibility of an ensemble of totally free electrons. We will not give the derivation, which the reader can find in text books on metal physics [2]. The result is very simple, and shows that "Landau diamagnetism" while being masked by "Pauli paramagnetism", still contributes to the latter a nonnegligible correction:

$$\chi_{dia} = -\chi_{Pauli}/3 \qquad (8.6)$$

Remark - *Formulas (8.5) and (8.6) are only applicable to the contribution that the valence electrons make to the susceptibility. It is necessary to remember that the electrons in the full orbitals of the ion also give rise to a diamagnetic contribution, as we have shown in chapter 4. The measured susceptibility is the sum of all these contributions.*

4. STONER'S MODEL OF ITINERANT FERROMAGNETISM

4.1. CRITERION FOR FERROMAGNETIC INSTABILITY

The description given in section 3 does not take into account the Coulomb repulsion between electrons. The problem of electrons moving in the potential created by all the other electrons is unsolvable using quantum mechanics, as it is far too complicated. However, it is not difficult to understand why a dissymmetry can be obtained between the electrons with the two spin directions, which then gives rise to ferromagnetism. We will begin from the Pauli principle, which postulates that the overall wavefunctions of all the electrons must be antisymmetric with respect to the interchange of two electrons. The consequence which interests us here is that two electrons with the same spin can never be in the same place at the same time. In contrast, nothing prevents the probability of finding two electrons with opposite spin in the same place from being nonzero. This means that two electrons with opposite spins will on average repel each other more than two electrons with the same spin, as the latter feel each other less because they can never be in the same place.

Stoner's model postulates a repulsion between electrons of opposite spins that is larger by a quantity I than that between electrons with the same spin direction. This is equivalent to a potential energy for the interaction between electrons in the form $IN_\uparrow N_\downarrow$, where $N_\uparrow$ and $N_\downarrow$ are the densities of electrons with the two spin directions. In this model, there are $N/2$ electrons with each spin direction.

Let us examine the stability of the paramagnetic state with respect to ferromagnetism when this interaction exists. We show that, in some cases, energy is minimized when electrons are transferred from one spin state to the other.

If one transfers a slice of thickness $\delta\varepsilon$ (see fig. 8.2), i.e. if one transfers $N(\varepsilon_F)\,\delta\varepsilon$ electrons, the kinetic energy of the electrons will increase by:

$$\Delta E_c = N(\varepsilon_F)(\delta\varepsilon)^2 \qquad (8.7)$$

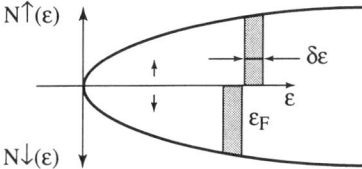

Figure 8.2 - Density of states for electrons with spins $\uparrow$ and $\downarrow$

The change in the interaction energy between electrons is then:

$$I\left[\frac{N}{2}+N(\varepsilon_F)\delta\varepsilon\right]\left[\frac{N}{2}-N(\varepsilon_F)\delta\varepsilon\right]-I\left(\frac{N}{2}\right)^2 = -IN^2(\varepsilon_F)(\delta\varepsilon)^2 \qquad (8.8)$$

The total change in energy is the sum of (8.7) and (8.8):

$$\Delta E = N(\varepsilon_F)(\delta\varepsilon)^2\,[1 - IN(\varepsilon_F)] \qquad (8.9)$$

The nonmagnetic state is therefore only stable if $IN(\varepsilon_F) < 1$. If this inequality is not satisfied, the minimal energy is obtained with a different number of $\uparrow$ and $\downarrow$ electrons, which implies ferromagnetism. The criterion for instability with respect to ferromagnetism, termed Stoner's criterion, is therefore:

$$IN(\varepsilon_F) > 1 \qquad (8.10)$$

The repulsion I introduced in this model is very difficult to evaluate, and in practice it is adjusted to fit experiment data. We note that this criterion shows that ferromagnetic metals should have a high density of states at the Fermi level.

4.2. MAGNETIC SUSCEPTIBILITY OF A METAL WITH INTERACTIONS

We will now calculate the susceptibility of a metal in Stoner's model when the nonmagnetic state is stable.

In the presence of an induction B, the energies of states k with moments in the direction of the field ($\uparrow$) are:

$$E_{k\uparrow} = \frac{\hbar^2 k^2}{2m} + IN_\downarrow - \mu_B B \qquad (8.11)$$

and for those states with opposed moments:

$$E_{k\downarrow} = \frac{\hbar^2 k^2}{2m} + IN_\uparrow + \mu_B B \tag{8.12}$$

The two bands are therefore shifted by an energy:

$$2\,\delta E = I(N_\uparrow - N_\downarrow) + 2\mu_B B \tag{8.13}$$

which, using the relation $N_\uparrow - N_\downarrow = 2N(\varepsilon_F)\,\delta E$, yields $\delta E = \mu_B B / [1 - IN(\varepsilon_F)]$.

The magnetization of the sample is therefore given by:

$$M = \mu_B (N_\uparrow - N_\downarrow) = \frac{2\mu_B^2 N(\varepsilon_F)}{1 - IN(\varepsilon_F)} B \tag{8.14}$$

which corresponds to a magnetic susceptibility:

$$\chi = \frac{2\mu_B^2 \mu_0 N(\varepsilon_F)}{1 - IN(\varepsilon_F)} \tag{8.15}$$

This susceptibility is larger than that of the system of noninteracting electrons (eq. 8.4) by a factor:

$$S = 1/[1 - IN(\varepsilon_F)] \tag{8.16}$$

termed Stoner's factor. S becomes infinite when the criterion for instability (eq. 8.10) is satisfied.

We note that *the magnetic moments predicted from equation (8.14) are not equal to an integral number of Bohr magnetons*, in agreement with the experiments made on the transition metal 3d series.

4.3. FERROMAGNETIC SOLUTION

Instability of the nonmagnetic solution can result either in *strong ferromagnetism*, where all the delocalized electrons have the same spin, or in *weak ferromagnetism*, where there is only an unbalance between the number of spins in the two directions (fig. 8.3). Which one occurs depends on the shape of the density of states, and the number of electrons per atom.

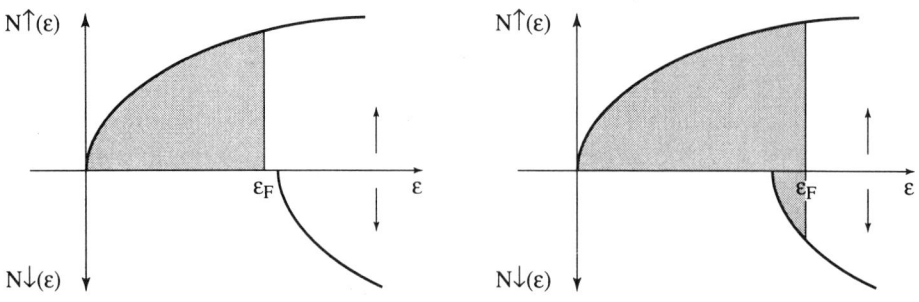

Figure 8.3 - Strong (on the left), and weak (on the right) ferromagnetism, represented schematically in the free electron limit

8 - MAGNETISM IN THE ITINERANT ELECTRON MODEL

Examination of the magnetic situation at nonzero temperature requires the solution to a self-consistent equation which is analogous to that of the molecular field, as we will emphasize later:

$$M = \frac{M}{\mu_B} = (N_\uparrow - N_\downarrow) = \frac{1}{N}\sum_k \left[f(E_{k\uparrow}) - f(E_{k\downarrow})\right] \qquad (8.17)$$

where M is the magnetization in μ_B per unit volume, and f(E) is the Fermi-Dirac function which determines the distribution of electrons over the different energy levels:

$$f(E) = \left[1 + \exp\left(\frac{E-\mu}{k_B T}\right)\right]^{-1}$$

μ represents the chemical potential of the electrons, and is equal to ε_F at zero temperature.

The detailed solution of this equation depends on the density of states, and we will not discuss it here. We will simply calculate the Curie temperature. This is given by equation (8.17) in the limit where m tends to zero:

$$1 = \frac{1}{N}\sum_k \frac{d}{dm}\left[f(E_{k\uparrow}) - f(E_{k\downarrow})\right]_{m=0} \qquad (8.18)$$

Using relations (8.11), and (8.12), (8.18) becomes $1 = \frac{I}{N}\sum_k \left[\frac{df(E_k)}{dm}\right]_{m=0}$, whence:

$$I\int N(\varepsilon)\frac{df}{d\varepsilon}d\varepsilon = -1 \qquad (8.19)$$

This equation determines the Curie temperature T_C, which tends to zero when $IN(\varepsilon_F)$ is equal to unity, as $(df/d\varepsilon) = -\delta(\varepsilon - \varepsilon_F)$ at zero temperature. It is useful to introduce an effective density of states at a finite temperature T defined by:

$$N(T) = -\int N(\varepsilon)\frac{df}{d\varepsilon}d\varepsilon \qquad (8.20)$$

It is simple to verify that this density of states is related to the Pauli susceptibility at temperature T in a metal without interactions:

$$\chi^0(T) = 2\mu_B^2 \mu_0 N(T) \qquad (8.21)$$

The Curie temperature is now given by: $IN(T) = 1$.

The susceptibility above T_C in the presence of interactions is given by a formula analogous to equation (8.15):

$$\chi(T) = \frac{2\mu_B^2 \mu_0 N(T)}{1 - IN(T)} \qquad (8.22)$$

While $\chi(T)$ does indeed become infinite at T_C, this formula does not give the Curie law at high temperatures as is observed in localized magnetism. Rather, the dependence on temperature stems from N(T), and therefore also from the form of the density of states.

4.4. APPLICATIONS

Stoner's theory has been widely used to explain the properties of ferromagnetic metals and alloys, and to qualitatively predict their properties. *Its great success was to explain why the magnetic moment cannot be expressed in terms of integral values of μ_B at zero temperature.* Unfortunately, the temperature dependence of the properties are poorly described. At low temperature, the observed variation of the magnetization does not follow the law predicted by Stoner's equation (8.17). More seriously, the critical temperatures calculated are generally an order of magnitude too large, and the Curie-Weiss law observed in many metals is not explained. We will return to these points later. The main quality of this theory is its great simplicity. Its failings result from the fact that this is a mean-field approximation, as we will now show.

5. GENERALIZATION OF STONER'S CRITERION

5.1. STONER'S THEORY AND MOLECULAR FIELD THEORY

Stoner's criterion can be generalized to instability with respect to forms of magnetic order other than ferromagnetism. To demonstrate this, we will first show that Stoner's theory is similar to the Weiss molecular field theory. The molecular field acts as an external field that enhances the applied field. It is supposed to describe the interactions between the magnetic moments, and to be proportional to the magnetization of the material. If χ^0 is the susceptibility of a material without interactions, and w is the molecular field coefficient, the magnetization in the presence of a small applied field H is given by: $M = \chi^0(H + wM) = \chi H$, from which one obtains:

$$\chi = \chi^0 / (1 - w\chi^0) \tag{8.23}$$

Comparison with equation (8.15), and the definition of χ^0, yield the molecular field coefficient which is the basis of Stoner's theory

$$w = I / (2\mu_0 \mu_B^2) \tag{8.24}$$

5.2. GENERALIZED SUSCEPTIBILITY

In order to be able to apply Stoner's theory to magnetic structures other than ferromagnetic, we first have to generalize the susceptibility for the case of a nonuniform magnetic field.

For any field $\mathbf{h}(\mathbf{r})$, a system invariant with respect to translation presents a magnetization that can be written: $\mathbf{M}(\mathbf{r}) = \int \chi(\mathbf{r} - \mathbf{r'}) \mathbf{h}(\mathbf{r'}) d\mathbf{r'}$. This function χ is the linear response to a nonuniform field, and is called the generalized susceptibility. It represents the magnetization due to a static localized magnetic field at a point $\mathbf{r_0}$. If $\mathbf{h}(\mathbf{r}) = \mathbf{h}_{\mathbf{r_0}} \delta(\mathbf{r} - \mathbf{r_0})$, we have therefore: $\mathbf{M}(\mathbf{r}) = \chi(\mathbf{r} - \mathbf{r_0}) \mathbf{h}_{\mathbf{r_0}}$.

Taking the Fourier transform, the susceptibility for wavevector **q** can be defined as the linear response to the magnetic field **h(q)** $e^{i\mathbf{qr}}$ by:

$$\mathbf{M(q)}\, e^{i\mathbf{qr}} = \chi(\mathbf{q})\, \mathbf{h(q)}\, e^{i\mathbf{qr}} \tag{8.25}$$

5.3. THE LOCAL MOLECULAR FIELD

The Weiss approximation, up to now, has only been applied to states with a uniform magnetization of the spins. In order to generalize Stoner's calculations, it is convenient to introduce the idea of a local molecular field. At any point **r** in a material with static magnetization **M(r)**, the static molecular field is defined by

$$\mathbf{H}_{mol}(\mathbf{r}) = \int w(\mathbf{r} - \mathbf{r'})\, \mathbf{M(r')}\, d\mathbf{r'} \tag{8.26}$$

where we postulate the existence of a molecular field coefficient w(**r**) with a certain characteristic range. Performing the Fourier transform, and using the translational invariance gives:

$$\mathbf{H}_{mol}(\mathbf{q}) = w(\mathbf{q})\,\mathbf{M(q)} = w(\mathbf{q})\,\chi(\mathbf{q})\,\mathbf{H}_{ext}(\mathbf{q}).$$

In this case, the response to an external sinusoidal field $\mathbf{H}_{ext}(\mathbf{q})$ will be:

$$\chi(\mathbf{q}) = \chi^0(\mathbf{q})[1 + w(\mathbf{q})\chi(\mathbf{q})] = \frac{\chi^0(\mathbf{q})}{1 - w(\mathbf{q})\chi^0(\mathbf{q})} \tag{8.27}$$

This equation is only valid when $w(\mathbf{q})\chi^0(\mathbf{q}) < 1$, and is a generalization of equations (8.15) and (8.23).

When $w(\mathbf{q})\chi^0(\mathbf{q}) = 1$ (equality instead of inequality) for some value of **q**, the system spontaneously magnetises, and adopts a periodic structure with wavevector **q**. The nonmagnetic state of the system is said to be unstable with respect to the formation of a spin density wave with wavevector **q**.

It is easy to see that, for the Heisenberg model, w(**q**) is proportional to the Fourier transform of J_{ij}, specifically to J(**q**) divided by $N_a(g\mu_B)^2$, where N_a is the number of atoms per unit volume. It can be shown that, for a metallic system, w(**q**) is independent of **q**, and is always given by equation (8.24).

5.4. DYNAMIC SUSCEPTIBILITY

In order to treat time dependent effects, it is useful to define a dynamic susceptibility. Application of a uniform magnetic field $h(t) = h(t_0)\,\delta(t - t_0)$ for a very short time at t_0 acts as an impulse, under the effect of which the system will have a magnetization M(t) that remains for a characteristic time after the end of the impulse. The time dependent susceptibility may be defined by the relation: $M(t) = \chi(t - t_0)\,h(t_0)$.

Causality demands that $\chi(t - t_0)$ be zero for $t < t_0$. Any time dependent magnetic field can be considered as a sum of impulses. If h(t) is always small, the magnetization of the system at time t is given by the sum of all the magnetizations that will have propagated until time t. For a time-dependent field that is uniform in space, the magnetization at time t is:

$$\mathbf{M}(t) = \int_{-\infty}^{t} dt' \, \chi(t-t') \, \mathbf{h}(t').$$

Imposing $\chi(t) = 0$ for $t < 0$, one can therefore write: $\mathbf{M}(t) = \int_{-\infty}^{+\infty} dt' \, \chi(t-t') \, \mathbf{h}(t').$

Defining the Fourier transform with respect to time as:

$$\mathbf{M}(\omega) = \lim_{\eta \to 0} \int_{-\infty}^{+\infty} dt \, \mathbf{M}(t) \, e^{-i\omega t - \eta t},$$

we obtain: $\mathbf{M}(\omega) = \chi(\omega) \, \mathbf{h}(\omega).$

The dynamic susceptibility characterizes the response to a field which was established adiabatically from time $t = -\infty$ until time $t = 0$.

If the magnetic field varies in both time and space, we can generalize the definitions, which yields:

$$\mathbf{M}(\mathbf{r},t) = \int d\mathbf{r}' \int_{-\infty}^{t} dt' \, \chi(\mathbf{r}-\mathbf{r}',t-t') \, \mathbf{h}(\mathbf{r}',t')$$

$M(\mathbf{q}, \omega) = \chi(\mathbf{q}, \omega) \, h(\mathbf{q}, \omega)$. $\chi(\mathbf{q}, \omega)$ characterizes the response of the system to a magnetic field $h(\mathbf{r}, t)$ where: $h(\mathbf{r}, t) = h(\mathbf{q}, \omega) \, e^{-i(\omega t + \mathbf{q} \cdot \mathbf{r})}$

5.5. THE INSTANTANEOUS LOCAL MOLECULAR FIELD

Let us make the following hypothesis: when there appears, at time t and point **r**, a magnetization **M**(**r**, t), there appears at the same time t at point **r'** a molecular field given by: $\mathbf{H}_{mol}(\mathbf{r}', t) = w(\mathbf{r} - \mathbf{r}') \, \mathbf{M}(\mathbf{r}, t).$

We generalize equation (8.26), but we neglect the possible effects of the delay in the propagation of the molecular field. This should be of little importance because the propagation takes place only over distances comparable to the range of w(**r**). This hypothesis is equivalent to that called the random phase approximation (RPA) [3, 4]. We have therefore:

$$\mathbf{H}_{mol}(\mathbf{q}, \omega) = w(\mathbf{q}) \, \mathbf{M}(\mathbf{q}, \omega) = w(\mathbf{q}) \, \chi(\mathbf{q}, \omega) \, \mathbf{H}_{ext}(\mathbf{q}, \omega).$$

Equation (8.27) can be generalized to:

$$\chi(\mathbf{q},\omega) = \chi^0(\mathbf{q},\omega)[1 + w(\mathbf{q})\chi(\mathbf{q},\omega)] = \frac{\chi^0(\mathbf{q},\omega)}{1 - w(\mathbf{q})\chi^0(\mathbf{q},\omega)}$$

The simplicity of these relations justifies the use of a dynamical **q**-dependent susceptibility. Moreover, if, for a frequency ω and wave vector **q**, $\chi(\mathbf{q}, \omega)$ is infinite, $M(\mathbf{q}, \omega)$ can still be finite even if $\mathbf{H}_{ext}(\mathbf{q}, \omega)$ is zero. The system then has an excited state with magnetization $\mathbf{M}(\mathbf{r}, t) = \mathbf{M}(\mathbf{q}, \omega) \exp{-i(\omega t + \mathbf{q}\mathbf{r})}$. A pole of $\chi(\mathbf{q}, \omega)$ is therefore an elementary excitation of the system. In fact the poles often have an imaginary component, i.e. they are in the complex plane, which gives a finite lifetime to these excitations. In the literature they are often called spin fluctuations or paramagnons [4].

6. TRANSITION METALS

Transition metals cannot be described by the free electron model used in section 3.1. The valence electrons are the s and d electrons. While the s electrons are completely delocalised, and behave like free electrons, their density of states is very weak, and they are therefore of little importance. It is the d electrons which are largely responsible for the magnetic properties of the transition metals. The two densities of states are schematically shown in figure 8.4. We now explain the origin of this large density of d states.

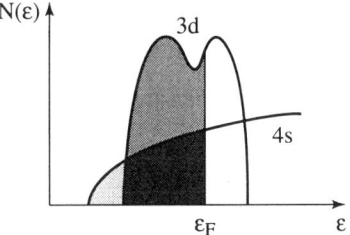

Figure 8.4 - Schematic density of states of an s and a d band

6.1. THE d BAND OF THE TRANSITION METALS

The d orbitals of the valence electrons of the transition metals are more localized than the s or p orbitals with the same energy. They are not very much affected by the lattice, but they overlap a little with the orbitals of neighboring atoms (see fig. 7.8). This is why the free electron approximation developed in preceding sections is not appropriate. Instead, another approach should be used, namely the *tight binding* approximation, which describes the wavefunction of the itinerant electrons as a linear combination of atomic orbitals. More precisely, the potential from ions is taken into account, and is written as the sum of atomic potentials centered on each site i of the lattice:

$$V = \sum_i V_i \tag{8.28}$$

For each lattice site, there are five d orbitals written $|i, m_l\rangle$ where m_l is the orbital quantum number of orbital l, and l varies from 1 to 5. These are eigenfunctions of the hamiltonian, and have energy E_0. For simplification, we will ignore the index m_l except when needed. The wavefunction can be approximated as a linear combination of the functions $|i\rangle$:

$$|\Psi(E)\rangle = \sum_i a_i |i\rangle \tag{8.29}$$

where:
$$(\hat{E}_c + \hat{V}_i)|i\rangle = E_0|i\rangle \qquad (8.30)$$

$\hat{E}_c$ and $\hat{V}_i$ being the kinetic and potential energy operators, respectively.

In this approximation, all the overlap integrals are assumed to be zero, $\langle i|j\rangle = 0$, and, among all the matrix elements of the potential, only the two center integrals that correspond to nearest, and sometimes also next-nearest neighbors, are kept. The coefficients a_i satisfy an infinite set of linear equations of the type

$$(E_0 + \alpha_i - E)a_i + \sum_{i \neq j} \beta_{ij} a_j = 0 \qquad (8.31)$$

with
$$\alpha_i = \langle i| \sum_{j \neq i} V_j |i\rangle \qquad (8.32)$$

$$\beta_{ij} = \langle i|V_j|j\rangle \qquad (8.33)$$

The integrals α simply act to displace the energy E_0 of the atomic levels $|i\rangle$ while the integrals β mix them into *molecular states that extend over the whole solid*. The solutions to this system of linear equations give the possible states and energies for the electrons. The resulting states can be classified according to the vector **k**, just as for plane waves (Bloch's theorem). For a simple cubic system with unit cell parameter a, where all the integrals β are equal (then the states $|i\rangle$ are the *s* states of the isolated atom), the energies ε_k of the states are:

$$\varepsilon_k = 2\beta(\cos k_x a + \cos k_y a + \cos k_z a) \qquad (8.34)$$

This dispersion relation replaces that of free electrons (eq. 8.1). In a simple cubic system the energy band therefore has a width of 12β (this is equal to $2z\beta$, where z is the number of nearest neighbors).

The width of the band depends therefore on the number of nearest neighbors of an atom, and on the value of the integral β, which itself depends very strongly on the interatomic distance (fig. 8.5). The boundary conditions quantize the values of k, and the number of states between ε and $\varepsilon + d\varepsilon$, i.e. the density of states, can be defined. It is useful to define the density of states $n(\varepsilon)$ per atom, and not per unit volume as we have done for free electrons. The two are related by $N(\varepsilon) = N_a n(\varepsilon)$, where N_a is the number of atoms per unit volume.

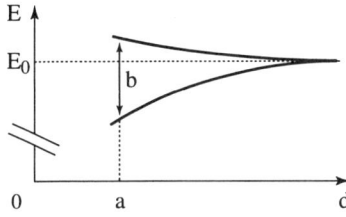

Figure 8.5 - Width b of the band as a function of interatomic distance

8 - MAGNETISM IN THE ITINERANT ELECTRON MODEL

We can have a very simple estimation of n(ε). For an energy band made up of s electrons we should have one state per atom and per spin direction. Therefore, if *the width of the band is b*, the density of states is of the order of 1/b. For the d electrons, the number of states per atom is 5, and therefore the density of states will be of the order of 5/b.

Numerical calculations made on the transition metals show that the bandwidths b are of the order of 5 to 10 eV. The shift of the band with respect to the atomic level is much smaller, and is of the order of 1 to 2 eV. The densities of states at the Fermi level will therefore be far greater than in the case of free electrons.

The larger the density of states, the smaller β. The latter will be small if the atomic wavefunctions are concentrated at the nucleus. In a series of transition metals, the extension of the atomic d orbitals reduces with increasing occupation, and correspondingly the largest densities of states are found at the end of the series. Orbitals of the second and third d series are more extended, hence the densities of state are smaller in these series. The largest densities of states are expected at the end of the first series, precisely where the magnetic materials are found. This is in perfect agreement with Stoner's criterion. Figure 8.6 shows an example of the density of states, that for pure nickel.

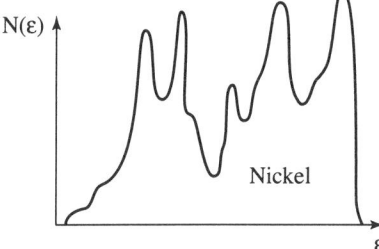

Figure 8.6 - Density of states (calculated) of nickel

6.2. ORIGIN OF MAGNETISM

We will now go deeper into the origin of the energy I that we have introduced in Stoner's model. In the Hartree approximation, the wavefunction of the electrons Ψ(1, 2, ... i, ... N) is given by the product of the wavefunctions Ψ_i of each electron i:

$$\Psi = \prod_i \Psi_i(i) \tag{8.35}$$

At this point, Ψ does not obey the Pauli principle, and so it is necessary to make it antisymmetric by constructing a determinant from the wavefunctions of all the electrons. Let us take the example of two electrons, 1 and 2, on two atomic orbitals centred at sites i and j. If their spins are parallel, the spatial wave function should be antisymmetric, i.e:

$$\Phi_A = \frac{1}{\sqrt{2}}\left[\Psi_{im}(1)\Psi_{jm'}(2) - \Psi_{im}(2)\Psi_{jm'}(1)\right] \tag{8.36}$$

where we signal the degeneracy of the d state through the orbital quantum number m. If we introduce the Coulomb repulsion between the two electrons, we add to the energy a term of the form:

$$<\Phi_A| \frac{e^2}{4\pi\varepsilon_0} \frac{1}{r_{12}} |\Phi_A> = U_{ij}^{mm'} - J_{ij}^{mm'} \quad (8.37)$$

with:
$$U_{ij}^{mm'} = <\Psi_{im}(1)\Psi_{jm'}(2)| \frac{e^2}{4\pi\varepsilon_0} \frac{1}{r_{12}} |\Psi_{im}(1)\Psi_{jm'}(2)> \quad (8.38)$$

$$J_{ij}^{mm'} = <\Psi_{im}(1)\Psi_{jm'}(2)| \frac{e^2}{4\pi\varepsilon_0} \frac{1}{r_{12}} |\Psi_{im}(2)\Psi_{jm'}(1)> \quad (8.39)$$

If the spins on the electrons are antiparallel, the Pauli principle requires that the spatial wavefunction is symmetric. This corresponds to replacing the sign − by the sign + in equation (8.37), and we obtain for the energy the result:

$$<\Phi_S| \frac{e^2}{4\pi\varepsilon_0} \frac{1}{r_{12}} |\Phi_S> = U_{ij}^{mm'} + J_{ij}^{mm'} \quad (8.40)$$

This gives rise to a difference in energy between the two spin configurations.

The integrals U are called the Coulomb terms, and J the exchange terms. It is evident that the interatomic terms ($i \neq j$) are smaller than the intra-atomic terms ($i = j$) because of the exponential fall-off of the atomic d functions. We will neglect them here, and keep only the intra-atomic terms. The largest exchange term is $J_{ii}^{mm} = U_{ii}^{mm} = U^m$.

Two electrons in the same orbital necessarily have opposite spins, and the energy is increased by the quantity U^m. If the two electrons are in different orbitals, the configuration with parallel spins has an energy that is reduced by $2J_{ii}^{mm'}$.

The intra-atomic terms therefore favor the situation where the two electrons are in different orbitals, and have parallel spins. This is simply an expression of Hund's rule. Numerical calculations give $U^m \sim 20$ eV, and $J_{ii}^{mm'} \sim 1$ eV. An average value U per pair of electrons and per atom can also be defined. This corresponds to the average energy gained by changing a pair of antiparallel spins to a pair of parallel spins. The 5 values of m being approximately equally represented at each energy, this estimation gives: $U = (U^m + 4J^{mm'})/5 = 5$ eV. The value of U is larger when the extension of the d orbitals is small, the same condition that gives a large density of states.

If we consider the quantities per atom, and not per unit volume, the density of states per atom $n(\varepsilon)$, and the susceptibility per atom, we have to replace in Stoner's calculation $N(\varepsilon)$ by $n(\varepsilon)$, and I by U.

The condition for instability with respect to ferromagnetism becomes:

$$U n(\varepsilon_F) > 1 \quad (8.41)$$

In this alternative expression of Stoner's criterion, the quantity I is replaced by one that is more physical, viz the repulsion between two electrons on the same site. This is just as difficult to calculate in practice because the s electrons of the transition metals can shield the repulsion between d electrons. Therefore, it usually remains a parameter that is adjusted to fit experiment. It does show that the magnetism arises from a local repulsion, on a site, which will be useful in the following sections.

This analysis is at the origin of the theoretical model known as Hubbard's hamiltonian [5]. It allows the study of the effects of Coulomb repulsion, which was ignored in the simple independent electron theories. This model introduces a repulsion only for two electrons on the same site. They must therefore have opposite spins. The effects are not limited just to magnetism, but can explain why some compounds are insulators (so-called Mott insulators) whereas band theory would lead us to expect a metal [6].

6.3. CRYSTALLINE ELECTRIC FIELD

In metals and metallic compounds based on the transition metals, the elements lose their valence electrons as they form the conduction band. This leads to a network of positive charges in a sea of conduction electrons. We saw in section 6.1 that these electrons are not totally free: they are influenced by the periodic potential of the metallic ions in the lattice, and its symmetry. As in the localized electron model, the 3d electrons, on a given atom, are in the electric field produced by the charges of the surrounding ions, with a shielding effect from the other electrons. The spherical symmetry of the isolated ion is broken, and is replaced by the local symmetry of the lattice. The orbitals to be considered are no longer the eigenfunctions of the component to the angular momentum, but linear combinations of them.

For example, in a cubic environment, the d orbitals which were given by equations (7.29) will now be decomposed into orbitals that are generally termed e_g ($x^2 - y^2$, and $3z^2 - r^2$), and t_{2g} (xy, yz, and zx). While these orbitals have the same energy in a system with spherical symmetry, their degeneracy is broken in a cubic environment, as shown in figure 8.7. This splitting can be simply explained if we suppose that the first neighbors of the atom represented in the figure are positive charges situated on the axes Ox, Oy, and Oz; the e_g orbitals with their negatively charged electronic clouds point in the direction of these positive charges, which is energetically favorable, while the t_{2g} orbitals do not point towards the positive charges: this situation corresponds to a state of higher energy.

There is therefore a coupling between the orbitals and the lattice. This effect is referred to as the crystalline electric field or more simply crystalline field.

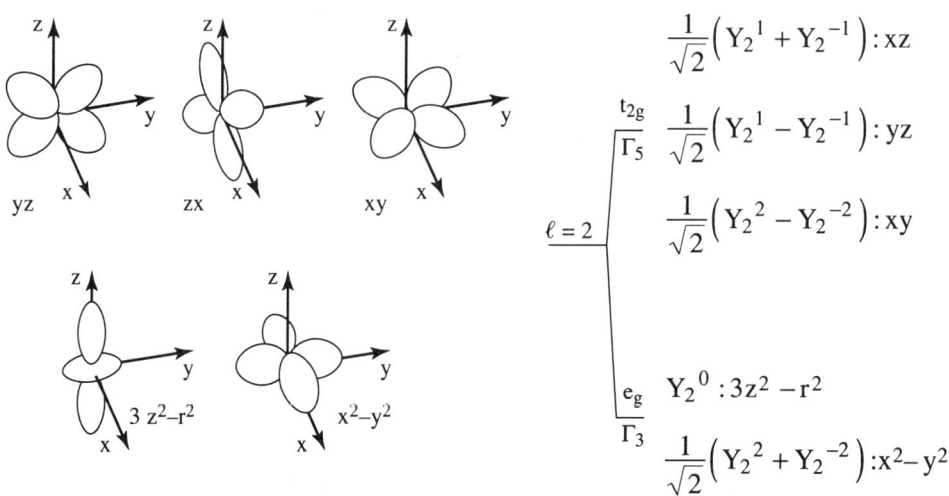

*Figure 8.7 - Shapes of the d orbitals (on the left),
and their splitting in a cubic environment (on the right)*

In the tight binding approach, this effect is described by the integrals α (see § 5.1). In the transition metals, this effect is very small. However, in compounds such as the perovskites, the crystal field can lift the degeneracy completely. We show in figure 8.8 the example of a copper based perovskite, as is the case in the new superconducting materials. The symmetry of the perfect perovskite lattice is cubic. In the superconductors, a crystallographic distortion makes the structure tetragonal, and breaks the degeneracy of the e_g orbitals. The Fermi level in these materials is situated within the band formed from the $x^2 - y^2$ states.

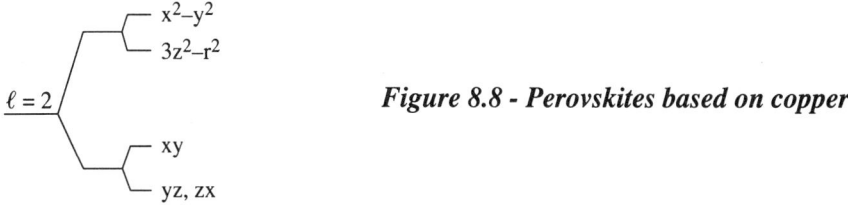

Figure 8.8 - Perovskites based on copper

6.4. MAGNETOCRYSTALLINE ANISOTROPY

The magnetocrystalline anisotropy depends both on the character of the orbitals at the Fermi level, and on the spin-orbit coupling. In metals, compounds, and alloys of transition metals, the spin-orbit coupling coefficient λ is always small compared to the crystal field parameters.

For the metals at the end of the first series, the crystal field is of the order of 1 eV, while the spin-orbit coupling, described in equation (7.20), is of the order of 0.01 eV. It can therefore be treated as a perturbation, and the anisotropy constants K_1, K_2, etc. can be determined from equations (3.3), and (3.4). One can shows that K_n decreases with n as a power of λ/b. As b, the width of the band, is of the order of 5 to 10 eV, the value of the anisotropy constants decreases very rapidly with n.

In cubic symmetry, the magnetic anisotropy energy is given mainly by the value of K_1 which is of the order of λ^4/b^3, i.e. 10^3 J.m^{-3}. This value is in agreement with the values measured for pure iron and nickel.

However, in the case of nickel, all the anisotropy coefficients, even those of high order, remain measurable at room temperature even though they vary strongly with temperature: this result is linked to the peculiarities of the Fermi surface in nickel [7], and demonstrates that it is more difficult to predict the magnetic properties of metals than those of insulators.

For a compound with hexagonal symmetry, K_1 is of the order of λ^2/b, i.e. 10^7 J.m^{-3}. The anisotropy energy is therefore far greater than in a cubic system. This is for example the order of magnitude for metallic cobalt.

7. LOCALIZED MAGNETIC MOMENT IN ITINERANT MAGNETISM

We have seen that a localized magnetic moment can exist in the rare earth metals because the very weak spatial extension of the f orbitals causes the electrons to remain localized in the solid. In transition metals, Stoner's theory does not require the magnetic moments to be localized on an atom, and the saturation magnetization at low temperatures is not submitted to the rules we have mentioned in chapter 4. Nevertheless, above the Curie temperature, the susceptibility practically obeys the Curie law, a characteristic of a localized moment.

We will now show how this is possible by studying the situations where the transition atom can be considered as carrying a magnetic moment, even though the d electrons are mobile.

7.1. MAGNETISM OF IMPURITIES

Let us begin by considering a single transition atom inside a nonmagnetic metallic matrix, for example in copper or aluminium. These metals are well described by the free electron theory developed in section 3. An electron in a d orbital of a transition atom does not remain bound to this atom because there is a probability that it will hop to an unoccupied state, with wavevector **k**, of the metallic matrix.

Conversely, an electron in the state **k** of the matrix can hop into the empty d state. This effect gives a finite lifetime to a d electron localized on the transition atom, and, according to quantum mechanics, results in a broadening of the d level. Rather than forming a bound state, a so-called "virtual" bound state is formed.

7.1.1. Local density of states

The idea of density of states can then be generalized to that on a given site, defined by:

$$n_i(\varepsilon) = \sum_k |\psi_k(i)|^2 \delta(\varepsilon - \varepsilon_k) \tag{8.42}$$

The density of states in the case of a virtual bound state is represented in figure 8.9. If Δ is the width in energy of the virtual bound state, the density of states on a transition atom will be $1/\Delta$. Δ is given by Fermi's golden rule; it is proportional to the product of the density of states at the Fermi level of the matrix by $|V_{kd}|^2$, where V_{kd} is the *hybridization* between a d state and a state k:

$$\Delta = \pi\, n(\varepsilon_F) |V_{kd}|^2 \tag{8.43}$$

In the limit where Δ is negligible, we recover a state which can be considered as bound. This is true for the f electrons of rare earths, where Δ is of the order of 0.01 eV. For a transition atom dissolved in copper, Δ is of the order of 1 eV.

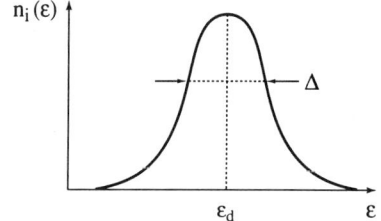

Figure 8.9 - Density of states for a virtual bound state

7.1.2. Magnetism of an impurity

We now introduce the Coulomb repulsion between two electrons on the site of the impurity. As two electrons of opposite spins repel each other more than two electrons with the same spin, we can define a quantity U analogous to that defined in section 5.2. This model is called Anderson's hamiltonian.

Following J. Friedel, we can make the same derivation as that made by Stoner, taking into account the density of states on the site of the impurity $n_i(\varepsilon)$. This then leads to a criterion for the formation of a magnetic moment on the site of the impurity.

$$U\, n_i(\varepsilon_F) > 1 \tag{8.44}$$

The impurity will or will not be magnetic depending on whether or not this criterion is satisfied. $n_i(\varepsilon_F)$ is inversely proportional to Δ. One sees immediately that the criterion will be the more difficult to satisfy the larger the density of states at the Fermi level of the metallic matrix. Aluminium has more electrons than copper, and correspondingly its density of states will be larger. It is therefore more difficult to have a magnetic moment on a transition atom dissolved in aluminium, which agrees well with experiment, as no transition atom is magnetic when dissolved in aluminium. In contrast, the metals in the middle of the first series are magnetic in matrices of copper, silver or gold.

7.2. THE KONDO EFFECT

If the (4f or 5f) impurity has a magnetic moment, or a spin, it will polarize the conduction electrons. We can define an exchange integral J_K which couples antiferromagnetically the impurity spin and that of a conduction electron. In the limit where U is far greater than Δ, it is given by:

$$J_K = 2|V_{kd}|^2/(\varepsilon_d - \varepsilon_F) \qquad (8.45)$$

This interaction is of the order of 1 eV, and can be larger than the direct exchange interaction, which is ferromagnetic. What is the effect of this antiferromagnetic interaction? It can reverse the magnetic moment of the impurity when coupling with a conduction electron occurs. The magnetic moment will therefore fluctuate with a characteristic time that we call τ. At T = 0 K, this will create a kind of singlet state between the spin of the impurity and those of the conduction electrons, and cause the impurity to appear nonmagnetic in measurements such as the susceptibility. This effect is termed the *Kondo effect*.

At finite temperature, thermal fluctuations normally lead to changes in the orientation of the magnetic moment with a characteristic time $\hbar/k_BT$. If this time is larger than τ, the impurity appears nonmagnetic, and if it is smaller, magnetic. There exists therefore a temperature, called the Kondo temperature, at which this inversion takes place. Above the Kondo temperature, the impurity is magnetic while below, it is nonmagnetic. The Kondo temperature is given by:

$$T_K = \frac{1}{n(\varepsilon_F)} \exp\left[\frac{1}{J_K n(\varepsilon_F)}\right] \qquad (8.46)$$

7.3. SPIN GLASSES, AND FRUSTRATION

If, rather than considering a single impurity, we consider a collection of impurities, they will interact with one another in a similar fashion to that observed in rare earths (which we describe in chap. 9), and the sign of the interaction will oscillate with the distance between the impurities.

Antiferromagnetic and ferromagnetic interactions will thus coexist. This situation leads to an effect called *frustration*: this behavior is schematically demonstrated in figure 8.10. In a triangle made up of three impurity atoms, if the three interactions are antiferromagnetic, they cannot all be satisfied at the same time. With a collection of impurities, a similar situation occurs many times. How will the system order at low temperature? Long-range order cannot occur. Rather, experiments show that, below a characteristic temperature, the magnetic moments will freeze with random orientations. The low temperature state therefore does not possess a global moment, and is said to behave like *a spin glass*.

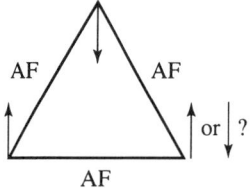

Figure 8.10 - Frustration of exchange interactions in a triangular lattice

There exist crystalline compounds with long range atomic order that contain triangles in their structure: frustration can also exist in these cases. In itinerant magnetism, there exists another possibility, beyond the one we have considered for disordered systems: the magnetic moment can simply disappear on a site which should be frustrated. The same atom, manganese for example, can have a magnetic moment on a favorable site, and be nonmagnetic on another site where it would be frustrated.

7.4. MAGNETIC MOMENTS AND THE TRANSITION METALS

We have seen that some transition metals, above the Curie temperature, display a behavior which strongly resembles that of a material with localized magnetism. We can now give an explanation for this. As we have seen when dealing with the problem of an impurity, the presence of a localized moment on an atom is not contradictory to the presence of itinerant electrons. What is true for an atom dissolved in a metallic matrix is also true for a collection of atoms. Transition metals can be considered in this way also.

In order to study the possibility of defining magnetic moments above T_C we will write the variation of free energy of the system in the presence of small magnetic fluctuations that act to create a distribution of small magnetic moments on each of the sites. We consider one Fourier component of this magnetization.

If the system has a small magnetization m(q) with wave vector q, the variation in the free energy density is given by:

$$\Delta F(q) = \frac{1}{2} \mu_0 \frac{[m(q)]^2}{\chi(q)} \tag{8.47}$$

For any fluctuations, we need to sum over all the values of q. Taking the inverse Fourier transform, the variation in the free energy can then be written:

$$\Delta F = \frac{\mu_0}{2N} \Phi(0) \sum_i m_i^2 + \frac{\mu_0}{2N} \sum_{i \neq j} \Phi(R_{ij}) m_i m_j \tag{8.48}$$

where:

$$\Phi(R_{ij}) = \frac{1}{N} \sum_k \frac{\exp(ikR_{ij})}{\chi_k} \tag{8.49}$$

This contains two terms: the first depends only on one site, and can be considered as the variation in energy corresponding to the creation of the magnetic moment, of value m_i, on site i; the second term depends on two sites, and can be considered as the interaction energy between the magnetic moments on the two sites. Several situations are therefore possible:

- This variation in energy is always negative, the system is unstable. In order to minimize its energy, it will create magnetic moments even in its disordered phase.
- This variation is negative only if the magnetic moments are ordered, i.e. the first term, which corresponds to the creation of a moment, is positive. The expression becomes negative only because of the magnetic order of these moments. Then the system does not have a magnetic moment in its disordered phase but it does have one in its ordered phase.
- This variation is always positive. Then the system is not magnetic, but the paramagnetic phase can be metastable.

This shows that the stability of the magnetic phases comes from both the stability of the magnetic moments m_i, and their magnetic coupling.

These three possibilities are observed in the transition metals. The size of the magnetic moment above the transition is not related to that observed at low temperature, and can even be zero. It can also vary with temperature. The function $\Phi(R_{ij})$ oscillates with distance. Making these considerations quantitative is very difficult, and the theory of these effects is complex. There does not actually exist a consensus of the type that prevails in localized magnetism.

8. MAGNETISM AND LOCAL ENVIRONMENT

We have noted that the magnetism of an impurity depends on the local density of states of the site it occupies. This density of states depends strongly on the environment of the atom: depending on its nature, and on the distance to the first neighbors, an atom may or may not be magnetic. We will give several examples of these situations.

While iron is magnetic in its normal body-centered-cubic structure, it is no longer magnetic, in the absence of stress, when it is prepared in the form of very thin films, in the unusual face-centred cubic structure.

If we consider a thin film of magnetic iron, the environment of a surface atom is not the same as that of an atom inside the film. The magnetic moment carried by an iron atom will therefore depend on its position in the film. Figure 8.11 shows, for a (110) film of Fe deposited on silver, the evolution of the hyperfine field B_{hf} (see chap. 23) as a function of the position of the individual layer in the film.

Because B_{hf} is proportional to the magnetic moment, it serves as a local probe which reveals the evolution of the moment at the position of the layer enriched in ^{57}Fe. With respect to an atom of iron in the bulk, Gradmann estimates that the moment of an iron atom increases by 32-38% at the surface of a (100) film, and by 19-26% at the surface of a (110) film [9].

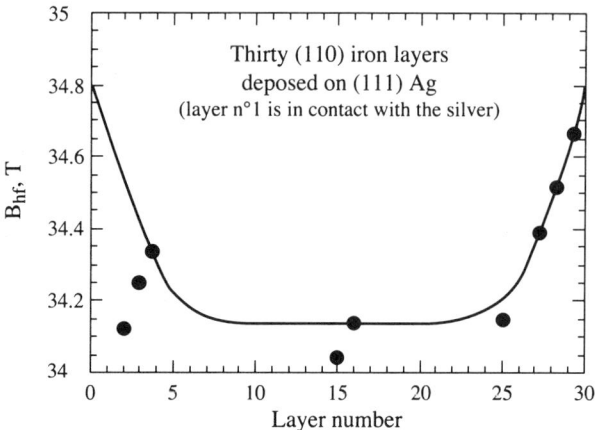

Figure 8.11 - Variation of the hyperfine field as a function of the position of the ^{57}Fe probe in the first 30 layers of Fe (110) deposited on Ag (111) at 4.2 K (layer N°1 is in contact with the silver) [8]

A nonmagnetic metal such as vanadium can become magnetic at the surface, because there the density of states is larger. Qualitatively, we have seen that the width of the density of states is proportional to the number of neighbors. Correspondingly, at a surface, this number is smaller, and therefore the density of states is higher. Stoner's criterion may thus be fulfilled by atoms at the surface only.

9. MAGNETISM OF TRANSITION METAL ALLOYS

The problem of alloys is far more difficult than that of an impurity. In fact, it exceeds the scope of this work. It is no more possible to calculate the electronic states, which are no more characterized by a wavevector **k** as in the case of free electrons or perfectly periodic solids (Bloch waves). Approximate methods have to be used to calculate the densities of states, either total, or local on a given atom. Theoretical predictions are therefore very difficult, but the evolution of the magnetic moment as a function of the number of s and d valence electrons per atom can be described by a universal curve called the *Slater-Pauling* curve (fig. 8.12). J. Friedel proposed a very simple interpretation for this curve.

In nickel and cobalt, the d band with the spin state ↑ is full. When an alloy is formed, all that changes is the number of electrons in the partially filled band with spin state ↓. To a first approximation, the change in the number of electrons in the sp band can be ignored because its density of states is far smaller at the Fermi level than that of the d band. Then the increase in the average magnetic moment (expressed in Bohr magnetons) will be equal to the average number of electrons removed from an atom. This explains the straight line, with a negative slope of 45°, observed for the alloys of nickel and cobalt with their neighboring elements.

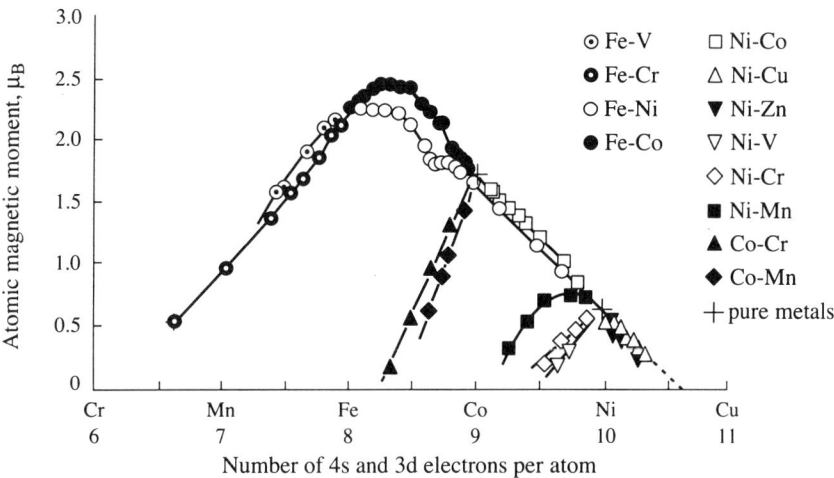

Figure 8.12 - Average atomic moment of binary alloys of elements of the iron group (Slater-Pauling curve)

For iron, where the two bands with different spin states are partially filled, the formation of the alloy should change the population of both bands. However, since the band with spin ↓ is assumed to have a small density of states, most of the electrons will be added or removed from the band with spin ↑. This explains the positive slope of approximately 45° observed for the alloys with elements neighboring iron.

For alloys between a metal at the beginning of the series with either nickel or cobalt, Friedel assumed that the repulsive potential due to the atoms at the start of a series pushes five d states from below the Fermi level to above. The five, initially full, states with spin ↑ empty themselves into states with spin ↓. This produces a sudden reduction in the magnetic moment, and so the different behavior shown by the Slater-Pauling curve.

Some authors explain the behavior in terms of magnetic valence. The chemical valence, Z, can be defined as $Z = N_\uparrow + N_\downarrow$, where $N_\uparrow$, and $N_\downarrow$ are the number of electrons of each spin state occupying the sp and d bands in a transition metal (Fe, Co, Ni). The magnetic moment in units of Bohr magnetons is then: $m = N_\uparrow - N_\downarrow = 2N_\uparrow - Z$.

In cobalt and nickel, the number $N_d^\uparrow$ of states per atom in the band with spin ↑ is five, and the band is completely full. The magnetic valence can then be defined by: $Z_m = 2N_d^\uparrow - Z$. The magnetic moment is therefore: $m = Z_m + 2N_{sp}^\uparrow$, where $N_{sp}^\uparrow = N^\uparrow - N_d^\uparrow$ is the number of electrons of spin ↑ belonging to the sp bands. This is of the order of 0.3 in the transition metals, and can be considered as being independent of the alloy in question.

With an alloy composed of a strongly ferromagnetic metal, each atom from the beginning of the series induces the suppression of five d states which are pushed above the Fermi level. If x is the concentration of transition atoms from the start of the series, the average magnetic valence becomes: $\overline{Z_m} = 2N_d^\uparrow(1-x) - Z_M(1-x) - xZ_A$, where Z_M (Z_A) is the magnetic valence of the magnetic (nonmagnetic) transition atom. The calculated variation is represented in figure 8.13.

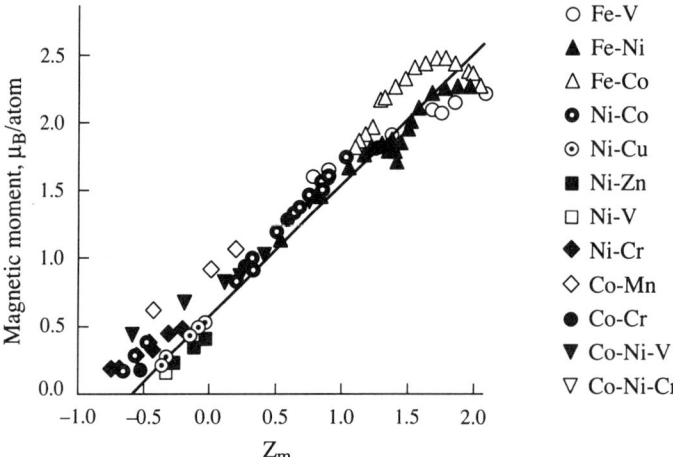

Figure 8.13 - Variation of the magnetic moment per atom as a function of the average magnetic valence in some series of 3d transition metal alloys [10]

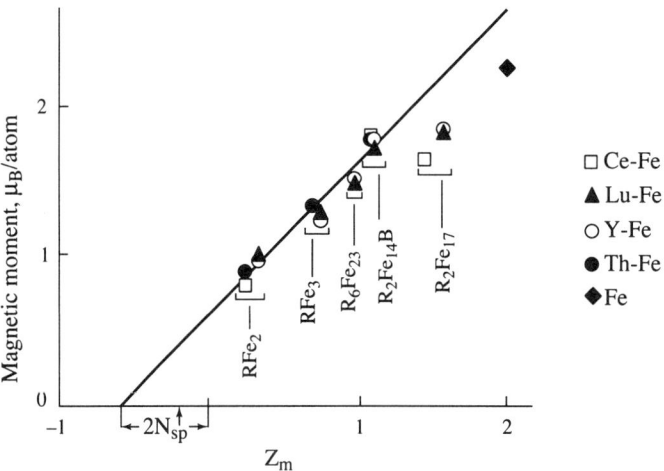

Figure 8.14 - Variation of magnetic moment per atom as a function of the average magnetic valence in the compounds R-Fe

Comment - *These considerations can be extended to analyze the magnetic moments with d character in rare earth-transition metal alloys. The rare earth ions (R), which have a configuration $5d^2 6s^1$, can be considered as transition metal elements (of the 4d series for Y, and 5d series for the other rare earths) at the start of their series. A d moment then results from the above arguments, considering a valence of $Z_m = 3$*

for the rare earth. In the compounds R-M, the 4f localized moment of the rare earth adds to this d moment. However, when the 4f shell is full (Lu) or empty (La), only d magnetism is present. The experimental moments can be directly compared to those deduced from the model of magnetic valence, as shown in figure 8.14. It is evident that these considerations are at best semi-quantitative as they do not take into account the effects of the local environment (see § 8).

10. CONCLUSION

The magnetism of metals arises from the Coulomb repulsion between two electrons present on the same site. This repulsion acts to push away electrons that have antiparallel spins. An atom carries a magnetic moment if Stoner's criterion is locally satisfied, i.e. if the local density of states is larger than a certain critical value.

The local density of states, and therefore the magnetism, is very strongly dependent on the environment of the atom. The reader will be able to learn more by consulting some of the authoritative works on this subject [11, 12, 13, 14].

REFERENCES

[1] D.P. YOUNG, D. HALL, M.E. TORELLI, Z. FISK, J.L. SARRAO, J.D. THOMPSON, H.R. OTT, S.B. OSEROFF, R.G. GOODRICH, R. ZYSLER, *Nature* **397** (1999) 412.

[2] N.F. MOTT, H. JONES, The theory of the properties of metals and alloys (1936) Dover Publications, New York.

[3] M. CYROT Ed., Magnetism of metals and alloys (1982) North Holland, Amsterdam.

[4] T. MORIYA, Spin fluctuations in itinerant electron magnetism (1985) Springer Verlag, Berlin.

[5] A. MONTORSI Ed., The Hubbard model (1992) World Scientific, Singapour.

[6] N.F. MOTT, Metal-Insulator Transition (1974) Taylor & Francis, G.B.

[7] R. GERSDORF, *Phys. Rev. Lett.* **40** (1978) 344.

[8] J. TYSON, A.H. OWENS, J.C. WALKER, G. BAYREUTHER, *J. Appl. Phys.* **52** (1981) 2487.

[9] U. GRADMANN, Magnetism in ultrathin transition metal films, in Handbook of Magnetic Materials, Vol. 7 (1993) K.H.J. Buschow Ed., North Holland, Amsterdam.

[10] A.R. WILLIAMS, V.L. MORUZZI, A.P. MALOZEMOFF, K. TERAKURA, *IEEE Trans. Magn.* **MAG 19** (1983) 1983.

[11] B. BARBARA, D. GIGNOUX, C. VETTIER, Lectures on modern magnetism (1988) Science Press, Springer Verlag, Berlin.

[12] B. COQBLIN, The electronic structure of rare earth metals and alloys (1977) Acad. Press, New York.

[13] J. FRIEDEL, The Physics of Metals (1969) Cambridge University Press, G.B.

[14] E.P. WOHLFARTH, K.H.J. BUSCHOW, Ferromagnetic Materials Vol. 1 to 6 (since 1980), North Holland, Amsterdam.

CHAPTER 9

EXCHANGE INTERACTIONS

(this chapter deals with fundamentals, and may be skimmed over during a first reading)

In the preceding chapters, we have shown that it is the exchange interactions between the spins of two atomic moments that are responsible for magnetic order. These interactions play a fundamental role, as the type of magnetic order that occurs depends on the sign of these exchange interactions, and the range over which they operate. The ordering temperature itself is generally of the same order of magnitude as the energy of the interactions. The aim of this chapter is to demonstrate the different microscopic mechanisms that are the origins of these exchange interactions.

1. MANY-ELECTRON WAVE FUNCTIONS

The existence of ordered states at high temperatures cannot be explained by classical physics using the dipolar magnetic interaction, as this interaction is at least 100 times weaker than required. Quantum mechanics has to be invoked to understand the origin of the exchange interactions that give rise to them.

The exchange interaction in a solid has the same quantum origin as the interaction between electrons within one atom (chap. 7): the correlations between two electrons lead to a difference in energy between configurations in which the spins are parallel and antiparallel. This originates in the Pauli principle, which stipulates that two electrons cannot occupy the same quantum state. One of the consequences of this principle is that many-electron wavefunctions must be antisymmetric with respect to the exchange of two electrons. We will treat here the case of a two-electron wave function $\Psi(1,2)$, where 1 and 2 describe both the spatial coordinates and the spins of the electrons. Application of the Pauli principle leads to the relation:

$$\Psi(1,2) = -\Psi(2,1) \tag{9.1}$$

The wavefunction $\Psi(1,2)$ can be written as a product of spatial $\varphi(1,2)$, and spin components $\chi(1,2)$. This leads to two types of wavefunctions depending on the symmetry of the functions $\varphi(1,2)$ and $\chi(1,2)$:

$$\begin{aligned}\Psi_I(1,2) &= \varphi_A(1,2)\chi_S(1,2) \\ \Psi_{II}(1,2) &= \varphi_S(1,2)\chi_A(1,2)\end{aligned} \tag{9.2}$$

Here S and A designate respectively symmetric and antisymmetric functions. In the case of two electrons, there is one antisymmetric wavefunction $\chi_A(1,2)$ that describes a singlet spin state: $\chi_A(1,2) = (1/\sqrt{2})(|1\uparrow, 2\downarrow> - |1\downarrow, 2\uparrow>)$, and three symmetric functions $\chi_S^m(1,2)$ (m = 0, ±1) that correspond to a state with total spin S = 1 (the triplet state): $\chi_S^1(1,2) = |1\uparrow, 2\uparrow>$, $\chi_S^{-1}(1,2) = |1\downarrow, 2\downarrow>$, and $\chi_S^0(1,2) = (1/\sqrt{2})(|1\uparrow, 2\downarrow> - |1\downarrow, 2\uparrow>)$. The energies of the two states can be calculated from the two-electron hamiltonian $\mathcal{H}$:

$$E_{I(II)} = \iint \varphi_{A(S)}(r_1, r_2) \mathcal{H}(r_1, r_2) \varphi^*_{A(S)}(r_1, r_2) dr_1 dr_2 \tag{9.3}$$

The energies of the spatial wavefunctions $\varphi_A(1,2)$ and $\varphi_S(1,2)$, E_I and E_{II} respectively, are different if the hamiltonian $\mathcal{H}$ contains interactions between electrons (such as Coulomb repulsion). It is this energy difference that is the origin of the exchange interaction between the spins of electrons 1 and 2. The exchange energy between the spins of two electrons can be defined by:

$$J_{12} = (E_{II} - E_I)/2 \tag{9.4}$$

The energy of any spin state can therefore be written as:

$$E(\mathbf{S_1}, \mathbf{S_2}) = E_0 - 2J_{12} \mathbf{S_1} \mathbf{S_2} \tag{9.5}$$

This is because, for 2 electrons, with spins $S_1 = S_2 = 1/2$, the scalar product $\mathbf{S_1}\mathbf{S_2} = \frac{1}{2}\left[(\mathbf{S_1}+\mathbf{S_2})^2 - \mathbf{S_1}^2 - \mathbf{S_2}^2\right]$ can take the values 1/4 (as $(\mathbf{S_1}+\mathbf{S_2})^2 = 2$ for a triplet state of total spin S = 1) or – 3/4 (as $(\mathbf{S_1}+\mathbf{S_2})^2 = 0$ for a singlet state with a total of spin S = 0).

The method we present has been successfully used to describe diatomic molecules such as molecular hydrogen H_2 [1]. In this example, the exchange interaction J_{12} can be written as a function of the Coulomb integral U, of the exchange integral V, and of the overlap between the two orbitals L. Calling $\phi_1(r)$ and $\phi_2(r)$ the respective one-electron wave functions centered on both hydrogen atoms:

$$J_{12} = \frac{V - UL^2}{1 - L^4}$$

with:
$$U = \int d^3r\, d^3r'\, H(r,r') |\phi_1(r)\phi_2^*(r')|^2$$
$$V = \int d^3r\, d^3r'\, H(r,r') \phi_1(r)\phi_1^*(r')\phi_2(r')\phi_2^*(r) \tag{9.6}$$
$$L = \int d^3r\, \phi_1(r)\phi_2^*(r)$$

H(r, r') being the Coulomb repulsion between electrons located at r and r'.

It can be seen that the exchange interaction arises as a direct consequence of the antisymmetry of the wave functions, and corresponds to the difference in energy between the symmetric and antisymmetric spatial wave functions; in the simple case presented here, its value depends on the Coulomb interactions (U and V) between

9 - EXCHANGE INTERACTIONS

electrons situated on neighboring atoms. However, applying this method to solids leads to a value for the exchange interaction due to the interatomic Coulomb repulsion (termed direct exchange) that is extremely weak, and often negative. Actually the exchange interaction in solids is more commonly related to the existence of indirect interactions between electrons situated on neighboring atoms. In the rest of this chapter, we describe the mechanisms of the most frequently observed interactions.

In chapter 4, the exchange interaction was introduced in a phenomenological fashion in order to describe the interactions between magnetic moments m_i (eq. 4.34). In the case of a magnetic moment arising from a spin of the type described above, the phenomenological interaction n_{ij} is related to the microscopic parameter J_{ij} by the relation $J_{ij} = 8\ \mu_0\mu_B^2 n_{ij}$. The exchange interaction J_{ij} therefore represents an interaction between spins, and from it the interaction n_{ij} between magnetic moments, in the form used in chapter 4, can always be deduced.

2. EXCHANGE INTERACTIONS IN INSULATORS

2.1. SUPEREXCHANGE [2]

In most magnetic insulators, the magnetic ions are separated by non-magnetic ions. This is the case in oxides, sulphides and halides of the transition metals. The magnetic ions are therefore situated at distances too great for their 3d wave functions to overlap. Instead, the exchange interaction is mediated by the non-magnetic ion (O^{2-}, S^{2-}, Br^-, Cl^-, F^-...), and it is the overlap between the 3d and p (2p, 3p or 4p) wave functions that is of importance.

The simplest example involves the hybridization of a single d electron of the transition metal (M) with the p orbitals of the intermediary non-magnetic atom, as shown in figure 9.1.

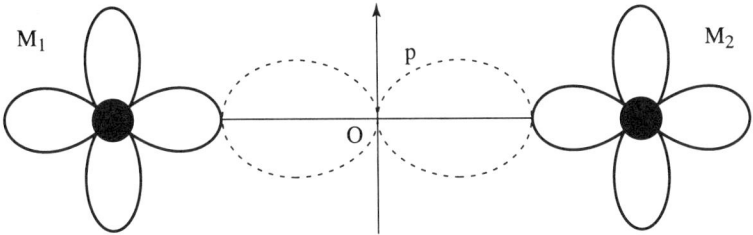

Figure 9.1 - M_1 and M_2: transition metal ions, O: O^{2-}
The p orbital oriented along the axis M_1-M_2 hybridises with the d orbitals of the M_1 and M_2 ions.

The description of the d band (chap. 8) needs to be modified, as the direct overlap integrals, β_{ij}, between the d orbitals are very weak. It therefore necessitates the inclusion of the overlap integrals between the p and d wavefunctions: $V_{pd}^{ij} = <\varphi_d^j | V | \varphi_p^i>$.

When this hybridization V_{pd}^{ij} is weak, the situation is the same as if the band was arising out of hybridised wave functions corresponding to a combination of d and p wave functions:

$$\psi_\sigma(r) = \sqrt{1-\varepsilon^2}\,\varphi_{d,\sigma}(r - R_1) + \varepsilon\varphi_{p,\sigma}(r - R_0)$$

instead of wavefunctions centred on each atom.

The mixed character of these wave functions indicates that the d or p electrons are no longer rigorously localized on a single ion, even though the system is an insulator. The energy associated with the delocalization of the p electrons on the M_1 and M_2 ions depends strongly on the relative orientation of the spins of the two d electrons. Figure 9.2 shows the two possible configurations for the case where only a single d orbital is available. Configuration (b) is the minimum energy configuration, as the reduction in kinetic energy, associated with the delocalization of p electrons, is the greater. In this situation, the sign of the exchange interaction is therefore negative.

Figure 9.2 - In configuration (a), only one of the p electrons of the O^{2-} ion (with ↓ spin) can delocalize, while in configuration (b) both electrons of the O^{2-} ion can delocalize, one on M_1, the other on M_2

The value and sign of this superexchange interaction depend strongly on the types of d orbitals (e_g or t_{2g}) involved, the number of electrons, and also the angle M_1-O-M_2. Superexchange interactions can lead to elevated ordering temperatures, up to nearly 900 K in the ferrites (T_C = 863 K for $NiFe_2O_4$). Semi-empirical rules allow us to estimate the sign of the superexchange interaction [3, 4]. Whilst it is commonly negative, ferromagnetic insulators such as EuO (T_c = 69 K) or $CrBr_3$ (T_c = 37 K) are known. Finally, this interaction is effective only over short distances, because it necessarily involves overlap between orbitals centered on neighboring ions.

2.2. ANTISYMMETRIC EXCHANGE

In some of the compounds which order antiferromagnetically, there exists a weak spontaneous moment at low temperature: the most well-known example is that of hematite, α-Fe_2O_3. This weak moment is attributed to the existence of antisymmetric exchange interactions of the form described by Dzyaloshinsky and Moriya [5]:

$$E_{12} = \mathbf{D}_{12}(\mathbf{S}_1 \times \mathbf{S}_2) \qquad (9.7)$$

It is clear from equation 9.7 that this antisymmetric exchange energy is only non zero if sites 1 and 2 are inequivalent, as may be the case when the environments of the two sites are different. For two equivalent sites, one should have $E_{12} = E_{21}$, and this gives $\mathbf{D}_{12} = 0$. The mechanism which gives rise to these antisymmetric interactions is analogous to superexchange, but requires the inclusion of spin-orbit coupling on the d electrons [5].

3. EXCHANGE INTERACTIONS IN METALS

This interaction favors non-colinear arrangements of moments. If one considers two sublattices of spins interacting with each other through both a negative exchange interaction J_{12} and an antisymmetric interaction $\mathbf{D}_{12}$, the two sublattices will no longer be exactly antiparallel, but instead they form an angle on the order of $\pi - |D/J|$ (fig. 9.3). The effect is commonly referred to as "canted antiferromagnetism" or "weak ferromagnetism", as the resulting moment is therefore very weak ($\sim 10^{-2} \mu_B$ per Fe^{3+} ion in $\alpha\text{-}Fe_2O_3$). It may be noted that the term "weak ferromagnetism" is also used in another context, that of itinerant magnetism, and refers e.g. to iron (see chap. 8).

Figure 9.3 - Configuration of the moments of two sublattices in the presence of antisymmetric interactions

3. EXCHANGE INTERACTIONS IN METALS

In a metal, it is the conduction electrons that are responsible for the exchange interactions. However, the cases of the 3d, and 4f metals are quite different: in the transition metals, the 3d electrons are the origin of both the magnetic moment and the exchange interactions (chap. 8, § 5). In the rare earth metals, the magnetic moment is that of the 4f electrons, and the exchange interactions are mediated via the conduction electrons (s, p, and d): (chap. 8, § 6).

3.1. INDIRECT INTERACTIONS BETWEEN 4f MOMENTS (THE RKKY INTERACTION) [6]

Traversing the fourteen rare earth elements corresponds to a filling of the 4f band. In most compounds, their s and d electrons are delocalized, and the ions of the rare earths are trivalent. The 4f electrons are very strongly bound, and the 4f orbitals are highly compact. Typically, their spatial extension is far less than the interatomic distances, and correspondingly there can be no direct interaction between the 4f electrons of different atoms. Rather, it is the conduction electrons which couple the magnetic moments: when a conduction electron is in the vicinity of a rare earth ion, it interacts locally with the 4f electrons of this ion. As before (§ 1), this interaction can be written in terms of the 4f spins localized on the ions, $\mathbf{S}_i$, and of the spins of the conduction electrons, $\boldsymbol{\sigma}(\mathbf{r})$:

$$E = -\sum_{\mathbf{R}_i, \mathbf{r}} J(\mathbf{R}_i - \mathbf{r}) \mathbf{S}_i \boldsymbol{\sigma}(\mathbf{r}) \quad (9.8)$$

This interaction is positive and highly localized, and correspondingly we can write it in the form $J(\mathbf{R}_i - \mathbf{r}) = J \delta(\mathbf{R}_i - \mathbf{r})$ (intra-atomic exchange). The magnitude of J is between 10^{-1} and 10^{-2} eV. Equation 9.8 contains the notion that a conduction electron sees a rare earth site i through a field created by the spin of the rare earth:

$\mathbf{h_i} = J\mathbf{S_i}/g\mu_B\mu_0$. This field polarises the conduction electrons and this polarization propagates through the lattice, creating at any point j a magnetization of the conduction electrons. This magnetization of the conduction electrons, associated with the local field $\mathbf{h_i}$ at another site j, is described by the generalised susceptibility χ_{ij}, introduced in section 7 of chapter 8, and can be written: $\mathbf{m_i} = \chi_{ij}\mathbf{h_j} = J\chi_{ij}\mathbf{S_j}/g\mu_B\mu_0$. This amounts to an indirect interaction between the spins (or the magnetic moments) on the two sites i and j. Using equation 9.8 again, this interaction energy can be written:

$$E_{ij} = -J\mathbf{m_i}\mathbf{S_i}/g\mu_B = -J^2\chi_{ij}\mathbf{S_i}\mathbf{S_j}/\mu_0(g\mu_B)^2 \qquad (9.9)$$

Equation 9.9 shows that the exchange interaction between spins is due to the polarization of the conduction electrons by these localized spins. The sign of the magnetic interaction depends only on the structure of the conduction band via the generalized susceptibility χ_{ij}. For free electrons, one finds that it acts at long distance, and for $\mathbf{k_F R_{ij}} \gg 1$, χ_{ij} varies as:

$$-[2\mathbf{k_F R_{ij}} \cos(2\mathbf{k_F R_{ij}}) - \sin(2\mathbf{k_F R_{ij}})]/(2\mathbf{k_F R_{ij}})^4.$$

The exchange therefore oscillates between positive and negative values as a function of the distance between spins, R_{ij} (fig. 9.4).

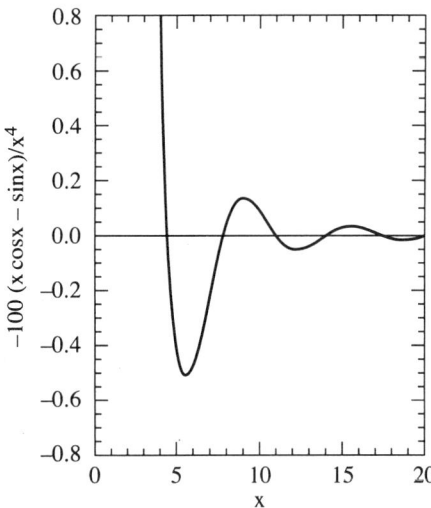

Figure 9.4 - Oscillations in the interaction between two magnetic moments

The wavevector at the Fermi level, $\mathbf{k_F}$, determines the wavelength of this oscillation. In general the RKKY interaction gives rise to ferromagnetism if k_F is small (this occurs when the band is nearly empty), and to antiferromagnetism when $k_F \sim \pi/a$ (half-filled band). The interaction can also lead to long period magnetic structures (helimagnets, etc.), that can be incommensurate with the lattice spacing.

These interactions are of the same order of magnitude in nearly all of the rare earths, but the ordering temperatures that they lead to can vary considerably (19 K for Nd,

289 K for Gd). This disparity is due to the variation in the magnitude of the magnetic moment with the different rare earths.

In chapter 20, section 4 we will show that the same long-range interactions are generally the origin of coupling between magnetic layers in multilayer systems. In these systems, it is possible to observe the oscillations in the RKKY exchange by varying the thickness of the non-magnetic layer which separates the two magnetic layers.

3.2. EXCHANGE INTERACTION IN THE 3d METALS

In chapter 8, we saw that, in the transition metals, the magnetic moment arises from a difference in the occupation of the bands with spin ↑ and ↓, and can therefore have non-integer values in terms of μ_B. This band-type magnetic moment appears when the intra-site Coulomb interaction, which tends to localize the electrons, is sufficiently large with respect to the kinetic energy of the d electrons. The interactions $U_{mm'}$ and $J_{mm'}$, previously defined (chap. 8), then lead to the stabilization of a localised magnetic moment on each site. This interaction between these moments takes place via overlap of the 3d wave functions of neighbouring sites (the transfer integrals β_{ij} defined in chap. 8). Notably, the s electrons, also present at the Fermi level, play a negligible role both in the stabilization of magnetic moments, and in the interaction between moments, as the polarization of the s band is negligible.

The overlap between wave functions has two contradictory effects: if $\beta \gg U$ and J, the band will not be magnetic because Stoner's criterion will not be satisfied, and, if $\beta \ll U$ and J, the intersite exchange interactions will be weak. Figure 9.5 shows how the interactions vary as a function of U/β. The strongest interactions are obtained for values of U/β slightly greater than that corresponding to the onset of magnetism.

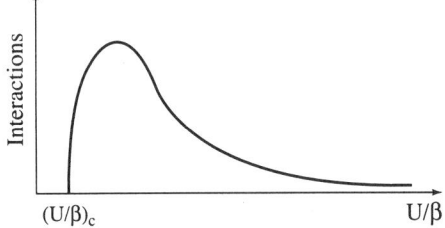

Figure 9.5 - Qualitative variation of the interactions between moments as a function of U/β

The critical value of $(U/\beta)_c$ corresponds to Stoner's criterion.

The sign of the interaction between 3d moments depends, as in the case of rare earths, on the filling of the band: it is ferromagnetic for a band that is nearly empty (or nearly full), and antiferromagnetic for a band that is half-filled.

It is possible to discuss very simply the sign of the exchange interaction for the transition metals. Consider two nearest-neighbor atoms with a small number of electrons (or holes). The ferromagnetic configuration will be preferred because it minimises the kinetic energy by permitting the electrons to delocalize from one atom to another (fig. 9.6), whilst still respecting Hund's rule. This is the situation in iron, cobalt, and nickel, where the 3d band is nearly full.

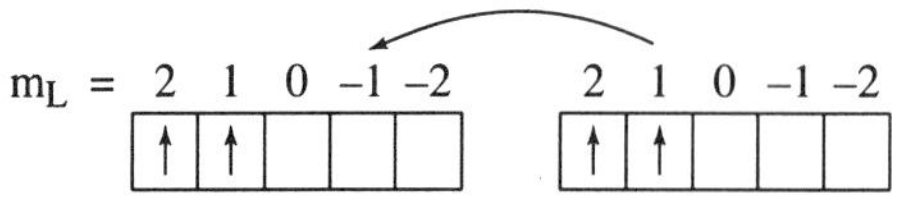

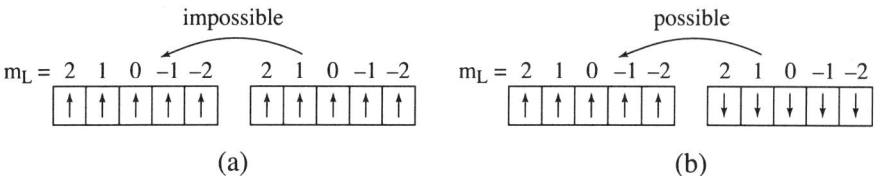

Figure 9.6 - *Ferromagnetism in a nearly empty band*

If the band is half-filled, the ferromagnetic configuration does not permit the electrons to be delocalized, as demonstrated in figure 9.7-a. However, the antiferromagnetic configuration does permit this, as shown in figure 9.7-b, and so it is the preferred configuration. This is the case for chromium and manganese. The antiferromagnetic structure in the first is sinusoidal, and in the second collinear.

Figure 9.7 - Antiferromagnetism in a half-filled band

The value of the exchange between 3d magnetic moments depends strongly on the electronic structure. The Curie temperatures of iron and of cobalt are very high (respectively 1,043 and 1,388 K) but, in the compounds close to Stoner's instability, they are far lower, e.g. 6 K for Sc_3In.

3.3. DOUBLE EXCHANGE [7]

In the transition metal oxides, we have seen that the exchange interaction is of the superexchange type. This is only true for the interactions between ions that are in a definite valence state.

In certain oxides, the transition ion can feature two valence states depending on the doping: for example this is the case in the substitutional series $La_{1-x}Ca_xMnO_3$ or $La_{1-x}Sr_xMnO_3$, in which the fraction $(1 - x)$ of Mn ions is in the configuration Mn^{3+} with a total spin $S = 2$, while the remainder of the Mn ions have the configuration Mn^{4+} ($S = 3/2$). Hybridization with the intermediate p orbitals of the O^{2-} ion in these manganites results in a ferromagnetic interaction between the Mn^{3+} and Mn^{4+} ions, called double exchange, as shown in figure 9.8. The ferromagnetic configuration (a) is stabilized by the reduction in the kinetic energy of the additional electron resulting from the delocalization associated with the hybridization.

Figure 9.8 - *Following Hund's rule, the extra electron of the Mn^{3+} ion can delocalize onto the site of a neighbouring Mn^{4+} only if the spins are parallel [configuration (a)]. In the case of configuration (b), the extra electron remains localized on the Mn^{3+}*

In fact, in the substitutional series $La_{1-x}Sr_xMnO_3$, both antiferromagnetic super-exchange interactions between ions of the same valence, and ferromagnetic double exchange between ions in different valence states are present. This leads to a complicated dependence of the type of magnetic order with variation in x. Depending on x, ferromagnetic, antiferromagnetic or "canted" states are stabilized. This last state results from a competition between ferromagnetic and antiferromagnetic exchange, and is characterized by two sublattices where the moments make an angle α between 0 and π (fig. 9.9).

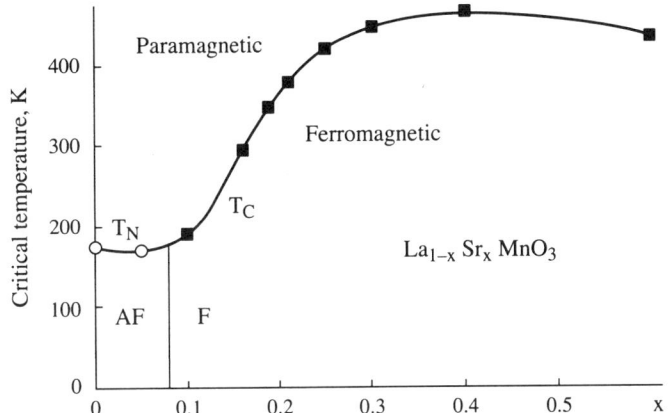

Figure 9.9 - Phase diagram of the magnetic alloys $La_{1-x}Sr_xMnO_3$
AF: antiferromagnetic phase, F: ferromagnetic or "canted" phase.

For more information on the manganites, the reader is advised to look up reference [8].

4. SUMMARY

The term "exchange interaction" covers many different mechanisms depending on whether the materials are insulators or metals, and on which magnetic ions (rare earths or transition metals) are involved. While the detailed nature of these microscopic mechanisms determines the sign, strength, and range of the interaction, the interaction energy can generally (except for the antisymmetric exchange § 2.2) be written as a phenomenological exchange energy between magnetic moments:

$$E = -\frac{1}{2}\mu_0 \sum_{ij} n_{ij} \mathbf{m}_i \mathbf{m}_j \quad (9.10)$$

REFERENCES

[1] L.P. LÉVY, Magnétisme et Supraconductivité (1997) CNRS Editions, Paris.
[2] P.W. ANDERSON, Magnetism, Vol. IIB (1962) G.T. Rado & H. Suhl Ed., Wiley, New York.

[3] J.B. GOODENOUGH, Magnetism and the chemical bond (1963) Wiley, New York.

[4] H.J. ZEIGER, G.W. PRATT, Magnetic Interactions in Solids (1973) Clarendon Press, Oxford, U.K.

[5] T. MORIYA, *Phys. Rev.* **120** (1960) 91.

[6] B. COQBLIN, The electronic structure of metals and alloys (1964) Acad. Press, New York.

[7] P.G. DE GENNES, *Phys. Rev.* **118** (1960) 141.

[8] A.P. RAMIREZ, *J. Phys. Condens. Mat.* **9** (1997) 8171; J.M.D. COEY, M. VIRET, S. VON MOLNAR, *Adv. Phys.* **48** (1999) 167.

CHAPTER 10

THERMODYNAMIC ASPECTS OF MAGNETISM

(this chapter deals with fundamentals, and may be skimmed over during a first reading)

This chapter applies thermodynamic methods to the analysis of magnetism in solids, and provides a brief introduction to critical phenomena.

1. BASIC THERMODYNAMICS: OUTSIDE MAGNETISM

We first recall some results on equilibrium thermodynamics in simple systems, in particular in the absence of an applied magnetic field. The presentation is based on the work by Callen [1], to which the reader is invited to refer both for a refresher and for derivations that are not presented here.

Classical thermodynamics deals with macroscopic systems *at equilibrium*, and describes the changes in characteristic quantities between equilibrium states. The simplest example is that of an isolated system made up of N particles. The internal energy U of this system, given by the sum of the kinetic and potential energies of the component particles, is then a constant, and its equilibrium state is described by the condition of maximum entropy S. A small change around this equilibrium state leads, for a single component fluid, to a change in the internal energy:

$$dU = T\,dS - p\,dV + \mu\,dN \tag{10.1}$$

For a mixture, it is necessary to introduce the sum $\mu_j\,dN_j$ over the various chemical species j. In a solid, the scalar quantities p and V have to be replaced by the stress and strain, integrated over the volume of the system.

The equilibrium state is quantitatively characterized by a small number of macroscopic variables, called the *state variables*. The pressure p, the temperature (absolute) T, and the chemical potential μ_j are *intensive* state variables. Their uniformity characterizes mechanical, thermal and diffusion equilibria, respectively. These variables appear, in nearly all thermodynamic equations, together with "conjugate" *extensive* variables: the volume V, the entropy S and the number of particles or moles N_j (depending on the the definition chosen for the chemical potential).

The fact that the differential of U involves the differentials of S, V and N_j implies that these are the *natural variables* for U. One can also formally consider the condition of equilibrium of an isolated system as a minimum of U, given the entropy, volume and number of particles. Thus U plays the role of a *thermodynamic potential*, just as the condition for equilibrium in a mechanical system is described by the minimum in its potential energy.

A system forced to remain at constant temperature T is no longer isolated. Its equilibrium state is then given by the minimum, consistent with the imposed constraints, of the *free energy* $F = U - TS$, the differential of which can be written:

$$dF = -SdT - pdV + \sum_j \mu_j dN_j \quad (10.2)$$

The natural variables for F are T, V and N_j. When temperature and pressure are the experimentally controlled variables, free enthalpy, or the Gibbs energy $G = U - TS + pV$, is involved. This has the differential:

$$dG = -SdT + Vdp + \sum_j \mu_j dN_j \quad (10.3)$$

The natural variables are then T, p and N_j.

Changing from one thermodynamic potential to another one is a simple procedure. The *Legendre transformations*, as from U to F or to G, correspond to the change in the physical variables that are experimentally controlled.

As $U(S, V, N_j)$, $F(T, V, N_j)$ and $G(T, p, N_j)$ are state functions, their mixed second derivatives are necessarily equal, i.e.

$$\left(\frac{\partial^2 F}{\partial V \partial T}\right)_{N_j} = \left(\frac{\partial^2 F}{\partial T \partial V}\right)_{N_j} \quad (10.4)$$

from which: $\left[\frac{\partial(\partial F/\partial T)}{\partial V}\right]_{T,N_j} = \left[\frac{\partial(\partial F/\partial V)}{\partial T}\right]_{V,N_j}$.

Or $\left(\frac{\partial F}{\partial T}\right)_{V,N_j} = -S,$ and $\left(\frac{\partial F}{\partial V}\right)_{T,N_j} = -p,$

which leads to: $$\left(\frac{\partial S}{\partial V}\right)_{T,N_j} = \left(\frac{\partial p}{\partial T}\right)_{V,N_j} \quad (10.5)$$

These thermodynamic identities are called *Maxwell relations*. We note that the variables involved (T, V, N_j in this example) appear explicitly at the bottom of these equations. They immediately indicate which thermodynamic function can lead to useful relations. The useful thermodynamic potential is that for which the natural variables are the relevant, i.e. the controlled variables, in the problem. In the above example, it is therefore $F(T, V, N_j)$.

Analysis of the stability conditions of a phase at equilibrium leads to the derivation of various inequalities between the response coefficients, such as:

$$(\partial S/\partial T)_{p,N_j} = \frac{C_p}{T} \quad \text{or} \quad (\partial S/\partial T)_{V,N_j} = \frac{C_V}{T}$$

C_p and C_V are the specific heats, or heat capacities, at constant volume and pressure, respectively. We note that these coefficients concern pairs of conjugate variables. It can be shown that C_p and C_V are always positive, and that $C_p > C_V$.

2. THERMODYNAMIC POTENTIALS FOR AN UNDEFORMABLE MAGNETIC SYSTEM [1, 2]

In the presence of magnetic fields $\mathbf{H}_0$, $\mathbf{H}$, $\mathbf{B}_0$, $\mathbf{B}$, and of magnetization $\mathbf{M}$, the situation becomes more complicated. Furthermore, we are usually interested in solids, hence p and V are no longer sufficient to describe the elastic effects. We therefore introduce, in section 4, the stress and strain tensors, each one involving up to 6 independent components.

For the moment, let us consider the case of an undeformable solid, i.e. the effects of the thermal variables (S, T) and of magnetic variables ($\mathbf{M}$, $\mathbf{H}_0$) only. We begin with the expression for the work associated with an elementary change. Legendre transformations will then provide the whole range of thermodynamic potentials. The equation for the magnetic work that we will use as a start is:

$$dW_{mag} = d\left(\frac{1}{2}\mu_0 \int \mathbf{H}_0^2 dV\right) + \mu_0 \int \mathbf{H}_0 d\mathbf{M} dV \qquad (10.6)$$

where μ_0 represents the magnetic permeability of vacuum, not a chemical potential.

The first term corresponds to the magnetostatic energy of the applied field stored over all space. It can often be omitted, which amounts to choosing a reference state where the applied field would exist, but where the sample would initially be in the state corresponding to zero field.

The second term describes the effect of the magnetic field on a sample with magnetization $\mathbf{M}$. The integration is automatically limited to the sample's volume because there is no magnetization outside the sample. This expression may be simplified if the applied field is uniform. The second term then becomes: $\mu_0 \mathbf{H}_0 d\mathbf{m}$, where $\mathbf{m}$ is the total magnetic moment of the sample: $\mathbf{m} = \int \mathbf{M} dV$, or again: $\delta W_{mag} = \mu_0 H_0 dm_{//}$. Here, $m_{//}$ depicts the component of the total magnetic moment parallel to the applied field.

Introducing this expression into the internal energy leads to a variety of thermodynamic functions, with various pairs of natural variables:

- with the natural variables S and $m_{//}$, a thermodynamic potential $U(S, m_{//})$ such that:

$$dU(S, m_{//}) = T\, dS + \mu_0 H_0\, dm_{//} \qquad (10.7)$$

- with T and $m_{//}$, a thermodynamic potential $F(T, m_{//})$ such that:

$$dF(T, m_{//}) = -S\, dT + \mu_0 H_0\, dm_{//} \qquad (10.8)$$

Use of a Legendre transformation then leads to functions:

- of S and H_0, thus a thermodynamic potential $U(S, H_0)$ such that:

$$dU(S, H_0) = T\, dS - \mu_0 m_{//}\, dH_0 \qquad (10.9)$$

- or of T and H_0, a thermodynamic potential $F(T, H_0)$ such that:

$$dF(T, H_0) = -S\, dT - \mu_0 m_{//}\, dH_0 \qquad (10.10)$$

The relation $F(T, \mathbf{H}_0) = F(T, \mathbf{M}) - \mu_0 \int \mathbf{H}_0 \mathbf{M} dV$ describes the conversion from $F(T, \mathbf{M})$ to $F(T, \mathbf{H}_0)$. This is in fact exactly the same procedure as that used in section 5.1.1 of chapter 5 when analysing the most favourable situation in terms of magnetic domains. The free energy $F(T, \mathbf{M})$ contains anisotropy, exchange and demagnetising field terms; the term $-\mu_0 \int \mathbf{H}_0 \mathbf{M} dV$ is the Zeeman term for the interaction with the applied field (symbolised by E_H in chap. 5, § 7.1.2).

Equilibrium corresponds to the minimum of $F(T, \mathbf{H}_0)$, with respect to the internal degrees of freedom. Here this corresponds to the distribution within the sample of $\mathbf{M}$, which in practice is its orientation.

We have chosen not to give a name to the many thermodynamic functions that can be devised. The important piece of information is the indication of the natural variables which follows the symbol U or F, i.e. their arguments.

While the above thermodynamic functions involving $\mathbf{H}_0$ are of particular interest because the applied field is frequently an experimentally controlled variable, they are not the only ones of interest.

The variable that is physically most meaningful is the macroscopic mean field, or induction, $\mathbf{B}$. It can be shown that the elementary work can also be expressed in terms of $\mathbf{B}$, with the total field: $\mathbf{H} = \mathbf{H}_0 + \mathbf{H}_{dem}$, as the conjugate variable. Here $\mathbf{H}_{dem}$ designates the field created, both inside the sample, and outside (we sometimes refer to the dispersion field, or stray field), by the distribution of magnetization in the specimen:

$$dW_{mag} = \int \mathbf{H} d\mathbf{B} dV = \int H dB_{//} dV) \qquad (10.11)$$

Neither $\mathbf{B}$ nor $\mathbf{H}$ are extensive variables, and an integration over all space is necessary as they are not limited to the interior of the sample.

These same procedures can lead to the four state functions: $U(S, B_{//})$, $F(T, B_{//})$, $U(S, H)$ and $F(T, H)$. Their respective differentials are written:

$$dU(S, B_{//}) = TdS + \int H dB_{//} dV \qquad (10.12)$$

$$dF(T, B_{//}) = -SdT + \int H dB_{//} dV \qquad (10.13)$$

$$dU(S, H) = TdS - \int B_{//} dH dV \qquad (10.14)$$

$$dF(T, H) = -SdT - \int B_{//} dH dV \qquad (10.15)$$

3. MAXWELL RELATIONS AND INEQUALITIES

The eight functions defined above provide, via the Maxwell relations which generalise equation (10.4), physically useful expressions for the heat involved, or the variation in temperature, associated with the application of a magnetic field. Thus, the iso-entropic change in temperature can be described in terms of a magnetothermal coefficient, which can be deduced from equation (10.9) associated with $U(S, H_0)$:

$$\left(\frac{\partial T}{\partial H_0}\right)_S = -\mu_0 \left(\frac{\partial m_{//}}{\partial S}\right)_{H_0} = -\mu_0 \left(\frac{\partial m_{//}}{\partial T}\right)_{H_0} \left(\frac{\partial T}{\partial S}\right)_{H_0} = -\frac{\mu_0 T}{C_{H_0}} \left(\frac{\partial m_{//}}{\partial T}\right)_{H_0} \qquad (10.16)$$

because $(\partial T / \partial S)_{H_0} = T / C_{H_0}$, where $C_{H_0} = T(\partial S/\partial T)_{H_0}$ is the heat capacity at constant applied field*. The determination of the finite change ΔT due to a finite change in applied magnetic field requires the use of an integration, and in the low-temperature regime it is definitely necessary to take into account the significant variation in C_{H_0} with magnetic field. This is in particular true for the most famous use of this technique: cooling by adiabatic suppression of the applied field, a technique (slightly improperly) called adiabatic demagnetization, which will be treated in an exercise at the end of chapter 11.

These relations lead to economy both in the concepts and in experiment. For example, relation (10.16) allows us to replace calorimetric measurements by isothermal magnetization measurements (see exercises).

In the same way that we have encountered, when dealing with the thermodynamics of simple systems, inequalities between the specific heats ($C_p > C_V$), we obtain inequalities between the coefficients that describe magnetic responses. In particular, $(\partial B / \partial H)_T$ and $(\partial B / \partial H)_S$, the isothermal and iso-entropic magnetic permeabilities $\mu_0 \mu_{r_T}$ and $\mu_0 \mu_{r_S}$ respectively, are always positive, with $\mu_{r_T} > \mu_{r_S}$.

If we also use this approach for the other pair of variables used, $\mathbf{H}_0$ and $\mathbf{M}$ or $\mathfrak{m}$, it appears to lead to the relation $(\partial M / \partial H_0)_{S \text{ or } T} > 0$. The fact that this relation is

---

* Not to be confused with the field coefficients of a coil (see Eq. 2.16).

obviously incorrect (as $\left(\frac{\partial M}{\partial H_0}\right)_{S \text{ or } T}$ is negative for diamagnetic materials, see chap. 3) is not a signal of the failure of thermodynamic arguments: it is simply associated with the fact that the expression for the work that led to this pair of variables was, as we have seen, artificially truncated.

Thermodynamics is a magnificent framework, but the physical meaning of any result obtained should never be forgotten. Let us return as an example to the procedure of cooling by adiabatic suppression of the applied field. During the suppression of the field, the loss of information about the orientations of the magnetic moments, or, in other words, the increase in their disorder, corresponds to an increase in their entropy. As, on the other hand, the entropy of the system consisting of the moments and of the vibrations of the atoms that carry them is held constant in an adiabatic quasi-static procedure (iso-entropic), the lattice has to lose some entropy, and therefore the sample cools down.

4. SITUATION OF A DEFORMABLE MAGNETIC SOLID

Introduction of the elastic variables, the symmetrical second rank stress σ_{ij} and strain η_{ij} tensors, does not simplify the calculations, as we must now take into account the elementary elastic work. The expression for this work is $\int \sigma_{ij} d\eta_{ij} dV$, where we adopt, as we will do from now on, *Einstein's convention* for summation over the repeated indices. Since the indices i, j, ... then describe the components with respect to a Cartesian reference frame, this expression means that:

$$\int \sigma_{ij} d\eta_{ij} dV = \int \sum_{i,j=1}^{3} \sigma_{ij} d\eta_{ij} dV$$

We thus face a large range of thermodynamic functions. We will call them G, by analogy with a fluid, when the natural thermal and elastic variables are temperature and stress, but there exists a large variety of them: $G(T, M_{//}, \sigma)$, $G(T, B_{//}, \sigma)$, $G(T, H, \sigma)$, $G(T, H_0, \sigma)$. The last of these is of particular interest as it describes the most common experimental situation, with temperature, applied field and stress fixed. In the same way, when the natural elastic variable is η, we will have four functions that describe U, and four others for F, for example $U(S, M_{//}, \eta)$ or $F(T, M_{//}, \eta)$. The essential point is again that the differentials can be immediately written, and that Maxwell relations follow naturally from them, for example:

$$dU(S, m_{//}, \eta) = T\,dS + \mu_0 H_0 dm_{//} + \int \sigma_{ij} d\eta_{ij} dV \qquad (10.17)$$

$$dG(T, H_0, \sigma) = -S\,dT - \mu_0 m_{//} dH_0 - \int \eta_{ij} d\sigma_{ij} dV \qquad (10.18)$$

We also have available a supplementary set of functions, with natural variables (S, σ and one of the magnetic variables) that are analogous to the enthalpy used in the thermodynamics of fluids. We will avoid denoting enthalpy with H, in order to prevent confusion with the magnetic field!

Use of the preceding relations requires the knowledge of the constitutive relations of the material to which we wish to apply them. The relations between **B** and **H**, and similarly those between $m_{//}$ and $\mathbf{H}_0$ are largely discussed in this book. In using these last relations, the effects of the demagnetising field are particularly important, except for the materials with weak susceptibility. This is a basic difficulty in the thermodynamic description of ferro- and ferrimagnetic materials: the field created outside the sample due to the magnetization in the material (this is sometimes called the dispersion field, or the stray field) is effective at large distances (the interaction is long ranged), and it depends on the shape of the sample.

5. COUPLING PHENOMENA

As well as the purely magnetic effects, it is also necessary to describe the coupling between magnetic properties and other effects, in particular the elastic deformations. This is obtained by introducing appropriate terms into the expressions for the thermodynamic potentials. We can thus describe the thermodynamic potential per unit volume of a *deformable* sample by

$$g(T, \mathbf{M}, \sigma) = g_0(T, \mathbf{M}) + S_{ijkl}\, \sigma_{ij}\, \sigma_{kl} + d_{ijk}\, \sigma_{ij}\, M_k + \lambda_{ijkl}\, \sigma_{ij}\, M_k M_l \quad (10.19)$$

again with Einstein's convention for the summation over repeated indices. $g_0(T, \mathbf{M})$ designates the thermodynamic potential per unit volume of an *undeformable* sample.

This relation introduces, in addition to the normal elastic energy, two very different coupling effects between stress and magnetization. The first, linear in magnetization, is called *piezomagnetic* coupling. This is often ignored, but it exists in some materials with special symmetries, in particular the antiferromagnetic fluorides MnF_2 and CoF_2 [3]. The second, bilinear in the magnetization, is always present, and corresponds to *magnetostriction*. It is developed in chapter 12. The relation $\eta_{ij} = (\partial g / \partial \sigma_{ij})_{T, \mathbf{M}}$, which follows from the differential equation for $F(T, \mathbf{M})$, implies that:

$$\eta_{ij} = S_{ijkl}\, \sigma_{kl} + d_{ijk}\, M_k + \lambda_{ijkl}\, M_k M_l \quad (10.20)$$

The first term corresponds to Hooke's law in its usual form. The other two terms, piezomagnetic and magnetostrictive, represent deformations associated with the magnetization. This formulation highlights the tensor nature of the various coupling effects.

Thus, when considering only the thermal, magnetic and elastic variables, the most general description of a magnetic system already requires 10 variables: $S = S/V$,

M_i (i = 1 to 3) and η_{ij} (i, j = 1 to 3) that are related to 10 conjugate variables T, H_i and σ_{ij} via a 10×10 matrix. This defines a total of 100 phenomenological coefficients, of which only 55 are linearly independent due to the Maxwell relations. We will show in the exercises what each one of these coefficients represents physically, and how best to treat them.

This formulation underlines certain intrinsic aspects of the symmetry of the tensors involved. Assuming that σ_{ij} and η_{ij} are symmetric ($\sigma_{ij} = \sigma_{ji}$ and $\eta_{ij} = \eta_{ji}$) implies that: $S_{ijkl} = S_{ijlk} = S_{jikl} = S_{jilk}$ and $d_{ijk} = d_{jik}$. The presence of the product $M_k M_l$ leads also to λ_{ijkl} having the same intrinsic symmetries as S_{ijkl}. In addition, such symmetry arguments also allow to predict, from knowledge of the crystal symmetry of a material, what effects are forbidden and which coefficients of the tensors are necessarily zero, or equal to each other. We also note that this reduction in the number of independent tensor components is based on the *magnetic symmetry* of the crystal, which is more complicated than its point symmetry because it involves the *time reversal* operator [4]. For an isotropic substance, or one with cubic symmetry, thermal expansion is described by a single coefficient, and so is magnetic susceptibility. Systems with hexagonal symmetry require two coefficients for thermal expansion, $\alpha_{//}$ and $\alpha_{\perp}$, and two magnetic permeabilities: $\mu_{//}$ along the c axis, and $\mu_{\perp}$ in the basal plane*.

All the thermodynamic "coefficients" thus defined depend strongly on the working point chosen, and even on the previous history of the material because of the hysteresis effects that we introduced in chapter 5. It is therefore essential to remember that they are not constants.

Other, less common, coupling effects can be described using the same formalism, with the introduction of additional variables. For example the linear magnetoelectric effect can be described in terms of $\alpha_{ij} E_i H_j$ using a free enthalpy density g(T, **E**, **H**) [5].

6. LANDAU-TYPE FREE ENERGY

Landau provided a different approach to thermodynamic potentials [2, 6]. His treatment of phase transitions profoundly marked physics, and remains a landmark even after the development of the more correct –but more complicated– approach to critical exponents based on the renormalization group theory. He introduced the concept of an *order parameter* and of the power expansion of a thermodynamic potential in terms of this order parameter. The order parameter is a function of temperature, defined as zero in the high symmetry phase, and nonzero in the lower symmetry ordered phase.

---

* When there exists a torque density, σ_{ij} and η_{ij} are no more symmetric, and rotational effects step in (see the end of § 2.2.1 of chap. 12).

The spontaneous magnetization is an excellent example. It is preferable to renormalise its value, and to use as an order parameter the reduced magnetization $m = M_s/M_{s0}$ where M_{s0} is the maximum value of M_s, that at 0 K. Very close to the transition, m will be small compared to unity. If the transition is continuous (hence corresponds to a second-type transition, or to a critical point), a power expansion of the energy in terms of m can be justified. The Landau expansion is written as:

$$f(T, m) = f_0(T) + a(T)\, m + b(T)\, m^2 + c(T)\, m^3 + d(T)\, m^4 + \ldots \qquad (10.21)$$

The essential difference with respect to a normal thermodynamic potential is that the first question is to find the value or values of m that minimise f at given temperature T (see fig. 10.1); if m_0 is such a value, the expression of f for *this* value of the order parameter plays the usual role of a thermodynamic potential.

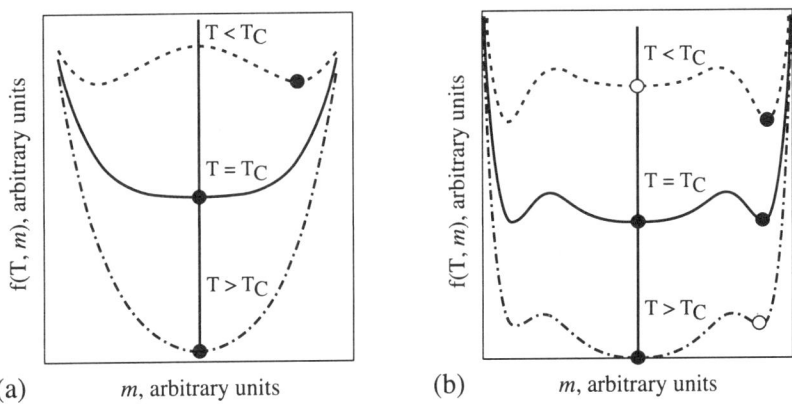

Figure 10.1 - Form of isothermal curves f(T, m):
(a) For a second-order transition - (b) For a first order transition

The simplest symmetry argument requires that a(T) and c(T) should be zero, because opposite directions of magnetization, and therefore opposite signs of m, should not affect the free energy. b(T) is then chosen so that the minimum in the function f(T) corresponds to $m = 0$ for $T > T_C$, where T_C is the transition temperature. The simplest form is $b(T) = b'(T - T_C)$ with $b' > 0$. Depending on the relative values and the sign of the higher-order terms in the expansion, one can observe a second-order transition (fig. 10.1-a and 10.2-a) or first-order transition (fig. 10.1-b and 10.2-b): in the latter case, one observes at $T = T_C$ a minimum for $m = 0$, and two others for $m = \pm m_0$.

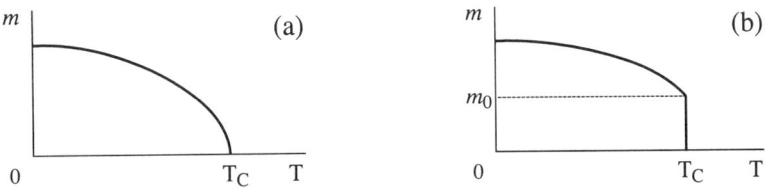

Figure 10.2 - Form of magnetization curves as a function of temperature for a second order transition (a), and a first order transition (b)

In figure 10.1-b, we note that, for $T = T_C$, the minima correspond to the same value of energy appearing for $m = 0$, and $m = \pm m_0$. Two phases then coexist, one is nonmagnetic, and the other carries a finite magnetization. This is the origin of the discontinuity in magnetization observed in figure 10.2-b.

The development of this model led to the quantitative description of the behavior of physical quantities close to a transition, for example magnetic susceptibility. It can be expanded upon by incorporation of terms in energy that correspond to spatial inhomogeneities of the order parameter, which permits the description of domain walls (regions where the order parameter is different, at least in terms of orientation).

The quantitative disagreement of this model with experiment, and specifically with the critical exponents as we will see, led to further development. The key point is the evidence for the essential role of fluctuations, that were previously ignored in this approach.

7. CRITICAL EXPONENTS AND SCALING LAWS

We have just seen that the simplest expression for the energy of a material featuring a second-order transition can be written using Landau's model as:

$$f(T, m) = f_0(T) + b'(T - T_C) m^2 + d(T) m^4 \qquad (10.22)$$

with b' and d > 0 (fig. 10.1). The value of m which minimises f is obtained by writing:

$$2b'(T - T_C) m + 4dm^3 = 0 \qquad (10.23)$$

which has the solutions: $m = 0$, valid above T_C, and:

$$m = [(b'/2d)(T_C - T)]^{1/2} \qquad (10.24)$$

which describes the spontaneous magnetization in the ordered phase. Experiments have shown that the reduced magnetization follows well, *close to T_C*, a power law in $T_C - T$:

$$m \sim (T_C - T)^\beta \qquad (10.25)$$

but that the exponent β is generally less than the value of 0.5 predicted for this model. In more general terms, all the physical qualities of a magnetic material feature, close to the critical temperature T_C, *critical behavior* which can be described in terms of the reduced variable:

$$t = (T - T_C)/T_C \qquad (10.26)$$

Thus, for example: $C_V \sim |t|^{-\alpha}$, $M \sim |t|^{-\beta}$, $\chi \sim |t|^{-\gamma}$, $\xi \sim |t|^{-\nu}$, where C_V is the specific heat, M the magnetization, χ the susceptibility and ξ the correlation length.

Near T_C, the magnetization also varies as a power of the field strength, which can be written: $M \sim |H|^{1/\delta}$, and the pair correlation function $G \sim \langle S_i S_j \rangle$ varies with distance r_{ij} as: $G \sim |r_{ij}|^{-(d-2+\eta)}$ where d represents (the dimensionality the dimension

10 - THERMODYNAMIC ASPECTS OF MAGNETISM

of the space considered). These power behaviors in terms of t, H or r are described as *scaling laws*, and the exponents, usually called *critical exponents,* are designated by a Greek letter. It was long believed that the critical exponents are not the same when T_C is approached from lower and higher temperatures. The exponents (α', γ', ν') and (α, γ, ν) were used for $T < T_C$ and $T > T_C$, respectively. These differences were in fact experimental artefacts. The theory of phase transition not only shows that $\alpha = \alpha'$, ... but also that there are relations between the different critical exponents. They are in fact thermodynamic inequalities valid at all temperatures, in particular far from T_C. For example, the inequality: $C_M = C_H - T[(\partial M/\partial T)_H]^2/(\partial M/\partial H)_T \geq 0$ leads to the Rushbrooke relation:

$$\alpha + 2\beta + \gamma \geq 2 \tag{10.27}$$

Similarly, one finds:
$$\alpha + \beta(1+\delta) \geq 2 \tag{10.28}$$

$$\gamma \geq \beta(\delta - 1) \tag{10.29}$$

and many other relations.

But, just as these exponents tend towards constant values very close to the ordering temperature, these inequalities become equalities. In a very general way, the knowledge of two critical exponents is thus sufficient to deduce all the others, which can be calculated from the following equalities [7, 8]:

$$\alpha + 2\beta + \gamma = 2 \tag{10.30}$$

$$\delta = 1 + \gamma/\beta \tag{10.31}$$

$$d\nu = 2 - \alpha \tag{10.32}$$

$$\nu = \gamma/(2 - \eta) \tag{10.33}$$

At the ordering temperature, the fluctuations have such an amplitude that the system becomes unstable: ξ then diverges as $1/(T - T_C)^\nu$, and, as a consequence, a number of physical quantities diverge as well, such as the magnetic susceptibility χ which follows $1/(T - T_C)^\gamma$.

Therefore, for a material with the values $\alpha = 0.125$ and $\beta = 0.3125$, the equality (10.30) yields $\gamma = 1.25$, the equality (10.31) gives $\delta = 5$ and (10.32) $\eta = 0$ for a space dimension d = 3. We therefore deduce that $\nu = 0.625$.

While the values for the critical exponents depend on the materials concerned, they depend essentially on the *dimensionality* of the space involved. Table 10.1 gives the values for these exponents, calculated by various theoretical models, and the values experimentally observed in real materials: magnetite [9] and nickel [10].

Souletie *et al.* [10] suggested that it was possible to apply the scaling laws to a far wider temperature range than the close neighbourhood of T_C under the condition that the reduced variable no longer corresponds to that defined by equation (10.26), but rather

$$t' = T - T_C/T \tag{10.34}$$

Magnetic susceptibility can then be expressed simply as a function of temperature by a relation that involves the critical exponent γ:

$$\chi = \frac{\mathscr{C}}{T}\left(\frac{T - T_C}{T}\right)^{-\gamma} \tag{10.35}$$

Table 10.1 - Some values for the critical exponents

Model	α	β	γ	δ	η	ν
Landau *	0	1/2	1	3	0	1/2
Ising (2 D)	0	1/8	7/4	15	1/4	1
Ising (3 D)	–1/8	–5/16	–5/4	–5	– 0	–5/8
Heisenberg	–0	–0.313	1.36 - 1.39	–5.25	–0	–0.64
Fe_3O_4	–0.16	0.405	1.35	4.33	1/8	0.72
Nickel	–0.1 ± 0.03	0.42 ± 0.04	1.316	4.2 ± 0.3	0.12	0.7

* These are also the exponents that are predicted by all the mean field models, such as the molecular field model. The reader will be surprised to find for the mean field models (Landau) a dimension of d = 4. The reason is that the molecular field model is exactly solvable only for d > 3. Detailed analysis of phase transitions and critical phenomena can be found in the classic works cited as [7] and [8].

This expression *diverges* (becomes infinite) at $T = T_C$ as it should, and reduces at high temperature ($T \gg T_C$) to the Curie-Weiss law: $\chi = \mathscr{C} / (T - \theta_p)$, where:

$$\theta_p = \gamma T_C \tag{10.36}$$

The molecular field model –as all the mean field models– predicts the equality of θ_p and T_C (fig. 4.11), i.e. the value of $\gamma = 1$ expected from Landau's model. The fact that γ exceeds unity simply indicates the presence, at high temperature, of fluctuations that are not totally random, but rather are correlated over a small volume of the material because of the exchange interactions. Short range order appears well above T_C, with a correlation length that increases as T is reduced until it diverges at $T = T_C$.

8. MAGNETIC ANOMALIES NEAR T_C

An important conclusion of this thermodynamic analysis is that all the physical properties of a magnetic substance present an anomaly close to T_C. Depending on whether the critical exponent associated with this property is positive or negative, the behavior will be similar to that of $m(T)$, with a substantial value at low temperatures, vanishing at or close to T_C (see for example fig. 3.19 and 3.20), or to that of $\chi(T)$, where a "lambda-type" divergence is observed near T_C (fig. 3.17 and 3.18).

9. THE MOLECULAR FIELD MODEL UNDER EXPERIMENTAL TEST

Experiments show that the coefficient β is always less than the value of 0.5 predicted by the molecular field model. This indicates that the spontaneous magnetization falls faster close to T_C than predicted by the molecular field model. In other words the magnetization near T_C is always larger than the value predicted by the molecular field model.

At low temperature, the situation is very similar. The experimental magnetization measured in real substances falls faster than predicted by the molecular field model towards 0 K; but this time, the experimental points are below the curve predicted by the molecular field theory, as demonstrated in figure 10.3 for pure metallic nickel.

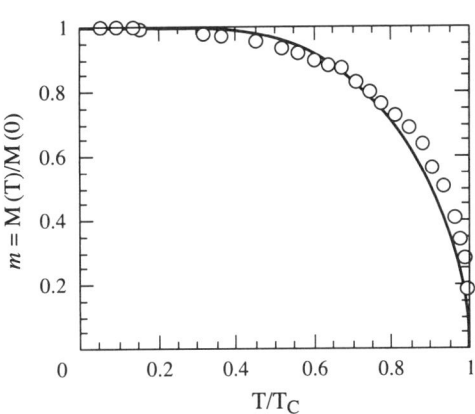

Figure 10.3 -
Spontaneous magnetization as a function of temperature

The data are plotted in reduced units. The curve predicted by the molecular field model for $S = 1/2$ (full line) is compared to the experimental points (circles).

The thermal variations at very low temperatures of the reduced magnetization can be explained in the following fashion: thermal agitation does not disorient the magnetic moments in a totally random fashion as was assumed in the molecular field model, but rather excites *spin waves*. Figure 10.4 illustrates the difference of behavior between random excitations (a) and excitations correlated by exchange interactions (b), in the simplified case of a linear chain of atomic moments. In both cases, the energy involved is exactly the same, but the variation in magnetization is far greater in case (b), as is shown by both classical calculations and quantum mechanical analysis.

Whereas the molecular field model predicts for *m* an exponential variation which begins very slowly, the *spin wave* excitations lead to a $T^{3/2}$ law:

$$m(T) = 1 - C\, T^{3/2} \qquad (10.37)$$

where C is the spin wave coefficient, and for a cubic face-centred lattice is given by:

$$C = \frac{0.0587}{2S}\left(\frac{k_B}{2JS}\right)^{3/2} \qquad (10.38)$$

A simple derivation is given by Kittel in his Introduction to Solid State Physics [11]. At higher temperature, in a metal, a contribution involving T^2 comes in and becomes predominant over the $T^{3/2}$ term.

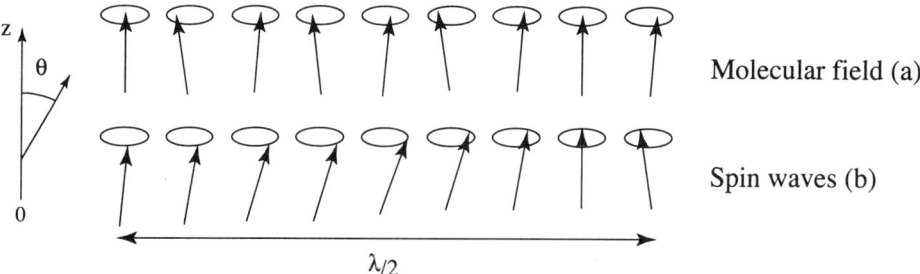

Figure 10.4 - Comparison between the random excitations considered by the molecular field model (a) and spin waves (b)

The exchange energy involved ($<J_{ij}S_iS_j> \sim <\cos \phi_{ij}>$, where ϕ_{ij} is the angle between two neighbouring magnetic moments) is the same in the two cases, but the resulting magnetization ($M_s<\cos \theta>$) is less in the case of spin waves.

EXERCISES

E.1 - Starting from the thermodynamic potential normalised to unit volume of a *deformable* sample g(T, **H**, σ), show that there are 100 second derivatives of g. Show that the Maxwell relations permit a reduction from 100 to 55 in the number of these second derivatives that are distinct.

E.2 - For virtual changes in temperature, applied field and stress, express the changes in $\mathscr{S}$ (entropy per unit volume), in the components of the induction, and in those of the strain tensor as a function of the 55 coefficients which can be identified with the 55 second derivatives of g. Describe their physical meaning. Show that both direct, and inverse coupling effects can be defined.

E.3 - In the case of a crystal with cubic symmetry, what happens to these 55 coefficients?

SOLUTIONS TO THE EXERCISES

S.1 - T is a scalar, and correspondingly has only one component, **H** is a vector, and has 3 components, and σ is a tensor with 6 symmetric components. G is thus expanded into 10 independent components, and so has 100 second derivatives such as $\partial^2 g/\partial T^2$, $\partial^2 g/\partial T\partial H_i$, $\partial^2 g/\partial H_i\partial \sigma_{jk}$, etc. The Maxwell relations tell us that the second derivatives of g are independent of the order in the derivatives, so that for example: $\partial^2 g / \partial T \partial H_i = \partial^2 g / \partial H_i \partial T$. The 10 second derivatives with only one variable $\partial^2 g / \partial T^2$, $\partial^2 g / \partial H_i^2$, $\partial^2 g / \partial \sigma_{jk}^2$ are not involved in these relations. Therefore only 90 of the mixed second derivatives will have their number halved, and so become 45. With 10 second derivatives of a single variable and 45 mixed second derivatives, we arrive at 55 distinct second derivatives.

S.2 - As $\partial g = -\mathscr{S} dT - B_i dH_i - \eta_{jk} d\sigma_{jk}$, we have: $\mathscr{S} = -\partial g / \partial T$, $B_i = -\partial g / \partial H_i$ and $\eta_{jk} = -\partial g / \partial \sigma_{jk}$. For a virtual change of the temperature dT, of the components of the field dH_i and of the components of stress $d\sigma_{jk}$, we have:

$$d\mathscr{S} = (\partial \mathscr{S} / \partial T) dT + (\partial \mathscr{S} / \partial H_j) dH_j + (\partial \mathscr{S} / \partial \sigma_{lm}) d\sigma_{lm},$$

and so: $d\mathscr{S} = (-\partial^2 g / \partial T^2) dT + (-\partial^2 g / \partial T \partial H_j) dH_j + (-\partial^2 g / \partial T \partial \sigma_{lm}) d\sigma_{lm}$.

In the same way:

$$dB_i = (-\partial^2 g / \partial T \partial H_i) dT + (-\partial^2 g / \partial H_i \partial H_j) dH_j + (-\partial^2 g / \partial H_i \partial \sigma_{jk}) d\sigma_{jk}$$

$$d\eta_{jk} = (-\partial^2 g / \partial T \partial \sigma_{jk}) dT + (-\partial^2 g / \partial H_i \partial \sigma_{jk}) dH_i + (-\partial^2 g / \partial \sigma_{jk} \partial \sigma_{lm}) d\sigma_{lm}$$

There are therefore 10 linear equations and 10 unknowns, with i, j = 1, 2, 3.

We now give the physical meaning of the coefficients of these equations.

♦ $C_{\sigma,H} = T(\partial \mathscr{S} / \partial T)_{\sigma,H} = -T(\partial^2 g / \partial T^2)_{\sigma,H}$ is the specific heat, which characterises the thermal properties of the sample,

♦ $(\chi_{ij})_{T,\sigma} = (\partial M_i / \partial H_j)_{T,\sigma} = (-\partial^2 g / \partial H_i \partial H_j)_{T,\sigma}$ is the susceptibility tensor, which has in the general case 3 diagonal and 3 off-diagonal components, and characterises its magnetic properties,

♦ and finally $(s_{ijkl})_{T,H} = (\partial \eta_{ij} / \partial \sigma_{kl})_{T,H} = -(\partial^2 g / \partial \sigma_{ij} \partial \sigma_{kl})_{T,H}$ the elastic compliance tensor which has, in the general case, 21 coefficients, 6 diagonal and 15 off-diagonal elements. It characterises the elastic properties.

The 27 remaining coefficients describe the *coupling effects* between the thermal, magnetic and elastic properties: in the general case there are 3 coefficients for the pyromagnetic coupling, $(i_j)_\sigma = \mu_0 (\partial M_j / \partial T)_{\sigma, H} = (\partial \mathscr{S} / \partial H_j)_{T, \sigma}$, which describes the effects of temperature on the magnetic properties, and inversely the effect of the magnetic field on the entropy (the magnetocaloric effect), six for the thermoelastic coupling, $(\alpha_{ij})_H = (\partial \eta_{ij} / \partial T)_{\sigma, H} = (\partial \mathscr{S} / \partial \sigma_{ij})_{T, H}$ which describes the thermal expansion and inversely the effect of stress on the entropy, and 18 for the piezomagnetic coupling, $(d_{ijk})_T = \mu_0 (\partial M_k / \partial \sigma_{ij})_{T, H} = (\partial \eta_{ij} / \partial H_k)_{T, \sigma}$ which describes the effects of mechanical stress on the magnetic properties, and inversely the effects of a magnetic field on the physical dimensions of the sample.

Thus, to each of these *direct coupling effects* there corresponds an *inverse coupling effect*.

S.3 - The symmetry properties of the cube requires that all the diagonal coefficients μ_{xx} are equal to μ_{yy} and to μ_{zz}, all μ_{xy} are equal to μ_{yx}, etc. as the exchange of x and of y cannot change the value of the energy. μ_{xy} is necessarily zero as exchange of x to $-$ x should not change the expression for the energy while it changes μ_{xy} into $-\mu_{xy}$. In a general way, any coefficient $A_{ijklmnop}$... should be even with respect to x, y and z, otherwise its value is necessarily zero, and its value must be conserved under all permutations of these three variables. In this way, there remain only one specific heat, one magnetic permeability, three elastic coefficients (c_{11}, c_{12} and c_{44}), one coefficient of pyromagnetic coupling, a single coefficient of thermal expansion and no piezomagnetic effect (but there remain three coefficients of magnetoelastic coupling that are responsible for magnetostriction).

REFERENCES

[1] H.B. CALLEN, Thermodynamics, and an introduction to Thermostatistics, 2nd ed. (1985) Wiley, New York. (Do not use printings earlier than the sixth).

[2] L.D. LANDAU, E.M. LIFSHITS, Electrodynamique des Milieux Continus (1969) Mir, Moscou.

[3] A.S. BOROVIK-ROMANOV, *J. Exp. Theor. Phys.* **36** (1959) 1954; *J. Exp. Theor. Phys.* **38** (1960) 1088; *Ferroelectrics* **162** (1994) 153.

[4] R.R. BIRSS, Symmetry, and Magnetism, *in Selected topics in Solid State Physics,* Vol. III (1964) E.P. Wohlfarth Ed., North Holland, Amsterdam.

[5] H. SCHMID, *Ferroelectrics* **161** (1994) 1.

[6] J.C. TOLEDANO, P. TOLEDANO, The Landau theory of phase transitions. Applications to structural, incommensurate, magnetic and liquid crystals systems (1987) World Scientific, Singapore.

[7] N. BOCCARA, Symétries brisées, théorie des transitions avec paramètre d'ordre (1976) Hermann, Paris.

[8] H.E. STANLEY, Introduction to phase transitions and critical phenomena (1971) Oxford University Press, New York.

[9] M. HAUG, M. FÄHNLE, H. KRONMÜLLER, *J. Magn. Magn. Mater.* **69** (1987) 163.

[10] E. CARRÉ, J. SOULETIE, *J. Magn. Magn. Mater.* **72** (1988) 29.

[11] C. KITTEL, Introduction to solid state physics, 7th ed. (1996) Wiley, London.

COUPLING PHENOMENA

CHAPTER 11

MAGNETOCALORIC COUPLING AND RELATED EFFECTS

We now tackle phenomena which arise due to coupling between magnetic properties, and other physical properties of a system. In this chapter we describe the simplest example, magnetocaloric coupling, as its analysis requires the introduction of only two scalars from outside the field of magnetism, namely temperature and entropy.

1. THERMAL VARIATIONS OF MAGNETIZATION AND SPECIFIC HEAT

The thermodynamical treatment of magnetism given in chapter 10 allows us to introduce the idea of coupling between the magnetic properties and other physical properties of a substance. Here we describe some effects that illustrate the coupling which can be observed between magnetic induction (or magnetization) and thermodynamic quantities, entropy, and temperature. In the first two parts of this chapter we consider equilibrium effects, and it is only in the third part that we consider losses associated with magnetic domains.

Experiment has shown that the spontaneous magnetization of a ferromagnetic substance decreases as its temperature is raised because of the increasing disorder of the magnetic moments (chap. 3, § 2.1). This variation in spontaneous magnetization is greatest near the Curie temperature (see fig. 3.8). The magnetization is also most sensitive to an applied magnetic field in this temperature region.

In the molecular field model (eq. 4.41) the exchange energy density is given by:

$$E_{ex} = -(1/2)\mu_0 w M^2 \tag{11.1}$$

and its variation with a change in temperature ΔT by:

$$\Delta E_{ex} = -\mu_0 w M \Delta M \tag{11.2}$$

where ΔM is the variation in magnetization associated with the increase in temperature ΔT. To increase the temperature of a sample by ΔT, it is necessary to supply a quantity of heat $\Delta Q + (\partial E_{ex}/\partial T)\, \Delta T$, where ΔQ alone is the "normal" quantity that would be needed if the sample were not magnetic. $(\partial E_{ex}/\partial T)\, \Delta T$ is positive (since $\partial M/\partial T$ is negative), has its maximum value at the Curie temperature, and is zero above this temperature. Thus, the specific heat of a ferromagnetic sample increases as T_C is approached. This is what is observed in figure 11.1, which shows the variation with temperature, under a weak magnetic field ($< 100\ A.m^{-1}$), of the specific heat of nickel [1].

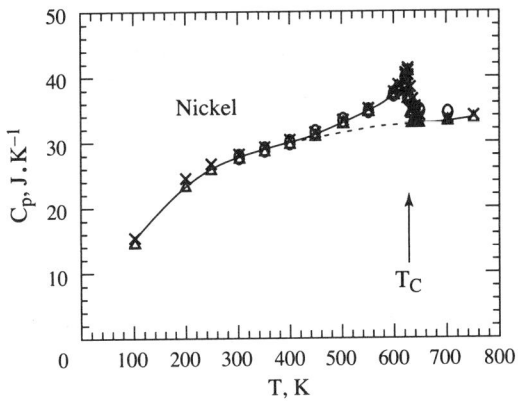

Figure 11.1 - Thermal variation of the molar heat capacity of nickel, after Ahrens (Δ), Grew (o), and Lapp ($\times$) [1]

However, contrary to the theoretical predictions of the molecular field model, this *"anomalous specific heat"* stretches somewhat above the Curie temperature, because of the persistence of short range order between the magnetic moments. A more rigorous statistical treatment, that takes account of the interactions between magnetic moments, predicts the magnetic contribution to the specific heat, within the Heisenberg model, to be:

$$C_v^{magn} = -NzJ_{AB}\, \frac{\partial}{\partial T} <S_A \cdot S_B> \qquad (11.3)$$

where N is the number of magnetic atoms per unit volume, z the number of nearest neighbours, and J_{AB} the exchange integral between atoms A and B. C_v^{magn} thus appears to be proportional to the derivative, with respect to temperature, of the *spin correlation function* $<S_A \cdot S_B>$. It is maximum at T_C, and then decreases rapidly above this temperature.

Evidently, similar effects can occur in ferri- and anti-ferromagnetic substances, but it must be then noted that it is the magnetization of each sublattice, and not the resultant magnetization, which contributes to the exchange energy. Thus, for a ferrimagnet of type "M" or "P" (see § 6 of chap. 4), while the resultant magnetization increases when the substance is heated at low temperature, this does not lead to a negative contribution to the specific heat. In all cases, this contribution presents a characteristic "lambda-shaped" behavior, as shown in figure 3.15 for the antiferromagnetic substance $ErGa_2$.

2. THE MAGNETOCALORIC EFFECT

The reversible heating of para- or ferromagnetic substances which occurs during the ordering of the magnetic moments is known as the magnetocaloric effect. It is the inverse of the effect we just discussed: unlike an increase in temperature, an increase in the magnetic field strength reduces the disorder of the magnetic moments. If the magnetization process is performed in an adiabatic way, that is without heat exchange, the total entropy of the substance will remain constant: the ordering of the magnetic moments will be compensated for by a greater disorder in the atomic arrangement. This corresponds to an increase in temperature, *heating by adiabatic magnetization*.

Conversely, if a substance is magnetised at a given temperature, and then thermally isolated, the adiabatic demagnetization of this substance will cause its temperature to decrease: this is known as cooling by *adiabatic demagnetization*, and, in theory, allows the attainment of very low temperatures.

A simple description of this effect may be given in the framework of the molecular field theory. The variation in energy ΔE_{ex} which accompanies the application of a magnetic field H is again given by equation (11.2), but this time ΔM is the variation in magnetization created by H, and the work of the magnetic field H necessary to bring about this variation in magnetization is:

$$\Delta W = \mu_0 H \, \Delta M \qquad (11.4)$$

The heat which is produced by this process will be equal to the difference between the work done by the field and the variation in exchange energy:

$$\Delta Q = \mu_0 (H + wM) \, \Delta M \qquad (11.5)$$

where wM is the molecular field. For temperatures *above the Curie temperature*, equations (4.50) and (4.51) allow us to write the molecular field as:

$$wM = H \, T_C / (T - T_C) \qquad (11.6)$$

which finally gives us the heat generated by magnetization reversal in the paramagnetic regime:

$$\Delta Q = \frac{1}{2} \frac{\mu_0}{w} \frac{T \, T_C}{(T - T_C)^2} \Delta(H^2) \qquad (11.7)$$

This generated heat will therefore raise the temperature of the sample by a proportional quantity, $\Delta T = \Delta Q / C_M$. This is the magnetocaloric effect. C_M is the specific heat at constant magnetization, and T is the initial temperature of the sample, in the absence of a magnetic field. Indeed, P. Weiss and R. Forrer observed, and precisely measured this heating in nickel; some of their results are reproduced in figure 11.2, which clearly shows that, above T_C, the heating varies as the square of the magnetic field, and decreases rapidly with temperature.

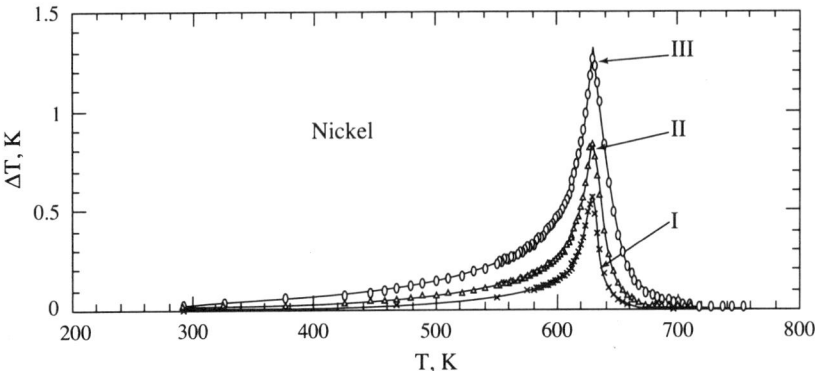

*Figure 11.2 - The magnetocaloric effect in nickel,
as a function of temperature, and magnetic field*
The strength of the magnetic field varies from 0 to $\mu_0 H = 0.6\,T$ (I), $1\,T$ (II), and $1.78\,T$ (III), after [2].

The effect also decreases at lower temperatures, as we move away from T_C, as the influence of the applied field quickly becomes negligible compared with the molecular field. Therefore, the effect shows a lambda-shaped maximum at T_C.

Using equation (11.6), we can express equation (11.7) in terms of the magnetization. We thus find that, *above the Curie temperature*:

$$\Delta T = (1/2C_M)\,\mu_0\,w\,\Delta(M^2)\,(T/T_C) \qquad (11.8)$$

Below the Curie temperature, where the applied magnetic field H becomes negligible compared with the molecular field, equation (11.5) simply gives:

$$\Delta T = (1/2C_M)\,\mu_0\,w\,\Delta(M^2) \qquad (11.9)$$

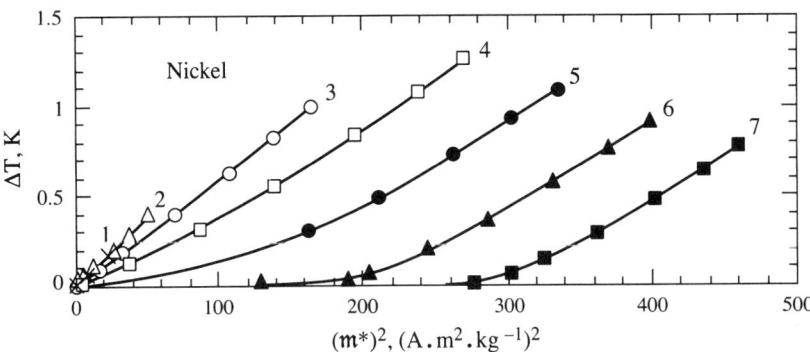

*Figure 11.3 - The magnetocaloric effect in nickel, as a function of the square
of the specific magnetic moment m^* at different temperatures, after [2]*
$T = 660.65\,K\,(1), 653.3\,K\,(2), 638.51\,K\,(3), 629.42\,K\,(4), 625.68\,K\,(5), 621.93\,K\,(6), 618.17\,K\,(7)$. Nickel is a very soft material: in the ordered state and on the scale considered, there is no noticeable heating during the technical magnetization process (movement of domain walls and rotation). Note that ΔT remains practically zero until the magnetization reaches M_S (spontaneous magnetization), and begins to increase only when the internal field becomes non-zero.

Weiss and Forrer also plotted the magnetocaloric effect as a function of the square of the magnetization, and found, beyond saturation, a series of parallel straight lines at low temperature (curves 7 to 4), and then straight lines of increasing slope as the temperature was raised above T_C (curves 3 to 1), as shown in figure 11.3.

3. IRREVERSIBLE THERMAL EFFECTS

Until now we have just considered the large reversible thermal effects associated with variations in entropy which accompany changes in the spontaneous magnetization of very soft materials. With harder materials, which show marked hysteresis, a complete magnetization cycle will be accompanied by a reversible magnetocaloric effect which is superimposed on the heating associated with hysteretic losses.

Figure 11.4 shows the result obtained for a sample of iron: the heating is plotted as a function of the applied magnetic field. The change in temperature is much smaller than in figure 11.2 because on the one hand the applied field is much weaker, and on the other hand, the measurement was made far below the Curie temperature.

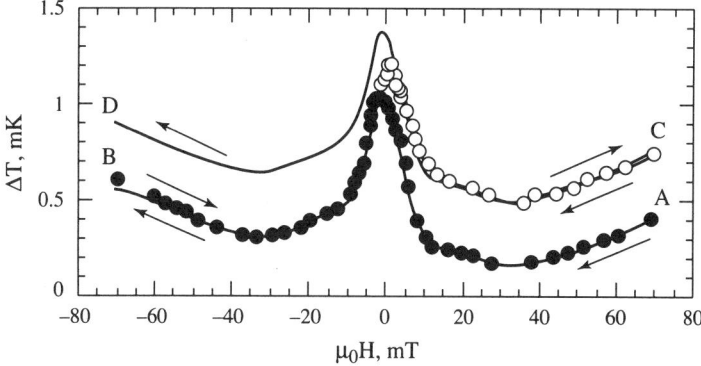

Figure 11.4 - Magnetocaloric effect along the hysteresis loop of a polycrystalline sample of iron measured at 300 K

We observe the superposition of the reversible magnetocaloric effect and of an irreversible dissipation of heat due to hysteretic losses [3]. An increase in temperature of about 0.2 mK takes place for every half cycle (moving from point A to B, from B to C, and so on).

The normal magnetocaloric effect is observed for magnetic fields greater than $\mu_0 H = 30$ mT, but an adiabatic cooling is also observed during magnetization reversal in the region 0-30 mT. This anomalous effect is explained by the fact that a reversible rotation of the magnetic moments away from the easy axis of magnetization occurs in low fields: the resulting angular distribution of the magnetic moments is wider than in the absence of a magnetic field, and therefore corresponds to an increase in the entropy of the system. This requires the absorption of heat which is observed in low field; a detailed analysis of this effect can be found in the work of S. Chikazumi [4]. Finally, an *irreversible* dissipation of heat occurs at low field (± 5 mT) where irreversible domain wall displacements occur. The total rise in temperature of the sample due to *hysteretic losses* during a magnetization cycle corresponds to the distance AC or BD in figure 11.4.

We can evaluate these losses by measuring the area of the hysteresis loop $\mu_0 m(H)$. Let us consider an elongated ellipsoidal ferromagnetic particle of magnetic moment m, placed at the centre of a solenoid carrying a current which is sufficient to saturate the sample, and oriented along the solenoid's axis. The demagnetising field of the sample is negligible. This situation corresponds to point A of the hysteresis loop (fig. 11.5).

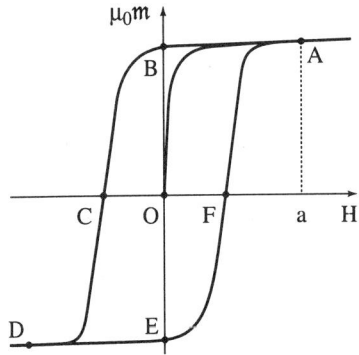

Figure 11.5 - Hysteresis loop

Let us extract the particle from the solenoid along its axis, to the point where the magnetic field is reduced to zero (point B of the loop). Let us now turn the particle through 180°, an action which requires no work (as H is zero), and then return the particle to the centre of the solenoid (point D of the loop). During this procedure, the magnetization is reduced to zero at point C of the loop. Extracting the sample from the solenoid a second time we arrive at point E, and finally after turning the sample once more we return to the initial situation (point A) after having changed the direction of magnetization a second time at point F.

The magnetic force acting on the ferromagnetic particle is zero at the centre of and far from the solenoid, but during the cycle, between these two equilibrium positions, the particle experiences a force $F_x = m_x \cdot \partial B / \partial x$, where the axis of the solenoid is the x axis. The work done to move the particle through dx is: $dW = F_x \, dx$, i.e. $dW = \mu_0 m \, dH$. Therefore, the work done to move the particle from the centre of the solenoid to infinity, i.e. the magnetic work needed to go from A to B, is equal to

$$W = \int_H^0 \mu_0 m \, dH = -\int_0^H \mu_0 m \, dH = -(\text{Area OaAB})$$

It is easy to see that the work done in making a complete cycle is zero if the material shows no hysteresis, while it is equal to the area of the loop for a real material. The work done is completely transformed into heat, since the system returns to its initial state. This is the origin of the heating observed in figure 11.4.

4. SIZE EFFECTS ASSOCIATED WITH THE MAGNETOCALORIC EFFECT

A solid usually expands when it is heated. The magnetocaloric effect may therefore be observed also by dilatometric techniques. Figure 11.6 shows the relative elongation of a ferromagnetic sphere observed near T_C. This elongation, plotted as a function of time, is the response to a step in the magnetic field. A significant elongation, followed by a contraction which is exponential in time, is observed upon application of a magnetic field.

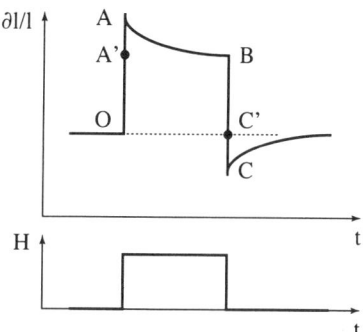

Figure 11.6 - Sample deformation in response to the switching on and switching off of a magnetic field

The sphere stays deformed even after having reached its thermal equilibrium state: this permanent deformation is called *magnetostriction*, and will be dealt with in the following chapter. When the magnetic field is removed, a strong contraction is observed, followed by a weaker expansion which returns the sphere to its initial size with an exponential time dependency. The sample is in thermal contact with a large sample holder, and the exponential thermal drifts which follow the switching on and the switching off of the magnetic field in fact correspond to the gradual thermalization of the sample, with the sample holder acting as a thermostat. OA'BC' represents the isothermal signal (due to magnetostriction), while OA and BC represent the signals obtained under adiabatic conditions. The lengths A'A and CC', which are given by $(\partial l/l)_A - (\partial l/l)_B$, represent the contribution of the magnetocaloric effect to the elongation.

If the linear thermal expansion coefficient of the substance, α_T, is known, it is then possible to deduce the magnetocaloric effect:

$$\Delta T = \frac{1}{\alpha_T}\left[\left(\frac{\delta l}{l}\right)_A - \left(\frac{\delta l}{l}\right)_B\right] \qquad (11.10)$$

Figure 11.7 shows the variations in the magnetocaloric effect of GdZn measured by this indirect technique, as a function of $(T - T_C)$, for an applied field of $\mu_0 H = 0.2$ T. The same behavior is always observed, viz a lambda-shaped peak close to the Curie temperature.

This effect is isotropic, within the precision of careful measurement: the heating is the same whether measured along the [100] direction or the [111] direction.

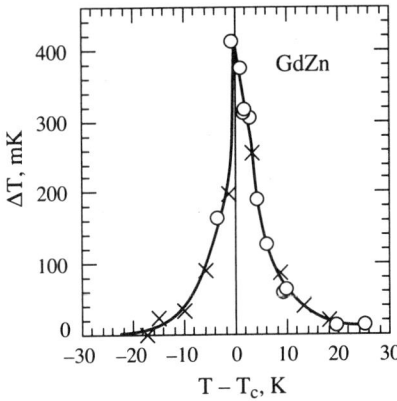

Figure 11.7 - Thermal variation of the magnetocaloric effect of a single crystal of GdZn, deduced from dilatometric measurements. Measurements made along [111]: × and along [001]: o [5]

5. APPLICATIONS OF MAGNETOTHERMAL EFFECTS

The most obvious technical application of the thermodynamics of magnetic materials is magnetic refrigeration. We have already mentioned and further address in exercise 2, the use of the adiabatic suppression of a magnetic field as a means of cooling a paramagnetic material in the low temperature region (< 1 K) using cerium magnesium nitride.

The magnetic cycling of materials to produce refrigeration (e.g. gadolinium for refrigeration near room temperature; $Gd_5(Si–Ge)_4$-type alloys for refrigeration in the broader temperature range 30-290 K), has been debated for some time now [6]. Nanostructured materials, which exploit the superparamagnetism of very small clusters, seem to be good candidates for this type of application [7].

EXERCISES

E.1 - Magnetocaloric effect: general relations

Consider a sample of a magnetic material, which has a sufficiently weak susceptibility, or alternatively an adequate shape, so that its demagnetising field can be neglected.

As we will not consider volume effects, it is convenient to consider a unit volume of the material. We denote the entropy density (entropy per unit volume) of the material as s, while c_H is its specific heat (per unit volume) at constant field H.

E.1.1 - State which shape the sample would need to have so that its demagnetising field may be neglected, even when it has a large susceptibility. If necessary, specify the direction in which the magnetic field must be applied.

E.1.2 - Using Nernst's principle, which states that entropy tends towards zero as temperature tends towards 0 K, determine:
- the limit, for $T \to 0$, of $(\partial s / \partial H)_T$
- $\lim (\partial M / \partial T)_H$ for $T \to 0$.
- Is this relation satisfied in the special case of a perfect paramagnet, defined as a material which obeys the *Curie law* $M = \mathscr{C} H / T$? What can you conclude about the validity of the Curie law (eq. 4.23)?

E.1.3
- Determine the expression for the variation in temperature resulting from an infinitesimal change in magnetic field, performed quasi-statically on a thermally insulated sample.

  - Determine the isothermal variation of entropy due to the application of a magnetic field $\Delta s = s(H, T) - s(0, T)$ using magnetic measurements only, i.e. without resorting to calorimetry. Clearly indicate how Δs can be graphically determined from isothermal magnetization curves measured at two temperatures, T_1 and T_2, close to T.

E.2 - Refrigeration at very low temperatures by "adiabatic demagnetization"

In the very low temperature regime (typically between some mK and 1 K), cerium magnesium nitrate (abbreviated to CMN) behaves like a perfect paramagnet, with a Curie constant $\mathscr{C} = 1.08 \times 10^{-2}$ K.

The lattice specific heat is negligible in this temperature range, and thus the total specific heat is almost entirely magnetic in origin.

Thus, in zero magnetic field, the specific heat at constant magnetic field has the form $c(T, 0) = A / T^2$, with $A = 0.137 \text{ J.K.m}^{-3}$.

E.2.1 - Calculate numerically the ratio $c(T, H) / c(0, 0)$ for $T = 1.0$ K and for a field $B = \mu_0 H = 0.45$ T. Recall that $\mu_0 = 4\pi.10^{-7}$ MKSA.

E.2.2
- Determine the differential expression for the change in temperature of a sample of CMN during a quasi-static adiabatic demagnetization process. This procedure involves reducing (not too rapidly) the applied magnetic field, while the sample is thermally insulated.

  - Determine numerically the final temperature T_f for an initial field $B_i = \mu_0 H_i = 0.45$ T, and an initial temperature $T_i = 1.0$ K, the final field being zero.

E.2.3 - The magnetic field is not completely suppressed, but reduced to a value H_f such that we can neglect the A / T_f^2 contribution to the specific heat.

Determine the ratio M_i/M_f of the initial magnetization, in a field H_i, and the final magnetization, in a field H_f, obtained under these conditions (i.e. a quasi-static process and a thermally insulated sample).

Have you any remarks to make on the use of the term "adiabatic demagnetization"?

SOLUTIONS TO THE EXERCISES

S.1 *S.1.1* A long cylinder, field parallel to the cylinder's axis; or a thin film, field parallel to the surface. We thus take H_0 to be equal to H. We also suppose that they are collinear, and thus restrict ourselves to scalar functions.

S.1.2 ♦ Nernst's principle indicates that $s \to 0$ for $T \to 0$ independently of H. Therefore, $\lim (\partial s/\partial H)_{T \to 0} = 0$. As a result, 0 K cannot be reached by adiabatic demagnetization (a feat equally impossible by any other method).

♦ T and H being the natural variables, we try

$$df(T, H) = -s\, dT - \mu_0 M\, dH.$$

The Maxwell relation gives $(\partial M/\partial T)_H = (1/\mu_0)(\partial s/\partial H)_T$. Thus the limit also equals 0 for $T \to 0$.

♦ No. The Curie law cannot remain valid at very low temperatures. This law implies that that the interactions between moments are negligible, but this means negligible compared with $k_B T$, where k_B is Boltzmann's constant.

S.1.3 ♦ $\left(\dfrac{\partial T}{\partial H}\right)_S = -\mu_0 \left(\dfrac{\partial m_{//}}{\partial S}\right)_H = -\mu_0 \left(\dfrac{\partial m_{//}}{\partial T}\right)_H \left(\dfrac{\partial T}{\partial S}\right)_H = -\dfrac{\mu_0 T}{C_H}\left(\dfrac{\partial m_{//}}{\partial T}\right)_H$

after (10.16).

♦ Equation $(\partial s/\partial H)_T = \mu_0 (\partial M/\partial T)_H$ leads to:

$$\Delta s = s(H, T) - s(0, T) = \mu_0 \int_0^H (\partial M/\partial T)_H\, dH = \mu_0 \int_0^H (\Delta M/\Delta T)_H\, dH$$

$$= \frac{\mu_0}{T_2 - T_1} \int_0^H [M(T_2, H) - M(T_1, H)]\, dH$$

$$= \frac{\mu_0}{T_2 - T_1} \left\{ \int_0^H [M(T_2, H)\, dH - \int_0^H [M(T_1, H)\, dH \right\}$$

The term in curly brackets is the area, in the M-H plane, between the isothermal magnetization curves M(H) measured at temperatures T_1 and T_2.

S.2 *S.2.1* $c(T, H) = T(\partial s/\partial T)_H$, from which:

$$(\partial c/\partial H)_T = T(\partial^2 s/\partial H \partial T) = \mu_0 T(\partial^2 M/\partial T^2)_H.$$

We deduce that: $c(T, H)/c(T, 0) = 1 + \mathscr{C}B^2/A\mu_0 = 1.27 \times 10^{-4}$ (enormous).

S.2.2
- $\delta T = T \mathscr{C} B \delta B / \{A\mu_0 + \mathscr{C}B^2\}$
- $T_f = T_i / \sqrt{1 + \mathscr{C}B^2/A\mu_0} = 8.9 \times 10^{-3}$ K

S.2.3 $T_f/T_i = B_f/B_i = H_f/H_i$, from which: $M_i/M_f = 1$.

We don't really demagnetise the sample!

REFERENCES

[1] P. PASCAL, Nouveau traité de chimie minérale, Vol. XVII (1963) 588, Masson, Paris.

[2] P. WEISS, R. FORRER, *Ann. Phys.* **V** (1926) 153.

[3] T. OKAMURA, *Sci. Reports Tohoku Univ.* **24** (1935) 745.

[4] S. CHIKAZUMI, Physics of Magnetism (1964) Wiley & Sons, New York.

[5] J. ROUCHY, P. MORIN, E. DU TRÉMOLET DE LACHEISSERIE, *J. Magn. Magn. Mater.* **23** (1981) 59.

[6] V.K. PECHARSKY, A. GSCHNEIDNER, JR. *Appl. Phys. Lett.* **70** (1997) 3299.

[7] R.D. MCMICHAEL, R.D. SHULL, L.J. SWARTZENDRUBER, L.H. BENNETT, R.E. WATSON, *J. Magn. Magn. Mater.* **111** (1992) 29.

CHAPTER 12

MAGNETOELASTIC EFFECTS

In a magnetic material, the interatomic distances can vary with the intensity and the orientation of magnetization: this is magnetostriction, or the direct magnetoelastic effect. Conversely, the magnetic state of the material is sensitive to mechanical influences. The hysteresis loop, for example, distorts under mechanical stress. This is called the inverse magnetoelastic effect. This chapter is devoted to a presentation of magnetoelastic coupling and to associated effects. The relevant materials and their industrial applications will be described in chapter 18.

1. THE MAIN MAGNETOELASTIC EFFECTS

1.1. ANOMALOUS EXPANSION
SPONTANEOUS AND FORCED EXCHANGE MAGNETOSTRICTION

Because exchange interactions are sensitive to interatomic distance, the order parameter $<\mathbf{S}_i \cdot \mathbf{S}_j>$ couples to the dimensions of a magnetic sample. This *magnetoelastic coupling* generates *exchange magnetostriction*, the spontaneous distortion which is observed below the ordering temperature.

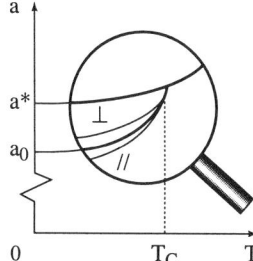

Figure 12.1 - Magnetostriction of nickel (§ 1.1, and 1.4)

Figure 12.1 shows the thermal variation of the lattice parameter of nickel. Above $T_C = 631$ K, the behavior is normal, and extrapolation can yield the value a^* of the lattice parameter which nickel would have at 0 K if it were not magnetic. An anomaly in thermal expansion is observed around T_C, as well as a relative decrease in the lattice parameter, which at low temperature reaches $(a_0 - a^*)/a^* = -4 \times 10^{-4}$, as was already noted for the compound $GdAl_2$ (fig. 3.19). This deformation *does not lower the symmetry* of the material. In isotropic and cubic materials, it can therefore be described just by the change in volume. In materials with lower symmetry, anisotropic strain, e.g. a change in the c/a ratio in hexagonal or tetragonal materials, can also occur.

Positive exchange magnetostriction can sometimes provide a thermal variation that is exactly opposite to the normal thermal expansion of the material. Then the material features zero thermal expansion through compensation. This is the *INVAR effect*, discovered by C.E. Guillaume in 1896 on iron nickel alloys. In the ordered state, exchange magnetostriction exhibits a weak linear dependence with magnetic field. This is called *forced isotropic magnetostriction*.

1.2. INFLUENCE OF HYDROSTATIC PRESSURE ON THE MAGNETIC PROPERTIES

Volume magnetostriction is due to exchange interactions. Conversely, hydrostatic pressure will change these exchange interactions, and consequently can modify the Curie temperature, and the magnetic moments. However, these variations are usually very small.

1.3. ANOMALIES IN THE ELASTIC CONSTANTS: THE MORPHIC EFFECT

Exchange magnetoelastic coupling is also responsible for some anomalies in the thermal variation of elasticity coefficients (fig. 3.20). Below the ordering temperature, a magnetic contribution to each of the elastic constants appears (the morphic effect). When this contribution is negative, its thermal variations can counterbalance the normal thermal variation of the elasticity coefficient. The result is then temperature independent over a range of temperatures: this is the ELINVAR effect, put to use around 1920 by C.E. Guillaume for precision watchmaking. A detailed study of this effect is outside the scope of this book.

1.4. ANISOTROPIC, HIGHLY MAGNETIC FIELD SENSITIVE, MAGNETOSTRICTION

If the expansion curve for pure nickel is blown up (fig. 12.1), it becomes clear that it is *sensitive to the application of an external magnetic field:* the latter induces an extra contraction (-36×10^{-6} at 300 K after saturation) along the field direction (//), and a twice smaller expansion ($+18 \times 10^{-6}$ at 300 K) in the plane perpendicular to the field ($\perp$). This is why this effect, discovered in 1842 by Joule on an iron rod, is called *anisotropic magnetostriction*.

The two magnetostriction effects we just introduced can be illustrated by figure 12.2. With respect to what it would be if it were not magnetic ($M_s = 0$), a spherical, isotropic sample suffers a relative change in volume $\delta V / V = \lambda^{\alpha,0}$ as soon as it is magnetic; this effect exists even if the sample is demagnetised, hence subdivided into magnetic (Weiss) domains ($H = 0$). $\lambda^{\alpha,0}$ is the *volume magnetostriction coefficient*.

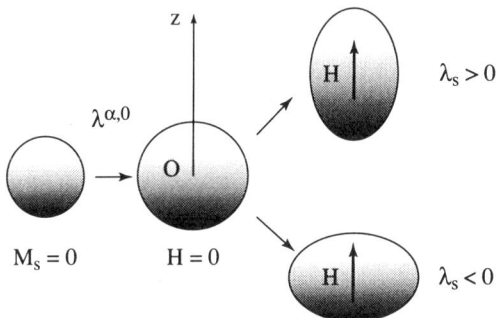

Figure 12.2 - The two principal magnetostriction modes observed in an isotropic material

This volume change is complemented, when a magnetic field $\mathbf{H}//Oz$ is applied, by an anisotropic deformation which changes the sphere into an ellipsoid of revolution around Oz, and occurs *at constant volume*. The relative variation in length measured along the field direction is called $\lambda_{//}$, and that observed along any direction perpendicular to Oz is called $\lambda_\perp$. If the starting demagnetised state is isotropic, the result obtained at technical saturation is $\lambda_{//} = -2\lambda_\perp = \lambda_s$. This defines the saturation magnetostriction constant λ_s of the isotropic material. The volume variation is indeed zero, as $\lambda_{//} + 2\lambda_\perp = 0$. Depending on the sign of λ_s, the sphere becomes a prolate ($\lambda_s > 0$) or oblate ($\lambda_s < 0$) ellipsoid. Below technical saturation, anisotropic magnetostriction is very sensitive to the magnetic field. Beyond saturation, a small variation of λ_s is sometimes observed: this is forced anisotropic magnetostriction $\partial \lambda_s / \partial H$, an effect often much smaller than the forced isotropic magnetostriction $\partial \lambda^{\alpha,0} / \partial H$ we already mentioned in section 1.1.

1.5. EFFECT OF A UNIAXIAL STRESS ON THE MAGNETIC PROPERTIES

A uniaxial stress applied to a magnetic material will generate an anisotropy along the direction $O\zeta$ of the stress, and the initial permeability will therefore be changed: this is the inverse effect to Joule anisotropic magnetostriction. Depending on the relative sign of the stress (tension or compression), and of λ_s, the magnetic moments will tend to align along or perpendicular to $O\zeta$. This effect plays an essential part in the performance of magnetic materials. It is usually a nuisance because it generally reduces the magnetic permeability, which is bad for soft materials. It is however put to use to make stress gauges, as we will see in chapter 18.

1.6. OTHER MAGNETOELASTIC EFFECTS

- Let us first mention a variant in Joule magnetostriction, observed when a helical magnetic field is applied to a ferromagnetic rod or wire, resulting in a twisting of the sample. This is the *Wiedemann effect*.

- Conversely, if a twist torque is exerted on a rod that is uniformly magnetised along its axis, the magnetization deviates from this axis. This is *the inverse Wiedemann effect*. An electric voltage difference appears between the ends of the rod, this is the *Matteuci effect*. These effects are used in many sensors.

- Applying a stress on a ferromagnetic rod of course deforms it according to linear elasticity. However this also changes its magnetization through the inverse magnetoelastic effect, and this induces a magnetostriction. Thus the apparent value of Young's modulus will be reduced: this is the *ΔE effect*.

- The *form effect* is a magnetostriction $\lambda^F(M^2)$ that is inhomogeneous even when magnetization is uniform, and which superimposes on the two principal modes we already introduced ($\lambda^{\alpha,0}$, and λ_s). This effect exists in all magnetic samples, even those with zero magnetostriction coefficients. It simply reflects the tendency to minimising the sum of the magnetostatic and elastic energies.

- Finally, *piezomagnetism* is an effect linear in the field, hence odd, whereas Joule magnetostriction is an even effect (an even function of the direction cosines of magnetization). This effect can be encountered only in some antiferromagnetic materials.

2. MICROSCOPIC ORIGIN OF THE MAGNETOELASTIC COUPLING

The effects we introduced can be explained by the very simple model developed by Néel [1] in the framework of localised electron ferromagnetism. The interactions (isotropic and anisotropic) responsible for magnetism are assumed to depend only on the direction of magnetization and on the distance r between two interacting atoms. The very symmetry of this problem (fig. 12.3) shows that the interaction energy between atoms A and B can be expanded into a series of Legendre polynomials:

$$\mathcal{E}_{AB}(r,\phi) = f(r) + g(r)\left(\cos^2\phi - \frac{1}{3}\right) + h(r)\left(\cos 4\phi - \frac{6}{7}\cos^2\phi + \frac{3}{35}\right) + ... \quad (12.1)$$

where ϕ is the angle between magnetization and the line AB.

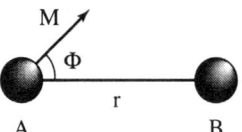

Figure 12.3 - A pair of atoms, and magnetization M

The first, isotropic, contribution, is just *Heisenberg's exchange interaction*. The other contributions describe *anisotropic exchange* interactions and *crystal field* interactions. The functions f(r), g(r), h(r) ... decrease rapidly with increasing distance. The special case of the dipole interaction, which decreases only as $1/r^3$, will be handled separately in section 4.7. The magneto-elastic coupling energy is obtained by summing (12.1) over all the nearest neighbours of each atom in a *deformed* magnetic solid. We will start out with the isotropic term, the treatment of which is simplest.

2.1. ISOTROPIC EXCHANGE MAGNETOELASTIC COUPLING

The exchange magnetoelastic coupling energy can be calculated by summing the isotropic exchange term, corresponding to the first term in expansion (12.1), over all the nearest neighbours:

$$\mathcal{E}_{ex} = f(r) = -J_{AB}(r) \langle S_A \cdot S_B \rangle \quad (12.2)$$

The contribution of further neighbours is assumed to be negligible, which is realistic because this interaction is short-ranged. J_{AB} is the Heisenberg exchange integral for a pair of nearest neighbour atoms, a distance r apart. For ferromagnetic materials, J_{AB} is positive. Expression (12.2) is scalar, and it involves only r. In a *cubic* crystal undergoing deformation, the magnetic energy density is found to be:

$$E_{magn} = -NzJ_{AB}(r) \langle S_A \cdot S_B \rangle$$

$$= -Nz \left[(J_{AB})_0 + r_0 \left(\frac{\partial J_{AB}}{\partial r} \right)_0 \frac{1}{3} \frac{\delta V}{V_0} + \frac{1}{2} r_0^2 \left(\frac{\partial^2 J_{AB}}{\partial r^2} \right)_0 \left(\frac{1}{3} \frac{\delta V}{V_0} \right)^2 \right] \langle S_A \cdot S_B \rangle \quad (12.3)$$

where N is the number of atoms per unit volume, z the number of nearest neighbours, and r_0 the interatomic distance which would be observed in the absence of magnetism. The elastic energy density is a quadratic function of $(\delta V/V_0)$, and it can be written, introducing the compressibility $\kappa = 3/c^\alpha$ of the material:

$$E_{el} = \frac{1}{2\kappa} \left(\frac{\delta V}{V_0} \right)^2 = \frac{c^\alpha}{6} \left(\frac{\delta V}{V_0} \right)^2 = \frac{c_{11} + 2c_{12}}{6} \left(\frac{\delta V}{V_0} \right)^2 \quad (12.4)$$

where the c_{ij} (c^α) are elastic constants (see the appendix at the end of this chapter).

The third term in (12.3), involving $(\delta V/V_0)^2$, thus represents a magnetic contribution to the elastic constants. This is one of the components of the morphic effect. We will not discuss it in detail because it is a difficult problem, involving both the theory of finite deformations and Lagrange elasticity tensors. For more details on this effect, the reader can look up reference [2]. Neglecting the small magnetic contribution to the c_{ij}, and minimising the sum $E_{el} + E_{magn}$ with respect to $\delta V/V_0$, it will be easy to calculate the volume magnetostriction. This will be done in section 3.

2.2. ANISOTROPIC MAGNETOELASTIC COUPLING

This coupling stems from anisotropic, short-range interactions with both magnetic and electrostatic origins such as anisotropic exchange (pseudo-dipolar and multipolar exchange), or crystal electric field effects. They all are rapidly varying functions of the distance between magnetic moment carriers. These interaction energies will, unlike exchange energy be sensitive to the orientation of the moments with respect to the crystal axes.

2.2.1. Principle of the coupling energy calculation

The anisotropic magnetoelastic coupling energy density can be obtained, to first order, by summing the second term in equation (12.1):

$$E_{anis} = [g(r_0) + (\partial g/\partial r)_{r_0} \delta r + ...][(\alpha_1 u + \alpha_2 v + \alpha_3 w)^2 - 1/3] + ... \quad (12.5)$$

over the nearest neighbours of each atom in a deformed material. The deformation is now anisotropic, and it is no longer described just by a change in the interatomic distance, hence in volume. It must be described with the elastic strain tensor ε_{ij}, which is defined in the appendix at the end of this chapter.

> *Warning:* the literature still contains far too often a description of strain through the Voigt components ($e_{ii} = \varepsilon_{ii}$, $e_{ij} = 2\varepsilon_{ij}$). As was pointed out forcibly by Callen *et al.* [3] in 1963, the e_{ij} do not form a tensor.

In equation (12.5), the α_i are the direction cosines of magnetization, and u, v and w those of the AB bond (fig. 12.3), referred *after deformation* to a set of axes defined in the *undeformed* material. δr, u, v and w can be expressed to first order as functions of the strains ε_{ij}, and of the β_i, the direction cosines of the AB bond *before* deformation, with respect to the same set of axes:

$$\delta r = r_0(\varepsilon_{xx}\beta_1^2 + 2\varepsilon_{yz}\beta_2\beta_3 + \text{circ. permut.}) \quad (12.6)$$

$$u = \beta_1[1 + \varepsilon_{xx}(1 - \beta_1^2) - \varepsilon_{yy}\beta_2^2 - \varepsilon_{zz}\beta_3^2 - 2\varepsilon_{yz}\beta_2\beta_3]$$
$$+ \varepsilon_{zx}\beta_3(1 - 2\beta_1^2) + \varepsilon_{xy}\beta_2(1 - 2\beta_1^2) - \omega_{zx}\beta_3 + \omega_{xy}\beta_2 \quad (12.7)$$

v and w can be deduced from the expression of u through circular permutation. The ω_{ij} are the components of the infinitesimal rotation. We will ignore it in what follows, because rotational effects only step in to second order, when describing the anisotropy torque [4] or to describe higher order magnetoelastic terms [2], due to the terms (...) which we ignored in equation (12.5).

2.2.2. The case of cubic symmetry

By summing equation (12.5), taking (12.6) and (12.7) into account, over crystals with simple cubic, body-centred cubic, and face-centred cubic lattices, the anisotropic magnetoelastic coupling energy, restricted to second order in α_i, can be written in all cases, per unit volume:

$$E_{mel} = B_1 \varepsilon_{xx}(\alpha_1^2 - 1/3) + 2 B_2 \varepsilon_{yz}\alpha_2\alpha_3 + \text{circ. permut.} \quad (12.8)$$

B_1 and B_2 are called the anisotropic magnetoelastic coupling coefficients. Table 12.1 shows the result of this calculation for these three lattices. N is the number of magnetic atoms per unit volume, $g = g(r_0)$, $g' = r_0(\partial g/\partial r)_0$ [1].

Table 12.1 - Magnetoelastic coupling coefficients for the three cubic lattices, in the pair model

	s.c	b.c.c.	f.c.c.
$B_1 = B^{\gamma,2}$	N.g'	(8/3) N.g	3 N.g + (1/2) N.g'
$B_2 = B^{\epsilon,2}$	2 N.g	(8/9) (N.g + N.g')	2 N.g + N.g'

Here again, the expression of magnetostriction will be found by minimising the sum of the elastic and magnetoelastic energy densities. However, the expression of the elastic energy as a function of the ε_{ij} is not diagonal, and it is simpler to use Callen's symmetrised notation. This consists in diagonalising the elastic energy, by introducing the deformation eigenmodes (see the appendix at the end of this chapter).

3. CALLEN'S SYMMETRISED NOTATION [3]

Using the identity:

$$ax + by + cz \equiv \frac{2}{3}\left(c - \frac{a+b}{2}\right)\left(z - \frac{x+y}{2}\right) + \frac{1}{2}(a-b)(x-y)$$

$$+ \frac{1}{3}(a+b+c)(x+y+z) \quad (12.9)$$

the first term in equation (12.8) can be altered, producing the expression of the magnetoelastic coupling energy density, to first order, and for a crystal with cubic symmetry:

$$E_{mel} = (B^\alpha/3)(\varepsilon_{xx} + \varepsilon_{yy} + \varepsilon_{zz})$$

$$+ B^{\gamma,2}\left[\frac{2}{3}\left(\varepsilon_{zz} - \frac{\varepsilon_{xx}+\varepsilon_{yy}}{2}\right)\left(\alpha_3^2 - \frac{\alpha_1^2+\alpha_2^2}{2}\right) + \frac{1}{2}(\varepsilon_{xx} - \varepsilon_{yy})(\alpha_1^2 - \alpha_2^2)\right]$$

$$+ 2 B^{\epsilon,2}(\varepsilon_{yz}\alpha_2\alpha_3 + \varepsilon_{zx}\alpha_3\alpha_1 + \varepsilon_{xy}\alpha_1\alpha_2) \quad (12.10)$$

We added the first term, describing the isotropic magnetoelastic coupling. This is the exchange magnetoelastic coupling (12.3) introduced in section 2.1. By matching the other two terms in expression (12.10) with equation (12.8), it is easy to see that $B^{\gamma,2} = B_1$ and $B^{\epsilon,2} = B_2$. These are the pseudo-dipolar and crystal field magnetoelastic coupling terms of order 2. Such *anisotropic* interactions can obviously not contribute to the *isotropic* coupling involving B^α.

The digit ^{2} that appears after γ and ε means these are coupling terms of second order with respect to the direction cosines of magnetization. Terms of order 4, 6 ... would appear if the expansion in (12.1) was extended to Legendre polynomials of order 4 [h(r)] or 6. Apart from a contribution to the magnetoelastic coupling (the coefficients $B^{\alpha,4}$, $B^{\gamma,4}$, $B^{\varepsilon,4}$, $B^{\alpha,6}$, $B^{\gamma,6}$, $B^{\varepsilon,6}$, $B^{\varepsilon,6'}$), we would also find an expression for the magnetocrystalline anisotropy energy involving $k^4(\alpha_2^2\alpha_3^2 + \alpha_3^2\alpha_1^2 + \alpha_1^2\alpha_2^2 - 1/5)$ for order 4, and $k^6[\alpha_1^2\alpha_2^2\alpha_3^2 - 1/11(\alpha_2^2\alpha_3^2 + \alpha_3^2\alpha_1^2 + \alpha_1^2\alpha_2^2 - 1/5) - 1/105]$ for order 6.

Thus, anisotropy and magnetoelastic coupling are two different manifestations of the same microscopic scale anisotropic couplings.

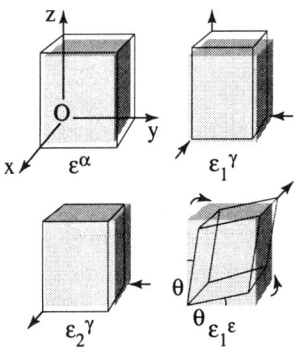

Figure 12.4
The normal modes of deformation

The symmetrised notation makes it possible to separate the deformation modes in cubic symmetry. The isotropic mode $\varepsilon^\alpha \sim (\varepsilon_{xx} + \varepsilon_{yy} + \varepsilon_{zz})$ does not lower the symmetry. The anisotropic modes:

$$\varepsilon_1^\gamma \sim [\varepsilon_{zz} - (\varepsilon_{xx} + \varepsilon_{yy})/2] \text{ and } \varepsilon_2^\gamma \sim (\varepsilon_{xx} - \varepsilon_{yy})$$

lower the symmetry by changing the lattice parameter, without changing the volume nor the angles, while the modes $\varepsilon_1^\varepsilon \sim \varepsilon_{yz}$, $\varepsilon_2^\varepsilon \sim \varepsilon_{zx}$ and $\varepsilon_3^\varepsilon \sim \varepsilon_{xy}$ lower the symmetry through shear, without altering the volume nor the lattice parameters (see fig. 12.4). Notation $^\alpha$, $^\gamma$, $^\varepsilon$, is based on symmetry group considerations, and refers to the irreducible representations of these groups [2, 3]. The nice feature in this notation is that it leads to anisotropy coefficients k_ℓ and to magnetoelastic coupling coefficients $B^{\mu,\ell}$ the thermal variations of which are simply expressed as a function of the reduced magnetization m of the material (see app. 6).

The formalism just sketched is based on the theory of symmetry groups developed by Bethe. It has the advantage of diagonalising the expression of Hooke's law. The elastic energy density is then expressed, for a crystal with cubic symmetry, in the form:

$$E_{el} = \frac{1}{2} c^\alpha \frac{1}{3} \left(\varepsilon_{xx} + \varepsilon_{yy} + \varepsilon_{zz} \right)^2 \quad (12.11)$$

$$+ \frac{1}{2} c^\gamma \left[\frac{2}{3} \left(\varepsilon_{zz} - \frac{\varepsilon_{xx} + \varepsilon_{yy}}{2} \right)^2 + \frac{1}{2} \left(\varepsilon_{xx} - \varepsilon_{yy} \right)^2 \right]$$

$$+ c^\varepsilon \left(\varepsilon_{yz}^2 + \varepsilon_{zx}^2 + \varepsilon_{xy}^2 \right)$$

where $c^\alpha = c_{11} + 2 c_{12}$, $c^\gamma = c_{11} - c_{12}$, and $c^\varepsilon = 2 c_{44}$ are elasticity coefficients.

This formalism has been extended to lower symmetries [2, 3], but it should be noted that it is then not possible to totally diagonalise the expression of the elastic energy. It also was extended to the case of isotropic materials [2], for which one can simply write: $c^\gamma = c^\varepsilon$ and $B^{\gamma,2} = B^{\varepsilon,2}$.

4. MAGNETOSTRICTION

4.1. EXCHANGE MAGNETOSTRICTION IN CUBIC SYMMETRY

Volume magnetostriction is found by minimising the sum $E_{el} + E_{magn}$ (eq. 12.3, and 12.4) with respect to $\delta V / V_0$

$$\delta V / V_0 = \lambda^{\alpha,0} = \frac{r_0 N z}{c^\alpha} \frac{\partial J_{AB}}{\partial r} <S_A \cdot S_B> \quad (12.12)$$

We note that, for a ferromagnetic material, $\lambda^{\alpha,0}$ is proportional to and has the same sign as the derivative $\partial J_{AB} / \partial r$. If there is no correlation between spins localised on the various sites, then $<S_A \cdot S_B>$ vanishes, and no exchange magnetostriction is observed, nor any anomaly in the elastic properties. This is the case of paramagnetism at very high temperature.

When cooling the material, the correlation function $<S_A \cdot S_B>$ grows while short range order generally appears above T_C. This is why the volume anomaly shown on figure 12.5 [5] for nickel already appears at 800 K, about 170 K above the Curie temperature $T_C = 631$ K. At low temperature, all the spins are ferromagnetically aligned, and the correlation function reaches its saturation value, $S_A S_B$, so that exchange magnetostriction and the morphic effects saturate.

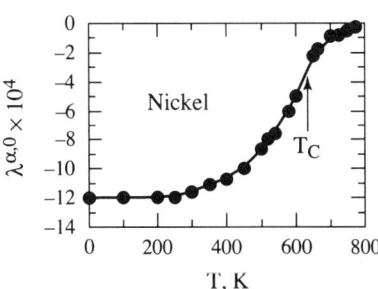

Figure 12.5
Exchange magnetostriction in metallic nickel [5]

Equation (12.12) is also valid for antiferromagnetic materials. Then, however, $<S_A \cdot S_B> = - S_A S_B$ at low temperature. This leads to negative volume magnetostriction if $\partial J_{AB}/\partial r$ is positive, contrary to the situation of ferromagnets. Figure 12.6 [6] shows the thermal variations of the thermal expansion coefficient for a series of oxides $U_xTh_{1-x}O_2$ (ThO_2 is diamagnetic and UO_2 is antiferromagnetic). The positive anomaly observed at the Néel temperature of UO_2 shows that this is where the thermal variation of $\lambda^{\alpha,0}$ is strongest.

Thus, this model gives a qualitative explanation for the thermal variation of volume magnetostriction. It therefore makes it possible to predict the order of magnitude and the sign of the magnetic anomalies in expansion, provided the variations with interatomic distance of the exchange integral are known.

Néel [7] discussed the variation of interaction energy with distance in detail (see fig. 4.9). It leads to a slightly negative volume magnetostriction for nickel, as we saw on figure 12.5, and a positive volume magnetostriction for nickel-iron alloys, in agreement with experiment. For iron, a negative volume magnetostriction would be expected. However, while $\lambda^{\alpha,0}(T)$ is strongly negative below 800 K, it becomes slightly positive near T_C. This shows the limitation of the localised electron model in the case of the 3d transition metals.

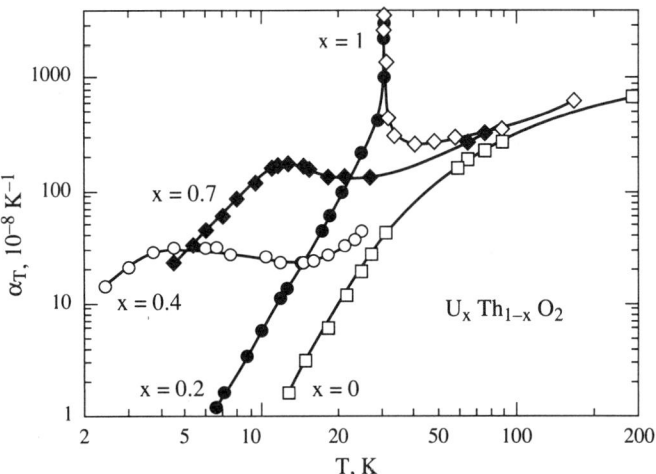

Figure 12.6 - Thermal variations of the linear thermal expansion coefficient in oxides $U_xTh_{1-x}O_2$ [6]

Forced volume magnetostriction is the variation of $\lambda^{\alpha,0}$ under a strong magnetic field. It consists in a change in volume due to the application of an external magnetic field. In the ordered state, applying a magnetic field on a ferromagnetic material will first orient all the spins in the same direction, without altering $<S_A \cdot S_B>$. Beyond technical saturation, the applied field, H, will strengthen the parallel orientation of S_A, and S_B, hence increase exchange magnetostriction. However, as this is but a small perturbation, the strain related to forced magnetostriction varies linearly with $|H|$:

forced magnetostriction is defined by $\partial \lambda^{\alpha,0}/\partial |H|$, which varies with temperature like $\partial <S_A \cdot S_B>/\partial |H|$. It therefore features maximum intensity (a lambda-shaped peak) near the ordering temperature, where the molecular field collapses while short range order remains appreciable. On the other hand, the deformation related to forced magnetostriction in the paramagnetic regime is a quadratic function of H, as is the magnetic energy density $\chi H^2/2$ from which it originates.

Figure 12.7 [8] illustrates this effect for a ferrimagnetic garnet. Forced magnetostriction remains very weak even at T_C (a few $10^{-12}/A \cdot m^{-1}$), as for most "normal" magnetic materials, oxides or metals. Comparing figures 3.17, 12.6 (for x = 1) and 12.7 shows the great similarity in behavior of the specific heat, of the magnetic contribution to the thermal expansion coefficient, and of the forced magnetostriction in the neighbourhood of magnetic transition temperatures.

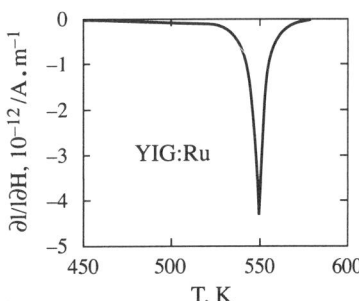

Figure 12.7
Forced magnetostriction of
$Y_3Fe_{4.976}Ru_{0.024}O_{12}$,
measured in 0.8 tesla [8]

We have discussed the exchange magnetoelastic effects in the simplest case, that of cubic materials. The approach can be generalised to other symmetries, but more parameters must be considered if symmetry is lower. Thus, there are two exchange integrals $J_{//}$ and $J_\perp$ in the case of materials with hexagonal or tetragonal symmetry, due to the different distances between nearest neighbours along the c axis and in the basal plane. For isotropic materials (glasses), the description could be expected to be very simple because of the high symmetry. Actually the analysis is made difficult by the fact that the sums over nearest neighbours must be replaced by integrals, because of the randomness in the geometrical positions of the nearest neighbours.

Finally, to conclude, the only technical applications of the exchange effects are related to a very special family of materials: the invar and elinvar families of alloys, characterised by strong exchange magnetostriction, both spontaneous and forced (a hundred times above normal). They will be dealt with in detail in chapter 18.

4.2. ANISOTROPIC MAGNETOSTRICTION IN CUBIC SYMMETRY

In Callen's formalism, magnetostriction is very easily calculated [3] by minimising the sum of the magnetoelastic (12.10) and elastic (12.11) energies. Magnetostriction is described by the six components of strain:

$$(\varepsilon_{xx} + \varepsilon_{yy} + \varepsilon_{zz}) = -B^{\alpha}/c^{\alpha}$$

$$\varepsilon_{zz} - \frac{\varepsilon_{xx} + \varepsilon_{yy}}{2} = -\frac{B^{\gamma,2}}{c^{\gamma}}\left(\alpha_3^2 - \frac{\alpha_1^2 + \alpha_2^2}{2}\right)$$

$$\varepsilon_{xx} - \varepsilon_{yy} = -\frac{B^{\gamma,2}}{c^{\gamma}}\left(\alpha_1^2 - \alpha_2^2\right) \quad (12.13)$$

$$\varepsilon_{yz} = -(B^{\varepsilon,2}/c^{\varepsilon})\alpha_2\alpha_3$$

$$\varepsilon_{zx} = -(B^{\varepsilon,2}/c^{\varepsilon})\alpha_3\alpha_1$$

$$\varepsilon_{xy} = -(B^{\varepsilon,2}/c^{\varepsilon})\alpha_1\alpha_2$$

Using identity (12.9), the relative expansion measured in a direction defined by its direction cosines β_i, which was classically expressed on the basis of expression (12.6), becomes:

$$\lambda = \frac{\delta l}{l} = \frac{1}{3}(\varepsilon_{xx} + \varepsilon_{yy} + \varepsilon_{zz}) + \frac{2}{3}\left(\varepsilon_{zz} - \frac{\varepsilon_{xx} + \varepsilon_{yy}}{2}\right)\left(\beta_3^2 - \frac{\beta_1^2 + \beta_2^2}{2}\right)$$

$$+ \frac{1}{2}(\varepsilon_{xx} - \varepsilon_{yy})(\beta_1^2 - \beta_2^2) + 2(\varepsilon_{yz}\beta_2\beta_3 + \varepsilon_{zx}\beta_3\beta_1 + \varepsilon_{xy}\beta_1\beta_2) \quad (12.14)$$

Taking equations (12.13) and (12.14) into account, the general expression for magnetostriction in cubic symmetry becomes:

$$\lambda = \frac{1}{3}\lambda^{\alpha,0}$$

$$+ \lambda^{\gamma,2}\left[\frac{2}{3}\left(\alpha_3^2 - \frac{\alpha_1^2 + \alpha_2^2}{2}\right)\left(\beta_3^2 - \frac{\beta_1^2 + \beta_2^2}{2}\right) + \frac{1}{2}(\alpha_1^2 - \alpha_2^2)(\beta_1^2 - \beta_2^2)\right]$$

$$+ 2\lambda^{\varepsilon,2}(\alpha_2\alpha_3\beta_2\beta_3 + \alpha_3\alpha_1\beta_3\beta_1 + \alpha_1\alpha_2\beta_1\beta_2) + \ldots \quad (12.15)$$

where: $\quad \lambda^{\alpha,0} = -B^{\alpha}/c^{\alpha}, \quad \lambda^{\gamma,2} = -B^{\gamma,2}/c^{\gamma}, \quad \lambda^{\varepsilon,2} = -B^{\varepsilon,2}/c^{\varepsilon} \quad (12.16)$

are the first magnetostriction coefficients of a cubic material. Higher order terms, associated with the coefficients $\lambda^{\alpha,4}$, $\lambda^{\gamma,4}$, $\lambda^{\varepsilon,4}$, $\lambda^{\alpha,6}$, $\lambda^{\gamma,6}$, $\lambda^{\varepsilon,6}$, $\lambda^{\varepsilon,6'}$ also exist [2], but they are usually small enough to be neglected, especially at room temperature: they very often decrease much faster with rising temperature than the coefficients of order 2. In what follows, we will ignore the isotropic exchange magnetostriction $\lambda^{\alpha,0}$ and the higher order coefficients. *Anisotropic Joule magnetostriction will therefore be essentially described by the two coefficients $\lambda^{\gamma,2}$ and $\lambda^{\varepsilon,2}$ for crystals with cubic symmetry.*

Thus, equation (12.15) describes the relative expansion ($\lambda = \delta l/l$) experienced by a crystal with cubic symmetry along a direction with direction cosines β_i when its magnetization is along the direction defined by the direction cosines α_i. If $\alpha_3^2 = \beta_3^2 = 1$, the magnetostriction measured in a parallel field along [001] can be written: $\lambda_{//} = (2/3)\lambda^{\gamma,2} = \lambda_{100}$, while magnetostriction $\lambda_{\perp}$ measured for $\beta_3 = 0$ will be $\lambda_{\perp} = -(1/3)\lambda^{\gamma,2}$, twice smaller, and with opposite sign. This is mode ε_1^{γ} in figure 12.4. λ_{100} denotes the relative expansion observed along [100] if the material is taken from the demagnetised state to saturation along this direction.

If magnetization is aligned along a three-fold symmetry direction [111], $\alpha_i^2 = 1/3$. Measuring the strain along this direction ($\beta_i = \alpha_i$) yields: $\lambda_{//} = (2/3)\lambda^{\epsilon,2} = \lambda_{111}$, while along a perpendicular direction ($\Sigma\alpha_i\beta_i = 0$) one finds: $\lambda_\perp = -(1/3)\lambda^{\epsilon,2}$. The sample then experiences the sum of three shears ($\epsilon_1^\epsilon + \epsilon_2^\epsilon + \epsilon_3^\epsilon$), i.e. a rhombohedral deformation. λ_{111} is thus the relative expansion observed along [111] when the material is taken from an isotropic demagnetised state to saturation along this direction. The coefficients λ_{100} and λ_{111} were the classical descriptors for magnetostriction before Callen's work [3].

In the demagnetised state, the moments are distributed among all the four-fold directions in the case $K_1 > 0$, and among all the three-fold directions if $K_1 < 0$. If the moments are equally distributed among the various easy magnetization directions, then the average deformation is zero, as shown by equation (12.15). This is referred to as an isotropic demagnetised state. The expressions we just calculated for $\lambda_{//}$ and $\lambda_\perp$ will only be provided by experiment if one starts out from such an isotropic demagnetised state, with no initial deformation of the sample.

4.2.1. *Measuring magnetostriction coefficients in cubic symmetry*

The measurement is performed by constraining the magnetic field to symmetry planes (001), (110) or (111) or to the principal directions perpendicular to these planes. The magnetic field is rotated within these symmetry planes, and the strains measured are analysed in terms of Fourier series in order to determine the magnetostriction coefficients through identification with equation (12.15). The relative strains can be measured by optical or capacitive dilatometry (sensitivity some 10^{-9}), or by extensometry, using strain gauges (sensitivity around 10^{-7}).

Figure 12.8 shows magnetostriction, measured along direction [1$\bar{1}$0] on a single crystal sphere of yttrium iron garnet ($Y_3Fe_5O_{12}$ or YIG). Under a magnetic field along [1$\bar{1}$0], $\lambda_{//}$ is negative. Under a perpendicular field, $\lambda_\perp$ depends on the orientation of the field in the (1$\bar{1}$0) plane.

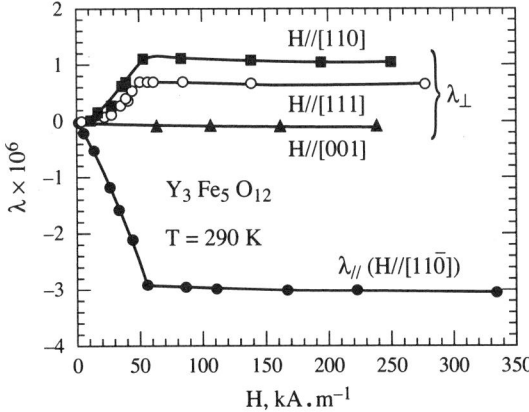

Figure 12.8 - Magnetostriction of a crystal of YIG measured along [1$\bar{1}$0]

Measuring the azimutal variation of $\lambda_\perp$ beyond saturation provides a sine curve (fig. 12.9) which shows the anisotropic character of magnetostriction. Saturation is reached for 50 kA.m^{-1}, exactly the value of the demagnetising field of the sphere, $M_s/3$, in this oxide with very weak magnetocrystalline anisotropy.

Note - *When the anisotropy is high, the magnetization may deviate from the applied field direction. The variations of magnetostriction must then be analysed as a function of the orientation of M, not of H.*

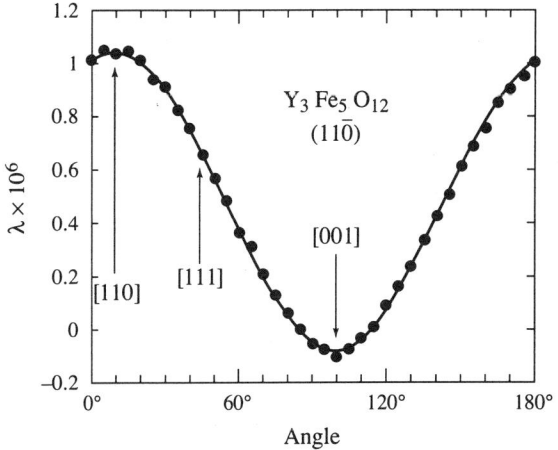

Figure 12.9 - Azimutal variations of the magnetostriction of a crystal of YIG, measured along [1$\bar{1}$0] with the magnetic field rotating in the plane perpendicular to this direction

In YIG, the fourth-order magnetostriction ($\lambda^{\alpha,4}$, $\lambda^{\gamma,4}$ and $\lambda^{\varepsilon,4}$), dipolar magnetostriction (shape effect: § 4.7) and forced magnetostriction (defined in § 1.1, and 1.4) are all negligible. Such situations are very frequent, and it is generally possible to restrict the expression of magnetostriction to the coefficients of order two ($\lambda^{\gamma,2}$, $\lambda^{\varepsilon,2}$).

4.2.2. Thermal variation of magnetostriction

Joule magnetostriction varies with temperature and vanishes at T_C. The model developed by Callen *et al.* under the localised moment assumption [3] predicts $B^{\gamma,2}(T)$ and $B^{\varepsilon,2}(T)$ to vary as $a\,m^2 + b\,\hat{I}_{5/2}[\pounds^{-1}(m)]$, where m is the reduced magnetization of the material, $\pounds(x)$ Langevin's function, and $\hat{I}_{5/2}(x)$ a modified Bessel function which varies like m^3 at low temperature, and $3/5\,m^2$ near T_C (see app. 6). This leads to an $a\,m^2 + b\,m^3$ regime at low temperature, tending to a purely m^2 regime at high temperature (T ≥ T_C) [2, 3].

The term involving "a" represents so-called *two-ion coupling*, or anisotropic exchange coupling. The term involving "b" describes *single ion* coupling, or the crystalline electric field effect. When coefficients a and b are opposite in sign, and of the same order of magnitude, the thermal variations can feature extrema or changes in sign. Generally, single-ion coupling coefficient is much larger than the other one, and, since the elasticity coefficients (c^γ, c^ε) generally have very little variation with temperature, one expects moderate thermal variations for the magnetostriction of order 2 of ferromagnets, with an approach to the Curie temperature linear in $(T - T_C)$, since magnetization varies as $(T - T_C)^\beta$ near T_C, while $\beta \sim 0.5$ with magnetostriction varying as the square of magnetization in this temperature zone.

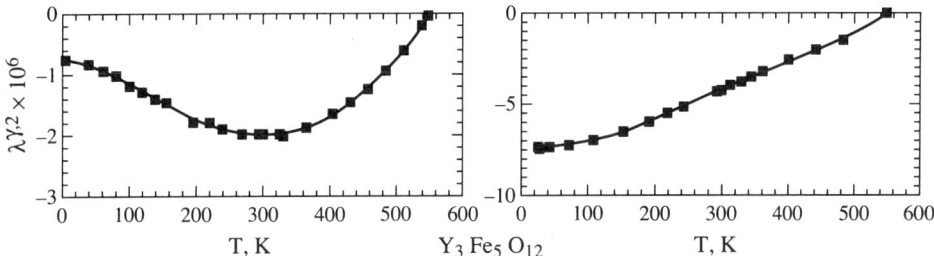

Figure 12.10 - Thermal variations of the magnetostriction coefficients of YIG: left, measured along [100]; right, measured along [111], after [9]

For ferrimagnets, Callen's model still applies. However, anomalies may occur, because of the competition between the sublattices, involving sometimes the cancellation and change in sign of magnetostriction. Thus, for the magnetostriction of YIG for which the thermal variations are given on figure 12.10, one should note the minimum around 300 K on the $\lambda^{\gamma,2}(T)$ curve. This is easily explained, as the thermal variations of the sublattice magnetizations are very different, and their contributions to the coefficient $B^{\gamma,2}$ are opposite in sign [9].

On the other hand, in the itinerant model of magnetism, any anomaly in the band structure at the Fermi level can totally disrupt the thermal behavior of magnetostriction, and no prediction is possible. As shown in figure 12.11 for iron, $\lambda^{\gamma,2}(T)$ shows a maximum at 800 K, and $\lambda^{\varepsilon,2}$ varies as m^{14}, which could not be predicted.

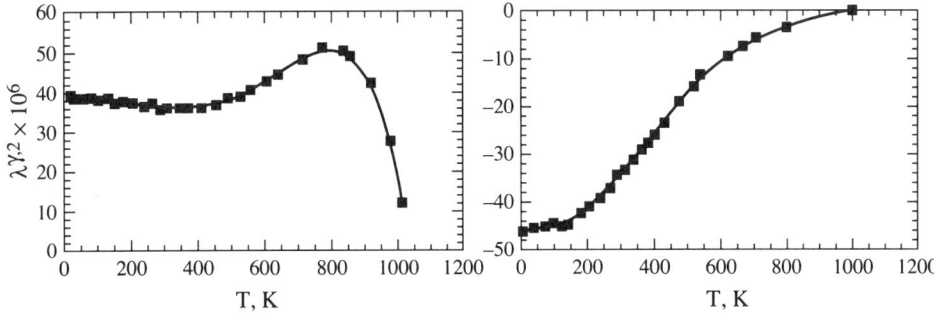

Figure 12.11 - Thermal variation of the magnetostriction coefficients of a single crystal of iron: left, measured along [100]; right, measured along [111], after [11]

4.2.3. Variation of anisotropic magnetostriction under magnetic field in cubic symmetry

The variations of magnetostriction under the influence of a magnetic field are closely related to the domain structure in the material. Consider a crystal of iron, starting from an isotropic demagnetised state and moving toward saturation along direction [001]. What will be the strain along this direction?

The magnetization can be expressed: $M_z = M_s <\alpha_3>$ and, since $\beta_3 = 1$, magnetostriction becomes: $\lambda_{//} = \lambda^{\gamma,2} (<\alpha_3^2> - 1/3)$. Since $<\alpha_3^2> \neq <\alpha_3>^2$, there is normally no relation between the variations under the influence of a magnetic field of these two quantities, contrary to what is sometimes mentioned in the literature. There can be a correlation only in the special case where $<\alpha_3> = \alpha_3$; this situation is possible if the magnetization process involving 180° wall displacement ends before the rotation process starts.

Let us look more closely at this possibility.

- Initially, we had an equal distribution of the moments among six families of domains (Néel's phases). Then $<\alpha_3> = 0$, $<\alpha_3^2> = 1/3$, $M_z = 0$, $\lambda_{//} = 0$.
- Under a weak magnetic field, the moments pointing towards [00$\bar{1}$] flip over to [001]. As one sixth of the moments flip from $-M_s$ to $+M_s$, we get $M_z = M_s/3$, and $\lambda_{//} = 0$ since λ is an even function of the α_i. There is then no correlation between $\lambda(H)$, and $M(H)$.
- On increasing the field, the two-thirds of the moments that were in the (001) plane will all rotate in phase towards direction [001]. During this process, the magnetization and magnetostriction are:

$$M_z = M_s \left(\frac{1}{3} + \frac{2}{3} x \right) \quad ; \quad \lambda_{//} = \frac{2}{3} \lambda^{\gamma,2} x^2 \qquad (12.17)$$

where x represents the α_3 for the four families of moments that are moving, and changes from 0 to 1 during this process. Then the variations of M_z and of $\lambda_{//}$ are correlated, as shown by the solid line in figure 12.12.

- At saturation, all the moments are parallel to [001], $M_z = M_s$, and $\lambda_{//} = 2/3\, \lambda^{\gamma,2}$.

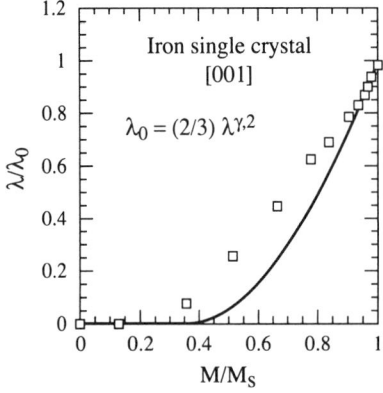

Figure 12.12
Variations in magnetostriction measured along [001] as a function of the reduced magnetization, for a single crystal of iron

The experimental points are from Webster [10]. The theoretical curve is given by equations (12.17) through elimination of x.

Far from saturation, the experimental points lie far above the theoretical curve, because the moment rotation process starts before all the 180° wall displacements are completed. Conversely, for $M \geq 0.9\,M_s$ all possible 180° wall displacements are seen to have occurred, since the experimental points lie on the theoretical curve. This example, in spite of its simplicity, shows it is futile to insist on relating the variations under a magnetic field of magnetostriction in a material with cubic symmetry with those of magnetization. This correlation will occur only in uniaxial crystals, and only when the magnetization process consists in coherent moment rotation (as e.g. if the field is applied along the c axis, when the basal plane is the easy magnetization plane).

4.3. MAGNETOSTRICTION IN HEXAGONAL CRYSTALS

Because hexagonal symmetry is lower than cubic symmetry, more coefficients are necessary to correctly describe the magneto-elastic effects (see chap. 10). Thus, the expression of magnetostriction for crystals with hexagonal symmetry can be written, to order 2:

$$\lambda = \frac{1}{3}\lambda_1^{\alpha,0} + \lambda_2^{\alpha,0}\left(\beta_3^2 - \frac{1}{3}\right) + \left[\frac{1}{3}\lambda_1^{\alpha,2} + \lambda_2^{\alpha,2}\left(\beta_3^2 - \frac{\beta_1^2 + \beta_2^2}{2}\right)\right]\left(\alpha_3^2 - \frac{1}{3}\right)$$

$$+ \lambda^{\varepsilon,2} \times \left[\frac{1}{2}\left(\beta_1^2 - \beta_2^2\right)\left(\alpha_1^2 - \alpha_2^2\right) + 2\beta_1\beta_2\alpha_1\alpha_2\right] \quad (12.18)$$

$$+ 2\lambda^{\zeta,2} \times \left(\beta_2\beta_3\alpha_2\alpha_3 + \beta_3\beta_1\alpha_3\alpha_1\right)$$

Note - *In the literature, modes ε and ζ are sometimes improperly designated as γ and ε, in disagreement with Bethe's group-theoretical notation. On the other hand, the expansion should be carried out to sixth degree terms in α in order to describe the azimuthal variation of λ in the basal plane. This effect is usually negligible.*

The expression of magnetostriction in equation (12.18) is based on the same definition of α_i and β_i as in cubic symmetry. $\lambda_1^{\alpha,0}$ is the volume magnetostriction, but now another mode of exchange magnetostriction appears, viz $\lambda_2^{\alpha,0}$, describing a change in the c/a ratio of the hexagonal structure. This deformation mode does not lower the symmetry, and that is why it is denoted with the superscript α. All the other terms are contributions to Joule anisotropic magnetostriction. We ignored higher order terms because, here too, the order 2 coefficients are dominant.

Figure 12.13 shows the thermal variations of the order 2 coefficients for the magnetostriction of metallic cobalt. They are seen to be an order of magnitutde higher than those of iron, with rather regular thermal variations, except for $\lambda_1^{\alpha,2}$ which changes sign near 500 K.

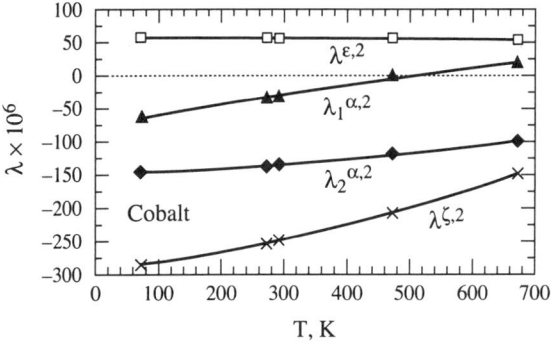

*Figure 12.13
Thermal variations
of the magnetostriction
coefficients of cobalt [12]*

4.4. MAGNETOSTRICTION OF ISOTROPIC MATERIALS

The symmetry of these materials is very high, and two coefficients, $\lambda^{\alpha,0}$ for exchange magnetostriction and λ_s for Joule magnetostriction, are enough to describe all the magnetoelastic effects *for an intrinsically isotropic material*. This means an amorphous alloy with no growth anisotropy (an isotropic metallic glass). Figure 12.2 shows the magnetostrictive deformation observed for an isotropic sphere. The calculation of magnetostriction from the pair model, as discussed in the case of cubic symmetry, would be very awkward. The sums over nearest neighbours would be replaced by integrals [2], as mentioned while discussing exchange magnetoelastic coupling at the end of section 4.1. An elegant and much simpler approach to the analytic expression of magnetostriction in isotropic materials consists in writing the condition or conditions for isotropy of magnetostriction for a crystallized material. In the case of cubic symmetry, for example, equation (12.15) can also be written, using identity (12.10):

$$\lambda = \frac{1}{3}\lambda^{\alpha,0} + \lambda^{\gamma,2}\left(\alpha_1^2\beta_1^2 + \alpha_2^2\beta_2^2 + \alpha_3^2\beta_3^2 - \frac{1}{3}\right) \\ + 2\lambda^{\varepsilon,2}\left(\alpha_2\alpha_3\beta_2\beta_3 + \alpha_3\alpha_1\beta_3\beta_1 + \alpha_1\alpha_2\beta_1\beta_2\right) \quad (12.19)$$

If $\lambda^{\gamma,2} = \lambda^{\varepsilon,2} = (3/2)\lambda_s$, this magnetostriction becomes isotropic, since the deformation is then *independent of the chosen measuring direction*. This yields:

$$\lambda = \frac{1}{3}\lambda^{\alpha,0} + \frac{3}{2}\lambda_s\left(\cos^2\theta - \frac{1}{3}\right) \quad (12.20)$$

where θ is the angle between the magnetization and the measuring direction. Starting from a hexagonal crystal, the isotropy conditions would be written as: $\lambda_1^{\alpha,0} = \lambda^{\alpha,0}$ and $\lambda_2^{\alpha,2} = \lambda^{\varepsilon,2} = \lambda^{\zeta,2} = (3/2)\lambda_s$, all the other coefficients being zero.

Note - *The case of materials that are isotropic through compensation, i.e. made up of a large number of randomly oriented crystallites (untextured polycrystalline materials) is completely different. Equation (12.20) does not provide a good description of their magnetostriction. The correct way of dealing with the magnetostriction of polycrystalline materials is to treat it as an average over the various crystallites. We return to this case in section 4.5.*

4.4.1. Magnetostriction curves for an intrinsically isotropic material

Magnetization is an odd function of the applied magnetic field H, but magnetostriction is an even function of H. The λ(H) hysteresis loop therefore has a butterfly shape, as shown in figure 12.14.

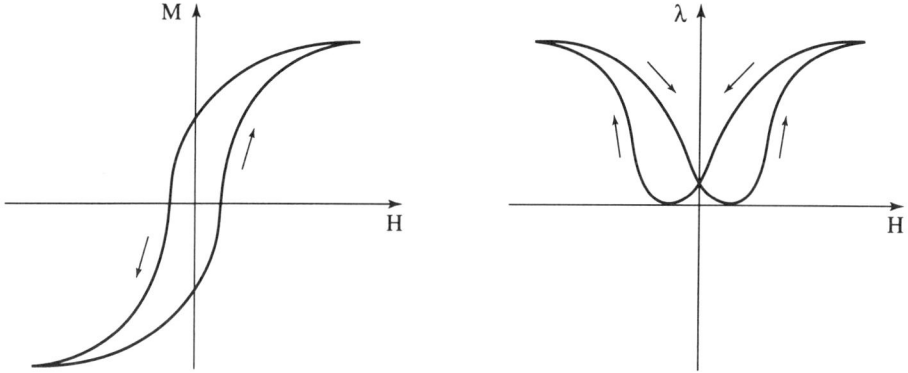

Figure 12.14 - Comparison between hysteresis loops for magnetization, and for magnetostriction. M(H) is odd, while λ(H) is even with respect to the applied field

The two loops have points in common: the loops are narrow and saturate quickly for soft materials, they are broad with difficult saturation for hard materials. It would however be wrong to believe that λ is a quadratic function of M: these quantities are correlated only in 90° wall displacement mechanisms, or for the form effect. Figures 12.15 and 12.16 show the variation, under the influence of a magnetic field, of magnetostriction, for two amorphous alloys, with compositions $Fe_{78}Si_9B_{13}$ and $Fe_4Co_{71}Si_{10}B_{15}$.

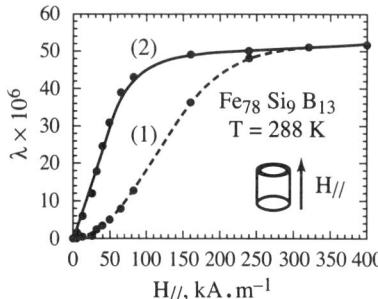 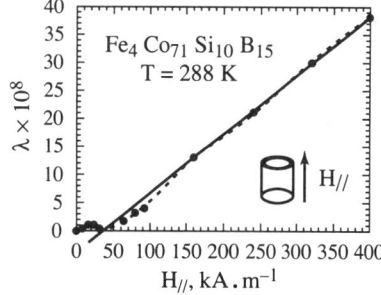

Figure 12.15 - Magnetostriction of an amorphous iron alloy: (1) as quenched, (2) annealed at 670 K

Figure 12.16 - Magnetostriction curve for an amorphous, as-quenched, cobalt-iron alloy

In both cases, the ribbon-shaped samples were wound and pasted. λ is measured in the field direction, along the axis of the cylinder thus obtained.

Annealing changes notably the aspect of the curve for the first alloy, which shows the influence of thermal and magnetic history on the performance of a given material. The second alloy is nominally a zero-magnetostriction material. Indeed, its Joule magnetostriction is two orders of magnitude smaller than for the first, which makes it possible to detect the effects of the dipole interaction (the shape effect, responsible for the little bump around 25 kA.m^{-1}), and of volume magnetostriction ($\lambda^{\alpha,0}$), which increases linearly with the applied field beyond the saturation magnetic field ($\approx$ 80 kA.m^{-1}): this is the *forced magnetostriction* effect ($\partial \lambda^{\alpha,0}/\partial H$) which we mentioned earlier. The value of the *saturation magnetostriction* must, when it is very small, be measured by extrapolating to zero internal field the linear part of the $\lambda(H_0)$ curve, in order to get rid of the effect of forced magnetostriction.

The magnetostriction that would have been measured perpendicular to the direction of the field would have shown, in the case of Fe$_{78}$Si$_9$B$_{13}$, the same shape, with opposite sign, and half the deformation because here the anisotropic Joule contribution is dominant and operates at constant volume. In the case of Fe$_4$Co$_{71}$Si$_{10}$B$_{15}$ it is the other way round: the dominant effect is volume magnetostriction, and, since this is isotropic, the two curves ($\lambda_{//}$ and $\lambda_\perp$) would have been practically identical. However, if dipole and volume effects are assumed negligible, $\lambda_\perp$ is equal to $-\lambda_{//}/2$ only if the initial strain is zero, i.e. if the orientation of the magnetic moments is randomly distributed in the demagnetised state.

4.4.2. Influence of the demagnetised state on the magnetostriction curve

Annealing in a magnetic field, or applying a uniaxial stress, can induce a magnetic anisotropy such that, in the demagnetised state, the moments are oriented preferentially in a given direction or plane. When a field is applied, the anisotropic magnetostrictive deformations are usually changed, and the ratio $\lambda_{//}/(-2\lambda_\perp)$ may become notably different from unity, the value expected for an isotropic demagnetised state.

Figure 12.17 shows how magnetostriction curves change as a function of the nature of the demagnetised state for an Fe$_{78}$Si$_9$B$_{13}$ alloy. With the measurement direction defined (Oz), the field is applied first along Oz ($\lambda_{//}$) then perpendicular to Oz ($\lambda_\perp$):

- in case (a), the result is $\lambda_{//} = -2\lambda_\perp$, because the demagnetised state is isotropic;
- in case (b), the easy magnetization direction happens to be perpendicular to the measuring direction for the strain. Thus all the moments rotate by 90° under a longitudinal magnetic field (θ goes over from $\pi/2$ to 0 in equation (12.20)) yielding $\lambda_{//} = (3/2)\lambda_s$, while $\lambda_\perp$ is zero because θ remains equal to $\pi/2$ in a transverse magnetic field;
- finally, in case (c), the easy magnetization direction is parallel to the strain measuring direction, and the situation is symmetrical with respect to the former case: $\lambda_{//}$ is zero while $\lambda_\perp = -(3/2)\lambda_s$.

The nature of the demagnetised state thus plays an essential role in the shape of the magnetostriction curves, and determining λ_s involves applying the magnetic field in succession along the measuring direction and along a perpendicular direction: $\lambda_{//} - \lambda_\perp = (3/2)\lambda_s$, in order to eliminate the effect of a doubtful initial state.

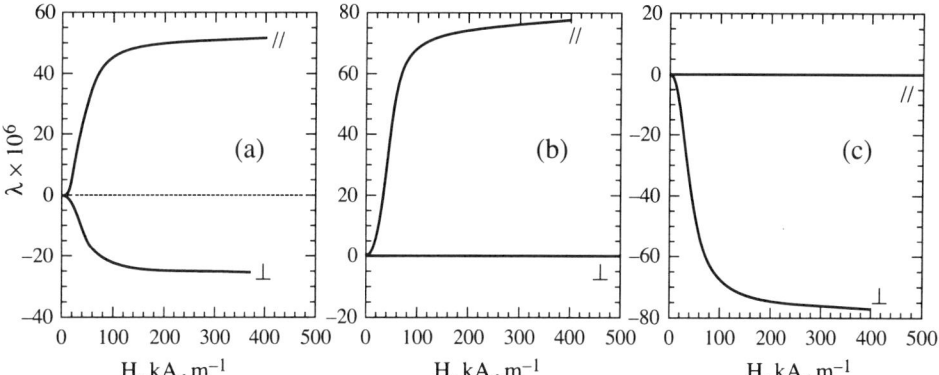

Figure 12.17 - Influence of the anisotropy of the demagnetised state on the magnetostriction curves of an amorphous ribbon
(a) Isotropy - (b) Easy direction perpendicular to the measuring direction
(c) Easy direction parallel to the measuring direction for the strain.
In all three cases, $\lambda_{//} - \lambda_{\perp} = (3/2)\lambda_s$.

We saw on figure 12.14 the "butterfly-shaped hysteresis loop" obtained when cycling the magnetic field alternatively from negative to positive values. There is another, just as basic, magnetostriction loop. This is the loop obtained by applying a magnetic field along the elongation measuring direction, taking this field to zero *without changing its sign*, and then applying, under the same conditions, a field perpendicular to the measuring direction.

Figure 12.18 shows the result: the remanent strain will not be the same depending on whether the saturated state corresponded to a perpendicular ($\lambda_R^{\perp}$) or a parallel ($\lambda_R^{//}$) field. We note that $\lambda_{//} - \lambda_{\perp}$ will be $(3/2)\lambda_s$ only if the same demagnetised state was used as the starting point.

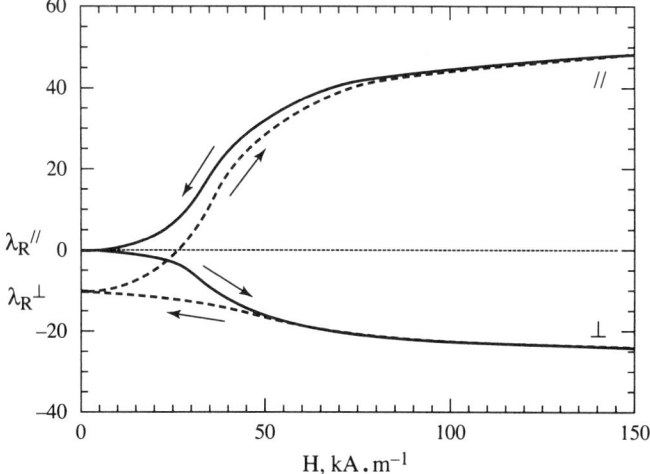

Figure 12.18 - Hysteresis effects on the magnetostriction curves for an amorphous alloy

4.4.3. Effect of temperature on magnetostriction

As for crystallized materials, the magnetostrictive characteristics of amorphous materials vary with temperature and collapse near T_C with the same rules. However, in these materials, there is also a gradual, irreversible change in the $\lambda(H)$ curves. This is due to the relaxation of stress and to the change in λ_s induced by structural relaxation or recrystallization.

Figure 12.19 [13] shows the thermal variation of the magnetostriction coefficient for two amorphous alloys. This variation is perfectly reversible, since it is related to the thermal variation of spontaneous magnetization.

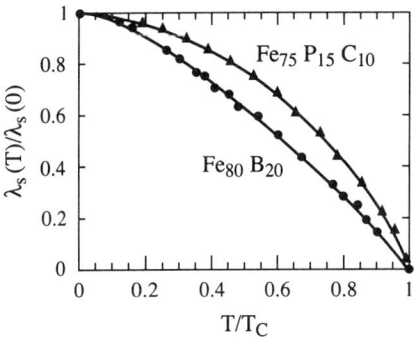

Figure 12.19 - Thermal variation of the reduced magnetostriction in two amorphous iron alloys, after [13]

Amorphous alloys are in a metastable state, and irreversible variations in magnetostriction are, as we saw, superimposed on the purely magnetic thermal variations. It is wise, when only the magnetic thermal effects are to be measured, to make a preliminary anneal of the sample at a temperature T_R slightly below the recrystallization temperature T_X in order to activate structural relaxation. Later measurements must not involve temperature above T_R.

4.5. MAGNETOSTRICTION IN POLYCRYSTALLINE MATERIALS

Most materials are used in industry as polycrystals, i.e. as aggregates or randomly oriented crystallites (or grains). Such materials are *isotropic by compensation*, or untextured polycrystals. The variations in magnetostriction they feature with magnetic field or temperature can involve anomalies, due to the *local anisotropic character* of each crystallite. Contrary to what can sometimes be read, it is *illegal* to treat it as an intrinsically isotropic material, and *to describe its magnetoelastic behavior through a single constant λ_s* for Joule magnetostriction. The behavior of each crystallite must be correctly analysed, and then an average can be taken over all the grains. Magnetostriction must then be calculated as the average of the deformation in each grain.

In the simple case where the parallel magnetostriction ($\lambda_{//}$) is required, i.e. the relative deformation of the material in the direction of magnetization, it is equivalent to consider one single grain and to calculate the average value of magnetostriction over

12 - MAGNETOELASTIC EFFECTS

all directions in space. For grains with cubic symmetry, ignoring volume magnetostriction, we get at saturation, letting $\alpha_i = \beta_i$:

$$\lambda_{//}^{sat} = \lambda^{\gamma,2}(<\beta_1^4> + <\beta_2^4> + <\beta_3^4> - \frac{1}{3})$$
$$+ 2\lambda^{\varepsilon,2}[<(\beta_2\beta_3)^2> + <(\beta_3\beta_1)^2> + <(\beta_1\beta_2)^2> - \frac{1}{3}] \quad (12.21)$$
$$= \frac{4}{15}\lambda^{\gamma,2} + \frac{2}{5}\lambda^{\varepsilon,2}$$

Comparing this result with equation (12.20) with the condition $\cos^2\theta = 1$ could lead us to believe that Joule magnetostriction is now described by a single coefficient, as in a truly isotropic material:

$$\lambda_s = (4/15)\,\lambda^{\gamma,2} + (2/5)\,\lambda^{\varepsilon,2} \quad (12.22)$$

But the approximation of isotropic magnetostriction is legitimate only when:
- the two coefficients $\lambda^{\gamma,2}$ and $\lambda^{\varepsilon,2}$ have the same sign and the same order of magnitude,
- the elastic constants also nearly satisfy the isotropy condition ($c^\gamma \approx c^\varepsilon$)
- and the polycrystalline material does not feature any crystallographic texture.

If these three conditions are not fulfilled, it is *very dangerous* to amalgamate a polycrystalline material with an intrinsically isotropic (glassy) material, as we now show.

Figure 12.20 [14] shows the variations with magnetic field of $\lambda_{//}$ for a polycrystalline, untextured disk of pure iron. This magnetostriction is, under a very weak stress, positive in a weak field (as was observed by Joule in 1842), it then decreases and changes sign at 50 kA.m^{-1} to reach a negative saturation value.

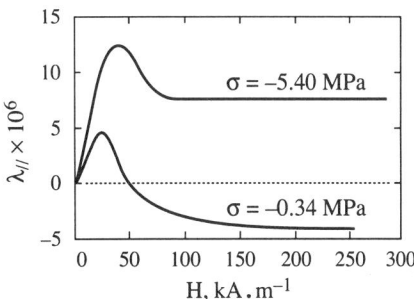

Figure 12.20 - Variations of the magnetostriction of an iron disk under magnetic field [14]

Under a stronger compressive stress, the curve moves upward, because the initial state is no more isotropic. This behavior is very different from the monotonous variation with field that would be expected from an isotropic material. It arises because the magnetostriction coefficients of iron have different signs ($\lambda^{\gamma,2} = +36 \times 10^{-6}$ and $\lambda^{\varepsilon,2} = -34 \times 10^{-6}$ at room temperature). With growing, but still small, magnetic field, the moments flip over from one easy direction to another one, nearer the direction of the field. This involves 90° wall displacement, hence, as shown by equation (12.15), implies constant $\lambda^{\gamma,2}$, and therefore positive magnetostriction. As saturation is approached, the moments are forced to align along the field direction. Some of them will have to rotate away from the four-fold axes, hence to rotate towards three-fold, hard magnetization, directions. Then the negative constant $\lambda^{\varepsilon,2}$ comes into action, which explains the change in sign of $\lambda_{//}$. The saturation value for the magnetostriction of polycrystalline iron is close to the prediction of equation (12.22): $\lambda_s = (4/15)\,\lambda^{\gamma,2} + (2/5)\,\lambda^{\varepsilon,2} = -4 \times 10^{-6}$.

Conversely, in the case of nickel for which both $\lambda^{\gamma,2}$ and $\lambda^{\varepsilon,2}$ are negative (at room temperature $\lambda^{\gamma,2} = -93 \times 10^{-6}$ and $\lambda^{\varepsilon,2} = -36 \times 10^{-6}$), the variations with applied magnetic field are close to those of an isotropic, untextured material, as shown by figures 12.21 (polycrystalline sphere), and 12.22 (a long polycrystalline wire); then the model of isotropic magnetostriction can be applied without too much danger. The wire ($N \approx 0$) magnetises in an external field a thousand times smaller than the sphere ($N = 1/3$).

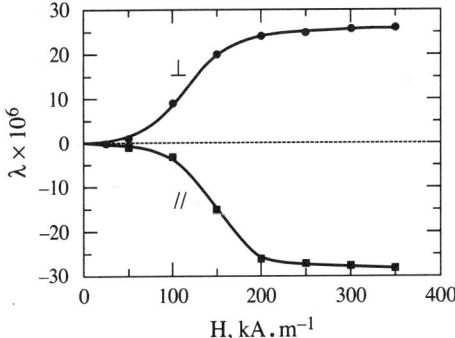

Figure 12.21 - Magnetostriction of a nickel sphere

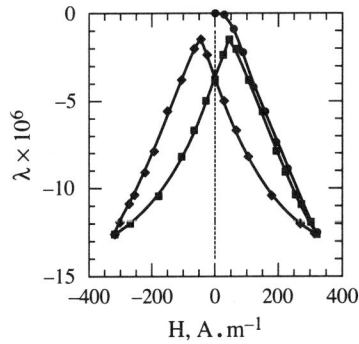

Figure 12.22 - Magnetostriction of a fine nickel wire [15]

4.6. MAGNETOELASTIC EFFECTS OBSERVED ON SURFACES AND IN THIN FILMS

Magnetostriction is sensitive to a material's symmetry, and the *symmetry is broken* at the surface of a solid. For example, symmetry becomes tetragonal near a (001) surface, and hexagonal near the (111) surface of a cubic crystal. As a result, a contribution from the surfaces superimposes on the bulk magnetostriction.

This surface contribution is negligible unless the number of atoms at the surface becomes sizeable compared to the total number of atoms in the sample. This is the case for thin films. Measuring magnetostriction for films with different thicknesses makes it possible to separate the contributions from the bulk and the surface, provided the density remains the same for all samples.

In the case of thin films, the problem is complicated by epitactic stresses. A thin film with only a few atomic layers is necessarily deposited on a rigid substrate. After deposition, the substrate is usually observed to bend, revealing the existence of huge stress (up to about ten GPa) in the thin film. This results, via the anharmonic effects related to these stresses, in a change in the magnetostriction coefficients. These strains furthermore generate a strong magneto-elastic anisotropy (§ 6.2) which can overcome the magnetocrystalline anisotropy, and thus change the easy magnetization directions.

The special interest of ultrathin films is that they make it possible to investigate the magneto-elastic coupling in new structures, which do not exist in bulk form. Examples are face-centred cubic iron, or artificial structures consisting of alternations of different, possibly non-miscible, metals. Such investigations are gaining momentum.

Because they are very thin, films are necessarily rigidly coupled to a much thicker substrate, and they cannot deform freely under the action of a magnetic field. The field will lead to the deformation of the *bimorph* system consisting of the film and its substrate. For example, an expansion along the field direction will occur together with a contraction in the perpendicular direction (see fig. 12.23-a). Thus the initially plane surface of the film will undergo an anticlastic deformation, i.e. take on a saddle shape.

In the case of bulk magnetostrictive materials, the component of displacement used is usually along the applied field. On the other hand, for thin film systems, figure 12.23-b shows that the useful displacement (here along Oz) is perpendicular to the plane of the bimorph, hence to the direction of the applied field (Ox).

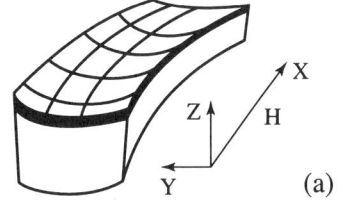
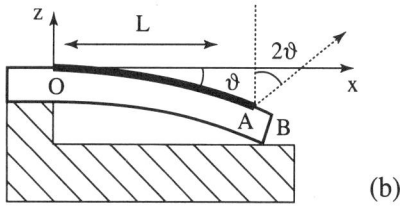

Figure 12.23 - Deflection under a magnetic field of a double strip consisting of a non-magnetic substrate and a ferromagnetic film: (a) saddle-like deformation of the bimorph; (b) longitudinal median section of the bimorph

The deformation suffered by a bimorph consisting of a non-magnetic substrate of thickness t_s, on which a very thin ferromagnetic film ($t_f \ll t_s$) with length L is deposited, can be expressed in terms of the shear modulus G_s of the substrate, of the Poisson coefficients (ν_f and ν_s) of the film and substrate and of the α_i, the direction cosines of magnetization. The angle ϑ measuring the deflection of the end of the double strip can be written as:

$$\vartheta = 3\frac{t_f}{t_s^2}\frac{L}{G_s}b(H) \qquad (12.23)$$

with: $\quad b(H) = \frac{1}{2}B^{\gamma,2}\left[\left(\alpha_1^2 - \alpha_2^2\right) - \frac{(1-\nu_s)(1+\nu_f)}{(1+\nu_s)(1-\nu_f)}\left(\alpha_3^2 - \frac{1}{3}\right)\right]$

This deformation can be measured by reflecting a laser beam on the end of the bimorph. The deflection (2ϑ) of the reflected beam can be detected using a photo-sensitive double diode [16].

We note that, while the magnetostriction coefficient $\lambda^{\gamma,2} = -B^{\gamma,2}/c^{\gamma}$ was the adequate parameter for describing magnetoelastic effects in the bulk, this is no more true for thin films. The deflection of the bimorph is expressed in terms of $-B^{\gamma,2}/G_s$. Hence the relevant parameter is now $B^{\gamma,2}$.

4.7. DIPOLAR MAGNETOSTRICTION, OR FORM EFFECT

When a sample is magnetised, it acquires a dipolar energy. It then tends to deform in order to minimise this energy, under the influence of *magnetic forces*. The analytic solution to this problem was given exactly only for a *spherical sample* submitted to a uniform magnetic field. Gersdorf showed that this sample undergoes surface stresses than can be expressed, at any point on the surface, through [17]:

$$\mathbf{F}_v = (1/2)\,\mu_0 M^2 (\cos^2 \psi) \mathbf{n} \tag{12.24}$$

where ψ is the angle between magnetization and the normal to the surface $\mathbf{n}$.

Such stresses create non-uniform stresses in the sphere. At the centre, they can be represented by an equation similar to that describing Joule magnetostriction.

Thus, for a crystal with **cubic symmetry** and **only in its centre**, the strain related to the form effect is described by equation (12.15) with three coefficients:

$$\lambda^{\alpha,0} = \frac{1}{2}\mu_0 M^2 s^{\alpha}$$

$$\lambda^{\gamma,2} = \frac{1}{2}\mu_0 M^2 s^{\gamma}(1+3A) \tag{12.25}$$

$$\lambda^{\varepsilon,2} = \frac{1}{2}\mu_0 M^2 s^{\varepsilon}(1+3B)$$

where:
$$A = \frac{s^{\alpha} - s^{\gamma}}{2s^{\alpha} + 13s^{\gamma} + 6s^{\varepsilon}} \quad B = \frac{s^{\alpha} - s^{\varepsilon}}{2s^{\alpha} + 4s^{\gamma} + 15s^{\varepsilon}} \tag{12.26}$$

Here, $s^{\alpha} = s_{11} + 2s_{12} = 1/c^{\alpha}$, $s^{\gamma} = s_{11} - s_{12} = 1/c^{\gamma}$ and $s^{\varepsilon} = (1/2)s_{44} = 1/c^{\varepsilon}$.

Away from the center of the sample, such a description is no more possible, the strain is variable. The solution consists in integrating all these strains to calculate the average strain of the sphere over a diameter. We give the result of the calculation, under the same assumption, viz a spherical crystal with cubic symmetry, when the magnetic field rotates from the measuring direction to a perpendicular direction, for all three principal symmetry directions. Using the same notations as before, we find [2]:

$$(\lambda_{//} - \lambda_{\perp})^F_{[001]} = \frac{1}{6}\mu_0 M^2 s^{\alpha}(1-2A) + \frac{1}{3}\mu_0 M^2 s^{\gamma}(1+4A) \tag{12.27}$$

$$(\lambda_{//} - \lambda_\perp)^F_{[111]} = \frac{1}{6}\mu_0 M^2 s^\alpha (1 - 2B) + \frac{1}{3}\mu_0 M^2 s^\varepsilon (1 + 4B) \qquad (12.28)$$

$$(\lambda_{//} - \lambda_\perp)^F_{[110]} = \frac{1}{6}\mu_0 M^2 s^\alpha \left[1 - \frac{1}{2}(A + 3B) + \frac{1}{2}(A + B)\cos 2\vartheta\right]$$

$$+ \frac{1}{12}\mu_0 M^2 s^\gamma [1 + A + 3B + (1 + A - B)\cos 2\vartheta] \qquad (12.29)$$

$$+ \frac{1}{4}\mu_0 M^2 s^\varepsilon [1 + A + 3B - (1 - 3A + 3B)\cos 2\vartheta]$$

In all three cases, the same expression for volume magnetostriction is found, and it is identical to that found at the center of the sphere:

$$(\lambda_{//} + 2\lambda_\perp)^F = \frac{1}{2}\mu_0 M^2 s^\alpha \qquad (12.30)$$

We note that the strain becomes isotropic (but still not uniform) when the elastic isotropy condition $s^\gamma = s^\varepsilon$ is fulfilled.

The form effect can be neglected in quite a number of materials with weak magnetization. Thus for nickel, with $M_s = 480$ kA.m^{-1}, $(\lambda_{//} + 2\lambda_\perp)^F = 0.26 \times 10^{-6}$, while $(\lambda_{//} - \lambda_\perp)^F$ is, in units of 10^{-6}, 0.8 along [001], 0.45 along [111], and $(0.4 + 0.12 \cos 2\vartheta)$ along [110]. These values are negligible compared to the Joule magnetostriction coefficients, $\lambda^{\gamma,2} = -95 \times 10^{-6}$ and $\lambda^{\varepsilon,2} = -42 \times 10^{-6}$.

For iron, the situation is different, because the Joule magnetostriction coefficients are weaker and the magnetization stronger. With $M_s = 1{,}710$ kA.m^{-1}, we obtain $(\lambda_{//} + 2\lambda_\perp)^F = 3.64 \times 10^{-6}$ while $(\lambda_{//} - \lambda_\perp)^F$ is, in units of 10^{-6}, 5.3 along [001] and 2.8 along [111], to be compared with $\lambda^{\gamma,2} = 36 \times 10^{-6}$ and $\lambda^{\varepsilon,2} = -34 \times 10^{-6}$. The shape effect reaches 15% of $\lambda^{\gamma,2}$, and this is no more negligible.

Finally, for the compound GdAl$_2$, coefficient $\lambda^{\gamma,2}$ is extremely small (0.5×10^{-6}) while the difference $(\lambda_{//} - \lambda_\perp)^F$ is nearly isotropic, with value 3.6×10^{-6}, i.e. 7 times the value of $\lambda^{\gamma,2}$. It is clearly essential not to forget it.

The reader will find details on the calculation of this effect in reference [2]. For other sample shapes and other symmetries, the calculations are even more complicated [18].

This effect is a nuisance in transformers because it contributes, along with Joule magnetostriction, to the vibration of the magnetic circuit under variable induction. We note that, during 180° wall displacement, the form effect still contributes to acoustic emission because it is proportional to magnetization squared, while Joule magnetostriction induces no strain. Even for a "zero magnetostriction" material, for which all the magnetostriction coefficients are zero, this effect remains a source of losses. It results from forces exerted by the magnetic charge distributions on the material, simply because it is magnetised.

4.8. NOTES ON THE MEASUREMENT OF MAGNETOSTRICTION

4.8.1. Internal stresses related to magnetostriction

As suggested by figure 12.24, the presence of 90° (or 71° and 109°) walls in a ferromagnetic crystal may, in some cases, generate internal stresses.

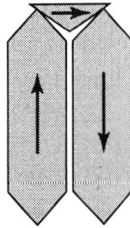

Figure 12.24 - Incompatibility for 90° domains

Assuming the magnetostriction to be positive, we see that, in a single crystal, each domain expands in the direction of the field and contracts along the perpendicular direction. This means there may occur an incompatibility between the strain in domains with perpendicular magnetization directions. Since the voids suggested in the figure cannot occur, the material would then be submitted to strong stresses at the walls to ensure the continuity of the material at all points. A non-negligible elastic energy density would thus be stored in the region near the walls, and would add to the exchange and dipolar energies in defining the most favorable domain structure.

Actually, this situation usually does not arise. The wall takes on, to within details that make the situation even more favorable, the (110) orientation if the domains magnetizations are along [100] and [010]. This is a Nye wall, the equivalent of a tilt boundary in metallurgy. Thanks to a tiny relative rotation, the lattice in both domains match perfectly, with no stress involved, and each domain can take on the spontaneous magnetostriction.

Another source for stress can be encountered in a polycrystalline material such as iron, in which the magnetostriction coefficients $\lambda^{\gamma,2}$ and $\lambda^{\varepsilon,2}$ have different signs. At saturation, all the moments are parallel to the direction Oz along which the field was applied. Consider two neighbouring grains, oriented along [001] and [111], respectively. One will tend to expand and the other to contract. The stresses generated by such a situation will somewhat alter the value of the "saturation magnetostriction" λ_s, and this change will depend on the material's elastic anisotropy [2]: the value measured will not exactly be the strain described by equation (12.22).

4.8.2. Accurate determination of Joule magnetostriction

We now analyse the components of the $\lambda(H)$ curve for a spherical sample, in which magnetization is homogeneous (fig. 12.25).

At low field ($H < H_m$), the magnetization process consists in the displacement of 180° walls, hence without Joule magnetostriction and with only the shape effect λ^F

(usually weak). Joule magnetostriction then develops when 90° (71° and 109°) walls move and when moments rotate, as H grows from H_m to H_D, with the internal field remaining zero. Joule magnetostriction saturates for $H > H_D$, when the field dominates over magnetic anisotropy. The internal field is no more zero and starts counteracting the thermal disorder of spins. This induces a forced magnetostriction, which can be expressed as: $(1/3) \partial \lambda^{\alpha,0}/\partial H + \partial \lambda_s /\partial H$ for $\lambda_{//}$, and $(1/3) \partial \lambda^{\alpha,0}/\partial H - (1/2) \partial \lambda_s /\partial H$ for $\lambda_\perp$. The straight lines representing forced magnetostriction intersect the vertical line through H_D at points A and B. The distance AB is exactly $(3/2) \lambda_s + (\lambda_{//} - \lambda_\perp)^F$, while $A_0 B_0$ would here provide a smaller value. Determining Joule magnetostriction thus implies extrapolating the experimental curves to the value H_D of the applied field, then subtracting out the shape effect, which can be calculated for a spherical sample. This determination will of course be correct only if both the $\lambda_{//}(H)$ and $\lambda_\perp(H)$ curves start from the same demagnetised state (see § 4.4.2).

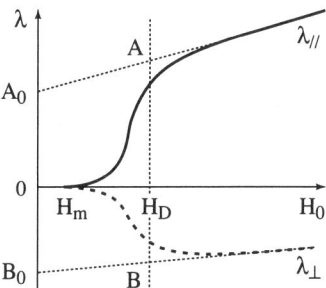

Figure 12.25 - Magnetostriction of an isotropic sphere as a function of applied field H//Oz

5. TWISTING OF A MAGNETISED WIRE CARRYING A CURRENT: THE WIEDEMANN EFFECT

Till now, we dealt with magnetostriction in a uniformly magnetised material, because this is the simplest case. Another geometry deserves attention, that of a ferromagnetic wire, or bar with circular cross-section, carrying a magnetization distribution with helical symmetry. This situation can be realised by superimposing on a uniform longitudinal magnetic field, created by a solenoid surrounding the sample, the circular field created by an electric current passing through the sample. This circular field is zero on the axis of the wire and maximum at its surface. The resulting field is helical and it twists the sample. The twist is zero on the axis of the wire and it grows with distance from the axis. This is an essentially non-uniform effect.

This twisting effect, called the "Wiedemann effect" for the physicist who discovered the phenomenon in 1862, is often used in a dynamical regime to generate shear waves in ferromagnetic bars, tubes or wires. For an isotropic wire, the amplitude of the effect will be described by the quantity $\xi = \phi/\ell$, the rotation per unit length of the wire. It can be shown [2] that:

$$\xi = 3 \lambda_s j / 2 H_{//} \qquad (12.31)$$

where j is the current density in the wire and $H_{//}$ is the magnetic field applied along the axis of the sample, *provided the magnetic field due to j remains small with respect to $H_{//}$*.

Figure 12.26 shows the elegant experiment performed in 1919 by Pidgeon on an annealed nicked wire, 1 mm in diameter, submitted to a magnetic field $H_{//} = 22.3 \, kA \cdot m^{-1}$.

The twist is described by equation $\xi = -6.74 \times 10^{-5} - 2.40 \times 10^{-9} j$ (rad.m^{-1}) if the current density j is expressed in A.m^{-2}. This yields for the magnetostriction coefficient λ_s the value -35.7×10^{-6}, in excellent agreement with the most recent results ($\lambda_s = -36 \times 10^{-6}$). This effect was also used to determine the magnetostriction coefficient for amorphous metallic ribbons: formula (12.31) also applies in this case.

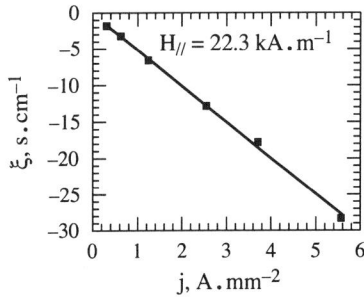

Figure 12.26 - Wiedemann effect for a nickel wire [19]

The twist ξ, expressed in arcseconds per centimetre, varies linearly with the current density.

6. INVERSE MAGNETOELASTIC EFFECTS AND THE ΔE EFFECT

After describing the direct magnetoelastic effects, we now show how the magnetic properties of a material can be altered through the application of mechanical stresses.

6.1. EFFECT OF HYDROSTATIC PRESSURE ON A MAGNETIC MATERIAL

Hydrostatic pressure is isotropic, hence it cannot lower the symmetry of a material. It will therefore change only the value of the Curie temperature and of the magnetic moment. Since the latter can be anisotropic in a material with uniaxial symmetry, the variation in moment under hydrostatic pressure can be different along the c axis and in the basal plane for such materials.

The variations in Curie temperature under pressure are usually small, except for materials featuring the Invar effect ($\partial T_C / \partial P = -35 \, K \cdot GPa^{-1}$ for $Fe_{64}Ni_{36}$ and $-24 \, K \cdot GPa^{-1}$ for the amorphous alloy $Fe_{80}B_{20}$) [20]. There exist fewer data on the pressure variations of magnetic moments, because these are difficult measurements.

Thermodynamic relations have been established between forced magnetostriction, the pressure variation of magnetic moment, and the pressure variation of T_C as expressed by the logarithmic derivative $\Gamma = \partial \ln(T_C) / \partial \ln(V)$. Thus [17]:

$$\mu_0 (\partial \mathfrak{m} / \partial p) = -(\partial V / \partial H) = -V(\partial \lambda^{\alpha,0} / \partial H) \quad (12.32)$$

$$\frac{\partial \lambda^{\alpha,0}}{\partial H}(T) = \frac{m \dfrac{\partial \lambda^{\alpha,0}}{\partial H}(0) - 3 \dfrac{\mu_0 T}{c^\alpha V} \left(\dfrac{\partial \mathfrak{m}}{\partial T} \right)_p \Gamma}{1 - 3\alpha_T T \Gamma} \quad (12.33)$$

where $m = M(T)/M(0)$ is the reduced magnetization at temperature T, and $\mathfrak{m}$ the magnetic moment associated to volume V of magnetised material; c^α is $c_{11} + 2 c_{12}$ and α_T is the linear thermal expansion coefficient. The first relation shows that forced volume magnetostriction is proportional to the pressure variation of magnetic moment. The second relation shows that it features a maximum around T_C (where $\partial \mathfrak{m} / \partial T$ becomes strongly negative), this maximum being the sharper the larger is Γ: Γ is less than one in most materials, it is 5 for the amorphous alloy $Fe_{80}B_{20}$ and 17 for the crystallised alloy $Fe_{64}Ni_{36}$ (Invar®).

6.2. EFFECT OF A UNIAXIAL STRESS ON THE MAGNETIZATION CURVE

We now discuss an essential effect, which has a strong influence on the performance of magnetic materials and has often been used to make sensors. A uniaxial stress deforms the material and this generates a magnetic anisotropy proportional to the strain and alters the magnetic permeability. Understanding the mechanism of this stress-induced anisotropy is essential for mastering its effects.

6.2.1. Material with cubic symmetry

The anisotropic part of the magnetoelastic coupling energy (12.10) can be rewritten, taking (12.16) into account, in the form:

$$E_{mel} = -\lambda^{\gamma,2} \sigma \left[\frac{2}{3} \left(\alpha_3^2 - \frac{\alpha_1^2 + \alpha_2^2}{2} \right) \left(\gamma_3^2 - \frac{\gamma_1^2 + \gamma_2^2}{2} \right) + \frac{1}{2}\left(\alpha_1^2 - \alpha_2^2\right)\left(\gamma_1^2 - \gamma_2^2\right) \right]$$

$$- 2 \lambda^{\varepsilon,2} c^\varepsilon (\varepsilon_{yz}\alpha_2\alpha_3 + \varepsilon_{zx}\alpha_3\alpha_1 + \varepsilon_{xy}\alpha_1\alpha_2) \quad (12.34)$$

We saw in section 3 that the formalism used here diagonalises Hooke's law, which results in a linear relationship between stress (σ) and strain (ε):

$$c^\gamma \left(\varepsilon_{zz} - \frac{\varepsilon_{xx} + \varepsilon_{yy}}{2} \right) = \sigma_{zz} - \frac{\sigma_{xx} + \sigma_{yy}}{2}$$

$$c^\gamma (\varepsilon_{xx} - \varepsilon_{yy}) = (\sigma_{xx} - \sigma_{yy}) \quad (12.35)$$

$$c^\varepsilon \varepsilon_{yz} = \sigma_{yz} \quad c^\varepsilon \varepsilon_{zx} = \sigma_{zx} \quad c^\varepsilon \varepsilon_{xy} = \sigma_{xy}$$

Since a uniaxial stress σ applied along a direction $[\gamma_1,\gamma_2,\gamma_3]$ can be described by the tensor $\sigma_{ij} = \sigma\gamma_i\gamma_j$, equation (12.34) can also be written as:

$$E_{mel} = -\lambda^{\gamma,2}\sigma\left[\frac{2}{3}\left(\alpha_3^2 - \frac{\alpha_1^2+\alpha_2^2}{2}\right)\left(\gamma_3^2 - \frac{\gamma_1^2+\gamma_2^2}{2}\right) + \frac{1}{2}\left(\alpha_1^2-\alpha_2^2\right)\left(\gamma_1^2-\gamma_2^2\right)\right]$$

$$- 2\lambda^{\varepsilon,2}\sigma(\alpha_1\alpha_2\gamma_1\gamma_2 + \alpha_2\alpha_3\gamma_2\gamma_3 + \alpha_3\alpha_1\gamma_3\gamma_1) \qquad (12.36)$$

This is the *analytic expression for the magnetic anisotropy induced by a uniaxial stress in a crystal with cubic symmetry.*

When σ is applied along a four-fold axis, e.g. [001], this anisotropy is uniaxial and can be written simply as:

$$E_{mel} = -\lambda^{\gamma,2}\sigma\left(\alpha_3^2 - 1/3\right) = K_u \sin^2\varphi + \text{Const} \qquad (12.37)$$

where φ is the angle between magnetization and the direction in which the stress is applied, with $K_u = \lambda^{\gamma,2}\sigma$.

In the same way, when σ is applied along a three-fold axis [111], this anisotropy is uniaxial. Setting $\alpha_1\gamma_1 + \alpha_2\gamma_2 + \alpha_3\gamma_3 = \cos\varphi$ and $K_u = \lambda^{\varepsilon,2}\sigma$, we see that:

$$E_{mel} = -\lambda^{\varepsilon,2}\sigma(\alpha_3^2 - 1/3) = K_u \sin^2\varphi + \text{Const} \qquad (12.38)$$

Outside these two very special cases, equation (12.36) does not usually reduce to uniaxial anisotropy. This is due to the tensor character of magnetostriction.

6.2.2. Isotropic material

In this case, the isotropy condition ($\lambda^{\gamma,2} = \lambda^{\varepsilon,2}$) together with identity (12.9) make it possible to rewrite equation (12.36) using $K_u = (3/2)\lambda_s\sigma$:

$$E_{mel} = -\frac{3}{2}\lambda_s\sigma\left(\cos^2\varphi - \frac{1}{3}\right) = K_u \sin^2\varphi + \text{Const} \qquad (12.39)$$

Here again, we have a uniaxial anisotropy induced by stress. We now discuss in detail the physical consequences of such an anisotropy.

When it is submitted to a magnetic field, an isotropic material with positive magnetostriction coefficient λ_s expands along the direction of the field, i.e. the direction in which the magnetic moments align. If it is submitted to a tensile mechanical stress ($\sigma > 0$), it elongates further, which favors the alignment of moments in the direction of the tension ($K_u > 0$).

Conversely, if λ_s is negative, a compressive stress ($\sigma < 0$) will give the same effect as the application of a field: here again the moments will align in the direction along which the stress is applied, but this time the stress will have the opposite sign compared to the tensile case.

Returning to the case of an isotropic material with positive λ_s, the application of a compressive stress ($\sigma < 0$) will tend to oppose the alignment of magnetic moments along the direction of the stress, and will therefore favor the settling of the moments in a plane perpendicular to this direction. Such an easy magnetization plane will also be favored by a tensile stress in the case where λ_s is negative.

Figure 12.27 schematically shows these various types of behavior. Starting from an isotropic distribution of magnetic moments in the absence of field or stress, we find an anisotropic distribution through the mere application of a stress, but the magnetization remains zero because there are as many moments along one direction as in the opposite direction. Applying a magnetic field will flip the moments by 180° in the "easy magnetization direction" situation, and rotate them gradually by 90° in the "easy magnetization plane" configuration.

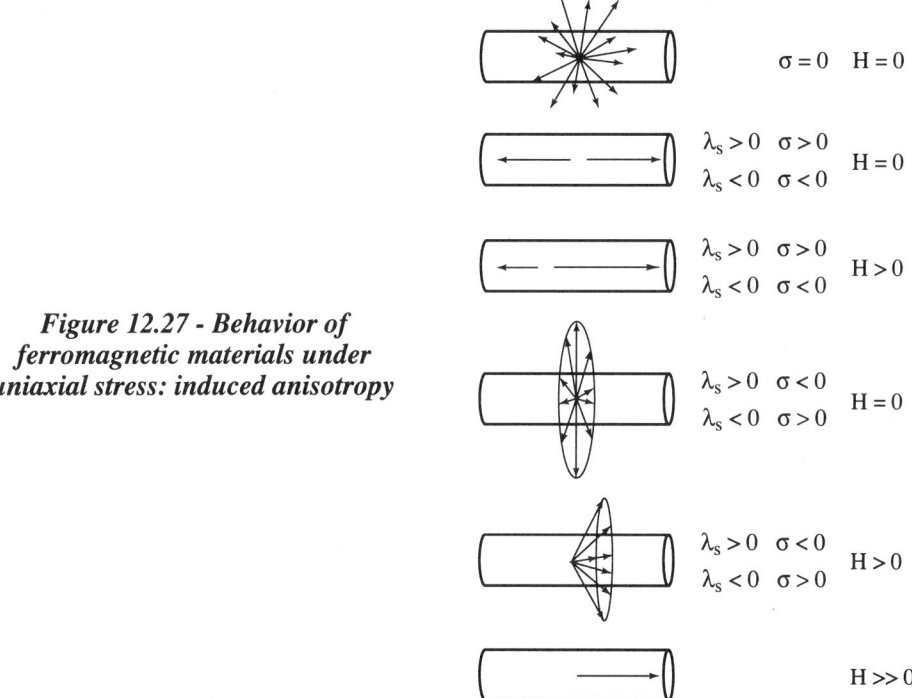

Figure 12.27 - Behavior of ferromagnetic materials under uniaxial stress: induced anisotropy

Depending on the configuration, saturation will thus be easier or more difficult to reach, because the stress will have created either a direction or a plane of easy magnetization. The magnetic permeability will hence be either enhanced or reduced depending on whether the sign of $\lambda_s \sigma$ is > 0 or < 0. The stress will have induced a magnetic anisotropy.

As an illustration, figure 12.28 shows the influence of tensile stress on the magnetization curve of an amorphous ribbon of the metallic glass $Co_{58}Fe_5Ni_{10}Si_{11}B_{16}$. Although its magnetostriction coefficient is very small, $\lambda_s = -0.17 \times 10^{-6}$, the effect is appreciable because the permeability of this metallic glass is very high under zero stress (because of the absence of anisotropy). The behavior is rather simple, because the material is intrinsically isotropic.

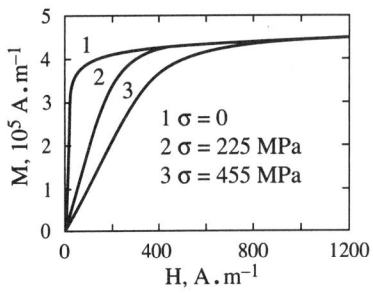

Figure 12.28 - Magnetization curves under tension of a metallic glass with negative magnetostriction

6.2.3. A polycrystalline material

For a polycrystalline material, even in the absence of texture, the problem becomes more complicated, because, as emphasised in section 4.5, the approximation of isotropic magnetostriction is not at all justified. Here again, it is useful to discuss the case of iron, because its magnetostriction is positive along the four-fold directions ($\lambda^{\gamma,2} > 0$) and negative along the three-fold directions ($\lambda^{\varepsilon,2} < 0$). In the demagnetised state, the moments are equally distributed along the various directions of easy magnetization. The application of stress will, for each crystallite, favor some of the six easy directions: the four-fold directions closer to the direction along which the strain is applied, or the plane perpendicular to this direction, depending on whether $\lambda^{\gamma,2}\sigma$ is positive or negative. As in the discussion above, the initial permeability will therefore be increased or reduced through the application of stress, depending on the sign of the product $\lambda^{\gamma,2}\sigma$. But, when the magnetic field and/or the stress is increased, the effect will reverse, i.e. $\partial\mu/\partial\sigma$ will change sign, because the moments will rotate away from the four-fold axes. This will signal that a combination of $\lambda^{\gamma,2}\sigma$, and of $\lambda^{\varepsilon,2}\sigma$ is coming into action, instead of only $\lambda^{\gamma,2}\sigma$.

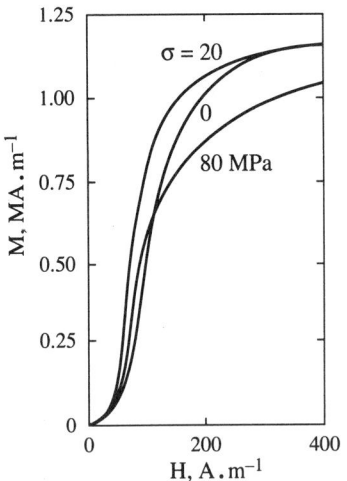

Figure 12.29 - Magnetization curves for iron under tensile stress [21]

Figure 12.29 [21] shows the magnetization curve of iron under zero stress, then for two different values of stress. Under 200 A.m^{-1}, a tensile stress of 20 MPa increases the magnetization, while a stress of 80 MPa reduces it. Beyond 300 A.m^{-1}, a tensile stress, however weak it is, always decreases the magnetization, while a weak tension increases the magnetization at low field. The sign reversal of $\partial\mu/\partial\sigma$ is known as Villari reversal. This complex behavior makes iron a doubtful candidate as the basis for linear stress sensors. Some industrial applications are possible, but under carefully determined conditions.

Hysteresis phenomena make the situation even more complicated. Thus, two very different magnetization curves under stress are observed depending on whether the stress is applied before the field (σ, H) or after (H, σ). For example, the magnetization of Permalloy (an alloy with 68% nickel in iron) can vary, under H = 3 A.m^{-1} and σ = 40 MPa, from 150 kA.m^{-1} (σ, H process) to 800 kA.m^{-1} (H, σ process). It is thus essential to specify the exact conditions of such an experiment. Finally, the effects we just described assume that the material is submitted to mechanical stress that remains below the elastic limit. Above this limit, the magnetization curve undergoes considerable irreversible changes.

Thus the performance of a material is very sensitive to the internal stress condition. If a material includes internal stresses with random orientation and amplitude, the resulting magnetic anisotropy will also be variable from point to point. It will then be more difficult to saturate the material: this is the negative aspect of anisotropy, which one tries to counteract, in metals and alloys, through annealing treatments designed to relax internal stresses and through the search for "zero magnetostriction" materials. On the other hand, it can be profitable, in some cases, to create a uniaxial anisotropy by applying for example a uniform stress along a given direction.

Finally, beyond the elastic limit, the introduction of plastic deformation into a ferromagnetic material also changes its magnetic properties, but irreversibly. A strain-hardening induced anisotropy is then observed.

6.3. EFFECT OF TWIST ON A MAGNETISED BAR (THE INVERSE WIEDEMANN AND THE MATTEUCI EFFECTS)

In section 5, we described the direct Wiedemann effect: a helical magnetic field applied to a magnetic sample twists the material.

Conversely, when a twisting torque is applied to a magnetised shaft, its magnetization is changed: this is the *inverse Wiedemann effect*. Figure 12.30 shows the distribution of stress in a cylindrical shaft submitted to twist: it is positive at +45°, negative at –45°. Through the inverse magneto-elastic effect, the permeability is reduced along +45° and increased along –45°, in the case $\lambda_s < 0$.

The difference in dynamic permeability between these two directions can generate, in a sensor, a signal proportional to the torque acting on the shaft, even if it rotates during the measurement. This is the principle of the contactless torquemeters based on the inverse Wiedemann effect. The direct Wiedemann effect is obtained by passing an electric current along a magnetised shaft. A second inverse effect could be anticipated: the appearence of an electric voltage between the ends of the twisted shaft. This is the *Matteuci effect*, discovered in 1852 by Wertheim and made popular in 1856 by Matteuci.

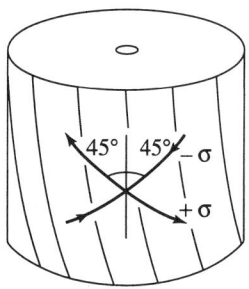

Figure 12.30 - Distribution of stress on a twisted shaft

6.4. DEVIATION FROM HOOKE'S LAW FOR FERROMAGNETIC SOLIDS: THE ΔE EFFECT

A bar oriented along direction Oz and submitted to a stress σ_{zz} deforms. Its relative length variation ε_{zz} along Oz will be proportional to σ_{zz}. This is Hooke's law:

$$\varepsilon_{zz} = \sigma_{zz}/Y \qquad (12.40)$$

This equation defines Young's modulus Y for the material. If the material is magnetic, applying stress will *change the magnetization* due to the inverse magnetoelastic effect we just described. And, due to the direct magnetoelastic effect, this stress-induced change in magnetization will induce a magnetostrictive deformation. The latter will add to the elastic deformation described by equation (12.40). This can be described as a change in the apparent value of Young's modulus, as shown in figure 12.31. This is called the ΔE effect because Young's modulus was long described by symbol E (in the present book, E designates an energy density).

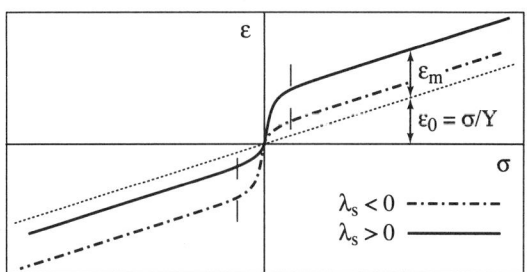

Figure 12.31 - Strain-stress curve for a ferromagnet

The deviation from Hooke's law is the ΔE effect. In all cases, the strain is larger than predicted by the linear law.

Using figure 12.27, it is easy to check that *Young's modulus is always decreased*. Starting from an isotropic distribution of the magnetic moments ($\sigma = 0$, H = 0), the application of a tensile stress on a material with positive magnetostriction aligns the spins along Oz, which induces an elongation which adds to the positive deformation described by Hooke's law. In the same way, applying a compressive stress on a material with $\lambda_s < 0$ also aligns the spins along Oz, which induces a shrinkage ($\lambda_s < 0$) which adds to the negative deformation from Hooke's law.

On the other hand, applying a stress such that $\lambda_s \sigma$ is negative puts the spins in the plane perpendicular to Oz. The measurement along Oz will yield not λ_s, but half of it *with reversed sign*, and we again check that the two deformations (elastic and magnetostrictive) have the same sign.

The "ΔE" effect is strongest in the low stress regime, where applying a stress can easily change the domain structure. At higher stress, the value Y_0 of Young's modulus that would prevail if the material were not magnetic is recovered. This effect is characterised by the ratio $\Delta Y / Y = (Y_0 - Y) / Y$, which can reach values above 10 in some materials. This is a first order magnetoelastic effect (related to the coefficient λ_s), and it should not be mixed up with second order effects mentioned in section 1.3 of the present chapter. Of course the "ΔE" effect varies with magnetic field, because it is very strongly correlated with the magnetization processes. It is therefore not necessarily a maximum at zero field (the same applies to maximum permeability), but it certainly tends to zero under a strong magnetic field.

The elastic properties of an isotropic material are characterised by two elastic constants: Young's modulus, defined by equation (12.40), and Poisson's ratio ν, generally near 0.3, which describes the transverse strain $\varepsilon_\perp$ associated with ε_{zz}: $\varepsilon_\perp = -\nu \varepsilon_{zz}$. Since $\varepsilon_\perp = -0.5 \varepsilon_{zz}$ for magnetostrictive strain, the ΔE effect generally entails a change in Poisson's ratio.

In a crystallised material, the coefficients c_{ij} are also observed to change under field, especially under weak magnetic fields, when Bloch walls are free to move. Therefore the true elastic constants of a (magnetically) soft ferromagnetic material should be measured under strong magnetic field.

7. MODELLING MAGNETOSTRICTIVE TRANSDUCERS

Actuators and sensors are the main two applications of magnetostrictive materials. Their operation is generally optimised by working in the most linear part of the characteristic curve, $\lambda(H)$ or $M(\sigma)$, for the material (fig. 12.32). To within the experimental accuracy, the $\lambda(H)$ and $M(\sigma)$ characteristics can be replaced by straight lines in the hatched areas.

In piezo-electric ceramics, the strain generated is an essentially linear function of the applied electric field, and a matrix formalism has been developed to relate elastic stress and strain to electric field and induction [22].

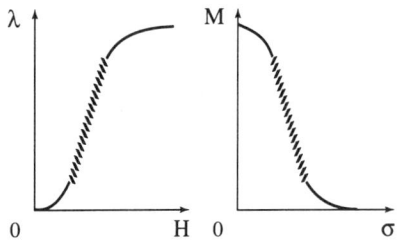

Figure 12.32 - Linearization of the direct and inverse characteristics of a magnetostrictive material

Isotropic Joule magnetostriction, being an even function of magnetic induction, is not naturally well suited to matrix representation. Nevertheless, it was desirable, towards modelling a material in view of technical applications, to transpose this simple matrix formalism to magnetostrictive materials. The technical solution consists in polarising the material through a static magnetic field H_0, for the curve $\lambda(H)$, and through a static strain Σ_0 for the curve $M(\sigma)$. H_0 and Σ_0 are at the inflection point of either curve. It is then possible to linearise the equations relating the magnetic (b and h), and mechanical (ε and σ) dynamic quantities:

$$\varepsilon = s^H \sigma + d^* h \qquad b = d\sigma + \mu^\sigma h \qquad (12.41)$$

To ensure the linearity of such a system, the excitation h (or σ) should in principle remain small enough compared to H_0 (Σ_0). However, Claeyssen *et al.* [23] showed it is possible, with a power sonar at resonance, to observe a load shift that is practically linear, even for high excitation and strain level (see fig. 12.33 below). Tensor s^H is called the *elastic compliance tensor* at constant field, μ^σ is the permeability tensor at constant strain, and d is the *piezomagnetic* tensor, with the convention $d_{ij}^* = d_{ji}$.

7.1. MAGNETOMECHANICAL COUPLING COEFFICIENT

The ability a magnetostrictive material has for transforming magnetic energy into elastic energy –or *vice versa*– can be characterised by a factor k, the "magnetomechanical coupling coefficient", defined as the ratio of the magnetoelastic energy to the geometric mean of the elastic and magnetic energies:

$$k = E_{mel} / \sqrt{E_{el} E_m} \qquad (12.42)$$

This factor can be expressed in terms of the elastomagnetic matrix coefficients. We will show this in the special case of a cylindrical bar, often used as an actuator in sonar (underwater acoustic radar) emitters as well as in the linear actuators which will be described in chapter 18.

Let the bar, with axis Oz, be submitted simultaneously to a compressive stress along its axis, σ_{zz}, and to a magnetic field, also along its axis, H_z. This field is often the superposition of a static field H_0 which polarises the material and of a dynamic field h_z, the excitation.

12 - MAGNETOELASTIC EFFECTS

The change in internal energy density E of the polarised material, under the effect of the excitation, can be expressed thermodynamically as:

$$dE = TdS + \sigma_{zz}d\varepsilon_{zz} + h_z db_z \tag{12.43}$$

where S is the entropy per unit volume.

The thermodynamic potential that should be used to analyse an experiment performed at constant temperature, under a given stress σ_{zz}, and under a magnetic field h_z, is the generalised Gibbs function (see § 4 of chap. 10):

$$G = E - TS - \sigma_{zz}\varepsilon_{zz} - h_z b_z \tag{12.44}$$

The differential of G at constant temperature is:

$$dG = -\varepsilon_{zz}d\sigma_{zz} - b_z dh_z \tag{12.45}$$

and, according to equation (12.41), we can express the components of strain and induction as functions of the components of stress and field:

$$\varepsilon_{zz} = s_{33}^H \sigma_{zz} + d_{33}h_z \qquad b_z = d_{33}\sigma_{zz} + \mu_{33}^\sigma h_z \tag{12.46}$$

from which we can rewrite (12.45):

$$dG = -d\left(\frac{1}{2}s_{33}^H \sigma_{zz}^2 + d_{33}h_z\sigma_{zz} + \frac{1}{2}\mu_{33}^\sigma h_z^2\right)$$
$$= -d\{E_{el} + 2E_{mel} + E_m\} \tag{12.47}$$

The magnetomechanical coupling coefficient in this experimental situation can thus be expressed as:

$$k_{33} = d_{33}\Big/\sqrt{s_{33}^H \mu_{33}^\sigma} = d_{33}\sqrt{Y^H/\mu_{33}^\sigma} \tag{12.48}$$

where s_{33}^H was replaced by Young's modulus Y^H. It could also be shown, using the above equations, that k_{33} can alternatively be expressed in terms of just Young's moduli, or just the permeabilities:

$$(k_{33})^2 = (Y^B - Y^H)/Y^B = (\mu_{33}^\sigma - \mu_{33}^\varepsilon)/\mu_{33}^\sigma \tag{12.49}$$

The second equality, for example, can be derived in the following way. From the first equation (12.46), we get the expression of σ_{zz} which is then inserted into the second equation:

$$b_z = \mu_{33}^\sigma h_z - (d_{33}^2/s_{33}^H)h_z + (d_{33}/s_{33}^H)\varepsilon_{zz} \tag{12.50}$$

When ε_{zz} is forced to vanish, this equation defines the permeability μ_{33}^ε, and, using equation (12.48), we recover the second equation (12.49):

$$\mu_{33}^\varepsilon = \mu_{33}^\sigma (1 - k_{33}^2) \tag{12.51}$$

Under an AC magnetic field h, the maximum magnetic energy localised per unit volume is $(1/2)\mu_{33}^{\sigma} h^2$ when the bar is free to deform under magnetostriction, and it is only $(1/2)\mu_{33}^{\varepsilon} h^2$ when it is stressed so that it cannot deform. In the latter case, no fraction of the magnetic energy is converted into mechanical energy. The difference $(1/2)(\mu_{33}^{\sigma} - \mu_{33}^{\varepsilon}) h^2$ thus represents the amount of magnetic energy that is transformed into elastic energy in the absence of any loss.

The square of the ratio k_{33} thus appears as the maximum fraction of the magnetic energy that can be converted into mechanical energy, and *vice versa*. It therefore expresses the overall balance of stored energies, with no consideration for possible losses in the process. It is not the efficiency, i.e. the ratio of the energies used to the energies provided, because this quantity involves the energy dissipation. The quantity analogous to the coefficient k_{33} in electrical engineering is the coupling coefficient k for a transformer, such that $k^2 = M_{12}^2 / L_1 L_2 = $ (mutual energy)^2 / [(primary energy)×(secondary energy)].

7.2. DYNAMIC ELASTIC ENERGY AWAY FROM RESONANCE

Engineers are interested in comparing the performance of a traditional motor, and of an actuator. They have to choose the best device in terms of the quality/cost ratio. An essential point in the comparison is the maximum elastic energy density that can be delivered by a material polarised by a magnetic field H_0. Returning to the example of the cylindrical bar actuator polarised along Oz, this energy density is expressed as $(1/2) Y^H \varepsilon_{zz}^2$. Its maximum value will be that corresponding to maximum strain $\varepsilon_{zz} = d_{33} H_0$, which assumes that the $\lambda(H)$ characteristic is linear down to zero field.

The maximum dynamic elastic energy density away from resonance is then:

$$E_{max} = \frac{1}{2} Y^H d_{33}^2 H_0^2 \qquad (12.52)$$

We will now see that resonance can improve this behavior.

7.3. MAGNETOELASTIC COUPLING AT RESONANCE

F. Claeyssen showed in his thesis that, at mechanical resonance, the maximum strain that can be observed on an actuator bar can be multiplied by the mechanical quality factor of the active material Q_m. Thus the maximum elastic energy density will be multiplied by the square of this factor [24]. Dynamical strains of $3,500 \times 10^{-6}$ were observed on rods of Terfenol-D with a static maximum strain of $1,600 \times 10^{-6}$, while using smaller fields than those needed to reach the maximum static strain.

Figure 12.33 shows that the strain-excitation current curves obtained with a sonar, the "tripod", based on this material near mechanical resonance ($F_0 = 1,200$ Hz) are fairly linear. A detailed analysis of these resonance effects is published in an excellent paper describing the actuators and transducers that use giant magnetostriction materials [23].

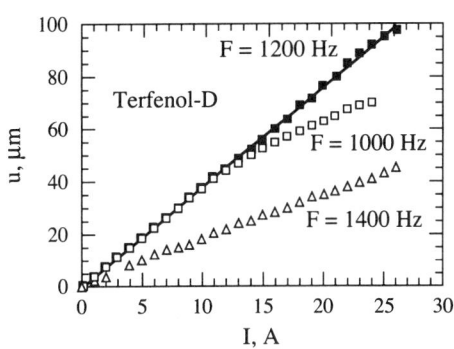

Figure 12.33
Displacement-current characteristics of a sonar [23]

Three basic books should be consulted for more information on magnetoelastic coupling:
- a theoretical book, dealing among others with the effects related to local rotations, and to non-uniform magnetoelastic effects [25]
- a work in Russian that discusses mainly the research performed in the Soviet Union [26]
- a review of 150 years of research on, and applications of magnetostriction in the world, since its discovery by Joule in 1842. It includes various subjects that are beyond the framework of the present textbook on magnetism (piezomagnetism, internal friction, parastriction, ...) [2].

APPENDIX: A REFRESHER ON LINEAR ELASTICITY

Elastic energy is a representation of the difficulty involved in deforming a solid by taking it out of its equilibrium configuration. It is therefore a measure of its cohesion.

In a first approximation, a spring is a one-dimensional object and its deformation is characterised by a single parameter, its relative elongation $\varepsilon = (x - x_0)/x_0$ where x is its length in the deformed state, and x_0 its length at rest. When it is deformed, its potential energy changes by the quantity:

$$\delta\mathscr{E} = (1/2)\,k\,(x - x_0)^2 = (1/2)\,(kx_0^2)\,\varepsilon^2 = (1/2)\,c\,\varepsilon^2.$$

A solid is generally a three-dimensional object, and its distorsion is defined by 3^2 parameters, the nine partial derivatives $\partial u_i / \partial x_j$ of the displacement vector **u** of a point linked to the solid, which is referred to an orthogonal normalised set of axes. The three parameters:

$$\varepsilon_{ii} = \partial u_i / \partial x_i \qquad (12.53)$$

measure the relative elongations, measured along three pependicular directions. The six other parameters, for $i \neq j$, are grouped into three symmetrical components:

$$\varepsilon_{ij} = \frac{1}{2}\left(\frac{\partial u_i}{\partial x_j} + \frac{\partial u_j}{\partial x_i}\right) \qquad (12.54)$$

and three antisymmetric components:

$$\omega_{ij} = \frac{1}{2}\left(\frac{\partial u_i}{\partial x_j} - \frac{\partial u_j}{\partial x_i}\right) \quad (12.55)$$

The ω_{ij} are the components of rotation and the ε_{ij} ($i \neq j$) are the three components of shear. Shear is the strain suffered by a square plate when one pulls on one of its diagonals (see fig. 12.4).

In the present work, we ignore rotations and only consider the six components of strain, *assuming they are all very small compared to unity*. Generalising the case of a spring, we will write the elastic energy density as an expression quadratic with respect to the strains:

$$E_{el} = \frac{1}{2} c_{ijkl} \varepsilon_{ij} \varepsilon_{kl} \quad (12.56)$$

The 36 *elastic coefficients* c_{ijkl} are thus the second derivatives of the potential energy density of the solid with respect to the six components of strain. Because of the Maxwell relations $\partial^2 E / \partial \varepsilon_{ij} \partial \varepsilon_{kl} = \partial^2 E / \partial \varepsilon_{kl} \partial \varepsilon_{ij}$, there are only 21 independent elastic coefficients, $c_{ijkl} = c_{klij}$. It is customary, following Voigt, to apply the rule of contraction over indices: $11 \to 1$, $22 \to 2$, $33 \to 3$, $23 \to 4$, $31 \to 5$, $12 \to 6$. Thus, c_{1112} is written as c_{16}.

Hooke's law expresses the proportionality of the strains to the stresses that produce them, within the framework of linear elasticity:

$$\sigma_{ij} = c_{ijkl} \varepsilon_{kl} \quad (12.57)$$

This law can also be written, using the *elastic compliances* s_{ijkl}:

$$\varepsilon_{ij} = s_{ijkl} \sigma_{kl} \quad (12.58)$$

The symmetry of a solid often makes it possible to notably simplify the expression of the elastic energy. It includes only five independent coefficients in hexagonal symmetry, three for cubic symmetry (c_{11}, c_{12} and c_{44}), and two for an isotropic solid, since $2 c_{44} = c_{11} - c_{12}$.

Hooke's law can be diagonalised in the case of *isotropic and cubic symmetry* materials. The eigenmodes of strain obtained are then:

$$\varepsilon_1 = \frac{1}{\sqrt{3}}\left(\varepsilon_{xx} + \varepsilon_{yy} + \varepsilon_{zz}\right)$$

$$\varepsilon_2 = \sqrt{\frac{2}{3}}\left(\varepsilon_{zz} - \frac{\varepsilon_{xx} + \varepsilon_{yy}}{2}\right) \quad \varepsilon_3 = \frac{1}{\sqrt{2}}\left(\varepsilon_{xx} - \varepsilon_{yy}\right) \quad (12.59)$$

$$\varepsilon_4 = \sqrt{2}\,\varepsilon_{yz} \quad \varepsilon_5 = \sqrt{2}\,\varepsilon_{zx} \quad \varepsilon_6 = \sqrt{2}\,\varepsilon_{xy}$$

The eigenmodes of stress are isomorphous to the above (the ε are substituted through the σ), and the elastic coefficients associated to the eigenmodes thus defined are:

for mode (1) $\qquad c_{11} + 2c_{12} = \dfrac{1}{s_{11} + 2s_{12}}$

12 - MAGNETOELASTIC EFFECTS

for modes (2) and (3) $\quad c_{11} - c_{12} = \dfrac{1}{s_{11} - s_{12}}$ (12.60)

for modes (4), (5) and (6) $\quad 2c_{44} = \dfrac{2}{s_{44}}$

In cubic symmetry, these three coefficients are respectively called c^α, c^γ and c^ε. For an isotropic solid, $c^\gamma = c_{11} - c_{12} = 2\, c_{44}$. The notations α, β, γ, ... refer to the irreducible representations Γ_1, Γ_2, Γ_3, ... of symmetry groups defined by Bethe and introduced into magnetoelasticity by Callen and coworkers [3].

So cursory a description of elasticity is valid only when the strains are very small and homogeneous. It would be insufficient to treat dynamical effects such as the propagation of spin waves coupled with elastic waves (magnetoelastic waves), which involve local rotations, or the elasticity of materials submitted to strong pre-stress. The general theory of elasticity, based on Lagrange's elasticity tensors and taking anharmonic effects (third order elastic constants) into account, is then mandatory [2, 24]. The interested reader is invited to refer to specialised texts on elasticity.

For lower symmetries, it is no more possible to diagonalise Hooke's law completely. But the complete description of magnetoelastic effects is already in the literature for hexagonal, tetragonal and orthorhombic symmetries. For still lower symmetries (rhombohedral, monoclinic and triclinic), the theory is not yet developed. It will have to take tensor *variance* into account.

Note - *Voigt's rule of index contraction simplifies the description of elasticity very nicely. Unfortunately, Voigt also defined the strains erroneously: $e_{ii} = \varepsilon_{ii}$, $e_{ij} = 2\varepsilon_{ij}$ ($i \neq j$). This is unacceptable because, with this definition, the components of strain do not make up a tensor any more. His expression for the shears is wrong. Unfortunately, many texts in magnetism continue, even today, using this bad definition.*

EXERCISES

E.1 - The free energy density of a ferromagnetic material can be written in the general form:

$$E = \sum_{i,j} k_{ij} \alpha_i \alpha_j + \sum_{i,j,k,l} k_{ijkl} \alpha_i \alpha_j \alpha_k \alpha_l + ...$$

$$+ \sum_{i,j} B_{ij} \varepsilon_{ij} + \sum_{i,j,k,l} B_{ijkl} \varepsilon_{ij} \alpha_k \alpha_l + \sum_{i,j,k,l,m,n} B_{ijklmn} \varepsilon_{ij} \alpha_k \alpha_l \alpha_m \alpha_n + ...$$

$$+ \frac{1}{2} \sum_{i,j,k,l} \left(C_{ijkl} + M_{ijkl} + \sum_{m,n} M_{ijklmn} \alpha_m \alpha_n + ... \right) \varepsilon_{ij} \varepsilon_{kl} + ... \quad (12.61)$$

where coefficients k, B, C are respectively anisotropy, magnetoelastic coupling and elasticity coefficients which, in a first approximation, depend only on temperature. The α_i are the direction cosines of magnetization, and the ε_{ij} are the components of the strain tensor, while the coefficients M describe the intrinsic magnetic contributions to the elastic constants.

E.1.1 - Starting from the above equation, show that the elastic and magnetoelastic energy density for a ferromagnetic crystal with *cubic symmetry* can be simply written, to second order in the directions cosines of magnetization:

$$E = \left(\frac{1}{2}c_{11}\varepsilon_{xx}^2 + c_{12}\varepsilon_{yy}\varepsilon_{zz} + 2c_{44}\varepsilon_{yz}^2\right) + \text{circ. permut.}$$
$$+ \left[B_0\varepsilon_{xx} + B_1\varepsilon_{xx}\left(\alpha_1^2 - \frac{1}{3}\right) + 2B_2\varepsilon_{yz}\alpha_2\alpha_3\right] + \text{circ. permut.} \quad (12.62)$$

E.1.2 - Show that this expression for the energy density can be recast into the form of equations (12.10) and (12.11).

E.2 - Determine the value of the magnetostriction coefficients for YIG from the experimental results of figures 12.8 and 12.9.

E.3 - Show that $(k_{33})^2 = \dfrac{Y_B - Y_H}{Y_B}$ (first term in eq. 12.49).

E.4 - Calculate how a long thin ferromagnetic metal tube, magnetised along its axis, and on the axis of which there is a wire carrying an electric current, will deform (this is a variant of the Wiedemann effect). Apply this to a nickel tube: $R = 5$ mm, $H = 1$ kA.m^{-1} and $i = 1$ A.

E.5 - In a localised model, Joule magnetoelastic coupling arises from "pair" interactions, i.e. from very short-range magnetic interactions between nearest neighbour interactions. Néel wrote them, in a classical approach for the crystal field term:

$$E = g(r)[\cos^2\phi - 1/3] \quad (12.63)$$

Restricting your consideration to interactions between nearest neighbours:

E.5.1 - Show that the magnetocrystalline anisotropy from these interactions is zero for a simple cubic crystal, and can be written $K(\cos^2\vartheta - 1/3)$ for a hexagonal crystal.

E.5.2 - Write the expression of the anisotropic magnetoelastic energy for a face-centred cubic crystal (terms $B^{\gamma,2}$ and $B^{\varepsilon,2}$ in eq. 12.10).

SOLUTIONS TO THE EXERCISES

S.1 **S.1.1** Use Maxwell's relations, $C_{ijkl} = C_{klij}$, with Voigt's index contraction rule, and write that energy must be invariant with respect to the symmetry operations allowed by cubic symmetry.

S.1.2 Diagonalise the expression of Hooke's law ($\sigma_{ij} = c_{ijkl}\varepsilon_{kl}$). This will define new components of strain which are linear combinations of the ε_{ii}.

S.2 In equation (12.15), set $\beta_1 = -\beta_2 = 1/\sqrt{2}$, then calculate the values of λ for the magnetization directions seen on figure 12.8, and take the differences $\lambda_{//} - \lambda_\perp$ for the three measured perpendicular directions. This yields three equations for two unknowns, $\lambda^{\gamma,2}$ and $\lambda^{\varepsilon,2}$. After correcting for the shape effect as given by equations (12.27) to (12.29), we check that these three equations are compatible and give: $\lambda^{\gamma,2} = -1.9 \times 10^{-6}$ and $\lambda^{\varepsilon,2} = -4.1 \times 10^{-6}$.

S.3 The derivation follows the same approach as for the second term of (12.49).

S.4 Consider a small square element of the cylinder, assume it to be parallel to the plane yOz.
- Calculate the magnetic field h_y due to i. Deduce from this the direction cosines of the resulting magnetic field $H + h_y$ as a function of the angle $\phi = \tan^{-1}(h_y/H)$. These will also be the direction cosines α_i of magnetization because the material is isotropic: $M // H$. ϕ will be taken as very small.
- Then calculate $\varepsilon_{yz} = (3/2)\lambda_s \alpha_2 \alpha_3$ in the presence of current i.
- Deduce the angle ϑ by which the generator of the cylinder has turned ($\vartheta = \varepsilon_{yz}$), and finally calculate the twisting of the cylinder, $\xi = \alpha/1 = \vartheta/R$.

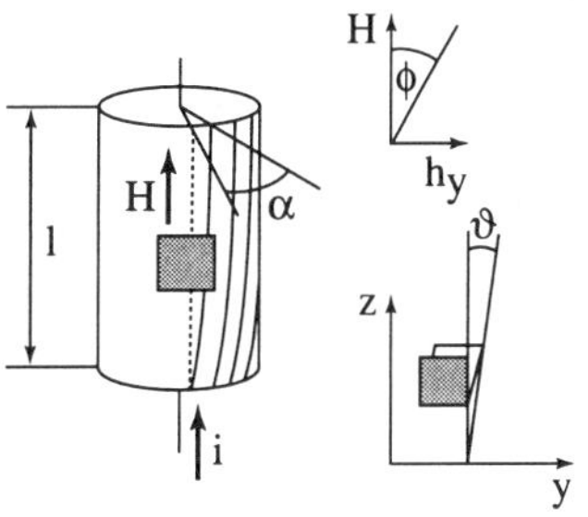

Note - ϑ *is not equal to* $\phi = \tan^{-1}(h_y/H)$*: it is much smaller.*
Numerically - *for nickel,* $\lambda_s = -36 \times 10^{-6}$ *hence:* $\xi = -0.34 \times 10^{-3}$ *rad. m*$^{-1}$.

Figure 12.34

S.5 *S.5.1* Cubic crystal case: $E = 2 g(r) \Sigma[\alpha_i^2 - (1/3)] = 0$; hexagonal crystal case: if $c/a < 1$, $E = 2 g(r) [\alpha_3^2 - (1/3)]$, and if $c/a > 1$, the sum must be taken over the 6 pairs of atoms in the basal plane, located at $(\pm a, 0, 0)$, $(\pm a\sqrt{3}/2, a/2, 0)$, and $(\pm a\sqrt{3}/2, -a/2, 0)$.

S.5.2 The strains in a crystal, referred to an orthogonal normalized set of axes, are represented by a tensor (ε_{xx}, ε_{yz}, ...) with six components; the relative elongation along any direction of space, $[\beta_1, \beta_2, \beta_3]$ is expressed as:

$$\partial r / r = (\varepsilon_{xx} \beta_1^2 + 2\varepsilon_{yz} \beta_2 \beta_3) + \text{circ. permut.}$$

and the direction cosines, for a fibre that was along $\beta_1, \beta_2, \beta_3$ before deformation, become after deformation:

$$u_1 = \beta_1 [1 + \varepsilon_{xx}(1 - \beta_1^2) - \varepsilon_{yy} \beta_2^2 - \varepsilon_{zz} \beta_3^2 - 2\varepsilon_{yz} \beta_2 \beta_3]$$
$$+ \varepsilon_{zx} \beta_3 (1 - 2\beta_1^2) + \varepsilon_{xy} \beta_2 (1 - 2\beta_1^2) - \omega_{zx} \beta_3 + \omega_{xy} \beta_2;$$

u_2, u_3: circ. permut.

In a fcc crystal, the twelve atoms that are nearest neighbours to the origin are at $(\pm a/\sqrt{2}, \pm a/\sqrt{2}, 0)$, $(0, \pm a/\sqrt{2}, \pm a/\sqrt{2})$, $(\pm a/\sqrt{2}, 0, \pm a/\sqrt{2})$, i.e. on the diagonals of the faces. Expression (12.50) must be summed over the twelve corresponding pairs.

The result is: $B^{\gamma,2} = (N/V_0)[(3\,g(r_0) + (1/2)\,r_0 \partial g/\partial r]$

and $B^{\epsilon,2} = (N/V_0)[(2\,g(r_0) + r_0 \partial g/\partial r]$

if one takes the number of magnetic atoms per unit volume to be N/V_0.

REFERENCES

[1] L. NÉEL, *J. Phys. Radium* **15** (1954) 225.

[2] E. DU TRÉMOLET DE LACHEISSERIE, Magnetostriction: Theory and Applications of Magnetoelasticity (1993) C.R.C. Press, Boca Raton (USA).

[3] E.R. CALLEN, H.B. CALLEN, *Phys. Rev.* **129** (1963) 578; *ibid. A* **139** (1965) 455.

[4] G. AUBERT, E. DU TRÉMOLET DE LACHEISSERIE, *J. Phys.* **48** (1987) 169.

[5] T.G. KOLLIE, *Phys. Rev. B* **16** (1977) 4872.

[6] G.K. WHITE, F.W. SHEARD, *J. Low Temp. Phys.* **14** (1974) 445.

[7] L. NÉEL, *Ann. Phys.* **8** (1937) 237.

[8] E. DU TRÉMOLET DE LACHEISSERIE, *J. Phys.* **37** (1976) 379.

[9] E.R. CALLEN, A.E. CLARK, B. DE SAVAGE, W. COLEMAN, H.B. CALLEN, *Phys. Rev.* **130** (1963) 1735.

[10] E. TATSUMOTO, O. OKAMOTO, *J. Phys. Soc. Jpn* **14** (1959) 1588; E. DU TRÉMOLET DE LACHEISSERIE, R. MENDIA MONTERROSO, *J. Magn. Magn. Mater.* **31-34** (1983) 837.

[11] W.L. WEBSTER, *Proc. Roy. Soc. London A* **109** (1925) 570.

[12] A. HUBERT, W. UNGER, J. KRANZ, *Z. Phys.* **224** (1969) 148.

[13] B.S. BERRY, W.C. PRITCHET, *Solid State Commun.* **26** (1978) 827.

[14] E. DU TRÉMOLET DE LACHEISSERIE, *J. Magn. Magn. Mater.* **13** (1979) 307.

[15] L.W. MCKEEHAN, *J. Franklin I.* **202** (1926) 737.

[16] J. BETZ, E. DU TRÉMOLET DE LACHEISSERIE, L.T. BACZEWSKI, *Appl. Phys. Lett.* **68** (1996) 132.

[17] R. GERSDORF, Thesis, Amsterdam (1961).

[18] H.D. BUTZAL, *Int. J. Magnetism* **3** (1973) 243.

[19] H.A. PIDGEON, *Phys. Rev.* **XIII** (1919) 209.

[20] G. BÉRANGER, F. DUFFAUT, J. MORLET, J.F. TIERS, Les Alliages de Fer et de Nickel (1996) Lavoisier, Tec & Doc. Paris.

[21] E. VILLARI, *Ann. Phys. Chem. (Ann. Poggendorf)* **126** (1865) 87.

[22] D.A. BERLINCOURT, D.R. CURRAN, H. JAFFE, Piezoelectric and Piezomagnetic Materials and their Function in Transducers, *in Physical Acoustics*, Vol. IA (1966) W.P. Mason Ed., Acad. Press, New York.

[23] F. CLAEYSSEN, N. LHERMET, R. LE LETTY, P. BOUCHILLOUX, *J. Alloy. Compd.* **258** (1997) 61.

[24] F. CLAEYSSEN, Thesis, INSA de Lyon, France (1989).

[25] W.F. BROWN JR., Magnetoelastic Interactions (1966) Springer Verlag, Berlin.

[26] K.P. BELOV, Magnetostrictive Phenomena and their Technical Applications (1987) Nauka, Glavnaïa Redactia Fiziko-Matematicheskoï Literatury, Moscow (*in Russian*).

CHAPTER 13

MAGNETO-OPTICAL EFFECTS

The magneto-optical effects were discovered a century and a half ago. They raised renewed interest over the last few years, and led to spectacular progress in physics as well as to many technological applications. After winning a place in consumer oriented magneto-optical recording for sound and video, they are now starting to conquer the computer market. The magneto-optical effects are widely used in the laboratory, both in the characterization of magnetic materials and in basic investigations of semiconductors. They are now being investigated in a spectral range to which they originally seemed completely foreign, that of synchrotron radiation X-rays.

We restrict the presentation to the magneto-optical effects that are most relevant within the context of the present book, centred on magnetic materials. A qualitative approach first describes the phenomena (§ 1). It is followed by a more formal phenomenological description (§ 2), while section 3 gives a simple physical point of view. Section 4 is devoted to the methods of measurement and to the orders of magnitude of magneto-optical phenomena. Section 5 gives an outline of their use in physics and of their technological applications. An appendix provides a short presentation of the optics of non-magnetic materials, as needed for this chapter, and in particular for section 2.

Several essential books or review papers on magneto-optical effects are among the references to this chapter. Sokolov's book [1] on the optics of metals deals at length with these effects. The very clear reviews by Dillon [2], as well as classic papers by Pershan [3] and Freiser [4], will be useful, provided caution is exercised toward the misprints which often mar the analytic expressions. Zvezdin & Kotov recently published a whole book devoted to magneto-optics [5], which is the work of reference in this area.

1. INTRODUCTION TO THE MAGNETO-OPTICAL EFFECTS

We will call magneto-optical effects the direct influence on light of the magnetic state, usually characterised by the magnetization, of the material in which the light propagates, or from which it is reflected. In magnetically ordered materials with non-zero spontaneous magnetization (ferro- and ferrimagnets), the relevant quantity is the mesoscopic magnetization **M** (see chap. 3, § 2).

Note - *We will not consider as a magneto-optical effect the (indirect) influence the deformation of the surface of a ferrofluid, or of a magnetostrictive solid sample, has on light under a magnetic field.*

These magneto-optical effects can show up through a change in the propagation direction, but more sensitively in the polarization, and / or the intensity of light. These changes are usually small, and it is difficult to separate out the magnetic effect from that of other physical properties of the sample. Detecting them will imply revealing variations in the properties of light when the magnetic state of the system is modulated, usually through a magnetic field. They will be evidenced, in ferro- and ferrimagnets, by reversing the magnetization of the sample. In para- and diamagnets, the behavior with and without an applied field can be compared.

The definition of the magneto-optical effects extends naturally to electromagnetic waves outside the visible range, including ultraviolet, infrared and microwaves. The extension can also cover X-rays: in this case, the distribution of magnetic moments on a microscopic scale can play a role, because the wavelength of the probe is then of the same order of magnitude as interatomic distances.

There often prevails an ambiguity in the terminology of magneto-optics. The qualifier "linear" is used both to say that an effect is different, in terms of the refractive index (birefringence) or of absorption (dichroism) depending on the rectilinear (or linear) polarization eigenmode the light is in. It also describes an effect that varies linearly with field or magnetization, as opposed to a quadratic variation. We will use the word "rectilinear" to refer to the polarization state. Thus we will term the Voigt or Cotton-Mouton effect a quadratic rectilinear birefringence. Further, we note that the effects that are termed "non-linear" (§ 1.3) are those where the frequency, hence wavelength, of the exiting light is not the same as that of the entering light.

1.1. TRANSMISSION EFFECTS

1.1.1. The Faraday effect

The Faraday effect was discovered in 1845. It remains the simplest to describe, and the easiest to observe. Consider a beam of light, initially with rectilinear polarization, propagating along a direction z in a material in which the magnetic induction **B** is parallel to z. Then the polarization of the beam will turn by an angle θ_F, proportional to the path length d in the material (fig. 13.1). Reversing the direction of **B** will lead to rotation the other way. If the material has non-zero absorption (which is always the case when magneto-optical effects exist), an ellipticity, increasing with d, will also appear.

The Faraday effect exists in "non-magnetic" materials, i.e. those with no spontaneous magnetization. It can then be described by $\theta_F = k_v B d$, where the proportionality constant k_v is called Verdet's constant. However we are more interested, in this chapter, in the case of ferro- and ferrimagnets. Then the effect of the magnetic field is mainly to steer the distribution, within the sample, of the domains, characterised by the different directions of the magnetization **M**. In general, the light exiting the specimen is in an inhomogeneous state, which makes it possible to observe the domains directly (see § 5.1.5 hereafter, and chap. 5, § 6.1.2). More generally, the effective component of **B** is that in the direction of propagation: there is no effect when **B** is perpendicular to the propagation direction.

A remarkable feature of the Faraday effect and more generally of the linear magneto-optical effects, is its *non-reciprocal* character. If, in figure 13.1, the direction of **B** is reversed, but the direction of propagation of light is unchanged, we expect the observer looking at the incoming beam to see the rotation reversed. Figure 13.2 shows the complementary situation: the magnetic field **B** is unchanged, but the propagation direction is reversed. For the observer (who is now placed on the left, and looks to the right) the rotation is clockwise, and the reverse of that shown in figure 13.1. This means that, with respect to the laboratory, the rotation takes place in the same direction on figures 13.1, and 13.2. Therefore, if the beam is passed successively in opposite directions through the same specimen, with length d, the rotation will be $2\theta_F$.

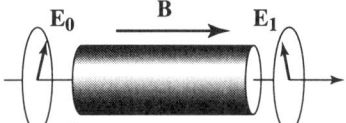
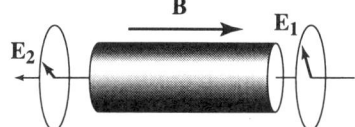

Figure 13.1 - Geometry of the Faraday effect   *Figure 13.2 - Non-reciprocal character of the Faraday effect*

Going through a distance d in a sample where the magnetic induction field is B entails for light a rotation of its plane of polarization by an angle θ_F, proportional to d. It also leads to an ellipticity (not shown here) of the outgoing wave. Figure 13.2 shows that going through the magnetic field region with reversed propagation direction leads to a rotation in the same direction as in figure 3.1 with respect to a fixed reference. Note however that an observer facing the incoming light in each case will describe the rotation as opposite for the situations of figures 13.1, and 13.2.

This distinguishes the rotation due to a magnetic field (magnetic gyrotropy) from that due to natural rotatory power, e.g. in quartz, where two passes in opposite directions would cancel out the rotation.

Both the rotation and the ellipticity produced by the Faraday effect can conveniently be described by treating the propagation of waves with circular polarization (right-hand and left-hand polarization). We will see later on that these modes are the eigenmodes, and that they are associated with different complex refractive indices.

The Faraday rotation results from the fact that one of the circular polarization states propagates a little faster than the other, as described by the difference in the real parts of these indices (circular magnetic birefringence). Circular dichroism, associated with the imaginary part of the indices, accounts for the ellipticity, because one of the circular commponents is more attenuated than the other.

1.1.2. Magnetic quadratic rectilinear birefringence

Liquids as well as solids feature a birefringence that affects the two perpendicular rectilinear polarization eigenmodes differently when the applied magnetic field, or the magnetization, is perpendicular to the propagation direction of light. This effect is quadratic with respect to field or magnetization.

This quadratic birefringence effect (fig. 13.3) is insensitive to a reversal of **B**. It encompasses both the Voigt effect and the Cotton-Mouton effect, due to anisotropic molecules getting aligned in liquids. There is an associated rectilinear dichroism. Note that this effect is unfortunately often called "linear birefringence" because it affects the two perpendicular eigenstates of rectilinear polarization differently (see § 1 above).

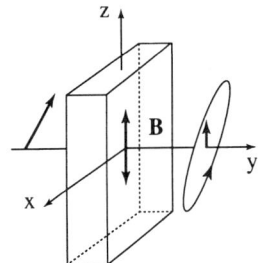

Figure 13.3 - The quadratic birefringence due to a field or magnetization along z leads to a phase difference between the components of E along x and along z

Light that is initially polarised along a direction which coincides neither with x nor with z becomes elliptically polarised after passage through the specimen. This is related to the fact that a material cannot remain isotropic when it features a magnetization, and formally this describes the non-equivalence between directions x or y on one hand, and z on the other hand, when the magnetization is along z.

1.1.3. Magnetic linear rectilinear birefringence

We again deal with birefringence related to the difference in refractive indices between the two perpendicular rectilinear polarization states (which distinguishes it from the Faraday effect, characterised by the difference in the indices describing the two circular polarization eigenstates). But this birefringence is linear (not quadratic as are the Voigt and Cotton-Mouton effects) with respect to the magnetic field, which now acts along the propagation direction. This effect is possible only in a few magnetic symmetry classes [6], in particular in magneto-electric materials and in antiferromagnetic transition metal fluorides (FeF_2, CoF_2, ...).

1.2. MAGNETO-OPTICS IN REFLECTION GEOMETRY

The magneto-optical effect in reflection geometry is called the magneto-optical Kerr effect. Its description involves three basic geometries.

1.2.1. The polar Kerr effect

A beam of light with rectilinear polarization falls under normal incidence on a surface which is magnetised in a direction perpendicular to the surface (fig. 13.4-a). After reflection, the polarization has turned by an angle θ_K, typically less than a degree, and some ellipticity has appeared. Reversing the magnetization leads to inverted rotation.

This case corresponds, in magnetostatic terms, to an unfavored situation (see chap. 2). It occurs only if the magnetization is forced to be perpendicular to the surface, under the effect either of a sufficiently strong field, or of "perpendicular" anisotropy.

1.2.2. The longitudinal Kerr effect

A beam strikes a surface which is magnetised in its plane under oblique incidence (fig. 13.4-b), the magnetization being in the plane of incidence. If the polarization (the electric field vector **E**) of the incident beam is perpendicular (s polarization) or parallel (p polarization) to the plane of incidence of the light, then the polarization of the reflected beam is slightly elliptical, with a major axis rotated with respect to the incident polarization by an angle of generally less than a degree. If the incident polarization is neither s nor p, then the phase shift due to ordinary reflection will make the magneto-optical effect much less noticeable. Here again, the rotation reverses when the magnetization is reversed.

1.2.3. The transverse Kerr effect

The geometry is the same as in the longitudinal case, except that the magnetization, although it is still in the plane of the specimen surface, is perpendicular to the plane of incidence (fig. 13.4-c). Now there is no rotation for s or p incident polarization, and in fact there is no effect at all for s polarization. However, reversing the magnetization leads, if the incident polarization is p, to a change in the intensity of reflected light.

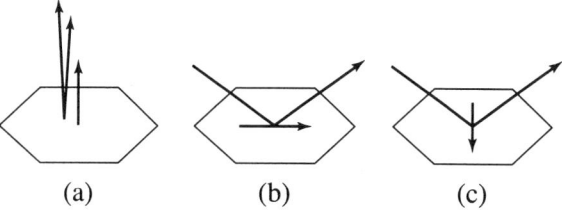

Figure 13.4 - The three basic configurations for the magneto-optical Kerr effect (a) polar - (b) longitudinal - (c) transverse

1.3. NON-LINEAR MAGNETO-OPTICAL EFFECTS

This corresponds to the situation of second harmonic generation, where a high intensity electromagnetic wave with angular frequency ω generates a wave with another frequency, in practice 2ω. The magneto-optical effects associated with this mechanism, which is usually very weak, are particularly interesting, precisely because they are forbidden by symmetry in centrosymmetric materials. They do occur in magneto-electric materials and at the surfaces of magnetic films, because there the symmetry is broken [7]. They therefore provide an elegant way of selectively probing surface and interface magnetism.

1.4. MAGNETIC DIFFRACTION AND MAGNETIC DICHROISM IN THE X-RAY RANGE

X-rays interact mostly with the electronic density in matter (Thomson scattering). However, there is also a small magnetic interaction term, and observing it experimentally, with a classical X-ray tube, was a *tour de force* [8]. The development of synchrotron radiation has made such experiments more commonplace. They make it possible to separately measure the orbital and spin contributions to magnetism. A resonant form of magnetic diffraction, in which the magnetic effect is enhanced for well chosen wavelengths, was discovered during the first synchrotron radiation measurements. The magnetism of atoms, when they are periodically arranged into a crystal, gives a contribution to the (Bragg) scattering associated with this regular arrangement. This effect is very strong [9] at the L edges, which however correspond, for the transition metals, to wavelengths too large for Bragg diffraction to occur in response to interatomic distances. However this situation can be used to investigate magnetic multilayers, in which the stacking period of different films can be adjusted at will. Just as for visible light, the absorption of X-rays, which is strong near special values of the photon energy (the absorption edges), also depends on the relative orientation of the direction of polarization or helicity of the X-rays (right-hand or left-hand circular polarization), and of the material's magnetization (parallel or antiparallel to the propagation direction). In the geometry of the transverse Kerr effect, at the 3p edge, hence in the range of very soft X-rays (60 eV), Hillebrecht *et al.* [10] observed the magnetic dichroism associated with rectilinear polarization, and obtained an image of domains on the basis of the local difference in the photoemission efficiency. Circular magnetic dichroism is also an important and blossoming basic research tool. In its application to imaging (see § 5.1.5 below), it provides images of the magnetic structure in multilayers (see chap. 20), with one or the other chemical element being selectively imaged depending on the chosen energy of the beam [11].

13 - MAGNETO-OPTICAL EFFECTS

2. PHENOMENOLOGICAL TREATMENT

2.1. PROPAGATION EQUATION. EIGENMODES

Consider first the propagation in a material of a plane electromagnetic wave, which we describe by $\mathbf{E} = \mathbf{E_0} \exp - j\,(\omega t - \mathbf{k}\cdot\mathbf{r})$. The meaning of the various quantities is indicated in the appendix to this chapter. They are related by Maxwell's equations, which can be written, considering that the material is electrically neutral:

$$\mathbf{curl\ E} = -\partial \mathbf{B}/\partial t \tag{13.1}$$

$$\mathbf{curl\ H} = \partial \mathbf{D}/\partial t + \mathbf{j}_{cond} \tag{13.2}$$

$$\mathrm{div}\,\mathbf{D} = 0 \tag{13.3}$$

$$\mathrm{div}\,\mathbf{B} = 0 \tag{13.4}$$

with $\qquad \mathbf{B} = \mu_0 \mu_r \mathbf{H} \quad$ and $\quad \mathbf{D} = \varepsilon_0 \boldsymbol{\epsilon}\,\mathbf{E} = \varepsilon_0 \mathbf{E} + \mathbf{P} \tag{13.5}$

where: $\qquad\qquad\qquad \mathbf{P} = \varepsilon_0 \chi_{el}\,\mathbf{E} \tag{13.6}$

is the electrical polarization, and χ_{el} the electrical susceptibility. $\boldsymbol{\epsilon}$ is the relative electrical permittivity tensor, $\boldsymbol{\mu}_r$ the relative magnetic permeability tensor for angular frequency ω. The charge displacement driven by the electric field induces a current density, the description of which involves, in the general case, the conductivity tensor $\boldsymbol{\sigma}$ and the electric susceptibility tensor χ_{el} of the materials

$$\mathbf{j}_{total} = \partial \mathbf{P}/\partial t + \boldsymbol{\sigma}\mathbf{E} = \partial \mathbf{P}/\partial t + \mathbf{j}_{cond} \tag{13.7}$$

The notation can be made more compact with the notations

$$\boldsymbol{\epsilon} = 1 + \chi_{el} \quad \text{and} \quad \boldsymbol{\epsilon}' = \boldsymbol{\epsilon} - \boldsymbol{\sigma}/j\omega\varepsilon_0 \quad \left(j = \sqrt{-1}\right) \tag{13.8}$$

The response to the magnetic field of the magnetization is restricted to frequencies of the order of the paramagnetic relaxation frequencies. Therefore, magnetization cannot follow variations at optical frequencies, and the relative magnetic permeability is always taken as equal to unity [12]. The magneto-optical effects thus involve only the effect of dc, or very low frequency, magnetic fields or magnetization on the dielectric permittivity. This approximation breaks down at very large wavelengths (the microwave regime).

Since $\boldsymbol{\epsilon}$ and $\boldsymbol{\sigma}$ are tensors, so is $\boldsymbol{\epsilon}'$ which can be expressed as:

$$\boldsymbol{\epsilon}' = \begin{bmatrix} \varepsilon'_{11} & \varepsilon'_{12} & \varepsilon'_{13} \\ \varepsilon'_{21} & \varepsilon'_{22} & \varepsilon'_{23} \\ \varepsilon'_{31} & \varepsilon'_{32} & \varepsilon'_{33} \end{bmatrix} \tag{13.9}$$

Onsager's principle in statistical physics states that the symmetry $\varepsilon_{ij} = \varepsilon_{ji}$ applies only for a system with invariance with respect to time-reversal. For a sample submitted to a magnetic field $\mathbf{H}$, or featuring a magnetization $\mathbf{M}$, invariance with respect to time-reversal requires that the direction of the field **H** or magnetization **M** is also reversed. Hence:

$$\varepsilon_{ij}(\mathbf{H}) = \varepsilon_{ji}(-\mathbf{H}) \tag{13.10}$$

Now consider, for simplicity, a ferromagnetic material which would be optically isotropic although it would bear a magnetization **M**. The optical anisotropy would then be due entirely to magnetism, and it would be described by non-diagonal terms in the permittivity tensor:

$$\varepsilon_{ij}(\mathbf{M}) = \varepsilon_{ji}(-\mathbf{M}) \tag{13.11}$$

Since these terms are zero in the absence of magnetization, they can be taken as linear with respect to **M**, which leads to:

$$\varepsilon_{ij}(\mathbf{M}) = -\varepsilon_{ji}(\mathbf{M}) \quad (i \neq j) \tag{13.12}$$

Thus the tensor $\boldsymbol{\varepsilon}'$ takes on the form:

$$\boldsymbol{\varepsilon}' = \begin{bmatrix} \varepsilon'_{11} & \varepsilon'_{12} & \varepsilon'_{13} \\ -\varepsilon'_{12} & \varepsilon'_{11} & \varepsilon'_{23} \\ -\varepsilon'_{13} & -\varepsilon'_{23} & \varepsilon'_{11} \end{bmatrix} \tag{13.13}$$

where the ε'_{ij} are generally complex.

If the direction of **M** coincides with one of the chosen coordinate axes, rotation around this axis has no effect on the tensor. The latter therefore features only two symmetrical non-diagonal terms, ε'_{ij} and $-\varepsilon'_{ij}$. Thus, for **M** along z:

$$\boldsymbol{\varepsilon}' = \begin{bmatrix} \varepsilon'_{11} & \varepsilon'_{12} & 0 \\ -\varepsilon'_{12} & \varepsilon'_{11} & 0 \\ 0 & 0 & \varepsilon'_{11} \end{bmatrix} \tag{13.14}$$

The fact that the non-diagonal terms ε_{ij} are proportional to magnetization is often expressed by introducing the magneto-optical coefficient Q, as in equation (13.15). We will make the form of the tensor $\boldsymbol{\varepsilon}'$ clear for the cases most frequently encountered in magneto-optics later.

$$\boldsymbol{\varepsilon}' = \begin{bmatrix} \varepsilon'_1 & jQ\varepsilon'_1 & 0 \\ -jQ\varepsilon'_1 & \varepsilon'_1 & 0 \\ 0 & 0 & \varepsilon'_1 \end{bmatrix} \tag{13.15}$$

The tensors $\boldsymbol{\varepsilon}$ and $\boldsymbol{\varepsilon}'$ are sometimes split into two parts, a symmetrical tensor and an antisymmetrical one, so that, for $\boldsymbol{\varepsilon}'$ for example:

$$\boldsymbol{\varepsilon}' = \begin{bmatrix} \varepsilon'_1 & 0 & 0 \\ 0 & \varepsilon'_1 & 0 \\ 0 & 0 & \varepsilon'_1 \end{bmatrix} + j \begin{bmatrix} 0 & jQ\varepsilon'_1 & 0 \\ -Q\varepsilon'_1 & 0 & 0 \\ 0 & 0 & 0 \end{bmatrix}$$

13 - MAGNETO-OPTICAL EFFECTS

A (complex) gyration vector **g** can be introduced by expressing **D** in the form: $\mathbf{D} = \varepsilon_0 \boldsymbol{\epsilon} \, \mathbf{E} = \varepsilon_0 \mathbf{E} + j \, (\mathbf{E} \times \mathbf{g})$. In an isotropic material, **g** is proportional to magnetization **M** ($\mathbf{g} = \alpha \mathbf{M}$), and the magnetic term in the polarization becomes: $\mathbf{P} = j \, \alpha \, (\mathbf{E} \times \mathbf{M})$. This term describes the gyrotropic effects that occur in configurations where the eigenmodes have circular polarization (the Faraday and circular dichroism effects). The above descriptions can also cover ferrimagnets, but then the resultant magnetization **M** is not sufficient to describe the state. In particular, **Q** and **g** must be expressed to first order as linear combinations of the sublattice magnetizations. When analysing the behavior of a plane wave in the magnetic material:

$$\mathbf{E} = \mathbf{E}_0 \exp - j \, (\omega t - \mathbf{k} \cdot \mathbf{r}) \tag{13.16}$$

$$\mathbf{H} = \mathbf{H}_0 \exp - j \, (\omega t - \mathbf{k} \cdot \mathbf{r}) \tag{13.17}$$

k is the wave vector in the material, such that $\mathbf{k} = N\mathbf{k}_0$ with $N = n + j\kappa$ the complex refractive index and $\mathbf{k}_0$ the wave vector in vacuum.

Maxwell's equations lead to the basic equation (13.18), which characterises the optical properties of an anisotropic material, with $\boldsymbol{\epsilon}'$ the tensor defined above and $\mathbf{u}_k$ the unit vector along **k**, with components x, y, z:

$$\boldsymbol{\epsilon}' \mathbf{E} = N^2 \, [\mathbf{E} - \mathbf{u}_k \, (\mathbf{E} \cdot \mathbf{u}_k)] \tag{13.18}$$

In the more general case of a material that has uniaxial anisotropy along Oz, this equation can be written $\mathbf{S} \cdot \mathbf{E} = 0$, with:

$$\mathbf{S} = \begin{bmatrix} \varepsilon'_1 - N^2(1 - x^2) & j\varepsilon'_1 Q + N^2 xy & N^2 xz \\ -j\varepsilon'_1 Q + N^2 yx & \varepsilon'_1 - N^2(1 - y^2) & N^2 yz \\ N^2 zx & N^2 zy & \varepsilon'_3 - N^2(1 - z^2) \end{bmatrix} \tag{13.19}$$

Non-trivial solution to this matrix equation exist only if the determinant of **S** is zero, i.e.: $|S| = 0$.

Solving this biquadratic equation provides the complex refractive indices and the associated propagation eigenmodes, hence the characteristic of the outgoing light produced by a given incident wave.

2.2. EFFECTS IN TRANSMISSION GEOMETRY

2.2.1. The Faraday effect

Light propagates along direction z, which also carries the magnetic field and the magnetization of the sample.

With $\mathbf{u}_k = (0, 0, 1)$ the biquadratic equation becomes:

$$N^4 - 2\varepsilon'_1 N^2 + \varepsilon'_1{}^2 (1 - Q^2) = 0 \tag{13.20}$$

The system of equations defined by $\mathbf{S} \cdot \mathbf{E} = 0$ thus has solutions:

$$N_+ = n_+ + j\kappa_+ \sqrt{\varepsilon'_1 (1+Q)} \qquad (13.21)$$

$$N_- = n_- + j\kappa_- \sqrt{\varepsilon'_1 (1-Q)} \qquad (13.22)$$

The corresponding electric fields are such that:

$$E_+ = (E_x, -jE_x, 0) \quad \text{and} \quad E_- = (E_x, jE_x, 0) \qquad (13.23)$$

which characterise right-hand and left-hand circularly polarised modes.

Incident light with rectilinear polarization can be considered as resulting from two circular vibrations, right-hand and left-hand. When it goes through a parallel-faced slab perpendicular to magnetization, with thickness e, there appears a phase difference ϕ between the circular components, such that $\phi = 2\pi(n_+ - n_-)e/\lambda_0$ where λ_0 is the wavelength in vacuum. The result is a rotation by θ_F of the plane of polarization of the transmitted light:

$$\theta_F = \pi(n_+ - n_-)e/\lambda_0 = \phi/2 \qquad (13.24)$$

The imaginary parts, κ_+ and κ_-, of the complex indices correspond to absorption in the material, and make the transmitted vibration elliptical. The ellipticity is expressed as:

$$\eta_F = -\tanh[\pi(\kappa_+ - \kappa_-)e/\lambda_0] \qquad (13.25)$$

Since the non-diagonal terms of tensor $\boldsymbol{\varepsilon}'$ are small with respect to the diagonal terms ($Q \ll 1$),

$$N_+ - N_- \sim (\varepsilon'_1)^{1/2} Q \sim \varepsilon'_1 Q/N \qquad (13.26)$$

with N the average complex refractive index.

The circular birefringence is thus proportional to Q.

2.2.2. The Voigt effect

Magnetization is parallel to the surface, along Oz, and the unit vector $\mathbf{u}_k$, perpendicular to the surface, coincides with Oy, hence $\mathbf{u}_k = (0, 1, 0)$. The biquadratic equation thus becomes:

$$(\varepsilon'_3 - N^2)\left[\varepsilon'_1(\varepsilon'_1 - N^2) - \varepsilon'^2_1 Q^2\right] = 0 \qquad (13.27)$$

The indices corresponding to polarization directions of the wave parallel and perpendicular to the magnetization are therefore:

$$N_z = (\varepsilon'_3)^{1/2} \qquad N_x = \left[\varepsilon'_1(1-Q^2)\right]^{1/2} \qquad (13.28)$$

The polarization eigenmodes follow:

$$E_{//}/E_0 = (0, 0, 1) \quad \text{and} \quad E_\perp/E_0 = (1, jQ, 0) \qquad (13.29)$$

with E_0 a normalization factor.

We see that the eigenmodes correspond to the electric field being parallel and perpendicular to magnetization, and that, in the latter case, the eigenmode has a weak longitudinal component because $Q \ll 1$.

The components along Ox and Oz of the incident wave's electric field suffer a phase difference as they go through the material. They therefore produce an outgoing elliptical vibration. Furthermore, since ε'_1 and ε'_3 are complex, there appears a rectilinear magnetic dichroism effect, corresponding to different values of the transmission factors along Ox and Oz. Thus, for an incident rectilinear vibration with arbitrary polarization direction, rotation of the polarization due to dichroism superimposes on the ellipticity resulting from birefringence. In either case the effect is even with respect to field or magnetization ($\propto H^2$ or $\propto M^2$).

2.3. THE KERR EFFECTS

The reflected amplitudes corresponding to polarization parallel (p) and perpendicular (s) to the plane of incidence are related to the incident p and s amplitudes by the Fresnel coefficients:

$$\begin{bmatrix} E_{p\,refl} \\ E_{s\,refl} \end{bmatrix} = \begin{bmatrix} r_{pp} & r_{ps} \\ r_{sp} & r_{ss} \end{bmatrix} \begin{bmatrix} E_{p\,inc} \\ E_{s\,inc} \end{bmatrix} \quad (13.30)$$

The coefficients r_{ij} thus correspond to special amplitude ratios: thus r_{ps} is the ratio $E_{p\,refl}/E_{s\,inc}$. In practice, the incident vibration is supposed to be polarised perpendicular to the plane of incidence (s), and the reflected (p) component is calculated.

The determination of the coefficients r_{ij} is rather cumbersome. The major steps involved are:

♦ determining the components of the electric field in each medium and for each propagation mode, as defined by a direction and a polarization (s or p) starting from the equation system $\mathbf{S}.\mathbf{E} = 0$,

♦ writing the continuity relations for the tangential components of the fields $\mathbf{E}$ and $\mathbf{H}$ at each interface. This is usually done in matrix form [13].

In the general case of oblique incidence in the yOz plane, with Oz perpendicular to the surface, and using again the notation:

$$\mathbf{k} = (N\omega/c)\mathbf{u}_k \quad (13.31)$$

with $\mathbf{u}_k = (0, \beta, \gamma)$, the result (neglecting Q^2 terms) is:

$r_{ss} = (\cos\theta - N\gamma)/(\cos\theta + N\gamma)$

$r_{ps} = [jNQ(m_y\beta - m_z\gamma)\cos\theta]/[\gamma(N\cos\theta + \gamma)(\cos\theta + N\gamma)]$

$r_{sp} = -j[NQ(m_y\beta + m_z\gamma)\cos\theta]/[\gamma(\cos\theta + N\gamma)(N\cos\theta + \gamma)]$

$r_{pp} = [(N\cos\theta - \gamma)/(N\cos\theta + \gamma)] + [j2m_xNQ\beta\cos\theta/(N\cos\theta + \gamma)^2] \quad (13.32)$

where m_x, m_y, and m_z are the direction cosines of magnetization **M**, or the reduced magnetization components along the x, y, z directions: $m_x = M_x/M$ etc. θ is the angle of incidence. These coefficients are complex in the case of a metallic material.

We note that r_{ss} is independent of Q, and hence of the magnetic state of the material, that r_{pp} depends on Q only in the case of the transverse effect, and that r_{sp} and r_{ps} are non-zero only in the cases of the polar and longitudinal Kerr effects. For Q = 0, the classical Fresnel coefficients r_{pp} and r_{ss} are recovered, and r_{ps} and r_{sp} are zero. This confirms the above-mentioned characteristics of the various Kerr effect configurations.

2.3.1. *The transverse Kerr configuration*

This corresponds to $m_x = \pm 1$. Then:
- there is no effect if the incident rectilinear polarization is s, or transverse electric (TE),
- the non-diagonal terms r_{ps} and r_{sp} are zero,
- p, or transverse magnetic (TM) polarization, and s (TE) are eigenmodes,
- a Q dependent term superimposes on the Fresnel term in r_{pp}. This term is usually small because $Q \ll 1$.

2.3.2. *The longitudinal Kerr configuration*

This corresponds to $m_y = \pm 1$. In this case:
- the coefficients r_{pp} and r_{ss} are independent of Q,
- the coefficients r_{ps} and r_{sp}, responsible for the rotation of the polarization of the reflected light with respect to the incident rectilinear light, have the same sign,
- normal incidence corresponds to $\beta = 0$; the non-diagonal terms r_{ps} and r_{sp} vanish, and there is no effect.

2.3.3. *The polar Kerr configuration*

This corresponds to $m_z = \pm 1$. In this case:
- normal incidence corresponds to $\beta = 0$ and $\cos\theta = \gamma = 1$; the terms r_{sp} and r_{ps} are equal, and they have maximum value,
- the rotation angle θ_K is maximum and is independent of the incident polarization direction.

In practice, one measures intensities. For instance, in the case of the transverse Kerr effect, the reflected intensity I_p is proportional to $r_{pp} r_{pp}^* (= R_p)$.

By noting by *Re* and *Im* the real and imaginary parts of a variable, the rotations θ_K and corresponding ellipticities η_K are such that, for oblique incidence:

$$\theta_{K\,s} = -Re(r_{ps}/r_{ss})$$

$$\theta_{K\,p} = Re(r_{sp}/r_{pp})$$

$$\eta_{K\,s} = Im(r_{ps}/r_{ss})\,Re(r_{ps}/r_{ss})$$

$$\eta_{K\,p} = Im(r_{sp}/r_{pp})\,Re(r_{sp}/r_{pp}) \qquad (13.33)$$

2.4. THIN FILM SYSTEMS

The technological importance of thin film and multilayer magnetic systems, for instance for recording applications, justifies that much work is performed in this area, and this is true for magneto-optics too [14]. Analysing such systems involves a combination of the interface effects (the Kerr effects) and of the effect of transmission through the films (Faraday effect). Matrix formulations, using the Fresnel coefficient matrices introduced above, are then very convenient.

3. PHYSICAL PRESENTATION

P.P. Ewald [15] first gave a microscopic theory of the refractive index. This approach, used in a very simplified form, provides a useful qualitative view of optical phenomena in matter. It consists in describing the effect of matter on propagation as the combination with the incident wave of that radiated by the distribution of oscillating dipoles induced in the material. In a continuous medium presentation, the distribution of dipoles is represented by the electric polarization **P**, the dipole moment per unit volume. **P** is then an oscillating quantity, with angular frequency equal to that of the incident wave. The (transverse) wave it radiates at a large distance r at time t will be proportional to the value of the transverse (perpendicular to the observation direction) component of the acceleration of the charge. This is equivalent to the second time derivative of **P**, but at the former time $(t - r/c)$, since account must be taken of the time for propagation r/c. In the optical range of frequencies and beyond, only electrons can contribute to the dipole moment because their mass, hence their inertia, is much small than that of the nuclei.

Let us try and imagine in simple terms the effect of a dc magnetic field **B** along z on a light wave in a material which, in the absence of **B**, would be as simple as possible, i.e. transparent and isotropic (only one real component of the dielectric constant ε).

Using a classical approach, we can describe the effect of **B** on the motion of electrons, with charge denoted $-e$, submitted to the effect of the ac electric field of the wave **E**, as a Lorentz force $-e\,\mathbf{v} \times \mathbf{B}$ (fig. 13.5). Thus an electric field with angular frequency ω, along x, will no longer exactly induce an oscillation along x as was observed in the absence of **B**. The Lorentz force is along y, and it will induce a component of the electron displacement, hence of the polarization, with direction y and with a phase shift $\pi/2$ with respect to the component due to **E**. The dielectric tensor must therefore feature a non-diagonal term ε_{yx}, as well as its companion ε_{xy} which would correspond to a situation where **E** would be along y and the Lorentz force along $-x$, with $\varepsilon_{yx} = -\varepsilon_{xy}$. The phase shift $\pi/2$ between **E** and **P** leads to the fact that both these terms are complex as indicated in equation (13.15).

But the acceleration of the electrons also acquires a component along y, also shifted by $\pi/2$ with respect to **E**.

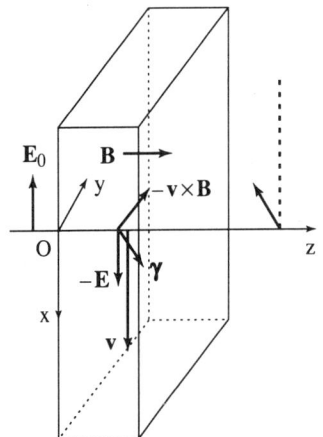

Figure 13.5 - Schematic representation of the Kerr effect in a classical approach

This shows the effect of a thin slab of magnetic material. The contributions $-e\mathbf{E}$ and $-e\mathbf{v}\times\mathbf{B}$ to the acceleration of the electrons would lead to a resulting acceleration $\boldsymbol{\gamma}$ that is not colinear to $\mathbf{E}$. The contribution from this slice to the wave radiated at large distance will therefore not be colinear to $\mathbf{E}$ either. This figure does not include the effect of the propagation time.

Thus the radiated wave will not be polarised along the x direction. In other words, rectilinear polarization along x or y is not an eigenmode for propagation in the material in the presence of **B**. Formal analysis shows that the eigenmodes which propagate without being altered in the material in the presence of **B** are the left- and right-hand circular polarizations L and R. Well-defined refractive indices, n_L and n_R, are associated with these eigenmodes.

When absorption is taken into account, the refractive index for the material and ε both are complex, even in the absence of magnetic field. The same is true for the non-diagonal terms of the permittivity tensor in the presence of **B**.

These are the two ingredients of the *Faraday effect*. The difference between the imaginary parts of the non-diagonal terms in $\boldsymbol{\varepsilon}'$ leads to the right-hand component E_R and the left-hand component E_L having a phase difference $2\pi d (n_L - n_R)/\lambda_0$, after travelling the same distance d, with λ_0 the wavelength in vacuum (circular birefringence). Their sum **E'** will thus, in the absence of absorption, be a rectilinear vibration rotated by an angle θ_F proportional to d (fig. 13.6).

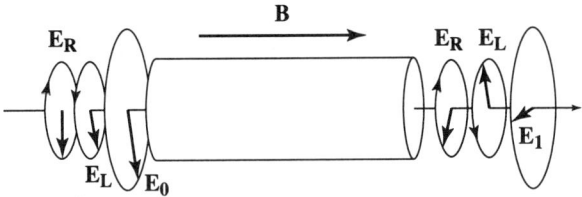

Figure 13.6 - The rotation associated with the Faraday effect is due to the difference in velocity of the right-hand E_R vibration and left-hand vibration E_L, hence of the indices n_R and n_L

The electric field vectors are shown before (E_0) and after (E_1) going through the magnetic sample. The arrows E_R and E_L in each group correspond to the same point and the same time, and the amplitudes of E_R and E_L correspond to their superposition, at the same points. Rotation of the polarization of light is due to the velocity for E_L being larger.

Absorption has the effect of further introducing (circular) dichroism. The emerging wave will actually be elliptical.

The *polar Kerr effect* can be qualitatively understood in at least two ways. Consider the situation of normal incidence, with magnetization perpendicular to the surface. The first approach, close to the Faraday effect, consists in noting that reflection is associated to a small path, first one way then the other, in the material. The non-reciprocal character of the Faraday effect shows that there will then be a resulting rotation θ_K of the light polarization and also ellipticity. We may however prefer to associate reflection with the actual surface, and describe it in terms of Fresnel equations. Then the different indices for right- and left-handed components, which remain the eigenmodes for propagation in the magnetised material, will again lead both to a difference in phase shift (hence a Kerr rotation) and a different reflected amplitude (ellipticity).

We saw, earlier on, that the magneto-optical effects can be interpreted, at least qualitatively and classically, in terms of the Lorentz force. It is worth-while discussing the Ewald approach on the same basis. In the polar Kerr effect, we thus understand the presence, in the polarization distribution induced in the material by the electric field of the incident wave, of a component perpendicular to the initial polarization. This component, due to the Lorentz force, leads to the radiated wave featuring a rotation and ellipticity.

It also shows, at least qualitatively, that there is no effect for an incident wave with electric field vector perpendicular to the plane of incidence (s polarization) in the *transverse Kerr effect* geometry. The vibration is then parallel to magnetization, and there is no Lorentz force. In a similar way, consider the p component (electric field parallel to the plane of incidence), whose change in amplitude on reflection is the transverse Kerr effect. It is easy to see that the Lorentz force is now in the plane of incidence and therefore cannot induce a perpendicular component. There is thus no rotation. The amplitude of the reflected wave results from interference between a contribution corresponding to metallic reflection E_{metal} and a magnetic component E_{mag}. The weakness of the Kerr effect results from the large difference in their amplitudes. Reversing the magnetization at the measured point ("up" to "down") produces a phase rotation by π for the magnetic component and a change in the amplitude E_{res} resulting from interference, as shown on the Fresnel diagram of figure 13.7.

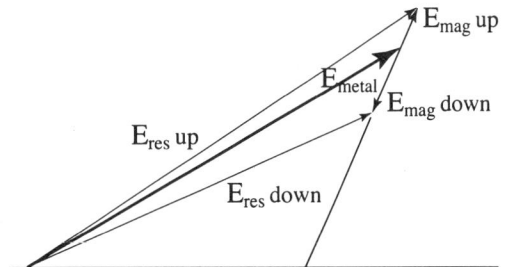

Figure 13.7 - Interference between the amplitude due to metallic reflection E_{metal} and the magnetic component E_{mag}, shown in Fresnel representation for the case of the transverse Kerr effect

This explanation of the magneto-optical coupling as due to the effect of the Lorentz force from the applied field on the electron distribution leads to fairly adequate orders of magnitude for dia- and paramagnets. The applied field is on the other hand blatantly insufficient in the case of ferromagnets. Agreement becomes satisfactory when the spin-orbit coupling, which acts on the electrons as would an effective field of the order of 10^2-10^3 teslas [5, 16], is taken into account.

Whether it applies to a rectilinear or a circular eigenmode, a refractive index will be magnetic field sensitive only if it is affected by a small variation in the angular frequency of the wave, of the order of the Larmor angular frequency $eB/2m$ of the electrons. The material must therefore feature non-negligible dispersion (a sizeable variation of the indices with the photon energy or wavelength). This again makes it mandatory to work in a spectral range where absorption lines due to magnetic-sensitive electronic levels or energy bands occur. This establishes the close connection between magneto-optical effects and absorption.

We can thus understand, semi-qualitatively, the fact that the 3d metals show, in the visible range, larger effects than the rare earth metals, the magnetism of which is associated with states further from the Fermi level.

The microscopic description of these effects should be based on that of solid state magnetism. Two simplified approaches are the localised electron model and the itinerant electron model, discussed in chapters 7 and 8.

In the localised electron model, which applies e.g. to the rare earth metals, the interactions in the material lift the degeneracy of electron energy levels. This leads to a representation in which successive splittings occur, with various energies (correlations between electrons, spin-orbit coupling, crystalline electric field, ...). Electromagnetic radiation will induce transitions with probabilities depending on the energy supplied, the polarization, the filling of the initial multiplet levels, as well as on that of the final levels following excitation. Thus the allowed transitions contribute to the effects with various weights [17]. As the states can be rather well defined, it is possible to build models accounting for the magneto-optical effects in those materials where the magnetic electrons are localised.

The itinerant electron model, which describes e.g. transition atoms in the metallic state, represents the electronic energy eigenstates on a density of states diagram. The allowed energies are distributed over one or several bands, for which the density of states diagram shows the profile. Under the action of an applied field, or in the ferromagnetic state, an imbalance appears in this diagram between the states with opposite spin direction, and the states are occupied up to the Fermi level. Electromagnetic radiation induces electronic transitions with probabilities that depend on the number of empty states among the final levels as well as on the polarization of the incident radiation. The induced transitions are different for the two spin populations, hence the effects are differential and generally have small amplitude.

4. MEASUREMENT AND ORDER OF MAGNITUDE OF THE MAGNETO-OPTICAL EFFECTS

Magneto-optical effects are measured during the exploration and magnetic characterization of a material. The sample can be bulk, a thin film, an ultra-thin film, ... For a given wavelength, the direct result of the measurement has the form of a curve representing the quantity or quantities characteristic of the measured effect as a function of applied field: $\Theta_F(H)$ for the Faraday effect, $\Theta_K(H)$ for the polar and longitudinal Kerr effects, $\Delta R / R\,(H)$ for the transverse Kerr effect, etc. Beyond this first approach, one is interested in how the characteristic quantity changes with the orientation of the sample, its temperature, the photon energy, etc.

The measurement is performed over a region of the sample that can be termed macroscopic. This results in an average value which is really significant only if the probed region has reached magnetic saturation. Values for intermediate field do however contain valuable information on the magnetization processes. Using one of the magneto-optical basic configurations provides information on the way a given component of magnetization changes. Depending on the effect used, the measurement will be:

- the determination of the rotation of a polarization direction: this is the case for the *Faraday effect,* and the *polar and longitudinal Kerr effects*,
- the ellipticity of the transmitted or reflected vibration: this is again the case for the *Faraday effect, and the polar and longitudinal Kerr effects*,
- the variation in the transmitted or reflected intensity: this is the *transverse Kerr effect, and circular and rectilinear dichroism*,
- the variation in the phase shift between perpendicular components of the field **E**, leading to the measurement of ellipticity and its variation: this is the case of the *Voigt effect*.

In practice, the three first points are, by far, the most frequent. The various effects can be measured using an ellipsometer. However, they are usually quite small, as the angles rarely exceed a degree and the relative intensity variations some 10^{-3}.

Therefore, in order to reach the necessary sensitivity in detection, and to make the operation simple, various special forms of ellipsometry have been developed. They involve modulation of the light signal, which improves the detection and amplification by eliminating unwanted frequency components and minimising drift.

We will briefly describe the basis of these techniques, and then discuss, with no claim to completeness, the experimental approaches used in the measurement of the various effects, viz polarization rotation, ellipticity, intensity variation and phase difference. We will in particular discuss differential methods, which have the advantage of always increasing the relative variation of the resulting signal so that the amplification possibilities in the measurement chain can be fully used.

4.1. MODULATION METHODS

4.1.1. Amplitude (or intensity) modulation

An amplitude modulator chops the beam at frequency $f = 1/T$ into square impulses or into a sine-like shape. The simplest example is the chopper. Call S(t) the time variation of the intensity of the unmodulated signal, as given by a photoelectric detector, due for example to the action of the applied field. After modulation into rectangular impulses with unit amplitude, and period $T = 2\pi/\omega$, the exit signal S'(t) can be decomposed into:

$$S'(t) = S(t)\left[\frac{4}{\pi}\left(\sin \omega t + \frac{1}{3}\sin 3\omega t + \frac{1}{5}\sin 5\omega t + ...\right) + \frac{1}{2}\right] \quad (13.34)$$

or, for sine modulation:

$$S'(t) = S(t)(\sin \omega t + 1) \quad (13.35)$$

After filtering at frequency $f = \omega/2\pi$, there remains:

$$S'(t) = S(t)\sin \omega t \quad (13.36)$$

A lock-in amplifier locked on frequency f gives around time t a resulting signal proportional to $\int_t^{t+\tau} S(t')\sin \omega t' dt'$, i.e. a sliding average of S(t). τ depends on the time constant of the lock-in amplifier.

4.1.2. Phase (and polarization direction) modulation

A phase modulator is an optical component operating in transmission, for which two perpendicular directions correspond to the propagation eigenmodes. Let Ox and Oy be these directions, with Oz the propagation direction of the beam. Rectilinear incident polarization can be decomposed along Ox and Oy. The modulator introduces a phase difference, generally sine-shaped with angular frequency ω, between the outgoing components, so that:

$$\phi_y - \phi_x = \phi_0 \sin \omega t \quad (13.37)$$

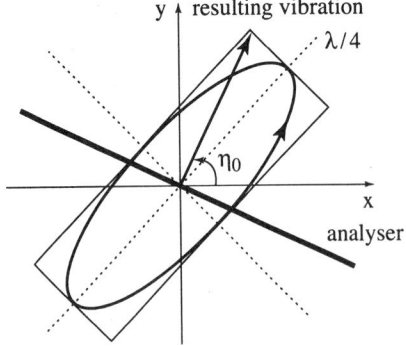

Figure 13.8 - Principle of polarization direction modulation

13 - MAGNETO-OPTICAL EFFECTS

The corresponding polarization is elliptical with variable ellipticity and, in the general case, with principal axes whose direction is modulated too. However, if the modulator axes are oriented at 45° to the azimut of the incident vibration, the outgoing vibration does have variable ellipticity but its principal axes have fixed directions (fig. 13.8).

Since the final detector is a photoelectric intensity detector, the phase variation must be changed into an intensity variation.

This will be realised by adding:
- a quarter-wave plate ($\lambda/4$) the neutral lines of which will be aligned with the axes of the ellipse so that the phase shift of $\pi/2$ between these axes is compensated for. This results in a rectilinear vibration, the direction of which is time-modulated with a sine variation in the xOy plane, around an average direction at angle η_0 with Ox:

$$\eta = \eta_0 + (\phi_0/2)\sin \omega t \qquad (13.38)$$

- an analyser with transmission direction at an angle Ψ to Ox. It converts the change in rectilinear polarization direction into a change in intensity:

$$I = I_0 \cos^2[\Psi - (\eta_0 + \phi_0/2)\sin \omega t)] \qquad (13.39)$$

We recall that:

$$\cos(\Phi_0 \sin \omega t) = J_0(\Phi_0) + 2\sum_{p=1}^{\infty} J_{2p}(\Phi_0)\cos 2\omega t \qquad (13.40)$$

$$\sin(\Phi_0 \sin \omega t) = 2\sum_{p'=1}^{\infty} J_{2p'-1}(\Phi_0)\sin[(2p'-1)\omega t] \qquad (13.41)$$

where the J_n are Bessel functions of the first type, of order n.

Expression (13.39) thus shows the decomposition into harmonics of the modulation frequency ω. There appear two solutions which cancel the ω component of the photoelectric signal: either $\Psi = 0$, i.e. the analyser is parallel to the average position of the rectilinear vibration; or $\Psi = \pi/2$ in which the analyser is perpendicular to this average position. It can easily be shown that the $\Psi = \pi/2$ configuration is better, as it provides better sensitivity, expressed by $\partial I/\partial \eta$, as well as a larger relative variation $\Delta I/I$.

4.2. MAGNETO-OPTICAL DETERMINATION OF THE EASY MAGNETIZATION DIRECTIONS

The easy magnetization direction(s) correspond to the orientation(s) of the magnetization vector in the absence of applied field. Knowing this direction is mandatory for the measurement of some effects. There are three ways of determining them:
- by looking for the sample orientation that corresponds to as rectangular a hysteresis loop as possible,
- by observing the domains and the preferential wall orientations directly,

◆ by measuring, with no applied field, the three components of the magnetization vector with respect to an orthonormal reference frame Oxyz [18], [19].

4.3. MEASURING THE ROTATION Θ
OF A RECTILINEAR POLARIZATION DIRECTION
(FARADAY EFFECT, POLAR AND LONGITUDINAL KERR EFFECTS)

The most straightforward way of determining the direction of a rectilinear vibration using a photoelectric detector consists in projecting it onto the transmission direction of a rectilinear polar (fig. 13.9) (*the term polar is recommended when the distinction between the polariser and the analyser is not clear or not necessary*).

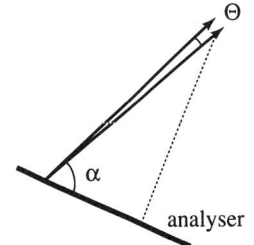

Figure 13.9
Polarization rotation in the polar Kerr effect

If $\alpha + \Theta$ is the angle between the vibration and this analyser, the emerging intensity is proportional to $\cos^2(\alpha + \Theta)$, where α is adjustable at will by the experimentalist, and Θ is the rotation to be measured. In general, the rotation angle due to the magneto-optical efect is small (polar Kerr effect), and the variation in transmitted or reflected intensity is thus proportional to $\Theta \sin 2\alpha + \Theta^2 \cos 2\alpha$. This variation reaches its maximum value Θ for $\alpha \# 45°$. The relative variation $\Delta I / I$ has its maximum around $\alpha = 90°$ because then the intensity is near zero while the absolute variation is proportional to Θ^2, hence very small. Thus the signal to noise ratio is not very favorable.

In an amplitude modulation regime, the photoelectric signal is proportional to the intensity, hence to $\cos^2 \alpha$. If G is the amplifier gain, the resulting signal is proportional to $G \cos^2 \alpha$ and the variation due to the magneto-optical effect to $G \sin 2\alpha$. Because the total exit signal cannot exceed the saturation threshold of the amplifier, a compromise must be found between the gain G and the angle α. It is reached around $\alpha = \pi / 2$ where G and the relative variation $\Delta I / I$ are maximum while the variation is equal to $G \Theta^2$. Preliminary calibration is required to provide Θ.

In order to enjoy both the maximum value of the variation ΔI and a large relative variation, making the use of a high gain G possible, a differential method is preferred. The incident rectilinear vibration is then physically decomposed into two perpendicular components, along Ox and Oy. These beams are collected by two distinct detectors, yielding signals S_x and S_y. By adjusting α to around 45°, these two components have the same amplitude. Then the difference is close to zero, while its variation between the two magnetization states, characterised by the rotation Θ, is proportional to $2\Theta \sin 2\alpha$, i.e. to 2Θ for $\alpha = 45°$. Θ can then be accurately determined from the ratio $(S_y - S_x)/(S_y + S_x)$ for either magnetization state (denoted $\uparrow$ and $\downarrow$) because:

$$\Theta = \frac{1}{4}\left[\sin^{-1}\left(\frac{S_y - S_x}{S_y + S_x}\right)_\uparrow - \sin^{-1}\left(\frac{S_y - S_x}{S_y + S_x}\right)_\downarrow\right] \quad (13.42)$$

In this case, no preliminary calibration is required.

In a polarization direction modulation regime of the form $\eta_1 \sin \omega t$ (see § 4.1.2 on phase modulation), the magneto-optical effect introduces a rotation Θ around the average orientation of the vibration. If the analyser is at angle ζ to this average position, the transmitted intensity is proportional to $\cos^2(\zeta + \eta_1 \sin \omega t)$. After filtering out the ω contribution, this results in a signal S proportional to $2J_1(2\eta_1) \sin 2\zeta$, where J_1 is the Bessel function of first type of order 1.

We note that S vanishes for $\sin 2\zeta = 0$, i.e. when the direction of the analyser either coincides, or is crossed, with the average orientation of the vibration. The crossed configuration is preferred, in order to enjoy a larger modulation depth and more sensitivity. Hence there are two methods for measuring Θ:

- locating, with respect to a fixed reference, such as Ox, the analyser positions α_1 and α_2 which cancel the signal S from the amplifier. They are such that $\alpha_2 - \alpha_1 = \Theta$, where α_1 and α_2 correspond to the two saturated states of the specimen;
- preliminary calibration of the variation of the signal S as a function of the average orientation of the vibration, i.e. $\partial S/\partial \zeta$, will give the value of Θ. Since the angle Θ is small, there will be no linearity problem ($\partial S/\partial \zeta$ = constant).

The angular resolution is high, and can reach a few microradians. The angles of interest are generally fairly small, and any method that can increase them or subtract parasitic rotations should be considered: antireflection coating [1, 20], automatic compensation of diamagnetic components [21].

4.4. MEASURING ELLIPTICITY

Ellipticity can be defined as the ratio of the a and b axes of the ellipse, viz $\tan \varepsilon = b/a$ where ε is the angle between the diagonal of the rectangle enclosing the ellipse and the latter's major axis. Measuring $\tan \varepsilon$ can be performed through the measurement of a and b. Intensity measurements only yield a^2 and b^2, i.e. $|\tan \varepsilon|$. The sign of $\tan \varepsilon$ gives the direction of rotation of the electric vector along the ellipse. In fact, determining $\tan \varepsilon$ completely implies a true ellipsometric measurement and requires the use of a compensator. This can be a Babinet type compensator [22, 23], with its neutral lines parallel to the incident rectilinear vibration, or Sénarmont-type (quarter-wave plate + analyser), with the neutral lines of the $\lambda/4$ plate parallel to the ellipse axes, the analyser being used to locate, via extinction, the outgoing rectilinear vibration.

Knowing the neutral lines (fast and slow axes) of the compensator (Babinet or $\lambda/4$) makes it possible to determine the direction of rotation on the ellipse.

4.5. MEASURING AN INTENSITY VARIATION (TRANSVERSE KERR EFFECT, CIRCULAR AND RECTILINEAR DICHROISM)

We deal with this measurement in the range of visible or near-visible wavelengths. The X-ray case, as used with synchrotron radiation, is special and will not be mentioned here.

We assume that the easy magnetization directions have been determined and that the experimentalist has selected the basic geometry for the effects to be measured. In all cases, the incident polarization corresponds to the propagation eigenmodes, viz p (or TM) for the transverse Kerr effect, right-hand circular then left-hand circular (or the other way round) for circular dichroism, rectilinear along two perpendicular directions successively for rectilinear dichroism. In practice, the sample is followed by a rectilinear analyser, possibly associated with a quarter wave plate, to make sure that the component that is measured is the correct one.

Measuring an intensity implies measuring a reference. As in the case of rotations, the procedure is based on the difference between the two saturation states of magnetization, in practice by cycling the field between the two saturations. The reason is that there is no reference state that would correspond to zero local magnetization, at least below the Curie point.

In the case of amplitude modulation, there is no special difficulty other than the weakness of the magneto-optical change in amplitude.

Phase modulation makes it possible, by adjusting the modulated phase, to go over alternatively from one polarization to the orthogonal one (right-hand to left-hand circular, or rectilinear along Ox to rectilinear along Oy), and thus to directly obtain the dichroism signal.

Here again, differential methods make it possible to increase the relative variation, and to enjoy more comfortable a signal. Thus, in the transverse Kerr effect setup, the constant part of the reflection factor R_p (sensitive to the sign of magnetization) can be compensated for through an intensity proportional to R_s, which is insensitive to the magnetization sign [24].

4.6. MEASURING THE PHASE DIFFERENCE AND ITS VARIATION

Once the easy magnetization direction is located, both the applied field and the projections on the sample of the compensator axes can be aligned with it. The phase difference, and its evolution with magnetization are related to the direction of the analyser placed downstream. A variation in phase by ϕ entails a rotation by $\phi/2$ of the emerging rectilinear vibration, hence of the crossed analyser.

4.7. SOME ORDERS OF MAGNITUDE

The results of measurements are expressed in different ways by different authors. The direct results correspond to the measurements for which we have just described the techniques. These are, for instance, the angle Θ_F and the ellipticity ε_F for the Faraday effect, Θ_K and ε_K for the polar or longitudinal Kerr effects, $\Delta R/R$ for the transverse Kerr effect. Another more elaborate formulation expresses the real and imaginary parts of the permittivity tensor [25]. Also encountered are the value of the magneto-optical constant Q, or the Verdet constant for dia- and paramagnets. For a material in a given form (bulk, thin film, ...) these values are represented as functions of the energy of the incident radiation, the angle of incidence, the temperature, etc.

We will just give a few orders of magnitude of the various effects for materials under well defined conditions.

4.7.1. Faraday effect

Diamagnets

The Verdet constant k_V for silicate glasses, at $\lambda = 632.8$ nm and at room temperature are between 4 (standard optical glasses of the SiO_2 and BK-7 types) and 27 rad.T^{-1}.m^{-1} (high refractive index and high dispersion glasses). Amorphous or crystalline oxides containing cations with easily polarisable outer electronic shells (Te^{4+}, Pb^{2+}, Ta^{5+}, Bi^{3+}, La^{3+}, ...) have large Verdet constants k_V, as do sulfur, selenium or arsenic chalcogenides, while oxygen-less fluorides have low Verdet constants. k_V is weakly temperature dependent in this case, and it decreases when wavelength increases.

Paramagnets

The Faraday rotations are induced by $(4f_n) \rightarrow (4f_{n-1}\ 5d)$ transitions of the paramagnetic rare earth ions. The ions with high 5d band energy (Gd^{3+}, Yb^{3+}) are transparent in the visible range and produce a weak rotation (less than 10 rad.T^{-1}.m^{-1}). The ions with low 5d band energy (Tb^{3+}, Ce^{3+}) give rise to high rotation, from about 20 to over 100 rad.T^{-1}.m^{-1}. The divalent ions such as Eu^{2+} have 5d bands with even lower energy. The Verdet constant is then even higher, 262 rad.T^{-1}.m^{-1}. k_V decreases as $1/\lambda^2$ with increasing λ.

Ferrimagnets

The ferrimagnetic garnets $R_3Fe_5O_{12}$ (where R is a rare earth or yttrium) gave rise to an extensive literature due both to the perspectives they were credited with in the area of data recording, particularly in the form of bubble domains and to their value in microwave applications. The high Faraday rotation angle for small sample thickness is one of the attractive properties of this family of materials.

At room temperature (300 K), in the visible and near infrared range corresponding to the spectral range in which they are transparent, the values of Θ_F for rare earth garnets range from $12°.\text{cm}^{-1}$ for $Yb_3Fe_5O_{12}$ to $-840°.\text{cm}^{-1}$ for $Nd_3Fe_5O_{12}$ at $\lambda = 1.06$ µm. Ytrium iron garnet gives a rotation Θ_F of $280°.\text{cm}^{-1}$ for the same wavelength, whereas, at $\lambda = 0.3$ µm, Θ_F is on the order of $10^5°.\text{cm}^{-1}$. Substituted garnets, of the type $(BiRGa)Fe_5O_{12}$, also give large rotations between 1 and $7 \times 10^{4}°.\text{cm}^{-1}$ in the visible range, and over 10^5 in the near ultraviolet. Other substitutions involving cobalt, lead or cerium give rotations of several thousand degrees per centimeter. This explains the attractiveness of garnets since, when epitactically grown as rather thick layers (several micrometers), they give rotation angles of several degrees.

Ferromagnetic metals and alloys

Considering the Faraday rotation, a transmission effect, makes sense in these materials only when the thickness is small, because absorption is generally high. The absorption coefficient of metals is usually between 10^5 and 10^6 cm^{-1} for metals. For *3d metals* (iron, cobalt and nickel), at saturation, for $\lambda = 546$ nm and at 300 K, the rotation angles are high, between 10^5 and $10^6°.\text{cm}^{-1}$ [26]. The rotation angles for the *4f rare earth metals* are much smaller in the visible range, because the electron transitions only play a part for wavelengths in the ultraviolet range. But then absorption is very high. Practically no data are available at room temperature, as the Curie temperatures are below.

The *R-M* alloys, where R is a rare earth and M a transition metal, are of interest toward applications insofar as their use as thin films or multilayers corresponds to small thicknesses, hence to the Faraday effect rather than the polar Kerr effect. Examples are the behavior of the R-M alloys (with R: Tb, Gd and M: Fe, Co) at 300 K and for $\lambda = 632.8$ nm. The maximum of Θ_F in the R-Fe system is between 1.8 and $2 \times 10^{5}°.\text{cm}^{-1}$ for 75 to 80% iron while the R-Co alloys have Θ_F growing toward the value for pure cobalt, viz around $4 \times 10^{5}°.\text{cm}^{-1}$ [27].

Ferromagnetic compounds can be *insulators*, as the chromium trihalogenides $CrCl_3$, $CrBr_3$ and CrI_3, which feature high Faraday rotations (between 10^3 and $10^5°.\text{cm}^{-1}$ around 1.5 K). They can also be *semiconductors* like the europium chalcogenides such as EuO or EuS, with a Curie temperature far below 300 K and high Faraday rotation for small thickness.

4.7.2. Polar Kerr effect

Plentiful data are available in the literature on the polar Kerr effect: see in particular Buschow [28]. This is mainly due to the potentialities for magneto-optical recording.

The *transition metals* have Kerr rotation angles at room temperature, in the visible range ($\lambda = 633$ nm), of $-0.30°$ for cobalt, $-0.41°$ for iron and $-0.13°$ for nickel. Buschow's reference book includes values of Θ_K for many *intermetallic alloys* of these metals in bulk and crystalline form for various compositions. For cobalt alloys, the values of Θ_K range from +0.01 for Co_7Gd_2 to $-0.40°$ for $Co_{70}Pt_{30}$. For iron alloys, they vary between $-0.14°$ for $Fe_{50}Pd_{50}$ and $-0.53°$ for $Fe_{85}Pt_{15}$. Pseudo-binary and ternary alloys have rather small values of Θ_K except in special cases ($-0.36°$ for $Fe_{2.7}V_{0.3}Ge$).

The *Heusler alloys* are ternary intermetallic ferromagnetic compounds with composition X_2YZ. The angles are generally small ($< 0.2°$), although there are high values for the alloys FeCoGa ($-0.41°$). The XMnY type alloys (PtMnSb, PdMnSb, NiMnSb, PtMnSn), related to the Heusler alloys and called semi-metals because they have a semiconductor-type gap in the minority spin band structure, can give rise to large Θ_K rotations ($-1.2°$ for PtMnSb at $\lambda = 750$ μm).

One of the criteria in choosing materials for magneto-optical recording is the room temperature value of the Kerr rotation for wavelengths available from laser diodes (near infrared or red). The materials are deposited as amorphous thin films or multilayers. They are *R-M films*, with R a rare earth (Tb, Gd, Dy) and M a 3d transition metal.

Θ_K at 300 K in the red range is about $0.4°$ for the $Gd_{1-x}Fe_x$, and $-0.1°$ to $-0.3°$ for the $Gd_{1-x}Co_x$ families as x increases from 0.6 to 1. The Kerr and Faraday rotations for these alloys are compared in a paper by Hansen *et al.* [27]; at 300 K, for $\lambda = 632.8$ nm in the case of $Gd_{0.15}Fe_{0.85}$, Θ_F is $2 \times 10^{5°} \cdot cm^{-1}$ whereas $|\Theta_K|$ is $0.4°$. The Kerr rotation thus corresponds to the Faraday rotation across 50 nm of the same material.

The TbFe alloys have been widely investigated. They give Θ_K around $0.2°$ at 600 nm and 300 K. Under the same conditions, the TbFeCo alloy used in commercial magneto-optical disks rotates the polarization by $\Theta_K = -0.17°$ whereas Θ_K is $-0.25°$ for GdFeCo.

The RFe_2 family (with R = Gd, Er, Ho, Dy, Tb) have been systematically investigated as a function of λ at 300 K [29]; the maxima of $|\Theta_K|$ occur around $\lambda = 300$ nm and range between $-0.4°$ for $TbFe_2$ and $+0.2°$ for $GdFe_2$.

4.7.3. Longitudinal Kerr effect

Although we discussed the longitudinal Kerr effect configuration as basic, it does not give unambiguous access to a well defined component of magnetization. It is impossible to make a clear-cut distinction between the longitudinal effect and the oblique incidence polar effect. Only measurements as a function of incidence angle can provide an answer. Furthermore, the in-plane components of magnetization are measurable through the transverse Kerr effect, which is quite widely used, especially for thin films and multilayers. This is why there are few results of longitudinal Kerr effect measurements in the literature, although this effect is broadly used for the observation of magnetic domains when the magnetization vector is roughly parallel to the surface, a very frequent situation, which is privileged by magnetostatic energy (see chap. 20) unless *perpendicular* anisotropy is present.

The values of the rotation angle and of ellipticity depend not only on wavelength and temperature, as for the polar Kerr effect, but also on the angle of incidence, which can vary between 0 and 90°. The rotation angle is zero for these extreme angles of incidence and it is maximum around the principal angle of incidence [30], for which the phase shift between the p and s components is $\pi/2$.

4.7.4. Transverse Kerr effect

Using the transverse Kerr effect has two advantages. It directly provides a quantity, ($\Delta R/R$), sometimes denoted as δ, which is proportional to the average value of magnetization over the measured area. Also it is simple to implement, and does not require an angle measurement because only the p or TM propagation eigenmode is involved. On the other hand, the value of $\Delta R/R$ depends on the angle of incidence. The maximum is generally around the principal refraction angle for the material, viz of the order of 65° to 80° for metals. $\Delta R/R$ also depends on λ and of course on T (hence M). The convenience of this effect has long made it a very handy tool [25], [31], [32]. Its use has sharply increased with the growing interest for magnetic thin films and, in general, for surfaces. It applies to ferro- and ferrimagnets.

The 3d metals were measured first. For thin cobalt films, $\Delta R/R$ goes through a maximum of 1.5×10^{-2} for $\lambda = 700$ nm and an angle of incidence of 71° [32]. For $\lambda = 632.8$ nm, at room temperature and for an incidence angle of 60°, a value of $\Delta R/R$ on the order of 1.1×10^{-2} was measured. Bulk cobalt shows identical behavior; infrared measurements show maxima in $\Delta R/R$ around $\lambda = 3$ µm, with values 1.6×10^{-2} and 3.4×10^{-2} for incidence angles of 75° and 85°, respectively [32].

For nickel, in the wavelength range between 0.7 and 5.8 µm, $\Delta R/R$ goes through extremal values around 1.5 µm, where $\Delta R/R$ is about -7×10^{-3}, and 5.5 µm where $\Delta R/R$ is around 2×10^{-3}, with a cancellation of the effect at 3.8 µm [32], [33]. Around 700 nm, $\Delta R/R$ is about -4×10^{-3}. The sign conventions are those of reference [32].

Iron shows extremal values of $\Delta R/R$ around $\lambda = 0.7$ µm and 3 µm with values respectively near 5.5×10^{-2} and 1.5×10^{-2} [32, 34]. The literature also contains many examples of transition metal and rare earth based multilayer systems which give a feeling for the sensitivity of this method (the atomic monolayer!) [35], [36].

5. USES AND APPLICATIONS OF THE MAGNETO-OPTICAL EFFECTS

We here distinguish the uses of the magneto-optical effects in physics, i.e. in experiments aiming at a better understanding of magnetic materials, from their application in devices.

5.1. USES IN PHYSICS

5.1.1. Magnetic characterization. Hysteresis loop determination

Under geometrically well defined conditions, involving a single component of induction **B** (i.e. in practice, in ordered magnetic materials featuring spontaneous magnetization, involving a single component of magnetization **M**), the sensitivity of the magneto-optical effects makes it possible to measure the variation in average magnetization as a function of the applied magnetic field, hence the hysteresis loop.

In reflection (Kerr effect), this measurement provides information about a shallow surface region of the material, with thickness on the order of the penetration depth, averaged over the illuminated region. The surface hysteresis loop thus obtained may be completely different from the loop corresponding to the bulk, as can be measured by classical methods. This difference is interesting because it reveals the importance of domains located inside the sample, where magnetization takes on a direction different from that in the closure domains which occupy the surface.

In the investigation of thin enough magnetic films, the magneto-optically measured loop is representative of the whole sample. Then the short measuring time, the possibility of measuring *in situ*, e.g. in an ultra-high vacuum chamber right after the specimen preparation and in parallel with other characterization methods, are highly appreciated. In particular, the hysteresis loop looks completely different depending on whether the field is applied in the easy direction or along a direction of difficult magnetization (see chap. 20). Comparing the results thus obtained makes it possible to determine the anisotropy of the films, as well as the values of the coercive field or of the remanent magnetization.

The development of the physics of magnetic multilayers and super-lattices, over the last years, has led to their magneto-optical investigation, both experimental and theoretical. The analysis of the results, on the basis of matrix formalisms such as that of Visnovsky [15], is particularly rich. Thus Penissard showed in his thesis that it should be possible to detect selectively the hysteresis loop of *one* of the films within a multilayer by appropriate choice of the photon energy. Hubert and Traeger [14] introduced *sensitivity functions* to deal with such systems.

5.1.2. Variation of magnetization in homogeneous systems Phase transitions

Magneto-optics provided a major contribution, in the 1970's, to the investigation of phase transitions and critical phenomena. The high sensitivity of optical measurements makes it possible to measure the variation of effects such as the Faraday rotation [37], [38] or rectilinear birefringence [39], [40]. Over a small temperature range (a few kelvin), the only quantity that varies strongly is often the magnetization or, in the case of antiferromagnets, the sublattice magnetization. Such measurements made it possible to determine a whole series of critical exponents, and to compare them to the predictions of the renormalization group theory (see chap. 10).

5.1.3. *Magneto-optics in diffraction*

Classically, magneto-optical effects deal with homogeneous systems. If a beam of visible light is sent on a spatially periodic object, with a period in the micrometer range, such as a grating made of magnetic material, or a two-dimensional periodic arrangement of magnetic islands, it is diffracted. Each of the diffracted beams then undergoes a magneto-optical effect, which can be described, for diffraction orders other than the zeroth, as non-specular [41].

5.1.4. *Resonances*

The presence, during the precession associated with a spin wave resonance, of a component of magnetization perpendicular to the direction of the applied dc magnetic field, can be detected through its effect on light passing between crossed polars [2]. This can provide an original way of analysing resonance modes in a transparent material. Here however we enter the area of inelastic magneto-optical effects (light scattering by spin waves), which we want to keep out of this chapter.

5.1.5. *Magnetic domain imaging*

Light will have different characteristics after it is reflected or transmitted by regions in which the orientation or the arrangement of magnetic moments are different, i.e. by different magnetic domains (see chap. 3). Its polarization direction may have turned one way or the other, or its intensity may be different. It will therefore be possible to image magnetic domains by detecting the inhomogeneity of the beam associated with local variations in magneto-optical behavior. We note that the effect of the domains leads to the macroscopic measurement of the hysteresis loop when integration is performed over the illuminated light instead of making the local use that corresponds to imaging.

The observation of *ferro- or ferrimagnetic* domains is based on the Kerr effects or, in transparent materials or very thin films, on the Faraday effect. The latter provides, for films a few µm in thickness of a magnetic garnet, grown for instance on a thick but transparent gadolinium gallium garnet (GGG) substrate, a comfortable rotation (a few degrees) of the polarization direction and an easily visible contrast, enhanced by colours. The observation of domains in a polarising microscope is then very pleasant, and indeed seeing them move under the influence of a magnetic field is a spectacular initiation to the physics of magnetic domains.

Conversely, the polar and longitudinal Kerr effects, the latter being applicable when the magnetization is in the surface plane, lead to very small rotation of the polarization plane of light (a few minutes). Observing the contrast from different domains thus leads to using an analyser almost crossed with the polariser, and the light intensity level becomes very low. Furthermore, use of the longitudinal mode requires oblique incidence, and any surface corrugation then produces, under crossed polars, violent contrast. In practice, this makes the demands on the surface polish more stringent.

Whatever the experimental method used, this surface must be free of a cold-worked layer, because otherwise the domains themselves will be altered through magnetostriction, and the observations lose significance. The surface must therefore be electrolytically polished or submitted to a controled etch. The supplementary demand from the optical side is that the surface be very plane.

Two auxiliary techniques can alleviate the difficulties. An electronic image processing device makes it possible to subtract the image of the sample in the magnetically saturated state, i.e. devoid of domains, and, if needed, to artificially enhance the contrast [42]. Also, depositing an interference film with high refractive index, such as ZnS, Si or SiO_2, with well-controled thickness, can increase the rotation for the longitudinal Kerr effect [43], [44]. The principle is to decrease, through destructive interference (fig. 13.10), the component of the electric field of the light wave that corresponds to ordinary reflection, without attenuating the small perpendicular component which can be associated with the magneto-optical effect.

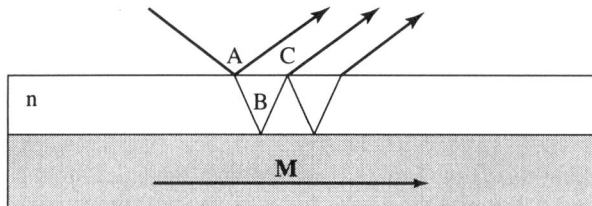

Figure 13.10 - Increasing the Kerr effect through a dielectric interference film. The waves reflected into vacuum from A, C etc. must have phase differences of π

The optical thickness (the thickness times the refractive index) of the film must therefore be near $\lambda/4$, so that the wave reflected on its surface be phase shifted by π with respect to that which comes out after going through the film twice, and acquiring its *magnetic* contribution on the magnetic surface. This method is also applicable to the polar Kerr effect [45] and to the transverse Kerr effect [46].

Magnetic domain imaging has recently started exploring a promising new path, with the advent of near-field microscopy [47] (see appendix, § A5). Apart from its interest in basic research and from its elegance, this approach holds very exciting possibilities for very high density magnetic recording. Much remains to be discovered on magneto-optical effects in the near field.

In type I superconductors, the Faraday effect made it possible to directly visualise the intermediate state (coexistence of the normal and superconducting phases, associated with the transition under an applied magnetic field), using a paramagnetic film with high Verdet constant [48]. In type II superconductors, magneto-optics also makes it possible to measure the hysteresis loop, although this is now related to changes in the configuration of the vortex lines (regions where **B** is non zero) in the mixed state. This type of investigation has been largely developed these last years, for high critical temperature superconductors [49]. The ferrimagnetic field detector invented by

Dorosinskii *et al.* [50], as mentioned in sections 5.1.6 and 5.2.2, is very helpful in these investigations, for which it was initially developed.

For the moment, we only discussed imaging in visible light. Nothing is changed in the principle if infrared light is used, e.g. to investigate domains in transmission in materials which are more transparent in the infrared than in the visible range. Obviously, however, this requires lenses that can operate in the required range, adequate polars, and a converter that makes it possible to see the image.

The development of synchrotron radiation sources that provide an extremely high flux of X-rays over a very broad range of energy recently made the investigation of magneto-optical effects in this range possible. The feasibility of the observation of domains through magnetic X-ray scattering was demonstrated [51] in the case of magnetite Fe_3O_4. The technique is then identical to Bragg diffraction imaging, often called X-ray topography [52]. However, in its usual form, where diffraction is due to the electronic charge (Thomson scattering), walls or domains contribute to contrast only through the variation in distortion they entail via magnetostriction (chap. 5, § 6.1.3). 180° walls and domains are then practically invisible. In magnetic scattering imaging, intensity varies from one domain to the other, in the same way as in neutron topography [51], but the contrast is weak.

Another family of approach appears promising: that associated with circular magnetic X-ray dichroism. This is the variation with the direction of magnetization in the material of the absorption of circularly polarised X-rays with energy near a well-chosen absorption edge. It could be used directly, in particular through a scanning imaging mode. An elegant variant was recently implemented; it uses the variations, related to those of absorption, in the number of photoelectrons ejected [53], by collecting these electrons through a system of (electron) lenses that produces an image. This process is intimately linked to the energy of the absorption edges, hence characteristic of one investigated element. It therefore becomes possible to selectively, and successively observe domains in one or the other films in a super-lattice or multilayer, hence to obtain very direct information on the way they are coupled. It is likely, as techniques for tuning X-ray polarization develop [54], that these approaches will blossom in the next years.

5.1.6. Magnetic field maps

The magneto-optical detector developed by Dorosinskii *et al.* [50], made up of a very soft transparent ferrimagnetic garnet film, grown on a non-magnetic garnet substrate, provides a higher sensitivity than paramagnetic films. The distribution of magnetization in the magnetic film largely follows the distribution of magnetic field which it encounters. When placed on a device or a specimen, and observed in a polarising microscope, it thus makes it possible, through the Faraday effect used in reflection, to visualise the distribution of magnetic field created by the sample.

5.2. APPLICATIONS

5.2.1. Optical isolators

Stimulated emission is generated in a laser through an external excitation which involves optical pumping in an active medium, called the amplifier, characterised by its gain. To ensure both stability of the emitted light and a thermal regime acceptable to the material, the light intensity in the active medium must remain below saturation. However, if part of the emitted light returns into the active medium through an unwanted reflection, it can generate instabilities in the emission, or even irreversible damage to the material. The purpose of an optical isolator is that of a photon valve, blocking any return of light toward the source.

The principle of an optical isolator is provided by the non-reciprocity property of the Faraday effect, as mentioned above. The beam is initially rectilinearly polarised (Pol$_1$ at 0°) at the exit of the laser. Then comes a component G such that its Faraday effect rotates the polarization direction by 45°, followed by a rectilinear polar (Pol$_2$ at 45°) passing this polarization with minimum attenuation. Any light that is reflected or back-scattered by a component downstream of the isolator suffers, as it returns toward the source, an extra rotation of its polarization plane by 45°. It is therefore blocked by the polar located on the source side.

Component G must induce a Faraday rotation Θ_F of 45°. It must therefore be made of a material that is transparent at the wavelength used (in the visible or near infrared range), with a thickness that must be consistent with the way the magnetic field H is produced.

The performance of an isolator operating at a given wavelength (for example 1.3 or 1.55 μm for fibre-optical communications) are expressed in terms of:
- the extinction ratio, or isolation given in decibels (dB), viz $-10 \log P_2/P_0$, where P_0 is the light flux coming directly from the source, and P_2 the flux returning to the source after crossing the isolator and being reflected by an ideal mirror. The isolation can reach 60 dB;
- insertion losses, also in decibels, i.e. $-10 \log P_1/P_0$ where P_1 is the transmitted, hence usable, flux. With good quality garnet, the losses can be less than 0.1 dB.

The best suited materials in terms of magnetic, magneto-optical [55], thermal and economic properties are found in the yttrium iron garnet ($Y_3Fe_5O_{12}$, YIG) family, thanks to appropriate substitutions. The bismuth doped garnets provide the highest rotations. The strong thermal dependence of Θ_F due to bismuth is alleviated by incorporating rare earth ions such as terbium, gadolinium, holmium, lanthanum or samarium. They bring the compensation temperature, for which $d\Theta_F/dT$ becomes very small, in the operating temperature range, typically 22°C [56].

The isolator configuration corresponds to magnetization perpendicular to the garnet, with typical thickness ranging from some hundreds of micrometers to a few millimetres [57].

The fields necessary to saturate the garnets range from 0.05 to 0.2 T. However, doping with gallium or aluminium leads to a weak permanent magnet type of material, which requires no applied field after it is saturated under a few hundreds of mT. This leads to a strong reduction in size and cost [56]. In this case the thermal dependence is less well compensated.

For isolators included in guided optical configurations [58], epitactically grown garnets are preferred. In this case the magnetization must be in the film plane. The isolator requires a smaller garnet thickness and a weaker applied field than in the description above. However it is mandatory to minimise rectilinear birefringence through annealing processes designed to relax the stresses.

5.2.2. *Magnetic field sensors*

A magnetic field sensor can be used as a tool for measuring the field (either its modulus, or the value of one component). In a different configuration, it can map its distribution, i.e. the spatial distribution of field lines. The latter application is described in the section devoted to non-destructive testing. The value of an electric current can be deduced from the magnetic field generated by the conductor.

The magnetic field tends to align the moments in a magnetic material. Some basic magneto-optical effects (Faraday, transverse and polar Kerr effects) are sensitive only to one component of this magnetization vector. By aligning the measuring device with respect to the magnetic field **H** so that the component of magnetization detected varies linearly with the amplitude of the field, at least in a given range, the local value of H can be deduced.

In a diamagnetic material with thickness and transmission coefficient such that light passes well, the Faraday rotation angle Θ_F is proportional to the applied field and such that $\Theta_F = k_V lB$, with k_V the Verdet constant, B the amplitude of the component along the light propagation direction of the field and l the thickness of the material. Good sensitivity will therefore be reached by choosing a diamagnetic material that associates a large Verdet constant and weak absorption at the wavelength used [59]. Applications include the measurement of an electric current through the Faraday rotation generated by the field created by the conductor in a single mode optical fibre wound around it. The small Verdet constant is then compensated by the large length l of the fibre.

For a ferromagnetic material with uniaxial anisotropy involving an easy magnetization direction Oz, the component of magnetization M_x perpendicular to Oz has, outside regions where the field is near the saturation value, an amplitude proportional to the applied field if the latter is along Ox. An odd magneto-optical effect (Faraday or Kerr) will then give a response proportional to the field.

Doped ferrimagnetic yttrium-iron garnets are here again frequently used [60] because S', the sensitivity per unit length, can reach high values, in particular near the compensation point [61]. This scaled sensitivity is independent of geometry, and it can be expressed as S' = N_{eff} S / l. Here N_{eff}, the effective demagnetising field factor, is a volume average of the geometrical factor N, which is usually expressed in tensor form, l is the length the beam goes through, and S = $d\Theta_F/dH = \Theta_{F,sat}/H_{sat}$ is the sensitivity. The sensitivities S, and S', the specific Faraday rotation (the rotation angle per unit length), the range of measurable fields and the temperature stability all depend on the garnet substitutions. In YIG, bismuth substitution increases the specific rotation, gallium reduces the saturation magnetization, and the thermal stability is improved by incorporating rare earths (see preceding section) such as gadolinium [62] or lanthanum into terbium iron garnet (TbIG). For these materials, the rotation angle per unit length can exceed $100°\,mm^{-1}$ while S' can be above $10°\,A^{-1}$. Furthermore, the frequency response involves no attenuation up to the MHz range, and it can even be acceptable up to the GHz range for some materials (YBiIG).

5.2.3. Non destructive testing

The sensitivity of optical methods is often used as a local or overall testing approach for materials or devices than affect the intensity, phase or polarization of light. It is generally assumed that the energy deposited by the probing beam does not alter the properties of the surface or volume probed. The magneto-optical effects are an extra non-destructive testing tool. They make it possible to directly visualise a magnetic state which can be affected by temperature or stress when they are used in the configuration described above.

They are also used to test the continuity of surfaces of magnetic materials [63] or the transport properties of non-magnetic materials. Any singularity or discontinuity in transport leads to an anomaly of the magnetic field generated by the electric current. A material featuring both high susceptibility and a strong magneto-optical coupling is used as a sensor of stray fields, and directly provides their spatial distribution. The ferrimagnetic garnets are once again used [64]. Such imaging systems, capable of detecting microcracks in metal sheets, are available commercially. There are many applications, ranging from tests of airplane bodies to those of metal plates in nuclear reactors. This type of control is part of a set of techniques for the inspection of metal surfaces and volumes.

5.2.4. Magneto-optical recording

Magnetic recording is one of the most important technologies, both for computer technology and in the arts involving sound and image [65]. It is discussed in chapter 21 of this book. Among the highways that have been explored over the last fifteen years is magneto-optical recording. It is now industrially implemented, and the magneto-optical compact disk has been present for several years on the consumer recording market (the Sony Minidisc®).

Appendix: Optics of non-magnetic materials

We briefly go over some basic notions of optics. Specialised books, such as those cited in the references, will of course have to be resorted to for more detail.

A1 - Angular frequency, wave vector, polarization

A light wave is characterised by its angular frequency ω, proportional to the energy $\hbar\omega$ of the associated photons, with $\hbar = 1.0546 \times 10^{-34}$ J.s Planck's constant divided by 2π; by its propagation direction represented, at least in simple cases, by its wave vector **k**; by its intensity, proportional to the modulus square of the amplitude of the electric field **E**; and by its polarization. In the simplest situation, that of propagation in vacuum, the wave is transverse, i.e. **E** is perpendicular to **k**. When the direction of **E** is constant, we refer to *rectilinearly polarised* light. During free propagation in vacuum, any two directions in the plane perpendicular to the propagation direction can be used as a reference frame to describe **E**. It is convenient, when dealing with refraction or reflection at an interface between two media, in oblique incidence, to refer the directions to the plane of incidence. This plane is defined as containing the perpendicular to the interface and the propagation direction of the incident beam. Light with its **E** vector perpendicular to the plane of incidence will be described by polarization s or σ, while the polarization is p or π if **E** is in the plane of incidence. These situations are also designated by the terms TE (**E** is Transverse with respect to the plane of incidence) and TM (the Magnetic field vector is Transverse with respect to the plane of incidence, hence **E** is in the plane of incidence).

But **E** can also turn, with angular velocity ω, during propagation. If it retains the same modulus, one speaks of *circular* polarization, right-hand or left-hand depending on the rotation direction. These descriptions refer to the curves which would be traced out by the end of **E** in the course of time. Conventionally [66], a right-hand circular wave is such that **E** seems to rotate clock-wise for an observer receiving the wave, hence looking at the source (fig. 13.11).

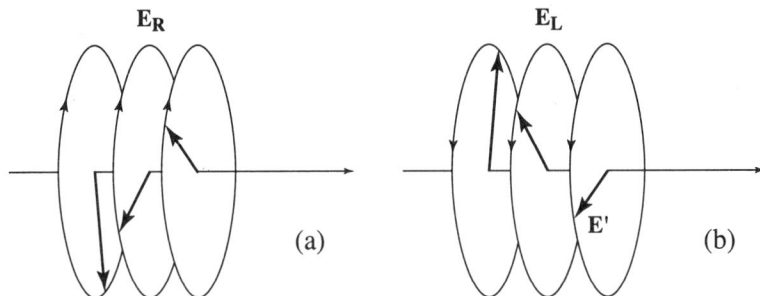

Figure 13.11- Right-hand (a) and left-hand (b) circular vibrations

*During propagation, the direction of the electric vector **E** rotates, and its end describes a circle. The vectors shown represent the field **E** at successive times, hence at shifted positions.*

If **E** changes modulus periodically, the polarization state is called *elliptical*, again right-hand or left-hand. Note that an elliptically polarised light can be considered as the coherent superposition of two circularly polarised waves, with opposite handedness and different amplitudes. It can also be described as the superposition of two rectilinearly polarised waves, with perpendicular polarization directions, with different amplitudes and a phase difference. *Coherent* superposition involves adding the amplitudes with the phases taken into account.

A beam of light is polarised only if it has undergone a preparation process, or if it is emitted by a laser. "Natural" light is unpolarised, i.e. the direction of **E** varies randomly. This situation can be thought of as resulting from incoherent superposition of two beams with equal intensities, polarised either rectilinearly or circularly in the two principal modes. *Incoherent* superposition consists in simply adding the intensities.

An *eigenmode* is a polarization regime that is conserved, i.e. is not altered, during propagation. A refractive index is relevant only for an eigenmode.

A2 - Refractive index, dielectric permittivity

Two related notions permeate the optics of materials: the refractive index and the dielectric permittivity tensor. In the visible range and even more so at higher frequencies (X-rays), the relative magnetic permeability μ plays no part, even in the magneto-optical effects. The magnetic moments cannot follow the extremely fast variations corresponding to these frequencies, hence one should take $\mu = 1$. All the effects are associated to the dielectric permittivity ε. This does not apply at lower frequencies.

The simplest expression for an electromagnetic wave in a material is for plane, monochromatic, rectilinearly polarised light. Its electric field is expressed as: $\mathbf{E} = \mathbf{E}_0 \cos(\mathbf{kr} - \omega t)$. This can be written in the form: $\mathbf{E} = \mathbf{E}_0 \exp[j(\mathbf{kr} - \omega t)]$ provided one keeps in mind the fact that, in this convention, only the real part of this complex expression is meaningful.

The modulus of the wave vector **k** is written as $k = nk_0$, where k_0 is the modulus of the wave vector in vacuum. This defines the refractive index of the material, n, and vacuum is associated with $n = 1$.

Since the phase velocity of the wave is $v_p = \omega/k$, we see immediately that, in a material with refractive index larger than 1, the phase velocity is divided by n with respect to its value in vacuum, c. When the refractive index changes with angular frequency ω, i.e. when there is dispersion, the group velocity is given by $v_g = d\omega/dk$. We can note that it is not shocking to encounter an index $n < 1$, which implies a phase velocity larger than the velocity of light in vacuum, c. The latter is a rigid upper bound to the propagation velocity, and therefore the group velocity must definitely satisfy $v_g < c$. It always does.

Experiment shows that, whether magnetic or not, many materials are *birefringent*. This means that the refractive indices associated with the two polarization eigenmodes are different. This can be described as
- *rectilinear* birefringence (the term linear is often used in the literature. We discussed this ambiguity at the beginning of this chapter) if the difference relates to the two rectilinear polarization eigenstates,
- *circular* birefringence if it is related to the circular polarizations (right- and left-hand).

The refractive index is then no more a characteristic of the material, a function only of the wavelength (or frequency, or angular frequency, or energy, it is all the same), and one is forced to use a more complicated description.

One of the convenient description modes is based on the dielectric permittivity tensor ε_{ij}, sometimes called the dielectric constant tensor. It is defined by the relation between the dielectric displacement **D** and the electric field **E**:

$$D_i = \varepsilon_0 \sum_j \varepsilon_{ij} E_j$$

ε_0 being the permittivity of vacuum, defined in chapter 2, with i and j designating the three components (for example x, y, z). It is customary to write this relation without explicitly mentioning the summation over repeated indices (this is the Einstein convention). The geometrical meaning is clear: for example ε_{21} or ε_{yx} describe the component of **D** along y due to the component along x of **E**.

The simplest case, that of a non-birefringent (optically isotropic) material, is then described by a diagonal tensor in which all non-zero elements are equal, viz:

$$\varepsilon_{ij} = \begin{pmatrix} \varepsilon & 0 & 0 \\ 0 & \varepsilon & 0 \\ 0 & 0 & \varepsilon \end{pmatrix} = \varepsilon \delta_{ij},$$

where δ_{ij} is Kronecker's symbol ($\delta_{ij} = 1$ if $i = j$; $\delta_{ij} = 0$ if $i \neq j$). In other words, $D_i = \varepsilon_0 \varepsilon E_i$, which means that, in this special case, **D** and **E** are colinear. This is the situation of amorphous materials (in particular glasses) and of crystallised materials with cubic structure.

In the general case, thermodynamic arguments (see § 5 of chap. 10) indicate that this tensor is necessarily symmetric: $\varepsilon_{ij} = \varepsilon_{ji}$, which reduces to 6 instead of 9 the number of values to be determined. Crystal symmetries can further lower the number of independent values, down to one in the cubic case we just encountered.

It should be noted that the form of the relation between **E** and **D** is the same in the dc (time independent) and the ac regimes, and in particular for light (very high angular frequency ω). However, the values of the ε_{ij} are often very different. A striking case is that of water, for which $\varepsilon \approx 80$ in the dc regime, but only $\varepsilon \approx 1.7$ (n = 1.33) at optical frequencies.

13 - MAGNETO-OPTICAL EFFECTS

It is easy to recover the relationship between ε and the refractive index n in the special case of an optically isotropic material. With $\mu = 1$ for the optical situation, we simply have $\varepsilon = n^2$, whatever the propagation direction (this is implied in isotropy). Any plane wave is an eigenmode, and there is free choice in the description of the basic polarization states. They can be chosen either as rectilinear, with again free choice of the two reference directions perpendicular to the propagation direction of the beam, or as circular.

In the general anisotropic case, the refractive index depends on the propagation direction. For anisotropic materials with rectilinear birefringence, the polarization directions of the eigenmodes are fixed (ordinary and extraordinary waves). For gyrotropic media and in particular in magneto-optics, the circular polarization modes are the eigenmodes for propagation along the magnetization direction.

A3 - Absorption. Complex refractive index and permittivity

The above description omits a central characteristic of most materials: they are not perfectly transparent, and metals are even very opaque. The formalism must therefore be improved, to include the *absorption* of light.

In an isotropic medium, the intensity transmitted after a beam has gone through a thickness z is $I = I_0 \exp(-\alpha z)$, with α the linear absorption coefficient. Translating this in terms of refractive index involves including in the amplitude a term $\exp - (\alpha z / 2)$. In the complex exponential form of the description of the field, this is equivalent to replacing $\exp[j(\mathbf{k} \cdot \mathbf{r} - \omega t)]$ by $\exp[j(k \cdot z + j\,\alpha z/2) - \omega t)]$ with the axis z taken along the propagation direction. The wave vector now becomes complex. This can be translated, in this simplified description, into a complex refractive index $N = n + j\kappa$ (see eq. 13.21), with n and κ real, and $\kappa = \alpha/2k_0 = \alpha \lambda_0 / 4\pi$, where λ_0 is the wavelength in vacuum. As an example [30], the values for bulk aluminium are, at wavelength 589.3 nm: $n = 1.44$; $\kappa = 5.23$. For evaporated iron, at the same wavelength, $n = 1.5$; $\kappa = 1.63$.

A4 - Transmission and reflection at an interface Inhomogeneous waves

One of the continuity conditions at an interface, for any kind of wave, is the conservation of the tangential component of the wave-vector. For transparent materials, this makes it easy to recover the Snell-Descartes law $n_1 \sin \theta_i = n_2 \sin \theta_t$. The continuity conditions also make it possible to express the ratio of the amplitudes of the reflected and transmitted waves to that of the incident wave (fig. 13.12).

These are the Fresnel coefficients. One of their forms is:

$$r_p = \frac{N_2 \cos \theta_i - N_1 \cos \theta_t}{N_2 \cos \theta_i + N_1 \cos \theta_t} \quad ; \quad r_s = \frac{N_1 \cos \theta_i - N_2 \cos \theta_t}{N_1 \cos \theta_i + N_2 \cos \theta_t}$$

$$t_p = \frac{2 N_1 \cos \theta_i}{N_2 \cos \theta_i + N_1 \cos \theta_t} \quad ; \quad t_s = \frac{2 N_1 \cos \theta_i}{N_1 \cos \theta_i + N_2 \cos \theta_t}.$$

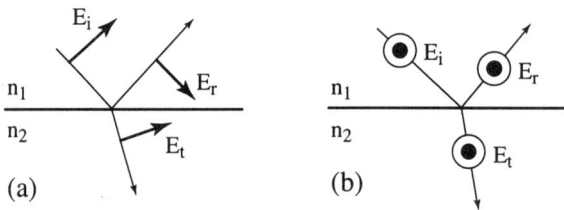

*Figure 13.12 - Refraction and reflection on a plane interface:
(a) p or TM polarization - (b) s or TE polarization*

Here, N_1 and N_2 are the complex refractive indices in the two media, (1) being that of the incident beam (incidence angle θ_i), and (2) that of the transmitted wave (refraction angle θ_t). r and t are respectively the ratios of the reflected and transmitted amplitudes to that of the incident wave, and the subscripts p and s refer to the polarization state. We note that here, in the absence of magneto-optical effects, there is no non-diagonal term such as r_{ps} in the matrix of Fresnel coefficients. The coefficients could be written r_{pp}, etc. An incident wave with p polarization gives neither a reflected nor a transmitted wave with s polarization.

When one of the media (for example medium 2) is absorbing, but isotropic, the amplitude ratios can still be obtained from the Fresnel relations by introducing its complex refractive index. This time the ratios are complex, i.e. they now carry also an information on the phase change, which is no longer just 0 or π (positive or negative sign of the Fresnel coefficient) as was the case for real refractive index. Reflection on an absorbing medium leads to reflection coefficients that differ both in amplitude and in phase for polarization states s and p. Reflection of a rectilinearly polarised wave, with components both along s and p, on an absorbing material (in particular a metal) at oblique incidence thus results in an elliptically polarised wave.

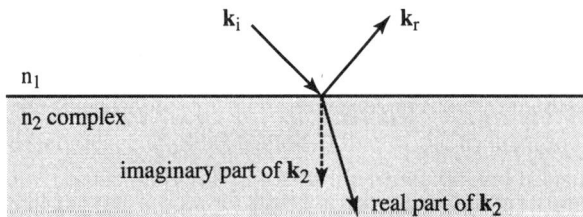

Figure 13.13 - Refraction into an absorbing material
The wave vector in the material is complex. Its real part has the same tangential component, on the interface plane, as the wave-vector incident from the non-absorbing material, and is therefore oblique. The imaginary part of the wave vector is perpendicular to the interface. The wave in the absorbing material is inhomogeneous.

A complication arises when considering the transition of a wave from a transparent medium (e.g. vacuum) into an absorbing material, under oblique incidence. In the transparent medium, the refractive index is real and so is the wave vector. There is no difficulty to find the real part of the wave vector in the absorbing material. This will describe the variation in phase of the wave. But the same condition of continuity requires that, since the imaginary part of the tangential component of the wave vector is zero in the transparent medium, it remains so in the absorbing material. This means that the imaginary part of the wave-vector, describing the spatial variation of the modulus of the wave, is necessarily perpendicular to the interface. The loci for equal phase (perpendicular to the real part of **k**) and for equal amplitude (parallel to the interface) are quite different (fig. 13.13). This situation corresponds to an inhomogeneous wave. Refraction becomes quite complicated.

A5 - Near-field microscopy

Standard microscopy involves the propagation of light over distances that are large with respect to the wavelength (the far field). It therefore uses propagating waves. Its resolution limit is about half the wavelength λ of the light (Rayleigh's criterion).

Near field microscopy makes it practically feasible to reach ten times better resolution [47, 67]. The basic idea is that information about small details (the large spatial frequencies) is contained in the components that are diffracted with large momentum transfer k_x. These components cannot propagate as soon as $k_x > \omega/c$ since the component of their wave vector perpendicular to the surface, $k_z = \sqrt{(\omega/c)^2 - k_x^2}$, is then imaginary: they are evanescent modes, localised in the vicinity of the surface. Near field microscopy requires collecting these modes, hence moving a receptor at distances to the surface much smaller than λ. It also involves turning them into propagating modes so they can then be used in a device, which implies using a probe with dimensions also much smaller than λ. It is fun to note that medicine has long been using the near field without saying so: the stethoscope provides a resolution limit much smaller than the wavelength of the sound waves emitted for example by the heart [68]. The present technology uses, as actuators for the displacements, the piezo-electric devices developed for scanning tunneling microscopy (STM), and the probe is often a tapered optical fibre or a metal tip.

A6 - X-rays

X-rays are electromagnetic waves, just as visible light, but with wavelength in the Ångström (10^{-10} m) range, and energy $\approx 10^4$ eV. They are produced by laboratory generators, or in synchrotron radiation sources, which are Large Facilities based on storage rings in which ultra-relativistic electrons circulate. These sources, which are more difficult to access but open to the research groups in their country or the group of countries cooperating for their construction, provide beams that are remarkable in their intensity as well as for their continuous spectrum (from infrared to hard X-rays), their polarization and their coherence. This is particularly true for the recent (third generation) sources, such as ESRF (European Synchrotron Radiation Facility) in Grenoble, France, APS (Advanced Photon Source) in Argonne, IL (USA), or Spring-8 in Japan.

A7 - Numerical values

Table 13.1 - Faraday rotation of some materials at 300 K

	Material	Θ_F	λ (nm)
Diamagnets	glasses; oxides; chalcogenides	4-27 rad$\cdot$T$^{-1}\cdot$m^{-1}	546
Garnets	$Yb_3Fe_5O_{12}$	12°$\cdot$cm^{-1}	1,060
Ferrimagnets	$Nd_3Fe_5O_{12}$	-840°$\cdot$cm^{-1}	1,060
" "	$(BiRGa)Fe_5O_{12}$	1-$7\cdot10^{4}$°$\cdot$cm^{-1}	"
Ferromagnets	Fe	10^5-10^{6}°$\cdot$cm^{-1}	546
"	Co	"	546
"	Ni	"	546
"	R-Fe	1.8-$2\cdot10^{5}$°$\cdot$cm^{-1}	632.8
"	R-Co	$4.5\cdot10^{6}$°$\cdot$cm^{-1}	632.8

Table 13.2 - Kerr rotation Θ_K for some materials at 300 K, mainly from [69]

	Material	Θ_K (degree)	λ (nm)
3d metals	Co	-0.35 (-0.39)	633 (830)
" "	Fe	-0.41 (-0.58 max.)	633 (1,100)
" "	Ni	-0.13	633
Binary compounds	CoGd	0.28	800
" "	FeCo	-0.55	830
" "	$Co_{70}Pt_{30}$	-0.4	633
" "	$Fe_{65}Pt_{35}$	-0.68	830
" "	$Gd_{1-x}Fe_x$	-0.1 to -0.3	750
" "	$Gd_{1-x}Co_x$	0.6 to 1	750
" "	TbFe	0.2	720
" "	$TbFe_2$	0.17 (-0.41 max.)	800 (270)
" "	$GdFe_2$	0.2	720
Ternary compounds	$Fe_{2.7}V_{0.3}Ge$	0.36	633
" "	Co_2FeGa	-0.48	830
" "	Co_2FeGe	-0.43	633
" "	TbFeCo	-0.23	670
" "	GdTbFe	-0.26	800
Heusler alloys	PtMnSb	-0.93	633
" "	NiMnSb	-0.30	830

REFERENCES

[1] A.V. SOKOLOV, Optical properties of metals (1961) State physical and mathematical Publication Company, Moscow (*in Russian*).

[2] J.F. DILLON JR., *J. Appl. Phys.* **39** (1968) 922.

[3] P.S. PERSHAN, *J. Appl. Phys.* **38** (1967) 1482.

[4] M.J. FREISER, *IEEE Trans. Magn.* **MAG 4** (1968) 152.

[5] A.K. ZVEZDIN, V.A. KOTOV, Modern magnetooptics and magnetooptical materials (1997) Institute of Physics, Bristol.

[6] N.F. KHARCHENKO, *Ferroelectrics* **162** (1994) 173.

[7] RU-PIN PAN, H.D. WEI, Y.R. SHEN, *Phys. Rev. B* **39** (1989) 1229; W. HÜBNER, K.H. BENNEMANN, *Phys. Rev. B* **40** (1989) 5973; J. REIF, J.C. ZINK, C.M. SCHNEIDER, J. KIRSCHNER, *Phys. Rev. Lett.* **67** (1991) 2878; TH. RASING, *J. Magn. Magn. Mater.* **175** (1997) 35.

[8] F. DE BERGEVIN, M. BRUNEL, *Phys. Lett. A* **39** (1972) 141; *Acta Cryst. A* **37** (1981) 324.

[9] K. NAMIKAWA, M. ANDO, T. NAKAJIMA, H. KAWATA, *J. Phys. Soc. Jpn* **54** (1985) 4099; D. GIBBS, D.R. HARSHMAN, E.D. ISAACS, D.B. MCWHAN, D. MILLS, C. VETTIER, *Phys. Rev. Lett.* **61** (1988) 1241.

[10] F.U. HILLEBRECHT, T. KINOSHITA, D. SPANKE, J. DRESSELHAUS, CH. ROTH, H.B. ROSE, E. KISKER, *Phys. Rev. Lett.* **75** (1995) 2224.

[11] J. STÖHR, Y. WU, B.D. HERMSMEIER, M.G. SAMANT, G.R. HARP, S. KORANDA, D. DUNHAM, B.P. TONNER, *Science* **259** (1993) 658.

[12] L. LANDAU, E. LIFSHITS, Electrodynamics of continuous media (1969) Mir, Moscow (and various translations).

[13] G. METZGER, P. PLUVINAGE, R. TORGUET, *Ann. Phys.* **10** (1965) 5; H.T. MINDEN, *Appl. Optics* **29** (1990) 3955; Z.J. YANG, M.R. SCHEINFEIN, *J. Appl. Phys.* **74** (1993) 6810; M. MANSURIPUR, *J. Appl. Phys.* **67** (1990) 6466.

[14] S. VISNOVSKY, *Czech. J. Phys. B* **36** (1986) 625; A. HUBERT, G. TRAEGER, *J. Magn. Magn. Mater.* **124** (1993) 185; K.R. HEIM, M.R. SCHEINFEIN, *J. Magn. Magn. Mater.* **154** (1996) 141.

[15] P.P. EWALD, *Ann. Phys.* **49** (1916) 117.

[16] P.N. ARGYRES, *Phys. Rev.* **97** (1955) 334.

[17] J.L. ERSKINE, E.A. STERN, *Phys. Rev. B* **8** (1973) 1239.

[18] J.M. FLORZAK, E. DAN DAHLBERG, *J. Appl. Phys.* **67** (1990) 7520.

[19] Z.J. YANG, M.R. SCHEINFEIN, *J. Appl. Phys.* **74** (1993) 6810.

[20] P.H. LISSBERGER, *J. Opt. Soc. Am.* **51** (1961) 957; *ibid.* **56** (1966) 192.

[21] Y. SOUCHE, P. PELLAT-FINET, A. SAADAOUI, *Optica Acta* **33** (1986) 327.

[22] S. HUARD, Polarisation de la lumière (1994) Masson, Paris.

[23] G. BRUHAT, A. KASTLER, Optique, 6e éd. (1965) Masson, Paris.

[24] O. CUGAT, D. GIVORD, J.P. REBOUILLAT, Y. SOUCHE, *J. Magn. Magn. Mater.* **104-107** (1992) 397.

[25] G.S. KRINCHIK, V.A. ARTEMJEV, *J. Appl. Phys.* **39** (1968) 1276.

[26] W. BREUER, J. JAUMANN, DO. SMITH, *Z. Phys.* **173** (1963) 117; T.R. MCGUIRE, M. HARTMANN, *IEEE Trans. Magn.* **MAG 21** (1985) 1644.

[27] P. HANSEN, C. CLAUSEN, G. MUCH, M. ROSENKRANZ, K. WITTER, *J. Appl. Phys.* **66** (1989) 756.

[28] K.H.J. BUSCHOW, Ferromagnetic materials, Vol. 4 (1988) 493, E.P. Wohlfarth and K.H.J. Buschow Eds, North Holland, Amsterdam.

[29] T. KATAYAMA, K. HASEGAWA, Proc. Int. Conf. Rapidly Quenched Metals IV (1981) T. Matsumoto, K. Suzuki Eds, Sendai.

[30] M. BORN, E. WOLF, Principles of Optics, 6th ed. (1993) Pergamon Press.

[31] D.H. MARTIN, K.F. NEAL, T.J. DEAN, *Proc. Phys. Soc.* **86** (1965) 605.

[32] R. CAREY, E.D. ISAAC, B.W.J. THOMAS, *Brit. J. Appl. Phys. (J. Phys. D)* **1** (1968) 945.

[33] G.S. KRINCHIK, G.M. NURMUKHAMEDOV, *Sov. Phys. JETP* **21** (1965) 22.

[34] P.E. FERGUSON, R.J. ROMAGNOLI, *J. Appl. Phys.* **40** (1969) 1236.

[35] D. KERKMANN, *Appl. Phys. A* **49** (1989) 523.

[36] S.D. BADER, *J. Magn. Magn. Mater.* **100** (1991) 440.

[37] J.T. HO, J.D. LITSTER, *Phys. Rev. Lett.* **33** (1974) 1576.

[38] J.A. GRIFFIN, S.E. SCHNATTERLY, *Phys. Rev. B* **2** (1970) 4523.

[39] A.S. BOROVIK-ROMANOV, N.M. KREÏNES, A.A. PANKOV, M.A. TALALAEV, *Sov. Phys. JETP* **37** (1973) 890.

[40] D.P. BELANGER, A.R. KING, V. JACCARINO, *Phys. Rev. B* **29** (1984) 2636.

[41] O. GEOFFROY, D. GIVORD, Y. OTANI, B. PANNETIER, A.D. SANTOS, M. SCHLENKER, Y. SOUCHE, *J. Magn. Magn. Mater.* **121** (1993) 516; N. BARDOU, B. BARTENLIAN, C. CHAPPERT, R. MÉGY, B. VEILLET, J.P. RENARD, F. ROUSSEAUX, M.F. RAVET, J.P. JAMET, P. MEYER, *J. Appl. Phys.* **79** (1996) 5848; A.A. STASHKEVICH, *Opt. Commun.* **178** (2000) 1.

[42] F. SCHMIDT, W. RAVE, A. HUBERT, *IEEE Trans. Magn.* **MAG 21** (1985) 1596; A. HUBERT, R. SCHÄFER, Magnetic domains (1998) Springer Verlag, Berlin.

[43] W. DRECHSEL, *Z. Phys.* **164** (1961) 308.

[44] J. KRANZ, A. SCHAUER, *Optik* **18** (1961) 186.

[45] K. NAKAMURA, T. ASAKA, S. ASARI, Y. OTA, A. ITOH, *IEEE Trans. Magn.* **MAG 21** (1985) 1654.

[46] R.P. HUNT, *J. Appl. Phys.* **38** (1967) 1215.

[47] J.K. TRAUTMAN, E. BETZIG, J.S. WEINER, D.J. DIGIOVANNI, T.D. HARRIS, F. HELLMAN, E.M. GYORGY, *J. Appl. Phys.* **71** (1992) 4659.

[48] H. KIRCHNER, *Phys. Status Solidi A* **4** (1971) 531.

[49] A. FORKL, T. DRAGON, H. KRONMÜLLER, *J. Appl. Phys.* **67** (1990) 3047.

[50] L.A. DOROSINSKII, M.V. INDENBOM, V.I. NIKITENKO, YU.A. OSSIP'YAN, A.A. POLYANSKII, V.K. VLASKO-VLASOV, *Physica C* **203** (1992) 149.

[51] H. KAWATA, K. MORI, *Rev. Sci. Instrum.* **66** (1995) 1407.

[52] M. SCHLENKER, J. BARUCHEL, X-ray diffraction topography, *in* Handbook of Microscopy, Vol. I (1997). S. Amelinckx, D. van Dyck, J. van Landuyt, G. van Tendeloo Eds, VCH, Weinheim.

[53] C.M. SCHNEIDER, R. FRÖMTER, H.P. OEPEN, J. KIRSCHNER, *J. Electron Spectroscopy and Related Phenomena* **84** (1997) 171.

[54] C. GILES, C. VETTIER, F. DE BERGEVIN, C. MALGRANGE, G. GRÜBEL, F. GROSSI, *Rev. Sci. Instrum.* **66** (1995) 1518.

[55] P. HANSEN, J.P. KRUMME, *Thin Solid Films* **114** (1984) 69.

[56] V.J. FRATELLO, S.J. LICHT, C.D. BRANDLE, *IEEE Trans. Magn.* **32** (1996) 4102.

[57] T. TAMAKI, K. TSUSHIMA, *Fiz. Nizk. Temp.* **18**, Suppl. S1 (1992) 415.

[58] R. WOLFE, J. HEGARTY, J.F. DILLON, L.C. LUTHER, G.K. CELLER, L.E. TRIMBLE, *IEEE Trans. Magn.* **MAG 21**, 5 (1985) 1647.

[59] H. TAKADA, S. MIYAMOTO, T. MITSUI, T. TOMIMASU, *Phys. E: Sci. Instrum.* **21** (1988) 371.

[60] K.B. ROCHFORD, A.H. ROSE, G.W. DAY, *IEEE Trans. Magn.* **32**, 5 (1996) 4113.

[61] M.N. DEETER, S. MILIÀN BON, G.W. DAY, G. DIERCKS, S. SAMUELSON, *IEEE Trans. Magn.* **30** (1994) 4464.

[62] O. KAMADA, H. MINEMOTO, N. ITOH, *J. Appl. Phys.* **75** (1994) 6801.

[63] N.F. KUBRAKOV, *SPIE Proc.* **1126** (1989) 85.

[64] N.F. KUBRAKOV, A.Y. CHERVONENKIS, *Fiz. Nizk. Temp.* **18**, Suppl. S1 (1992) 361.

[65] M. KRYDER, *MRS Bull.* **21** (1996) 17 (and the other papers in this issue).

[66] E. HECHT, Optics, 2nd ed. (1988) Addison-Wesley, Reading, MA, USA.

[67] D. COURJON, C. BAINIER, *Rep. Prog. Phys.* **57** (1994) 989.

[68] D.W. POHL, W. DENK, M. LANZ, *Appl. Phys. Lett.* **44** (1984) 651.

[69] K.H.J. BUSCHOW, P.G. VAN ENGEN, R. JONGEBREUR, *J. Magn. Magn. Mater.* **38** (1983) 1.

CHAPTER 14

MAGNETIC RESISTIVITY, MAGNETORESISTANCE, AND THE HALL EFFECT

The effects of magnetic fields, or of magnetization, on electrical or thermal transport are grouped under the label of galvanomagnetic effects. As is the case for the resistivity, galvanomagnetic effects behave very differently in a semiconductor and in a metal. Only electrical transport will be considered here. For information on thermal effects see Seeger [1] or Gratz and Zuckermann [2].

1. DEFINITIONS

1.1. ELECTRICAL RESISTIVITY

The current density in a material, $\mathbf{j}$, is, in the weak field limit, proportional to the electric field $\mathbf{E}$: $\mathbf{j} = \sigma \mathbf{E}$, where σ is the conductivity. The resistivity ρ is the reciprocal of the conductivity: $\mathbf{E} = \rho \mathbf{j}$. This equation is in fact a tensor relation. The field and the current are not always colinear, and the conductivity is a function of direction in a material with low crystal symmetry. For a cubic material, the conductivity is isotropic in the absence of a magnetic field; it is related to the density of free carriers n, their effective mass m*, their charge $q = \pm e$, and to the mean time between collisions τ, by:

$$\sigma = n e^2 \tau / m^* = n q \mu \qquad (14.1)$$

μ is the mobility of the carriers, i.e. the ratio of their mean velocity to the electric field. The mean free path is $\lambda = v_F \tau$, where v_F is the average velocity at the Fermi level. A description of scattering mechanisms is given in chapter 20.

1.2. MAGNETORESISTANCE

In an applied field, or in a material where a magnetic induction $\mathbf{B}$ exists, the current carriers (electrons or holes) are subjected to the Lorentz force $\mathbf{F} = q \mathbf{v} \times \mathbf{B}$ (eq. 2.57).

The electronic trajectories are modified, and so are the band energies and the scattering probabilities, which depend on the orientation of the electronic spins. Magnetoresistance (MR) is the modification of the resistivity by the magnetic field in the material $\Delta\rho/\rho\,(B=0)$, expressed by the variation of the diagonal components of the resistivity tensor, e.g. $\rho_{xx}(B)$. This is an even function of the field. The longitudinal resistivity (or MR) is that in the direction of the applied field ($\mathbf{B}\,//\,\mathbf{j}$), whereas the transverse resistivity is that perpendicular to the applied field ($\mathbf{B}\perp\mathbf{j}$).

1.3. THE HALL EFFECT

Consider a rectangular plate with small thickness d, carrying a current along the length of the plate I_x, with a field B_z perpendicular to the plate (fig. 14.1). A carrier (electron or hole) moves with average velocity v_x, and is subjected to the Lorentz force. This force has the effect of deviating the electrons in the y direction, resulting in an accumulation of charge on the lateral faces. A Hall voltage V_y appears between the faces such that the resulting electric field E_y compensates the Lorentz force.

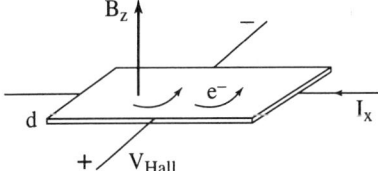

Figure 14.1 - Hall voltage V_y

If n is the volume concentration of carriers, the current density j_x is $j_x = n q v_x$, and the electric field at equilibrium is $E_y = B_z j_x / nq$. The Hall effect thus corresponds to the component of electric field perpendicular to the plane containing the magnetic field and the current; it comes from the non diagonal part of the tensor ($\rho_{xy}...$), and is an odd function of field. The Hall resistivity $\rho_H = E_y / j_x$ is often proportional to the field, with the slope defining the Hall coefficient:

$$R_H = E_y / j_x B_z = V_y d / I_x B_z = 1/nq \qquad (14.2)$$

This coefficient is negative for electrons ($q = -e$), and positive for holes, and is expressed in $\Omega.m.T^{-1}$ or equivalently in $m^3.C^{-1}$. For situations where there is only one type of carrier, the ratio R_H/ρ gives their mobility.

1.4. THE PLANAR HALL EFFECT

If the magnetic field is parallel to the plate, a component of electric field can exist within the field-current plane, perpendicular to the current. This planar Hall effect [3] is due to the distortion of the current lines by the field, and is related to the anisotropy of the magnetoresistance: $E_p = g\,I\,B^2 \sin(2\Psi)$, where Ψ is the angle between $\mathbf{I}$ and $\mathbf{B}$.

2. TRANSPORT IN MAGNETIC METALS

2.1. MAGNETIC RESISTIVITY

The general problem of resistivity in metals is considered by Schroeder [4], and Rossiter [5]. The resistivity has contributions from atomic disorder ρ_i, phonons ρ_{ph}, and, in a metal with magnetic atoms, magnetic scattering ρ_m. When the scattering probabilities are small (when the mean free path is large compared to interatomic distances), the resistivity is the sum of the three contributions (Matthiessen's law).

$$\rho(T) = \rho_i + \rho_{ph}(T) + \rho_m(B, T) \qquad (14.3)$$

The magnetic resistivity ρ_m depends on the magnetic disorder, and thus on the magnetic entropy. It is a function of temperature and field, and its derivative $d\rho_m/dT$ is similar to the magnetic contribution to the specific heat (fig. 3.17 and 3.18).

2.2. LOCALISED MOMENTS

De Gennes and Friedel [6] showed that the magnetic resistivity is related to the spin correlation function $<S_i.S_j>$ by:

$$\rho_m(B, T) = \rho_\infty [S(S+1) - <S_i.S_j>_{B,T}] \qquad (14.4)$$

In the paramagnetic state, the magnetic resistivity is constant if the correlations between the moments vanish. For a ferromagnet, the resistivity varies approximately as $M_0^2 - M_s^2(T)$ in the region where it is ferromagnetically ordered, provided the Fermi surface is only weakly modified by the ordering (fig. 14.2).

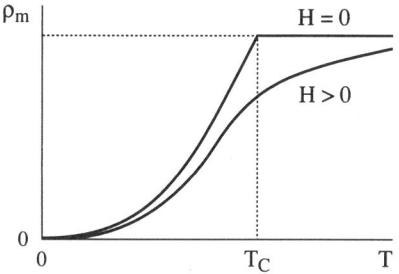

Figure 14.2 - Magnetic resistivity in a ferromagnet, and its evolution in a magnetic field

In rare earth compounds [7, 8], the high temperature resistivity tends towards the value ρ_∞:

$$\rho_\infty \approx (g-1)^2 J(J+1),$$

where J is the total angular moment, and g the Landé factor.

The crystalline electric field often lifts the degeneracy of the *4f* levels (see chap. 7). The resistivity in the paramagnetic state depends on the number of occupied levels, and on the probability of scattering from one state to another. The resistivity is no longer constant, but increases as a function of the population of the levels [9].

In an antiferromagnetic compound, two terms contribute to the resistivity: the spin disorder terms as in the previous case, but also the change in the number of carriers due to the existence of magnetic Brillouin zones which alter the Fermi surface. This extra term generally results in a decrease in the number of carriers, and hence an increase in resistivity. A typical case is that of the heavy rare earth metals [10, 11, 12], where the structure is modulated, and where the resistivity along the six-fold axis rises below the Néel temperature T_N.

2.3. ITINERANT MAGNETISM, ALLOYS, AND TRANSITION METALS

In 3d transition metals, the electrons responsible for magnetism are delocalised to varying degrees, and form bands which participate in conduction (see chap. 8). Transport phenomena are much more complex than for localised magnetism.

2.3.1. A magnetically ordered metal

The electrons are separated into two half bands, with spins parallel and anti-parallel to the magnetization ($\uparrow$ and $\downarrow$ respectively), characterised by densities $n\uparrow$ and $n\downarrow$ at the Fermi level. One can consider that these two bands carry current in parallel, with resistivities $\rho\uparrow$ and $\rho\downarrow$, and a probability of spin flip during scattering which gives a contribution $\rho\uparrow\downarrow$. The total resistivity is thus [13]:

$$\rho = [\rho\uparrow\rho\downarrow + \rho\uparrow\downarrow(\rho\uparrow + \rho\downarrow)]/(\rho\uparrow + \rho\downarrow + 4\rho\uparrow\downarrow) \tag{14.5}$$

One can evaluate the contribution of dilute impurities in a given matrix, for each spin direction, by measuring the resistivity as a function of concentration.

The resistivity per impurity atom is a function of the difference in charge ΔZ between the matrix and the impurity; it is larger for magnetic impurities which have virtual bound states [14], i.e. 3d states which are localised on the impurity, with an energy corresponding to the conduction band (see § 7.1 of chap. 8). An example is that of chromium impurities in Ni.

The angular dependence of the resistivity with respect to the magnetization gives rise to an anisotropy of the magnetoresistance: the resistivity is different depending upon whether the current is parallel ($\rho_{//}$) or perpendicular to the magnetization ($\rho_\perp$).

$$\Delta\rho/\rho_o = (\rho_{//} - \rho_\perp)/\rho_o \approx \gamma(\alpha - 1) \tag{14.6}$$

where ρ_o is the resistivity of the demagnetised state, α is the ratio $(\rho\uparrow/\rho\downarrow)$, and γ depends on the spin orbit coupling [13, 5]. This property was used in the first magnetoresistive sensors (permalloy).

2.3.2. Spin fluctuations, and the Kondo effect

Fluctuations and instabilities of the magnetic moment, resulting from hybridization between moment carriers and conduction electrons, give rise to strong scattering. Two situations can arise:

3d metals with spin fluctuations

In this case, the electrons responsible for transport are also the moment carriers (transition metals with magnetic *3d* impurities: Mn, Fe in Pd or Ni, compounds on the verge of being magnetic such as YCo_2). The electronic orbits and the moments tend to align in a coherent manner at low temperature. The residual resistivity is thus due only to scattering by the difference in potential between the matrix and the impurities: $\rho_m \approx |V|^2 n(E_F)$, where V is the scattering potential. Increasing the temperature gives rise to fluctuations of the moments, and the resistivity goes up as $(T/T_{sf})^2$, T_{sf} being the spin fluctuation temperature. At high temperature, ρ_m tends towards a constant which is characteristic of scattering by a disordered magnetic system.

The Kondo effect

The most mobile *s* electrons hybridise with "magnetic" electrons (*d* or *f*), and are temporarily scattered into these localised orbits, for which the spin fluctuation time is relatively long (cerium atoms in a lanthanum compound, iron in copper). This resonance between two states at the Fermi level gives rise to the Kondo effect [15, 16]: the antiferromagnetic interaction between the local moment and the spin of the conduction electron leads to a scattering probability which gets larger the lower the temperature. For dilute magnetic impurities, the magnetic resistivity decreases linearly with the logarithm of temperature:

$$\rho = A[V_0^2 + f(\lambda) J_K^2 + g(\lambda) J_K^3 n(E_F) \ln(k_B T / \Delta)] \tag{14.7}$$

Δ is the width of the band, $n(E_F)$ is the density of states at the Fermi level, and J_K is the Kondo (negative) *s-f* or *s-d* interaction: $J_K = |V_{sf}|^2 / (E_{4f} - E_F)$. f and g are functions of the degeneracy λ of the *3d* or *4f* level. V_0 is the part of the scattering potential that is independent of the spin.

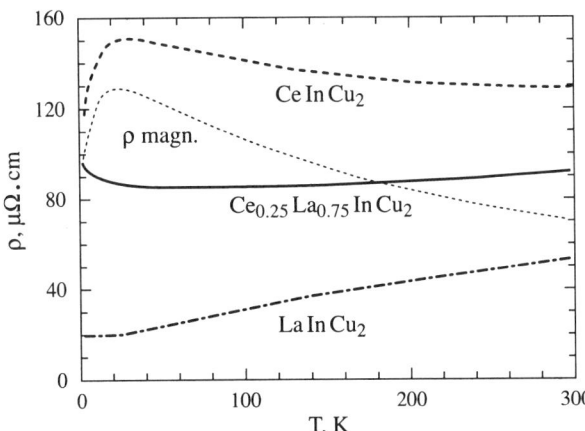

Figure 14.3 - Resistivity of the compounds $CeInCu_2$ (Kondo), $LaInCu_2$ (non magnetic), and the solid solution $(Ce,La)InCu_2$
The difference between Ce- and La-$InCu_2$ (dashed line) represents the magnetic resistivity of cerium.

Below the Kondo temperature (see eq. 8.46), the regime is different, and the resistivity tends progressively towards a constant. Taking the resistivity of the phonons into account, the total resistivity shows a minimum.

In the case of an ordered cerium lattice, the correlations between sites reduce the resistivity per magnetic ion, and at low temperatures one obtains a coherent behavior where the resistivity goes initially as T^2. The resistivity thus displays a maximum and a minimum as a function of temperature. In some cases, the increase in resistivity due to the phonons can be compensated over a large range of temperatures by a reduction in the magnetic contributions, giving a nearly constant resistivity. Note that the thermoelectric power often reaches a high value in Kondo systems; this property is exploited in some thermocouples.

2.4. MAGNETORESISTANCE

The distinction is made between those metals devoid of moment (cyclotron effect), and those with magnetic atoms (spin disorder).

2.4.1. Cyclotron effect

Under the influence of the Lorentz force, the electrons precess around the field at the cyclotron (Larmor) frequency $\omega_c = eB/m^*$, which increases their total path relative to that travelled parallel to the electric field. The electrons thus undergo more collisions with impurities or phonons, which increases the resistivity (positive magnetoresistance).

The average velocity parallel to the electric field is less affected if the magnetic and electric fields are parallel. The longitudinal MR is thus, in general, weaker than the transverse one (fig. 14.4).

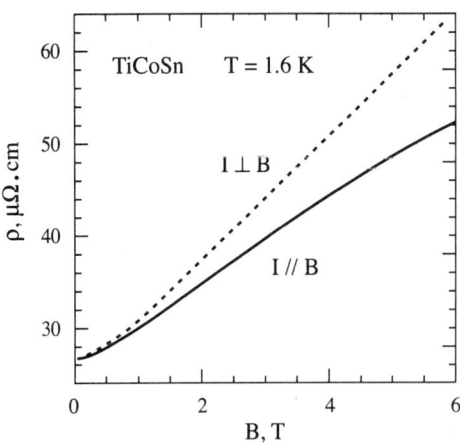

Figure 14.4 - Cyclotron type magnetoresistance for B//I, and B⊥I

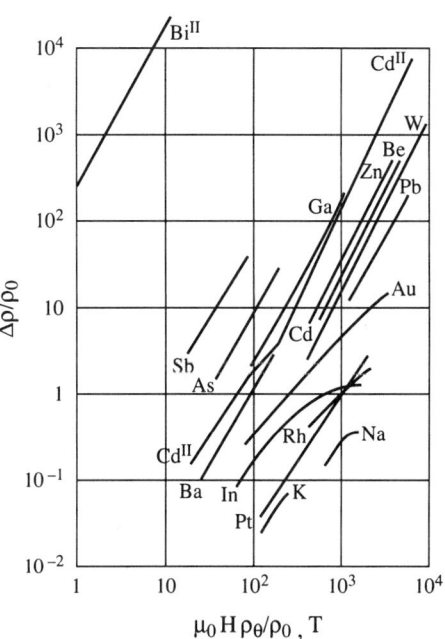

Figure 14.5 - Kohler's law in some metals [17]

The relative increase $\Delta\rho/\rho_0$ is larger when the initial resistivity ρ_0 is weak, i.e. at low temperature. For simple metals it follows Kohler's law: $\Delta\rho/\rho_0 \approx f(B/\rho_0)$, as shown in figure 14.5. The changes in amplitude are biggest when the metal no longer conforms to the free electron model, the record being held by bismuth, which is a semi-metal.

$\Delta\rho$ varies as B^2 in weak fields, and then either becomes linear or saturates in higher fields. The modified electronic trajectories being constant energy curves near the Fermi energy, MR reflects the symmetry of the Fermi surface, and can be strongly anisotropic, particularly when open orbits exist.

2.4.2. Magnetization dependent terms - Spin disorder

In the paramagnetic domain, the magnetoresistance (which is negative) reflects the variations in the correlation function between spins (eq. 14.4) under an applied field (i.e. it goes as M^2). Its initial dependence on the field is quadratic. In the ferromagnetic range, the magnetoresistance initially changes linearly with field as a result of the molecular field H_{mol} (see fig. 14.6).

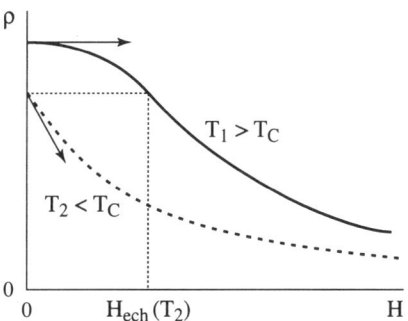

Figure 14.6
Magnetoresistance versus field
in the para- or ferromagnetic ranges

A more rigorous study [18] shows it is possible to decompose the magnetoresistance into two terms, arising from the transverse and longitudinal fluctuations of the magnetization. One is linked to the change in magnitude of the magnetization, while the other is linked to the change in the mean angle between the moments and the field. For an antiferromagnet, magnetoresistance depends on the angle between the moments in different sublattices, and can display different behaviors depending on the field direction with respect to the direction of antiferromagnetism, and on the ratio between anisotropy and exchange.

2.4.3. Anisotropy of magnetoresistance

MR often shows an anisotropic behavior in magnetic systems, even for cubic paramagnets or ferromagnets. A first origin is the anisotropy of the band structure and of the Fermi surface, as in the case of non magnetic systems (§ 2.4.1). A second one is the anisotropic coupling of the current carriers with the moment carriers. This may arise from spin-orbit couplings within the conduction band (eq. 14.6) as well as from the interaction with an anisotropic magnetic shell. In the case of rare earth compounds, current carriers include *6s5p* electrons from the rare earths, which are coupled to the anisotropic *4f* shell, characterized by spin, orbital, and total moments S, L, J. As a first approximation, the anisotropy of MR is proportional [19] to the electric quadrupole of the *4f* shell: $(\rho_{//} - \rho_{\perp})/\rho_0 \approx \alpha\,[3 <J_z^2> - J(J+1)]$, where α is the Stevens coefficient.

2.5. THE HALL EFFECT IN A MAGNETIC MATERIAL

The Hall effect contains two contributions: the normal contribution linked to the direct effect of the field $\mathbf{B} = \mu_0(\mathbf{H}_0 + \mathbf{M})$ on the carriers, and a second (anomalous) term arising from the coupling between the current carriers and the magnetization carriers. One can distinguish the effects of coupling with the spin and with the local orbital moment, which give rise to *skew* scattering and a side jump effect, respectively [20].

Taking the effect of the demagnetisating field $\mathbf{H}_d = -N_d\mathbf{M}$ into account, the Hall resistivity can be written, in the case of a thin plate, as a function of the applied field H_0 and the magnetization:

$$\rho_H = R_0 B + R_a \mu_0 M = R_0 \mu_0 H_0 + \mu_0 [R_0 (1 - N_d) + R_a] M \qquad (14.8)$$

In the paramagnetic state, the anomalous Hall effect goes as the susceptibility χ, and can show a Curie-Weiss like behavior.

In the ferromagnetic state, a term linked to the spontaneous magnetization appears. The variation as a function of field resembles a magnetization curve. The anomalous Hall resistivity is a function of the magnetization (fig. 14.7), but the coefficient R_a changes with temperature [20], and contains terms proportional to ρ_{xx} and ρ_{xx}^2.

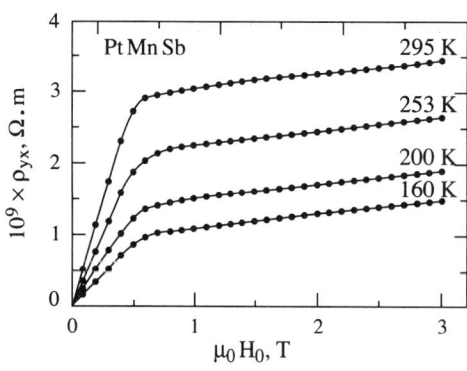

Figure 14.7
Isothermal variations in the Hall resistivity of PtMnSb [21]

3. MAGNETOTRANSPORT IN SEMICONDUCTORS

3.1. NON MAGNETIC SEMICONDUCTORS

Many books cover the Hall effect and the magnetoresistance of semi-conductors [1, 3, 4, 22, 23]; it is impossible here to present the variety of observed behaviors. For doped semi-conductors, or at finite temperature, both positive and negative carriers can exist (electrons and holes). Assuming the bands to be independent, the overall conductivity results from the sum of the two currents:

$$\sigma = e(n_e\mu_e + n_p\mu_p) \qquad (14.9)$$

where n_e, n_p are the carrier densities and μ_e, μ_p are their mobilities, while the Hall coefficient is given by:

$$R_H = (\mu_p^2 n_p - \mu_e^2 n_e)/e(\mu_p n_p + \mu_e n_e)^2 \qquad (14.10)$$

As with resistivity, the Hall effect has an intrinsic regime, in which the number of carriers is governed by a thermal activation law:

$$n_e = n_p = CT^{3/2} \exp(-E_g/2k_BT) \qquad (14.11)$$

where E_g is the width of the forbidden band. There is also an extrinsic regime linked to the number of impurities (donors and acceptors), where the mobility of the carriers plays a large role.

3.2. MAGNETIC SEMICONDUCTORS

These semi-conductors contain magnetic ions which interact with the carriers via a $J\mathbf{S}\cdot\mathbf{s}_c$ type coupling, and among themselves via super-exchange (see chap. 9).

Doping modifies the interactions: an additional electron polarises the moments of the neighbouring ions, which can give rise to local ferromagnetic order within a globally paramagnetic system. This phenomenon is called a *magnetic polaron*. Carriers "dressed" with the spins of the ions have a large effective mass, and therefore are not very mobile. They contribute to conduction via hopping or diffusion.

These polarons give rise to ferromagnetic or non-collinear order at lower temperature (EuO, EuS, $Gd_{3-x}S_4$ [24], manganites). The activation laws are no longer valid, and the system can display giant magnetoresistance around the Curie point (see § 4.2.5).

4. METAL-INSULATOR TRANSITIONS IN MAGNETIC MATERIALS

In many transition metal compounds, the metals can have more than one valence state. Most of the oxides for example are insulators, but some are metallic ($LaNiO_3$, TiO...). The delocalization or polarization of the electrons responsible for magnetism can lead to a metal-insulator transition. Mott has given a general review on metal-insulator transitions, and their relationship to magnetism [25].

4.1. SEMICONDUCTING TRANSITION METAL COMPOUNDS

What factors determine whether a compound is metallic or insulating, and how do the energy levels of the magnetic electrons influence these transitions?

In a semiconductor, the width of the forbidden band (*gap*) between the valence band and the conduction band is linked to the possible excitations of the system. The ground state of the metal is, for example, the configuration $3d^n$, the anion (oxygen) having its $2p$ shell filled by the s electrons of the metal.

Two types of excitations are important [26]:
- An electron is transferred from a transition metal i to its neighbour j: $d_i^n\, d_j^n \to d_i^{n+1}\, d_j^{n-1}$ costing the energy U of the intra-atomic Coulomb interaction.
- An oxygen gives up an electron to a neighbouring metal, leaving a hole L: $d_i^n \to d_i^{n+1}\, L$. This costs the energy Δ of the charge transfer.

If U and Δ are big compared to the width of the $3d$ band (W), the compound will either be a charge transfer insulator ($\Delta < U$), or a Mott-Hubbard insulator ($\Delta > U$). When U is small, the electrons propagate from one $3d$ cation to another (d type metal), and, when Δ is small, the electrons hop from oxygens to cations (p type metal).

4.2. TYPES OF METAL-INSULATOR TRANSITIONS

There are several types, for example:

4.2.1. The "classic" Mott transition

When atoms get closer together in a solid, the atomic levels broaden, and hybridise. At a critical distance, the bands overlap, and give rise to a metallic state. This explains qualitatively some metal-insulator transitions under pressure (hydrogen under very high pressure).

4.2.2. Mott-Hubbard transitions

The d electrons occupy a lower Hubbard band of width B_1, separated from excited levels forming the upper band, of width B_2, by the intra-atomic Coulomb energy U (see chap. 9); a metal-insulator transition occurs when $U \approx (B_1 + B_2)/2$. As an example, V_2O_3, doped with titanium or under pressure, undergoes a transition from an insulating antiferromagnetic state to a magnetically disordered metallic state.

4.2.3. Band crossing transitions (EuO at the Curie point)

EuO orders ferromagnetically at $T_C = 69$ K. It is a semiconductor in the paramagnetic state. As a result of the polarization below T_C, the energies of the valence electrons of europium cross the trap levels due to vacancies on the oxygen sites. The transfer of electrons gives rise to a transition to a metallic state.

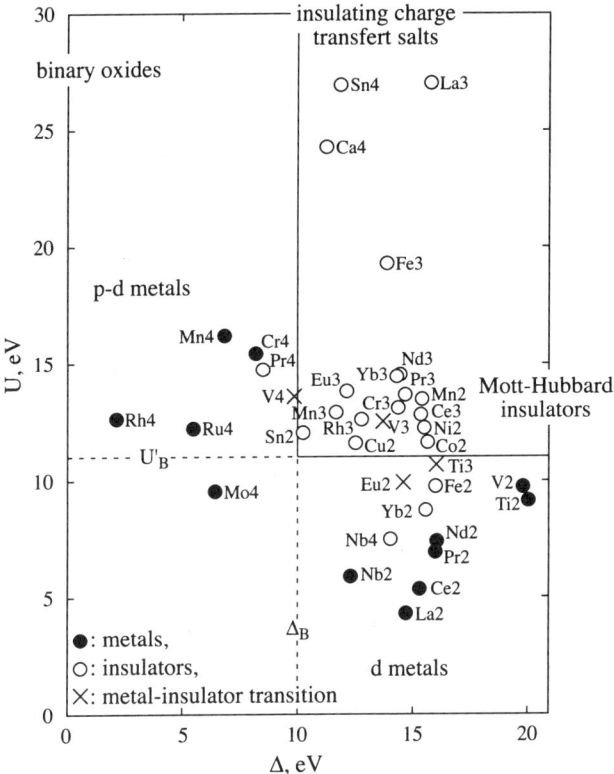

Figure 14.8 - Classification of 38 binary oxides from ref. [27] using the notation of Zaanen, Sawatzky, Allen [26]

4.2.4. Anderson transitions in disordered systems

In heavily doped semi-conductors, electrons are trapped on sites distributed in a random manner, within potential wells of variable depth. Conduction proceeds by *hopping* [28]. Below a certain energy, the mobility edge, the electrons can only leave their site by hopping to a distant site having the same energy. On the other hand, above this limit the carriers are mobile. The conductivity does not exactly follow an activation law, but can be expressed by:

$$\rho = \rho_0 T^s \exp(T_0/T)^p \qquad (14.12)$$

Depending on the situation, one observes: $s = 1/2$, $p = 1/4$ or $s = 1$, $p = 1/2$. T_0 is the Mott temperature. The conduction will thus be a function of the disorder (that can be of magnetic origin), and of the localization. The location of the Fermi level with respect to the mobility edge is determined by the level of doping, magnetic field, etc. A magnetic splitting may induce a transition to a metallic state, by driving the mobility edge for one spin subband close to the Fermi level.

4.2.5. *An example: giant magnetoresistance in manganites*

While it was known since 1954, this MR was extensively studied recently in thin films [29], because it may have applications as magnetoresistive sensors. $LaMnO_3$ and $MMnO_3$ (M = Ca, Sr, Ba) are antiferromagnetic semiconductors. Solid solutions with 1/3 of alkaline earth metal are ferromagnetic, and display a semiconductor-metal transition around their Curie temperature. Applying a magnetic field aligns the moments, and provokes a *colossal* negative magnetoresistance [30]. Several mechanisms are likely to be involved: conduction by hopping requiring magnetic *polarons* in the semiconducting paramagnetic domain, a decrease in spin disorder that reduces scattering, and a transition to the metallic state. The latter can occur either by crossing of bands, when the polarization of the bands below the Curie point becomes larger than the semi-conductor gap, or when this decoupling brings the Fermi energy close to the mobility edge.

4.3. HALF-METALS

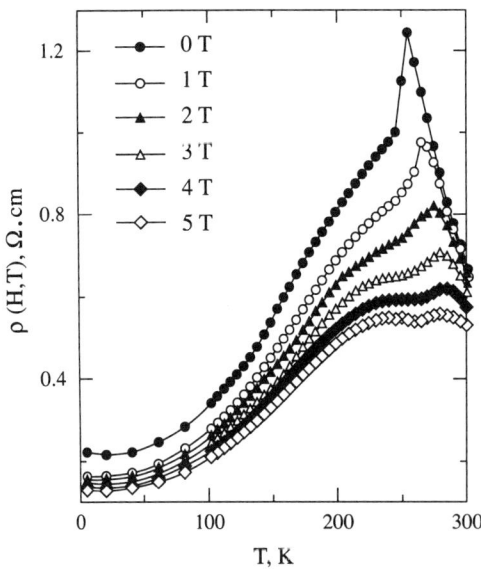

Figure 14.9
Resistivity of a sintered sample of $La_{0.67}Ca_{0.33}MnO_3$ as a function of temperature and magnetic field, from ref. [32]

A very special case of the interaction between ferromagnetism and the conduction properties is that of half-metals [31]. This situation sometimes arises due to the magnetic decoupling of the two spin subbands (↑ and ↓ spins) when the band structure contains energy gaps. In this case, the Fermi level for the up-spin band may fall in a gap of the down-spin band. These materials are then metals for the up-spin band, but semiconducting for the down-spin band. Half metals should not be confused with semi-metals (paramagnetic metals with slightly overlapping bands like Bi). Half metallic states are encountered for NiMnSb or PtMnSb semi-Heusler phases [31], CrO_2, and some ferromagnetic manganites. For a stoichiometric compound, the magnetic moment per formula unit should be an integer (4.0 μ_B for NiMnSb), as the filled down-spin band contains an integral number of electrons. These systems may have interesting properties concerning magneto-optics (PtMnSb) or spin-dependent transport (injection of spin-polarized carriers, spin valves, field sensitive transistors) [30].

5. QUANTUM EFFECTS

5.1. THE SHUBNIKOV-DE HAAS EFFECT

Classically, electrons precess around a field with the cyclotron frequency ω_c (§ 2.4.1). The quantization of energy states of the bands [33, 34, 35] manifests itself when the electrons can complete several turns without being scattered by phonons or impurities (when $\omega_c \tau \gg 1$), i.e. under intense magnetic field, and at low temperature. For nearly free electrons, the energy levels in an applied field along Oz are given by:

$$E = (n + 1/2)\,\hbar\omega_c + (\hbar^2 k_z^2 / 2m^*) + (1/2)\,g\mu_B B_0 \qquad (14.13)$$

where m* is the effective mass of the electrons, and n is an integer. The parabolic band is decomposed into sub-bands separated by energy $\hbar\omega_c$ (Landau levels). The distribution of the electrons in k space will condense into cylinders whose cross-section perpendicular to the field is quantised: $A_n = (n + 1/2)(eB_0/\pi\hbar)$. The section of the orbits is modified when the field is increased such that those near the Fermi energy periodically depopulate, the electrons moving to lower levels.

The free energy, and all of the properties which depend on it, have a pseudo-periodic behavior with $1/B_0$, the amplitude of the oscillations being strongly reduced with temperature as a result of the broadening of the Landau levels. In the same way as the susceptibility (de Haas-van Alphen effect), the observed resistivity oscillates (Shubnikov-de Haas effect), as was first observed in semi-metals (Bi) and semiconductors.

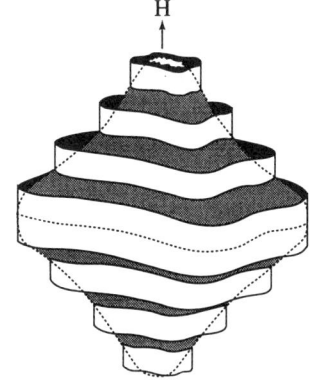

Figure 14.10 - Quantization of orbit areas (after Ziman)

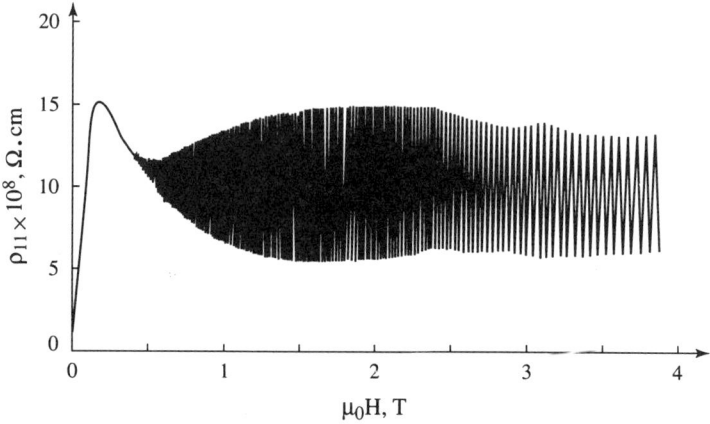

Figure 14.11 - Shubnikov-de Haas effect in Mg, from ref. [36]

5.2. THE QUANTUM HALL EFFECT

This effect (QHE) was discovered in 1980 by K. von Klitzing *et al.* [37]. It appears at very low temperatures for a layer of electrons confined to two dimensions (at the junction between two semiconductors). Unlike the preceding case, there is no z degree of freedom ($k_z = 0$ in eq. 14.13). In contrast to the classical Hall effect, the Hall voltage increases in steps as a function of field; these steps correspond to a reduction or even cancellation of the longitudinal resistivity [35, 38]. Each plateau of the Hall conductance corresponds to a multiple (integral QHE) or sub-multiple (fractional QHE) of the *conductance quantum* e^2/h, independently of the nature of the sample under consideration: $\rho_H = h/\nu e^2$. The values of the resistance plateaus are the same, to better than 10^{-7}, as the calculated theoretical values. This has allowed a new standard of resistance to be defined: the ratio (h/e^2), or the klitzing, is 25,812.8 ohms.

What do these steps correspond to? The Hall resistivity represents the ratio of the field to the electronic density: $\rho_H = B/ne$. The flux in the sample being a multiple of the elementary flux quantum Φ_0, the steps appear for definite values, integral or fractional, of the *filling factor*, the ratio between the number of electrons and the number of flux quanta. The integral QHE corresponds to the occupation of an integral number of Landau levels. The existence of a *gap* in the excitation spectrum makes the electron gas incompressible. More generally, the repulsive interactions between electrons stabilise a new type of quantum fluid, where the electrons are linked to the flux quanta. The phenomenon can thus be interpreted as a condensation of fermions, under the influence of an intense field, into bosons subjected to a reduced field [39]. This condensation into bosons explains the cancellation of the longitudinal resistance, and has analogies with superconductivity.

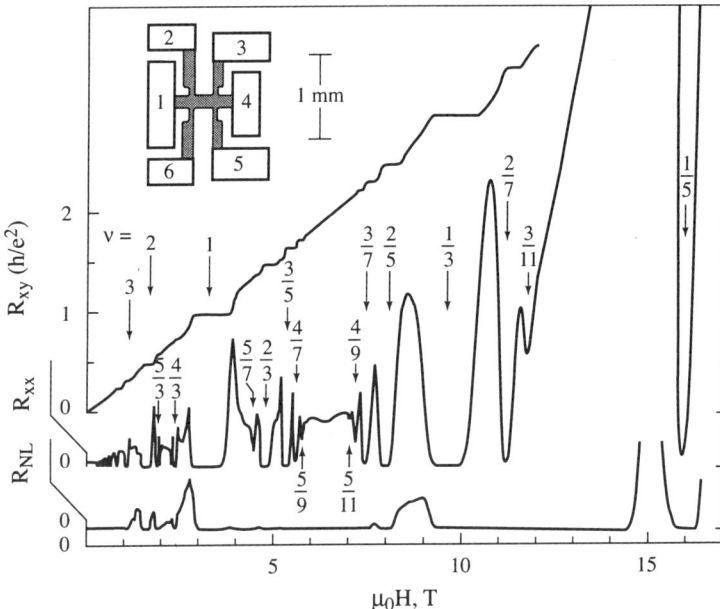

Figure 14.12 - Experiment showing the quantum Hall effect
From top to bottom: Hall resistance V_{26}/I_{14}, longitudinal resistance V_{23}/I_{14}, and the non-local resistance V_{26}/I_{35}. $T = 40$ mK (from V.J. Goldman, unpublished, cited in [39]).

6. APPLICATIONS

While magnetic scattering contributes to temperature-independent resistors (manganin, constantan), and to large thermoelectric effects (chromel, rhodium-platinum), its main applications are magnetic sensors (magnetoresistance, Hall effect). New applications could emerge from spin polarised transport, for example the injection of polarised carriers from a half-metal.

6.1. HALL EFFECT FIELD SENSORS

The material is chosen from doped semiconductors (InAs, InSb) whose Hall resistivity changes minimally with temperature. Sensors in thin film form give a large signal as a result of their small thickness. The Hall voltage does not always change linearly with the field, but this can be corrected for by shunting the Hall signal via an adaptation resistor.

6.2. MAGNETORESISTIVE SENSORS

The highest performance sensors in low fields are multi-layer spin valve systems (see chap. 20). The first magnetoresistive hard disk read heads (1994) used the anisotropy of the magnetoresistance of permalloy $Ni_{80}Fe_{20}$, where the anisotropy is around 2-3% in fields less than 0.1 tesla. The resistivity of a rod is a function of the angle θ that the field makes with the current:

$$\rho(\theta) = \rho_\perp + \Delta\rho \cos^2\theta \qquad (14.14)$$

Taking a sensor where the current is at 45° to the easy axis (*Barber pole* geometry), the weak field response can be linearised.

FINAL NOTE

The various effects linked to galvanomagnetic coupling have briefly been presented; certain aspects will be revisited in much greater detail in what follows, in particular giant magnetoresistance, which is the subject of very intense research related to read head applications (see chap. 20, § 5, and 21 § 5.2).

REFERENCES

[1] K. SEEGER, Semiconductor Physics, 2nd ed. (1982) Springer Verlag, Berlin.

[2] E. GRATZ, M.J. ZUCKERMANN, *in* Handbook on the Physics and Chemistry of Rare Earths, Vol. 5 (1982) 117, K.A. Gschneidner, L. Eyring Eds, North Holland, Amsterdam.

[3] E.H. PUTLEY, The Hall effect and related phenomena (1960) Butterworths, London.

[4] K. SCHROEDER, Electronic, magnetic and thermal properties of solid materials (1978) M. Dekker Inc., New York.

[5] P.L. ROSSITER, The electrical resistivity of metals and alloys (1991) Cambridge University Press.

[6] P.G. DE GENNES, J. FRIEDEL, *J. Phys. Chem. Solids* **4** (1958) 71.

[7] B. COQBLIN, The electronic structure of metals and alloys (1977) Acad. Press, New York.

[8] J.M. FOURNIER, E. GRATZ, *in* Handbook on the Physics and Chemistry of Rare Earths, Vol. 17 (1993) 409, K.A. Gschneidner, L. Eyring Eds, Elsevier Science Publ.

[9] T. VAN PESKI-TIMBERGEN, A.J. DEKKER, *Physica* **29** (1963) 917.

[10] H. MIWA, *Prog. Theor. Phys.* **29** (1963) 477.

[11] R.J. ELLIOTT, F.A. WEDGWOOD, *Proc. Phys. Soc.* **81** (1963) 846.

[12] K.A. McEWEN, *in* Handbook on the Physics and Chemistry of Rare Earths, Vol. 1 (1978) 469, K.A. Gschneidner, L.R. Eyring Eds, North Holland, Amsterdam.

[13] A. FERT, I.A. CAMPBELL, *J. Phys. F.: Met. Phys.* **6** (1976) 849; I.A. CAMPBELL, A. FERT, *in* Ferromagnetic Materials, Vol. 3 (1982) 747, North Holland Publ.

[14] J. FRIEDEL, The theory of magnetism in transition metals (1968) Acad. Press, New York.

[15] B. COQBLIN, *in* Magnetism of metals and alloys (1982) 295, M. Cyrot Ed., North Holland Publ., Amsterdam.

[16] E. BAUER, *Adv. Phys.* **40** 1991) 417.

[17] M. KOHLER, *Ann.Phys.* **6** (1949) 18 107.

[18] H. YAMADA, S. TAKADA, *J. Phys. Soc. Jpn* **34** (1973) 51.

[19] A. FERT, P.M. LEVY, *Phys. Rev. B* **16** (1977) 5052.

[20] A. FERT, *Physica B* **86-88 B** (1977) 491, and references therein.

[21] M.J. OTTO, Thesis (1987) University of Groningen.

[22] High magnetic fields in semiconductor physics III (1992) G. Landwehr Ed. Springer Series in Solid State Sciences 101, Springer Verlag, Berlin.

[23] T. DIETL, *in* Handbook on semiconductors, Vol. 3-b (1994) 1251, T. Moss and S. Mahajan Eds, North Holland.

[24] S. VON MOLNAR, T. PENNEY, *in* Physics of disordered materials, Vol. 3 (1985) 183, D. Adler *et al.* Eds, Plenum Press, New York.

[25] N.F. MOTT, Metal-insulator transitions, 2nd ed. (1990) Taylor and Francis, London.

[26] J. ZAANEN, G.A. SAWATSKY, J.W. ALLEN, *Phys. Rev. Lett.* **55** (1985) 418.

[27] J.B. TORRANCE, P. LACORRE, C. ASAVAROENGCHAI, R.M. METZGER, *Physica C* **182** (1991) 351.

[28] M. POLLAK, B. SHKLOVSKII, Hopping transport in solids (1991) North Holland, Amsterdam.

[29] R. VON HELMOLT, J. WECKER, B. HOLZAPFEL, L. SCHULZ, K. SAMWER, *Phys. Rev. Lett.* **71** (1993) 2331.

[30] A. RAMIREZ, *J. Phys: Cond. Matter* **9** (1997) 8171, and references therein; J.M.D. COEY, M. VIRET, S. VON MOLNAR, *Adv. Phys.* **48** (1993) 167.

[31] R.A. DE GROOT, F.M. MÜLLER, P.G. VAN ENGEN, K.H.J. BUSCHOW, *J. Appl. Phys.* **51** (1995) 1077.

[32] A. NOSOV (1997), private communication.

[33] A.B. PIPPARD, D. SCHÖNBERG, R.G. CHAMBERS, *in* The Physics of metals, Electrons (1969), J.M. Ziman Ed., Cambridge University Press.

[34] G. LANDWEHR, *in* Physics of solids under intense magnetic fields (1969) 415, E.D. Haidemenakis Ed., Plenum Press, New York.

[35] L.P. LÉVY, Magnétisme et supraconductivité, Chapter 2 (1997) Editions du CNRS, Paris.

[36] L.M. FALICOV, R.W. STARK, *Prog. Low Temp. Phys.* **5** (1967) 235.

[37] K. VON KLITZING, G. DORDA, M. PEPPER, *Phys. Rev. Lett.* **45** (1980) 494.

[38] J.K. JAIN, *Adv. Phys.* **41** (1992) 105.

[39] S. KIVELSON, D-H. LEE, S-H. ZHANG, *Sci. Am.* **274** (1996) 64.

APPENDICES

APPENDIX 1

SYMBOLS USED IN THE TEXT

a : Lattice parameter ; ka, Ma: 10^3, 10^6 years (Chap. 24). $\mathcal{A}$: Surface area.

A : Number of nucleons ; A_{ex}: Exchange constant; **A**: Magnetic vector potential.

B : Magnetic induction; B_1, B_2, ... $B^{\mu,\ell}$: Magnetoelastic coupling coefficients.

$\mathcal{B}$: Brillouin function.

c : Speed of light; lattice parameter; c_{ij}, c_{ij}^{μ} : Elastic moduli.

C : Capacitance of a condenser, specific heat; $\mathcal{C}$: Curie constant.

d : Density, distance.

e : emf; $-e$: Electron charge; $\mathfrak{e}$: Electric potential (voltage).

E : Energy density; E_F: Fermi energy. **E**: Electric field.

$\mathcal{E}$: Magnetomotive force; energy.

f : Function, frequency.

F : Free energy density; **F**: Force.

$\mathcal{F}$: Free energy.

g : Acceleration of gravity; gap (airgap); g, g_J: Landé factor.

G : Shear modulus, free enthalpy density; $\mathcal{G}$: Free enthalpy (Gibbs function).

h : Planck's constant $\{\hbar = h/2\pi\}$.

H : Magnetic field; H_C: Coercive field; H_{crit}, H_c: Critical field; H_A: Anisotropy field; H_d: Demagnetising field; H_0: Applied field (external); H_{mat}: Field produced by matter; H_{mol}: Molecular field.

$\mathcal{H}$: Hamiltonian.

i : Instantaneous electric current; I: Effective electric current.

j : Electric current density; $j = \sqrt{-1}$.

J : Total angular momentum operator (in units of $\hbar$); **J** $(= \mathbf{B} - \mu_0\mathbf{H})$: Magnetic polarisation.

$\mathcal{J}$: Exchange integral; J, j_i: Total angular momentum quantum number.

k	:	Wave vector; k_B: Boltzmann's constant.
K_i	:	Magnetocrystalline anisotropy constants.
L	:	Self inductance; L, l_i: Orbital angular momentum quantum number.
L, $\boldsymbol{\ell}$	:	Orbital angular momentum operator (in units of $\hbar$).
$\boldsymbol{\mathcal{L}}$	:	Angular momentum ($= \mathbf{r} \times \mathbf{p}$).
$\mathcal{L}$	:	Langevin function.
m	:	Mass; m: Reduced magnetisation $\{= M(T, H) / M_0\}$.
$\mathfrak{m}$	:	Magnetic moment; $\mathfrak{m}_{eff}, \mathfrak{m}_o, \mathfrak{m}_s$: Effective, orbital, spin magnetic moment. $\mathfrak{m}$: Magnetic moment (modulus).
M	:	Mutual inductance; M_J, M_L, M_S, m_i: Magnetic quantum number.
M	:	Magnetisation ($\mathbf{M} = d\mathfrak{m}/dV$); M_s: Spontaneous magnetisation; M_0: M_s (T = 0 K).
n, n_i	:	Principal quantum number; n_{ij}: Molecular field coefficient.
N	:	Demagnetising field coefficient; Avogadro's number; $N(\varepsilon)$: density of states.
p	:	Pressure.
p	:	Momentum ($= m\mathbf{v} + q\mathbf{A}$).
P	:	Power.
q	:	Electric charge; q_m: magnetic mass.
Q	:	Quantity of electricity, quality factor, magneto-optic coefficient.
R	:	Electric resistance; R_H: Hall coefficient.
$\mathcal{R}$	:	Reluctance; $\mathcal{R}_{ij}$: Components of rotation ($i \neq j$).
$[s_{ij}]$	:	Elastic compliance tensor (in reduced notation).
S	:	Entropy density; Surface area.
S	:	Spin operator (in units of $\hbar$); oriented surface area; S, s_i: Spin quantum number.
t	:	Time.
T	:	Period, temperature; T_N: Néel temperature; T_C: Curie temperature.
T_{crit}	:	Critical temperature (e.g. superconducting); T_{comp}: Compensation temperature.
u	:	Potential difference (voltage); U: Effective potential difference.
u	:	Displacement vector; $u_{i,j}$: Distortion.
U	:	Internal energy density; U: Magnetic potential difference.
$\mathcal{U}$	:	Internal energy.
v	:	Velocity.
$\hat{\mathbf{v}}$	:	Unit vector (for instance : $\hat{\mathbf{r}} = \mathbf{OM} / |\mathbf{OM}|$).

APPENDIX 1 - SYMBOLS USED IN THE TEXT

V	:	Volume, electric potential.
$\mathcal{V}$	:	Magnetic potential.
w	:	Molecular field coefficient.
W	:	Work.
Y	:	Young's modulus.
Z	:	Electrical impedance, atomic number.
α	:	Damping coefficient; α_T: Linear thermal expansion coefficient.
α_i	:	Direction cosines of magnetisation.
β_i	:	Direction cosines of the observation direction.
γ_i	:	Direction cosines of direction of application of a stress.
γ	:	Gyromagnetic ratio; γ: Wall energy per unit area.
$[\gamma_{ij}]$	:	Electric conductivity tensor.
Γ	:	Torque, moment of a force.
δ	:	Wall thickness. δ_{ij}: Kronecker symbol.
ε	:	Permittivity, energy; ε_0: Permittivity of free space; $\boldsymbol{\epsilon}$, $[\varepsilon_{ij}]$: Permittivity tensor, strain tensor.
η_p	:	Viscous friction coefficient.
Θ_p	:	Paramagnetic Curie temperature.
κ	:	Compressibility.
λ	:	Wavelength, spin-orbit coupling parameter; elongation; λ_s, λ^α, $\lambda^{\mu,\ell}$: Magnetostriction coefficient ($\mu = \gamma, \delta, \varepsilon, \zeta ...; \ell = 2, 4, ...$).
Λ	:	Permeance.
μ	:	Magnetic permeability; μ_0: Permeability of free space; μ_r: Relative permeability.
μ_B	:	Bohr magneton.
ν	:	Poisson's coefficient; frequency.
ξ	:	Torsion; coherence length.
ρ	:	(Mass) density, resitivity; ρ_e: Charge density; ρ_m: Magnetic mass density.
$[\rho_{ij}]$	:	Electrical resistivity tensor.
σ	:	Surface electric charge density; specific magnetic moment ($\sigma = \mathfrak{m}/m$); spin quantum number.
σ_m	:	Surface magnetic mass density.
$[\sigma_{ij}]$	:	Mechanical stress tensor; $\boldsymbol{\sigma}$: Conductivity tensor.
τ	:	Relaxation time.

Φ   : Magnetic flux.

χ   : Magnetic susceptibility (χ_p, χ_d: paramagnetic, diamagnetic).

ψ(**r**) : Wave function for the electron {$\psi_k(\mathbf{r}) = u_k(\mathbf{r}) \exp(i\mathbf{k}.\mathbf{r})$ for a Bloch wave}.

ω   : Angular frequency, relative volume change; ω_L: Larmor (angular) frequency.

ω : Angular velocity.

Vectors and tensors are printed in bold type. For instance : $\mathbf{r}_{IJ}$ = vector $\overrightarrow{IJ}$.

The modulus of a vector **V** is written : |**V**|.

Unit vectors are printed with a circonflex accent (example : $\hat{\mathbf{r}}$ = **OM** / |**OM**|).

The cross product is written : $\mathbf{A} \times \mathbf{B}$.

APPENDIX 2

UNITS AND UNIVERSAL CONSTANTS

1. CONVERSION OF MKSA UNITS INTO THE CGS SYSTEM, AND OTHER UNIT SYSTEMS OF COMMON USE

Length (metre) : 1 m = 10^2 cm = 39.37 " (inch) = 10^{10} angström (Å).

Force (newton) : 1 N = 10^5 dyn = 0.102 kgf.

Energy (joule) : 1 J = 10^7 erg = 0.7243×10^{23} K = 0.6241×10^{19} eV.

Energy density : 1 J.m^{-3} = 10 erg.cm^{-3}.

Power (watt) : 1 W = 10^7 erg.s^{-1} = 1.359×10^{-3} CV = 1.340×10^{-3} hp.

Pressure (pascal) : 1 Pa = 10 baryes = 10^{-5} bar = 1.02×10^{-5} kgf.cm^{-2}
 = 7.49×10^{-3} torrs = 1.45×10^{-4} psi.

Magnetic induction (tesla): $\quad\quad\quad\quad\quad\quad\quad$ 1 T = 10^4 gauss (= 1 Wb.m^{-2}).

 B can be called magnetic induction, magnetic induction field or magnetic flux density.

Magnetic field (ampere / metre): $\quad\quad\quad\quad$ 1 A.m^{-1} = $4\pi \times 10^{-3}$ œrsted.

 The magnetic field **H** is often expressed in units of $\mu_0 H$, hence in tesla (T) or its submultiple, the gamma (γ) which is equal to 10^{-9} T. A field of 1 A.m^{-1} corresponds to 1.2566 µT.

Magnetisation (ampere / metre): $\quad\quad\quad\quad\quad\quad$ 1 A.m^{-1} = 10^{-3} emu.cm^{-3}.

Magnetic moment (ampere-square metre or joule per tesla): 1 A.m^2 = 1 J.T^{-1} = 10^3 emu.

Specific magnetic moment: $\quad\quad\quad\quad\quad\quad\quad\quad$ 1 A.m^2.kg^{-1} = 1 emu.g^{-1}.

 Note that, sometimes, magnetisation (**M**), magnetic moment (𝔪) and specific magnetic moment (σ) are expressed in tesla, Weber-meter and Weber-meter per kilogram, respectively : the reason is that we have adopted as definition of magnetisation (d𝔪/dV): **M** = **B**/μ_0 – **H**, whereas some authors call "magnetisation" the quantity **B**$_i$ = **J** = **B** – μ_0**H** which is usually called "(magnetic) polarisation".

Temperature : to know the temperature in degrees Celsius or centigrades (°C), subtract 273.15 from the value in kelvin (K).

2. SOME FUNDAMENTAL PHYSICAL CONSTANTS

Name and symbol	Numerical value
Speed of light in vacuum, c	2.9979×10^8 m.s^{-1}
Permeability of vacuum, μ_0	$4\pi \times 10^{-7}$ H.m^{-1}
Permittivity of vacuum, $\varepsilon_0 = 1/c^2\mu_0$	8.8542×10^{-12} F.m^{-1}
Planck's constant, h	6.6261×10^{-34} J.s
$\hbar = h/2\pi$	1.0546×10^{-34} J.s
Acceleration of gravity, g	9.8066 m.s^{-2}
Electron rest mass, m	9.1094×10^{-31} kg
Electron charge (absolute value), e	1.6022×10^{-19} C
Bohr magneton, $\mu_B = e\hbar/2m$	9.2742×10^{-24} A.m^2
Flux quantum, $h/2e$	2.0678×10^{-15} Wb
Avogadro's number N	6.0221×10^{23} mol^{-1}
Boltzmann's constant, k_B	1.3807×10^{-23} J.K^{-1}

APPENDIX 3

PERIODIC TABLE OF THE ELEMENTS

In the cells of the table below, for each element, the first line indicates its symbol, with its atomic number as superscript, the second one its atomic weight, the third one its spectroscopic ground state in the neutral atomic state, and finally the fourth one gives the electronic configuration of the external shells, again in the neutral atomic state.

Legend (example cell):
- element — H^1 — atomic number
- atomic mass — 1.008
- $^2S_{1/2}$ — spectroscopic state
- $1s^1$ — configuration of external shells

H^1 1.008 $^2S_{1/2}$ $1s^1$																	He^2 4.003 $^1S_0$ $1s^2$
Li^3 6.940 $^2S_{1/2}$ $2s^1$	Be^4 9.012 $^1S_0$ $2s^2$											B^5 10.811 $^2P_{1/2}$ $2p^1$	C^6 12.011 $^3P_0$ $2p^2$	N^7 14.007 $^4S_{3/2}$ $2p^3$	O^8 15.999 $^3P_2$ $2p^4$	F^9 18.998 $^2P_{3/2}$ $2p^5$	Ne^{10} 20.180 $^1S_0$ $2p^6$
Na^{11} 22.990 $^2S_{1/2}$ $3s^1$	Mg^{12} 24.305 $^1S_0$ $3s^2$											Al^{13} 26.982 $^2P_{1/2}$ $3p^1$	Si^{14} 28.086 $^3P_0$ $3p^2$	P^{15} 30.974 $^4S_{3/2}$ $3p^3$	S^{16} 32.060 $^3P_2$ $3p^4$	Cl^{17} 35.453 $^2P_{3/2}$ $3p^5$	Ar^{18} 39.948 $^1S_0$ $3p^6$
K^{19} 39.100 $^2S_{1/2}$ $3d^04s^1$	Ca^{20} 40.08 $^1S_0$ $3d^04s^2$	Sc^{21} 44.95 $^2D_{3/2}$ $3d^14s^2$	Ti^{22} 47.90 $^3F_2$ $3d^24s^2$	V^{23} 50.941 $^4F_{3/2}$ $3d^34s^2$	Cr^{24} 51.996 $^7S_3$ $3d^54s^1$	Mn^{25} 54.938 $^6S_{5/2}$ $3d^54s^2$	Fe^{26} 55.847 $^5D_4$ $3d^64s^2$	Co^{27} 58.933 $^4F_{9/2}$ $3d^74s^2$	Ni^{28} 58.70 $^3F_4$ $3d^84s^2$	Cu^{29} 63.546 $^2S_{1/2}$ $3d^{10}4s^1$	Zn^{30} 65.38 $^1S_0$ $3d^{10}4s^2$	Ga^{31} 69.72 $^2P_{1/2}$ $4p^1$	Ge^{32} 72.59 $^3P_0$ $4p^2$	As^{33} 74.922 $^4S_{3/2}$ $4p^3$	Se^{34} 78.96 $^3P_2$ $4p^4$	Br^{35} 79.904 $^2P_{3/2}$ $4p^5$	Kr^{36} 83.80 $^1S_0$ $4p^6$
Rb^{37} 85.47 $^2S_{1/2}$ $4d^05s^1$	Sr^{38} 87.62 $^1S_0$ $4d^05s^2$	Y^{39} 88.906 $^2D_{3/2}$ $4d^15s^2$	Zr^{40} 91.22 $^3F_2$ $4d^25s^2$	Nb^{41} 92.906 $^6D_{1/2}$ $4d^45s^1$	Mo^{42} 95.94 $^7S_3$ $4d^55s^1$	Tc^{43} ~98 $^5F_5$ $4d^55s^2$	Ru^{44} 101.07 $^5F_5$ $4d^75s^1$	Rh^{45} 102.91 $^4F_{9/2}$ $4d^85s^1$	Pd^{46} 106.4 $^1S_0$ $4d^{10}5s^0$	Ag^{47} 107.87 $^2S_{1/2}$ $4d^{10}5s^1$	Cd^{48} 112.40 $^1S_0$ $4d^{10}5s^2$	In^{49} 114.82 $^2P_{1/2}$ $5p^1$	Sn^{50} 118.69 $^3P_0$ $5p^2$	Sb^{51} 121.75 $^4S_{3/2}$ $5p^3$	Te^{52} 127.60 $^3P_2$ $5p^4$	I^{53} 126.90 $^2P_{3/2}$ $5p^5$	Xe^{54} 131.30 $^1S_0$ $5p^6$
Cs^{55} 132.91 $^2S_{1/2}$ $5d^06s^1$	Ba^{56} 137.34 $^1S_0$ $5d^06s^2$	La^{57} 138.91 $^2D_{3/2}$ $5d^16s^2$	Hf^{72} 178.49 $^3F_2$ $5d^26s^2$	Ta^{73} 180.95 $^4F_{3/2}$ $5d^36s^2$	W^{74} 183.85 $^5D_0$ $5d^46s^2$	Re^{75} 186.21 $^6S_{3/2}$ $5d^56s^2$	Os^{76} 190.2 $^5D_4$ $5d^66s^2$	Ir^{77} 192.22 $^4F_{3/2}$ $5d^76s^2$	Pt^{78} 195.09 $^3D_3$ $5d^96s^1$	Au^{79} 196.97 $^2S_{1/2}$ $5d^{10}6s^1$	Hg^{80} 200.59 $^1S_0$ $5d^{10}6s^2$	Tl^{81} 204.37 $^2P_{1/2}$ $6p^1$	Pb^{82} 207.2 $^3P_0$ $6p^2$	Bi^{83} 208.98 $^4S_{3/2}$ $6p^3$	Po^{84} ~209 $^3P_2$ $6p^4$	At^{85} ~210 $6p^5$	Rn^{86} ~222 $^1S_0$ $6p^6$
Fr^{87} ~223 $6d^07s^1$	Ra^{88} 226.03 $^1S_0$ $6d^07s^2$	Ac^{89} ~227 $^2D_{3/2}$ $6d^17s^2$															

Lanthanides:

Ce^{58} 140.12 $^3H_4$ $4f^1$ $5d^16s^2$	Pr^{59} 140.91 $^4I_{9/2}$ $4f^3$ $5d^06s^2$	Nd^{60} 144.24 $^5I_4$ $4f^4$ $5d^06s^2$	Pm^{61} ~145 $4f^5$ $5d^06s^2$	Sm^{62} 150.35 $^7F_0$ $4f^6$ $5d^06s^2$	Eu^{63} 151.96 $^8S_{7/2}$ $4f^7$ $5d^06s^2$	Gd^{64} 157.25 $^9D_2$ $4f^7$ $5d^16s^2$	Tb^{65} 158.92 $^8H_{17/2}$ $4f^8$ $5d^16s^2$	Dy^{66} 162.50 $^7J_{10}$ $4f^9$ $5d^16s^2$	Ho^{67} 164.93 $^6K_{10}$ $4f^{10}$ $5d^16s^2$	Er^{68} 167.26 $^5K_{10}$ $4f^{11}$ $5d^16s^2$	Tm^{69} 168.93 $^2F_{7/2}$ $4f^{13}$ $5d^06s^2$	Yb^{70} 173.04 $^1S_0$ $4f^{14}$ $5d^06s^2$	Lu^{71} 174.97 $^2D_{5/2}$ $4f^{14}$ $5d^16s^2$

Actinides:

Th^{90} 232.04 $^3F_2$ $5f^0$ $6d^27s^2$	Pa^{91} 231.04 $^4F_{3/2}$ $5f^2$ $6d^17s^2$	U^{92} 238.03 $^5L_6$ $5f^3$ $6d^17s^2$	Np^{93} 237.05 $5f^4$ $6d^17s^2$	Pu^{94} ...	Am^{95}	Cm^{96}	Bk^{97}	Cf^{98}	Es^{99}	Fm^{100}	Md^{101}	No^{102}	Lw^{103}

APPENDIX 4

MAGNETIC SUSCEPTIBILITIES

We list the magnetic susceptibilities (MKSA) of some so-called "non magnetic" substances: pure elements in the solid and liquid state, and materials of common use: alloys, plastics, glasses, and ceramics. The values usually were published in cgs units of specific susceptibility (χ_{cgs}). We therefore give the value of the density which was used to convert the susceptibility into MKSA units (χ is dimensionless in this system with our convention for the magnetisation: it is simply given by: $\chi = 4\pi . d . \chi_{cgs}$ where d is the density expressed in g/cm^3).

We begin with diamagnetic substances, and then move on to paramagnetic substances.

Table A 4.1 - Susceptibility of some diamagnetic pure elements ($\times 10^6$)

Element	d g.cm^{-3}	$-\chi$	Element	d g.cm^{-3}	$-\chi$	Element	d g.cm^{-3}	$-\chi$
Ag	10.492	25	Au	18.88	34	B	2.535	20
Bi	9.78	165	C-diamond	3.52	22	Cd	8.65	19
Cu	8.933	11.8	Ga	5.93	23	Ge	5.46	7.25
H$_2$	0.0763	2.6	Hg	14.2	29	I	4.94	22
In	7.28	51	red P	2.20	20	Pb	11.342	16
S (α)	2.07	12.6	S (β)	1.96	11.4	Sb	6.62	67
Se	4.82	19	Si	2.42	3.4	grey Sn	7.30	29
Te	6.25	24	Tl (α)	11.86	37	Zn	6.92	15

All these elements were studied in their solid phase except for hydrogen and mercury, which are in the liquid state (in grey). The data were obtained at room temperature, except for hydrogen, measured at 13 K. The data are mostly taken from reference [1]. For the case of copper, for instance, one can check [2] that the thermal variations of χ are very weak from 4.2 K to 300 K.

Table A 4.2 - Susceptibility at room temperature of some diamagnetic materials of common use ($\times 10^6$)

Material	d g.cm^{-3}	$-\chi$	Material	d g.cm^{-3}	$-\chi$	Material	d g.cm^{-3}	$-\chi$
Cu Sn 5	8.9	11	Cu Sn 12	8.9	14	Cu Sn 20	8.8	19
Cu Zn 10	8.9	12	Cu Zn 20	8.9	16	Cu Zn 40	8.9	24
Corning7052	2.27	8	Zerodur	2.52	12	Plexiglas	1.19	9
Polyethylene	0.923	10	Polyimide	1.43	9	PTFE	2.15	10
Alumina	3.87	12	Macor	2.52	11	Silica	2.21	12

The convention for the two first lines is that, for example Cu Sn 5 is a tin-copper alloy with 5 wt% tin. The absolute values of the susceptibility of bronzes (first line), and brasses (second line) increases when the copper content decreases [1]. The third line shows two glasses [3], and Plexiglas [4]. The fourth line presents three plastics (PTFE is often called Teflon), and the fifth line, two ceramics and fused silica (amorphous) [2].

Table A 4.3 - Paramagnetic susceptibility at room temperature of some pure elements ($\times 10^6$)

Element	d g.cm^{-3}	χ	Element	d g.cm^{-3}	χ	Element	d g.cm^{-3}	χ
Al	2.70	21	Ba	3.5	6.6	Ca	1.55	20
Cs	1.873	5.2	Hf	13.3	70	Ir	22.42	38
K	0.87	5.8	La	6.174	66	Li	0.534	14
Lu	9.842	>0	Mg	1.74	5.5	Mo	9.01	105
Na	0.9712	8.5	Nb	8.4	232	Os	22.5	15
Pd	12.16	815	Pt	21.37	278	Rb	1.53	38
Re	20.53	97	Rh	12.44	170	Ru	12.1	65
Sc	2.992	263	white Sn	7.30	2.4	Sr	2.60	34
Ta	14.6	177	Th	11.0	79	Ti	4.5	180
U-α	18.7	404	V	5.87	370	W	19.3	78
Y	4.478	120	Yb	6.959	126	Zr	6.44	108

All the elements given in this table are solid are room temperature. Magnesium was found diamagnetic by Pascal [5], and paramagnetic by Foëx [1] and then by Thomas, and Mendoza [6]. The reported susceptibility is that given by the latter. One can notice that the weakest susceptibilities are observed for alkali metals.

APPENDIX 4 - MAGNETIC SUSCEPTIBILITIES

Table A 4.4 - Paramagnetic susceptibilities of some alloys of common use, at various temperatures ($\times 10^6$) [7]

	304 (N)	304 L	316 (LN)	90Cu 10Ni
T = 300 K	3.1	2.9	3.2	210
T = 77 K	6.6	7.6	8.7	...
T = 4.2 K	9.0	11.3	14.2	280

Note that stainless steels (304 and 316) exhibit paramagnetic susceptibilities comparable to those of alkali metals, i.e. at least 20 times smaller than for other weakly magnetic alloys, even at low temperature. This is why they are widely used in magnetic instrumentation, e.g. for the manufacturing of cryostats for superconducting coils.

There exist several tables of numerical value where engineers can find complementary information on paramagnetism in solids, e.g. that edited by Landolt et Börnstein [8]. *Important comment: some tables still give numerical values in cgs units, and the reader should thus be careful, and not confuse them with data in SI / m^3, which can be of the same order of magnitude!*

Note that a ferromagnetic impurity can strongly disturb the magnetic properties of very weakly magnetic substances, even for concentrations of the order of a ppm: the perturbation will generally be temperature dependent, and be larger at low temperature.

On the other hand, special care must be taken in processing materials expected to exhibit very weak magnetism: brazing or soldering under argon atmosphere is for instance liable to precipitate a strongly magnetic phase in a weakly magnetic alloy. Stainless steels that are well known to be very weakly magnetic, are then seen to become slightly ferromagnetic! Special care is therefore required, and it is important to check the magnetic properties of an allegedly non magnetic alloy before using it, and after thermal or mechanical processing. An extremely simple method to check the magnetism of a metallic part consists in hanging a strong magnet (Nd-Fe-B or SmCo$_5$) at the end of a 1 metre long thin nylon thread, and bringing it near the part to be checked: if the slightest attraction is observed, the substance is already slightly ferromagnetic or at least very strongly paramagnetic.

REFERENCES

[1]   Tables de constantes et Données numériques: 7. Constantes sélectionnées: G. FOËX, Diamagnétisme et paramagnétisme; C.J. GORTER, L.J. SMITS, Relaxation paramagnétique (1957) Masson & Cie, Paris, 317 p.
[2]   C.M. HURD, *Cryogenics* **6** (1966) 264.
[3]   P.T. KEYSER, S.R. KEFFERTS, *Rev. Sci. Instrum.* **60** (1989) 2711.

[4] PH. LETHUILLIER, private communication.

[5] P. PASCAL, Nouveau traité de chimie minérale, Tome IV (1958) 143, Masson, Paris.

[6] J.G. THOMAS, E. MENDOZA, *Phil. Mag.* **43** (1952) 900.

[7] Handbook on materials for superconducting machinery, metals and ceramics (1974) Battelle Information Center, Columbus (Ohio).

[8] LANDÖLT, BÖRNSTEIN, Zahlenwerte und Funktionen aus Physik. Chemie. Astronomie. Geophysik. Technik II - 9 - Magnetische Eigenschaften 1, Section 29.1111 (1962) Springer, Berlin.

APPENDIX 5

FERROMAGNETIC MATERIALS

We summarize the structural physical properties, the linear thermal expansion coefficient, the elastic properties (tab. A 5.1) and the intrinsic magnetic properties (tab. A 5.2) of some common strongly magnetic substances.

Table A 5.1 - Physical properties of some strongly magnetic substances, measured at room temperature

Substance	Molecular weight g	Structure	a nm	c/a	Density g.cm^{-3}	α_T 10^{-6} K^{-1}	c_{11} GPa	c_{12} GPa	c_{44} GPa
iron	55.847	bcc	0.287	1	7.85	12.1	237	140	116
cobalt	58.933	hcp	0.251	1.622	8.84	12.4	307*	165*	75.5*
nickel	58.70	fcc	0.352	1	8.90	12.8	250	160	118.5
gadolinium	157.25	hcp	0.363	1.591	7.90	-	67.8*	25.6*	20.8*
Fe_3O_4	231.54	spinel	0.839	1	5.19	-	273	106	97
$CoFe_2O_4$	234.63	spinel	0.838	1	5.29	-	-	-	-
$NiFe_2O_4$	234.39	spinel	0.834	1	5.38	7.5	220	109.4	81.2
$Y_3Fe_5O_{12}$	737.95	garnet	1.238	1	5.17	-	269	107.7	76.4
Fe-80%Ni	(58.10)	fcc	0.354	1	8.65	12.0	-	-	-
Cu_2MnAl	209.01	bcc	0.596	1	6.55	-	135.3	97.3	94
$BaFe_{12}O_{19}$	1111.5	h	0.589	3.937	4.5	-	-	-	-
$SmCo_5$	445.01	h	0.500	0.794	8.58	-	-	-	-
Sm_2Co_{17}	1302.7	h	0.838	0.973	?	-	-	-	-
$Nd_2Fe_{14}B$	1081.1	tetra.	0.879	1.389	7.60	-	-	-	-

* For hexagonal substances, two additional elastic constant have to be considered: for cobalt, they are $c_{13} = 103$ GPa and $c_{33} = 358$ GPa; for gadolinium, $c_{13} = 20.7$ GPa and $c_{33} = 71.2$ GPa.

Table A 5.2 - Magnetic properties of some strongly magnetic substances

Substance	T_C K	M_s kA.m^{-1}	$\mu_0 M_s$ T	σ_s A.m^2.kg^{-1}	K_1 kJ.m^{-3}
iron	1043	1720	2.16	218	48
cobalt	1394	1370	1.72	162	530
nickel	631	485	0.61	56	– 4.5
gadolinium	289	2117	2.66	268	(at 0 K)
Fe_3O_4	858	477	0.60	91.0	– 13
$CoFe_2O_4$	793	398	0.50	75.2	180
$NiFe_2O_4$	858	271	0.34	50.4	– 6.9
$Y_3Fe_5O_{12}$	553	139	0.17	26.1	– 2.5
Fe-80%Ni	595	828	1.04	95.7	– 2*
Cu_2MnAl	610	560	0.70	85.5	– 0.47
$BaFe_{12}O_{19}$	723	382	0.48	84.9	250
$SmCo_5$	995	836	1.05	97.4	17 000
Sm_2Co_{17}	1190	1030	1.29	101	3 300
$Nd_2Fe_{14}B$	585	1280	1.61		4 900

The values of magnetization (M_s and T_s) and of anisotropy (K_1) are for room temperarure (300 K), except in the case of gadolinium, for which the values at 0 K are given.

* K_1 *is very sensitive to thermal treatments for Permalloy (Fe-80%Ni), and changes from – 2 kJ.m^{-3} to a value ten times smaller after quenching.*

Note - *The magnetisation M of a substance, the magnetic moment of which is expressed in μ_B per molecule $\sigma(\mu_B)$, is obtained in A/m through the relation:*

$$M = 5.585 \times \frac{\sigma(\mu_B) \times d(kg.m^{-3})}{M_{mol}(kg)} = 5.585 \times 10^6 \times \frac{\sigma(\mu_B) \times d(g.cm^{-3})}{M_{mol}(g)}.$$

APPENDIX 6

SPECIAL FUNCTIONS

We recall the definition of some special mathematical functions (spherical harmonics, Legendre polynomials, Langevin function, Brillouin functions, and derivatives) which are used several times in this book. A detailed presentation of their properties can be found in many textbooks on mathematics. Here we only give the analytical expression of these functions, we illustrate them with some figures, and we recall their most common uses in magnetism.

1. SPHERICAL HARMONICS

Any function defined in real space can be expressed as a function of the three variables: r, θ, φ in spherical coordinates. A function independent of r can be expressed using only the angular variables θ and φ, and an expansion into a series of orthogonal functions which are solutions of Laplace's equation. They are called spherical harmonics, and denoted $Y_\ell^m(\theta, \phi)$, where l refers to the degree in θ, and m the degree in φ.

$$Y_0^0 = \sqrt{\frac{1}{4\pi}} \qquad Y_1^0 = \sqrt{\frac{3}{4\pi}}\cos\theta \qquad Y_1^{\pm 1} = \mp\sqrt{\frac{3}{8\pi}}\sin\theta\, e^{\pm i\phi}$$

$$Y_2^0 = \sqrt{\frac{5}{16\pi}}(3\cos^2\theta - 1) \quad Y_2^{\pm 1} = \mp\sqrt{\frac{15}{8\pi}}\sin\theta\cos\theta\, e^{\pm i\phi} \quad Y_2^{\pm 2} = \sqrt{\frac{15}{32\pi}}\sin^2\theta\, e^{\pm 2i\phi}$$

$$Y_3^0 = \sqrt{\frac{7}{16\pi}}(2\cos^3\theta - 3\sin^2\theta\cos\theta) \qquad Y_3^{\pm 1} = \mp\sqrt{\frac{21}{64\pi}}\sin\theta(5\cos^2\theta - 1)e^{\pm i\phi}$$

$$Y_3^{\pm 2} = \sqrt{\frac{105}{32\pi}}\sin^2\theta\cos\theta\, e^{\pm 2i\phi} \qquad Y_3^{\pm 3} = \mp\sqrt{\frac{35}{64\pi}}\sin^3\theta\, e^{\pm 3i\phi}$$

These functions are normalized to unity by the relationship: $\int_0^{2\pi}d\phi\int_0^\pi Y_l^m \sin\theta\, d\theta = 1$.

They are used to describe the orbital contribution to the s (Y_0^0), p (Y_1^m), d (Y_2^m), f (Y_3^m) ... wavefunctions of free electrons [1].

In the absence of crystalline electric field and magnetic field, the probability of occupation of the 2m + 1 orbitals describing the state of the level m is the same: therefore, one easily shows that the spatial distribution of the electron is spherical, because: $|\sum_{m} Y_{\ell}^{m}|^2 = \text{Constant}$. On the other hand, when the electron is submitted to the periodic potential of a crystal lattice, the 2m + 1 orbitals are no more energetically equivalent. The orbital contribution to the wavefunctions of the electron is then described with linear combinations of spherical harmonics.

For instance, in cubic symmetry, the five d orbitals are distributed over two different energy levels that are usually called e_g (or Γ_3), which is twice degenerate and associated with the two orbital functions Y_2^0 and $(Y_2^2 + Y_2^{-2})$, and t_{2g} (or Γ_5), which is three-fold degenerate, and associated with the three orbital functions $(Y_2^1 + Y_2^{-1})$, $(Y_2^1 - Y_2^{-1})$, and $(Y_2^2 - Y_2^{-2})$.

As Cartesian coordinates can be expressed as functions of spherical coordinates: $x = r \sin\theta \cos\phi$, $y = r \sin\theta \sin\phi$, and $z = r \cos\theta$, one easily obtains that $Y_2^0 \sim 3z^2 - r^2$, $(Y_2^2 + Y_2^{-2}) \sim x^2 - y^2$, whereas the three functions associated with the t_{2g} level are proportional to yz, zx, and xy, respectively.

The probability for an electron to be present at a given distance r from a nucleus will thus depend on the direction in space. The polar representation shows, for given r, this variation with direction. The distance from the origin to any point on the surfaces shown is proportional to the modulus square of the wave function, hence to the probability, in the direction of the line from the origin to this point. This representation brings out lobes which show the directions of higher probability, as well as sets of directions (planes, cones, ...) along which the electron cannot be found. Figure A 6.1 shows such "orbitals" for a 3d electron.

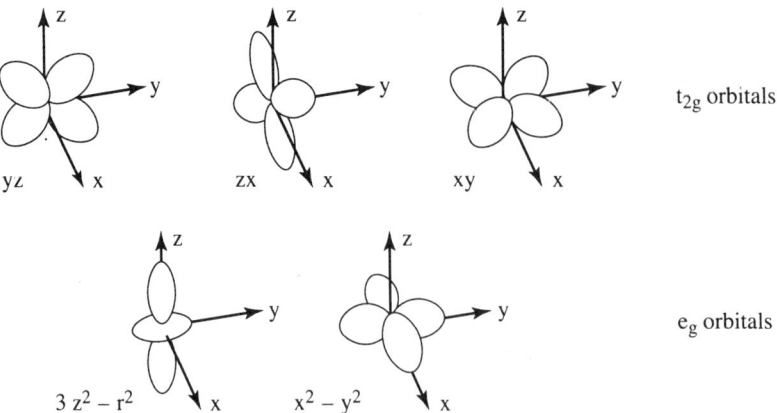

Figure A 6.1 - Orbitals of 3d electrons in cubic symmetry

APPENDIX 6 - SPECIAL FUNCTIONS

2. LEGENDRE POLYNOMIALS

These polynomials were introduced in order to describe physical systems with cylindrical symmetry. The axis Oz is chosen along the symmetry axis, and the only relevant variable is θ. Any function can be expanded into a series of the Y_1^0's, called the *Legendre polynomials*. The Legendre polynomials most used in magnetism are the following:

$P_0 = 1$ $\quad\quad$ $P_1 = \cos \theta$ $\quad\quad$ $P_2 = (3/2)(\cos^2 \theta - 1/3)$

$P_3 = \cos^3 \theta - (3/2) \sin^2 \theta \cos \theta$ $\quad\quad$ $P_4 = (1/8)(35 \cos^4 \theta - 30 \cos^2 \theta + 3)$

These polynomial are normalized so that $P_n (\theta = 0) = 1$.

Figure A 6.2 shows the variations of these polynomials $P_n (\theta)$.

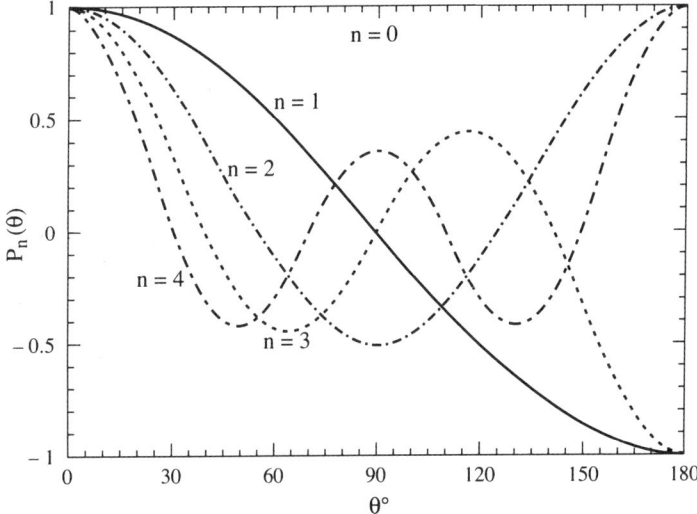

Figure A 6.2 - θ dependence of some Legendre polynomials

L. Néel expanded into a series of even Legendre polynomials the pair interaction energy (eq. 12.1) that allowed him to write the expressions of the magnetocrystalline anisotropy and of the magnetoelastic coupling. The geomagnetic potential of the earth is also expanded (eq. 24.2) into even and odd Legendre polynomials.

3. THE LANGEVIN FUNCTION

The Langevin function $\mathcal{L}(x) = \coth x - 1/x$ has been introduced to account for Curie paramagnetism (eq. 4.21), but it is mainly used to describe superparamagnetism, where the magnetic moment carriers involve so many elementary magnetic moments $\mathfrak{m}$ that quantum effects are irrelevant. The variable x is then equal to:
$x = \mu_0 \mathfrak{m} H / k_B T$.

At high temperature (x << 1), one can restrict the expansion to the first terms:

$$\mathcal{L}(x) \sim \frac{x}{3}\left[1 - \frac{1}{15}x^2 + \frac{2}{315}x^4 - \frac{2}{1575}x^6 + O(x^8)\right]$$

whereas when x >> 1 (very intense fields), this function is close to: $\mathcal{L}(x) \sim 1 - 1/x$.

4. BRILLOUIN FUNCTIONS

Brillouin functions are a generalization of the $\mathcal{L}(x)$ function:

$$\mathcal{B}_J(x) = \frac{2J+1}{2J}\coth\left(\frac{2J+1}{2J}x\right) - \frac{1}{2J}\coth\left(\frac{1}{2J}x\right)$$

They were introduced in order to better account for Curie paramagnetism in the presence of quantum effects (small atomic magnetic moments, J = 1/2, 1, ...). The Langevin function is the limit of the Brillouin functions when J tends toward infinity.

When x is much smaller than 1, i.e. at high temperature, one has:

$$\mathcal{B}_J(x) \sim \frac{x}{3}\frac{J+1}{J}\left[1 - \frac{1}{15}x^2\left(1 + \frac{1}{J} + \frac{1}{2J^2}\right) + O(x^4)\right]$$

and when x is much larger than 1 (very intense fields), this function tends toward 1 much faster than the Langevin function.

Figure A 6.3 shows the x dependence of two Brillouin functions, which are compared to the Langevin function.

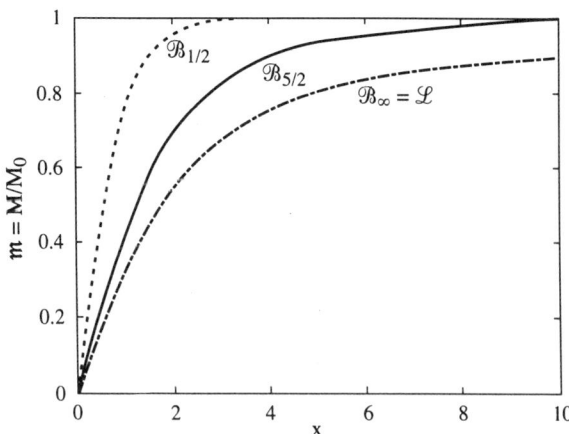

Figure A 6.3 - The functions $\mathcal{B}_{1/2}(x)$, $\mathcal{B}_{5/2}(x)$, and $\mathcal{L}(x)$

APPENDIX 6 - SPECIAL FUNCTIONS

5. MODIFIED BESSEL FUNCTIONS

The three first modified (or hyperbolic) Bessel functions of the first kind with half-integer indices (ref. [2], p. 384) are:

$$I_{1/2}(z) = \sqrt{\frac{2}{\pi z}} \sinh z \qquad I_{3/2}(z) = \sqrt{\frac{2}{\pi z}} \left(\cosh z - \frac{\sinh z}{z} \right)$$

$$I_{5/2}(z) = \sqrt{\frac{2}{\pi z}} \left[\left(\frac{3}{z^2} + 1 \right) \sinh z - 3 \frac{\cosh z}{z} \right]$$

One easily notices that the Langevin function is none else than the ratio:

$$\mathcal{L}(x) = \hat{I}_{3/2}(x) = \frac{I_{3/2}(x)}{I_{1/2}(x)} = \coth x - \frac{1}{x}.$$

It is possible to define the generalized Langevin functions $\hat{I}_{\ell+1/2}(x)$. The functions $\hat{I}_{\ell+1/2}(x)$ with even degree ℓ are mostly used in magnetism:

$$\hat{I}_{5/2}(x) = 1 - (3/x)\mathcal{L}(x) = \frac{3}{x^2} + 1 - \frac{3}{x}\coth(x)$$

$$\hat{I}_{9/2}(x) = 1 + \frac{35}{x^2} - \left(\frac{105}{x^3} + \frac{10}{x} \right) \mathcal{L}(x) = \frac{105}{x^4} + \frac{45}{x^2} + 1 - \left(\frac{105}{x^3} + \frac{10}{x} \right) \coth(x)$$

Callen et al. showed [3, 4] that, *in a model of localized ferromagnetism*, the field or temperature dependence of the anisotropy coefficients $K_\ell(T, H)$ or of the magneto-elastic coupling coefficients $B_\ell(T, H)$, associated with harmonic polynomials of degree ℓ, follows the modified Bessel function $\hat{I}_{\ell+1/2}(x)$. They proposed to relate these variations to those of the reduced magnetisation $m = M(T, H)/M_o$, given by the function $m = \mathcal{L}(x)$, by writing: $K_\ell(T, H)/K_\ell(0, 0) = \hat{I}_{\ell+1/2}[\mathcal{L}^{-1}(m)]$. The field and temperature dependences of the anisotropy (and of the magnetoelastic coupling) coefficients can thus be implicitly expressed, *in this model*, as a function of the reduced magnetisation of the substance by means of functions which generalize the Langevin function.

The function $\hat{I}_{5/2}(x)$ has to be used to treat the second order coefficients (the K_1 constant of a uniaxial crystal or the $B^{\gamma,2}$ and $B^{\epsilon,2}$ coupling coefficients of a cubic crystal). The $\hat{I}_{9/2}(x)$ function is used for the fourth order coefficients (for instance the K_1 constant of a cubic crystal), and so on.

We note that, *only at low temperature*, the functions $\hat{I}_{\ell+1/2}[\mathcal{L}^{-1}(m)]$ vary as $m^{\ell(\ell+1)/2}$. This is easily verified on the equations which define the functions with $\ell = 1, 2$ and 4: at very low temperature, x tends toward infinity, $\mathcal{L}(x)$, i.e. m, tends toward $1 - 1/x$, and accordingly m^n tends toward $1 - n/x$. We see that $\hat{I}_{5/2}(x)$ tends toward $1 - 3/x$, and $\hat{I}_{9/2}(x)$ toward $1 - 10/x$, which proves our statement.

On the other hand, at *high temperature*, the functions $\hat{I}_{\ell+1/2}[\mathcal{L}^{-1}(m)]$ vary as m^ℓ: around and above T_C, $\mathcal{L}(x) = m \sim x/3$, $\hat{I}_{5/2}(x) \sim x^2/15$, and $\hat{I}_{9/2}(x) \sim x^4/945$, which leads to $\hat{I}_{5/2}(x) \sim (3/5) m^2$, and $\hat{I}_{9/2}(x) \sim (3/35) m^4$.

Figure A 6.4 shows the variations of the functions $\hat{I}_{\ell+1/2}(x)$ for $\ell = 1, 2,$ and 4, the curve $\mathscr{L}(x)$ being associated with $\ell = 1$. Knowing the experimental value of $m(T, H)$, say 0.9, corresponding to point A on the curve $\ell = 1$, we can right away deduce the values of $K_\ell(T, H) / K_\ell(0, 0)$ or those of $B_\ell(T, H) / B_\ell(0, 0)$: we draw a vertical straight line through A; the intercept of this line with the curve associated with $\ell = 2$ (B), and $\ell = 4$ (C) immediately gives the result.

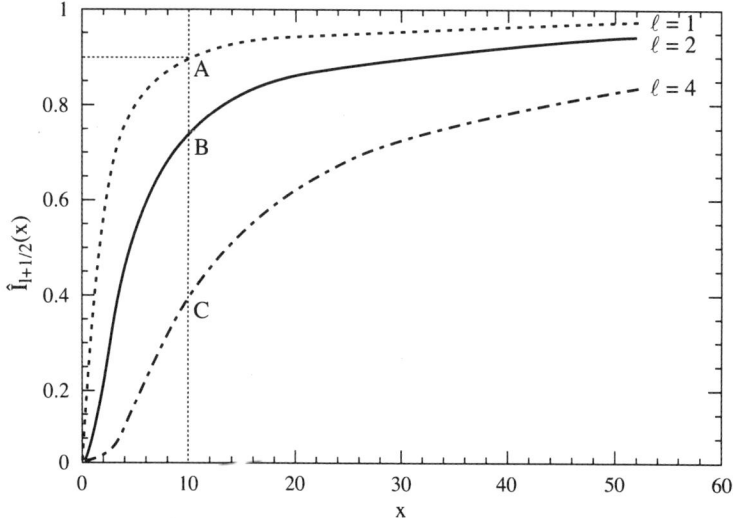

Figure A 6.4 - The functions $\mathscr{L}(x)$, $\hat{I}_{5/2}(x)$ and $\hat{I}_{9/2}(x)$

These figures explain very well why the approach to saturation of the magnetostriction curves is slower than for magnetisation curves: when magnetisation reaches 90% of the saturation value, magnetostriction –which is proportional to $B_2(T, H)$– is only around 70% of its saturation value.

We note that the same approach as that generalizing the Langevin function could be developed for the Brillouin functions; the results [5, 6] are more accurate, but at the cost of a larger analytical complication.

REFERENCES

[1]   C. COHEN-TANNOUDJI, B. DIU, F. LALOË, Mécanique Quantique (1977) Hermann, Paris, 2 volumes. In English: Quantum Mechanics (1977) Wiley

[2]   A. ANGOT, Compléments de mathématiques (1957) Editions de la revue d'Optique, Paris, 836 p.

[3]   E.R. CALLEN, H.B. CALLEN, *J. Phys. Chem. Solids* **16** (1960) 310-328.

[4]   E. CALLEN, *J. Appl. Phys.* **39** (1968) 519-527.

[5]   Y. MILLEV, M. FÄHNLE, *J. Magn. Magn. Mater.* **135** (1994) 284-292.

[6]   Y. MILLEV, M. FÄHNLE, *J. Phys. Condens. Mat.* **7** (1995) 6909-6923.

APPENDIX 7

MAXWELL'S EQUATIONS

curl E = $-\partial \mathbf{B}/\partial t$ $E_{T1} = E_{T2}$ (time independent case)
div **D** = ρ $D_{N2} - D_{N1} = \sigma$
div **B** = 0 $B_{N1} = B_{N2}$
curl H = **j** + $\partial \mathbf{D}/\partial t$ $H_{T2} - H_{T1} = \mathbf{j_S} \times \mathbf{n}$ (time independent case)

In vacuum:

$\mathbf{D} = \varepsilon_0 \mathbf{E}$ $\mathbf{B} = \mu_0 \mathbf{H}$ $\varepsilon_0 \mu_0 c^2 = 1$

In matter:

$\mathbf{B} = \mu_0 (\mathbf{H} + \mathbf{M})$ if $\mathbf{M} = \chi \mathbf{H} \Rightarrow$ $\mathbf{B} = \mu \mathbf{H}$ with: $\mu = (1+\chi)\mu_0 = \mu_0 \mu_r$
$\mathbf{D} = \varepsilon_0 \mathbf{E} + \mathbf{P}$ if $\mathbf{P} = \chi_e \varepsilon_0 \mathbf{E} \Rightarrow$ $\mathbf{D} = \varepsilon \mathbf{E}$ with: $\varepsilon = (1+\chi_e)\varepsilon_0 = \varepsilon_0 \varepsilon_r$
$\mathbf{j} = \gamma \mathbf{E}$ $\varepsilon \mu v^2 = 1$

Equations for the current density:

div **j** + $\dfrac{\partial \rho}{\partial t} = 0$ $j_{N1} - j_{N2} = \dfrac{\partial \sigma}{\partial t}$

Force acting on an electric charge:

$\mathbf{F} = q\,(\mathbf{E} + \mathbf{v} \times \mathbf{B})$

Scalar potential V, and vector potential A:

B = curl **A** $\Box \mathbf{A} + \mu \mathbf{j} = 0$
E = $-$ grad V $- \partial \mathbf{A}/\partial t$ $\Box V + \rho/\varepsilon = 0$

where the D'Alembertian operator $\Box$ can be expressed as a function of the Laplacian operator Δ as:

$$\Box = \Delta - \frac{1}{v^2}\frac{\partial^2}{\partial t^2}$$

We propose in the following page some useful formulae of vector algebra.

SPHERICAL COORDINATES

$$\mathbf{grad}\, U = \frac{\partial U}{\partial r}\mathbf{e}_r + \frac{1}{r}\frac{\partial U}{\partial \theta}\mathbf{e}_\theta + \frac{1}{r\sin\theta}\frac{\partial U}{\partial \varphi}\mathbf{e}_\varphi$$

$$\text{div}\,\mathbf{A} = \frac{1}{r^2}\frac{\partial}{\partial r}(r^2 A_r) + \frac{1}{r\sin\theta}\frac{\partial}{\partial \theta}(\sin\theta\, A_\theta) + \frac{1}{r\sin\theta}\frac{\partial A_\varphi}{\partial \varphi}$$

$$\mathbf{curl}\,\mathbf{A} = \frac{1}{r\sin\theta}\left[\frac{\partial}{\partial \theta}(\sin\theta\, A_\varphi) - \frac{\partial A_\theta}{\partial \varphi}\right]\mathbf{e}_r$$

$$+ \frac{1}{r}\left[\frac{1}{\sin\theta}\frac{\partial A_r}{\partial \varphi} - \frac{\partial}{\partial r}(rA_\varphi)\right]\mathbf{e}_\theta$$

$$+ \frac{1}{r}\left[\frac{\partial}{\partial r}(rA_\theta) - \frac{\partial A_r}{\partial \theta}\right]\mathbf{e}_\varphi$$

$$\Delta U = \frac{1}{r}\frac{\partial^2(rU)}{\partial r^2} + \frac{1}{r^2\sin^2\theta}\frac{\partial^2 U}{\partial \varphi^2} + \frac{1}{r^2\sin\theta}\frac{\partial}{\partial \theta}\left(\sin\theta\frac{\partial U}{\partial \theta}\right)$$

CYLINDRICAL COORDINATES

$$\mathbf{grad}\, U = \frac{\partial U}{\partial r}\mathbf{e}_r + \frac{1}{r}\frac{\partial U}{\partial \theta}\mathbf{e}_\theta + \frac{\partial U}{\partial z}\mathbf{e}_z$$

$$\text{div}\,\mathbf{A} = \frac{1}{r}\frac{\partial}{\partial r}(rA_r) + \frac{1}{r}\frac{\partial A_\theta}{\partial \theta} + \frac{\partial A_z}{\partial z}$$

$$\mathbf{curl}\,\mathbf{A} = \left[\frac{1}{r}\frac{\partial A_z}{\partial \theta} - \frac{\partial A_\theta}{\partial z}\right]\mathbf{e}_r + \left[\frac{\partial A_r}{\partial z} - \frac{\partial A_z}{\partial r}\right]\mathbf{e}_\theta + \frac{1}{r}\left[\frac{\partial}{\partial r}(rA_\theta) - \frac{\partial A_r}{\partial \theta}\right]\mathbf{e}_z$$

$$\Delta U = \frac{\partial^2 U}{\partial r^2} + \frac{1}{r}\frac{\partial U}{\partial r} + \frac{1}{r^2}\frac{\partial^2 U}{\partial \theta^2} + \frac{\partial^2 U}{\partial z^2}$$

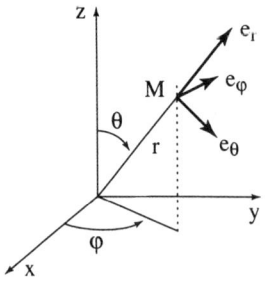

Spherical coordinates

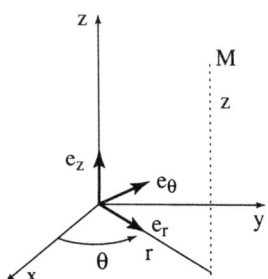

Cylindrical coordinates

GENERAL REFERENCES

AHARONI A. - *Introduction to the theory of ferromagnetism.* Clarendon Press, Oxford, 1996.

BLUNDELL S. - *Magnetism in condensed matter.* Oxford University Press, 2001.

BRISSONNEAU P. - *Magnétisme et matériaux magnétiques pour l'électrotechnique.* Hermès, Paris, 1997.

BUSCHOW K.H. & WOHLFARTH E.P. Eds - *Ferromagnetic Materials.* North Holland, Amsterdam (a multi-volume series).

CHIKAZUMI S. - *Physics of magnetism.* Wiley, New York, 1964; *Physics of ferromagnetism.* Clarendon Press, Oxford, 1997.

COEY J.M.D. Ed - *Rare earth permanent magnets.* Clarendon Press, Oxford, 1996.

CULLITY B.D. - *Introduction to magnetic materials.* Addison-Wesley, Reading, 1972.

DURAND E. - *Magnétostatique.* Masson, Paris, 1968.

HERPIN A. - *Théorie du magnétisme.* Presses Universitaires de France, Paris, 1968.

HUBERT A. & SCHAEFER R. - *Magnetic domains. The Analysis of Magnetic Microstructures.* Springer, Berlin-Heidelberg-New York, 1998.

LANDAU L.D., LIFSHITZ E.M. & PITAEVSKII L.P. - *Electrodynamics of Continuous Media.* Elsevier, 1997.

LANDAU L.D. & LIFSHITZ E.M. - *Classical theory of fields.* Butterworth-Heinemann, 1997.

MATTIS D.C. - *Theory of magnetism.* Harper & Row, New York, 1965.

MORRISH A.H. - *The physical principles of magnetism.* Wiley, New York, 1965.

NEEL L. - *Oeuvres scientifiques.* Editions du CNRS, 1978.

O'HANDLEY R.C. - *Modern magnetic materials: principles and applications.* Wiley, New York, 2000.

SKOMSKI R. & COEY J.M.D. - *Permanent magnetism.* Institute of Physics Publishing, Bristol and Philadelphia, 1999.

INDEX BY MATERIAL

Alnico	II - 13, 17, 19, 43, 68, 491
Amorphous alloy	II - 114-115, 118
Amphibole	II - 397
$BaFe_{12}O_{19}$	I - 475-476 • II - 487-488
$BaO-6Fe_2O_3$	II - 43, 48, 51
$Bi_2Ca_2Sr_2Cu_3O_{10}$	II - 251
$(BiRGa)Fe_5O_{12}$	I - 422, 438
$CaCO_3$	II - 393
Calcite	II - 393
Carbonates	II - 393
Carbonyl iron	II - 145
$CeCl_3$	I - 267
CeF_3	I - 267
Cerium magnesium nitrate (CMN)	I - 347
Co	I - 140, 162, 284, 301, 367, 438, 475-476 • II - 487-488
CoF_2	I - 181, 402
Co_2FeGa	I - 438
Co_2FeGe	I - 438
$CoFe_2O_4$	I - 475-476 • II - 487-488
CoGd	I - 438
Co_7Gd_2	I - 423
$Co_{70}Pt_{30}$	I - 423, 438
Cobalt based amorphous alloy	I - 369-370
$CrBr_3$	I - 314, 422
$CrCl_3$	I - 422
CrI_3	I - 422
$CrK(SO_4)_2 \cdot 12H_2O$	I - 113
Cu	II - 162
Cu_2MnAl	I - 475-476 • II - 487-488
$CuS_4C_4(CF_3)_4$	II - 426
Dextran ferrite	II - 430
Dupeyredioxyl	II - 426
Dy	I - 269
Dy_3Al_2	I - 189
$DyNi_2Si_2$	I - 134
DyZn	II - 217
Elinvar	I - 352 • II - 215
Er	I - 269
$ErGa_2$	I - 100
EuO	I - 314, 422, 451-452
EuS	I - 422, 451

Entry	Reference
Fe based amorphous alloy	II - 145
Fe/Cr multilayer	II - 286
α-Fe	I - 85, 243
γ-Fe	I - 85
$FeCO_3$	II - 393
$FeCl_2$	I - 266
FeCo	I - 438
$Fe_{0.75}Co_{0.25} / Tb_{0.27}Co_{0.73}$	II - 231
FeCoGa	I - 423
Fe-Co	II - 108, 111, 114, 147
$Fe_{49}Co_{49}V_2$	II - 147
$Fe_{73.5}Cu_1Nb_3Si_{13.5}B_9$	II - 118, 121
FeF_2	I - 181, 266, 402
Fe_3N	II - 347
$Fe(NH_4)_2 (SO_4)_2 \cdot 12H_2O$	I - 113
FeNi	II - 108
$Fe_{20}Ni_{80}$	II - 146
$Fe_{35}Ni_{65}$	II - 146
$Fe_{45}Ni_{55}$	II - 146
$Fe_{50}Ni_{50}$	II - 145, 147
$Fe_{65}Ni_{35}$	II - 213
$Fe_{15}Ni_{80}Mo_5$	II - 147
$Fe_{17}Ni_{81}Mo_2$	II - 145
Fe-Ni	I - 234 • II - 108, 114, 145-147, 213, 493
Fe-80%Ni	I - 475-476 • II - 487-488
FeO	I - 266
$FeO-TiO_2-Fe_2O_3$	I - 4
α-Fe_2O_3	I - 266 • II - 394-395
γ-Fe_2O_3	I - 4, 85, 266 • II - 314, 347, 394-395
Fe_3O_4	I - 4, 266, 475-476 • II - 191, 217, 339, 394, 428, 487-488
α-Fe_3O_4	II - 395
FePd	I - 179
$Fe_{50}Pd_{50}$	I - 423
$Fe_{65}Pt_{35}$	I - 438
$Fe_{85}Pt_{15}$	I - 423
FeS	I - 266 • II - 396
Fe-Si	I - 99, 173, 175 • II - 93, 97, 99, 101, 114, 147, 493
Fe-3%Si / Goss	II - 147
$Fe_{2.7}V_{0.3}Ge$	I - 438
Ferrinox	II - 114
Ferrite	II - 17, 68, 194, 491
Ferritin	II - 427
Ferrofluid	II - 337
Ferroxyhite	II - 395
Fibrin	II - 428
Gadolinium Gallium Garnet	see $Gd_3Ga_5O_{12}$
Garnet	II - 197, 318, 344
Gd	I - 269, 317, 346, 475-476 • II - 487-488
$GdAl_2$	I - 101, 352, 377

INDEX BY MATERIAL

Gd-Co	I - 423
GdFe$_2$	I - 438
Gd-Fe	I - 423
Gd$_3$Ga$_5$O$_{12}$	I - 426
Gd$_2$O$_3$	I - 267
Gd$_{3-x}$S$_4$	I - 451
Gd$_2$(SO$_4$)$_3$.8H$_2$O	I - 113
Gd$_5$ (Si-Ge)$_4$	I - 346
GdTbFe	I - 438
GdZn	I - 101
GGG	see Gd$_3$Ga$_5$O$_{12}$
Goethite	II - 395
Greigite	II - 395-396, 428
Haemoglobin	II - 427
Haemosiderin	II - 427
Hard ferrite	II - 19, 48
Hematite	see α-Fe$_2$O$_3$
Heusler alloy	I - 423
Hexaferrite	II - 49, 193, 197, 493
HgBa$_2$Ca$_2$Cu$_3$O$_{8+\delta}$	II - 251
High T$_c$ superconductor	II - 251
Ho	I - 269
HoCo$_2$Si$_2$	I - 139
Ilmenite	II - 397
Interstitial compound	II - 63
Invar	I - 352 • II - 213-215
Iron	I - 165-166, 195, 224, 366, 373, 377, 384, 438, 475-476 • II - 145, 162, 395, 487-488, 492
Iron based amorphous alloy	I - 369, 372 • II - 118, 145
Iron-cobalt alloy	II - 111
(La$_{1-x}$Ba$_x$)$_2$CuO$_4$	II - 251
La$_{1-x}$Ca$_x$MnO$_3$	I - 318
La$_{1-x}$Sr$_x$MnO$_3$	I - 318
Lodestone	see magnetite
Lu$_3$Co$_{7.77}$Sn$_4$	I - 136
Maghemite	see γ-Fe$_2$O$_3$
Magnetite	see Fe$_3$O$_4$
Manganese-zinc ferrite	see Mn$_x$Zn$_{1-x}$Fe$_2$O$_4$
Manganite	I - 451, 454
Metglas	II - 226, 227, 494
MgCO$_3$	II - 393
MnF$_2$	I - 181
MnO	I - 10
Mn$_x$Zn$_{1-x}$Fe$_2$O$_4$	II - 112, 162
Nb$_3$Ge	II - 251
Nb$_3$Sn	II - 251, 459
NbTi	II - 251, 459
Nd	I - 316
Nd$_2$Fe$_{14}$B	I - 285, 475-476 • II - 41, 43, 51, 63, 68, 487-488
Nd-Fe-B	I - 13, 186, 242 • II - 17, 19, 491

$Nd_3Fe_5O_{12}$ .... I - 422, 438
$NdFe_{11}TiN_{1.5}$ .... II - 63
Nd_2Se_3 .... I - 267
Ni .... I - 215, 301, 351-352, 359, 374, 377, 380, 422, 438, 475-476 • II - 215, 217, 487-488
$Ni_{80}Co_{20}$ / Ru / $Ni_{80}Co_{20}$ .... II - 275
Nickel-zinc ferrite .... see $Ni_xZn_{1-x}Fe_2O_4$
$NiCl_2$ .... I - 267
NiF_2 .... I - 267
$NiFe_2O_4$ .... I - 314, 475-476 • II - 217, 487-488
NiMnSb .... I - 423, 438
NiO .... I - 180
$Ni_xZn_{1-x}Fe_2O_4$ .... II - 112, 162
Olivine .... II - 397
Oxyhydroxides .... II - 393, 396
PdMnSb .... I - 423
Permalloy .... I - 385 • II - 110, 217
Permanent magnet .... II - 3
Permendur .... II - 217
Phosphates .... II - 393
Phyllosilicates .... II - 397
PtMnSb .... I - 423, 438, 450
PtMnSn .... I - 423
Pyrite .... II - 396
Pyroxene .... II - 397
Pyrrhotite .... II - 395-396
Quartz .... II - 393
R .... often stands for a rare earth element
RAl_2 .... I - 268
RCo_5 .... I - 279 • II - 55, 59
R_2Co_{17} .... II - 61
R-Co .... I - 422, 438
RCu_2 .... I - 268
RFe_2 .... I - 423
$RFe_{10.5}Mo_{1.5}$ .... II - 63
$RFe_{11}Ti$ .... II - 63
$RFe_{10}V_2$ .... II - 63
R_2Fe_{17} .... II - 62
$R_2Fe_{14}B$ .... I - 279 • II - 55
$R_2Fe_{17}N_y$ .... II - 63
$R_3Fe_5O_{12}$ .... I - 268, 421 • II - 318, 493
R-Fe .... I - 308, 422, 438
R-FeCo amorphous alloy .... II - 317
RNi_2 .... I - 268
Rare earth garnet .... see $R_3Fe_5O_{12}$
RIG .... see $R_3Fe_5O_{12}$
Siderite .... II - 397
Silicates .... II - 393, 397
Silicate glass .... I - 421
$SmCo_5$ .... I - 186, 234, 285, 475-476 • II - 19, 41, 43, 51, 63, 68, 487-488

INDEX BY MATERIAL

$SmCo_{3.5}Cu_{1.5}$	I - 243
Sm_2Co_{17}	I - 475-476 • II - 19, 43, 63, 68, 72, 487-488
Sm-Co	II - 17, 491
Sm-Co phase diagram	II - 58
$Sm_2Fe_{17}C_{1.1}$	II - 63
$Sm_2Fe_{17}N_{2.3}$	II - 63
$SmFe_{11}Ti$	II - 63
Soft ferrite	II - 112
Spinel	II - 191, 193, 197
$SrO\text{-}6Fe_2O_3$	II - 43, 48
Sulphates	II - 393
Sulphides	II - 396
Tanol suberate	II - 425
Tb	I - 269
$TbCo_5$	II - 60
TbFe	I - 423, 438
$TbFe_2$	I - 438
TbFeCo	I - 423, 438
$Tb_3Fe_5O_{12}$	I - 431
$Tb_{75}Gd_{25}$	I - 100
Tb_2O_3	I - 267
TbIG	see $Tb_3Fe_5O_{12}$
Terfenol	II - 217
Terfenol-D	I - 390 • II - 216, 219, 221, 494
Tetrathiafulvalene	II - 426
$ThMn_{12}$	II - 63
ThO_2	I - 360
Titanohematite	II - 395
Titanomagnetite	II - 395
$Tl_2Ca_2Ba_2Cu_3O_{10}$	II - 251
Tm	I - 269
Tm_2O_3	I - 267
Transition metals	I - 422
Troilite	II - 396
UO_2	I - 360
$YBa_2Cu_3O_7$	II - 251
YCo_5	II - 60
YFe_2	I - 85
YFe_2Si_2	I - 85
$Y_3Fe_5O_{12}$	I - 184, 266, 361, 363-365, 430, 475-476 • II - 183, 190, 344, 487
Y-Co/Gd-Co/Y-Co	II - 276
$Yb_3Fe_5O_{12}$	I - 422, 438
YIG	see $Y_3Fe_5O_{12}$
Yttrium-iron garnet	see $Y_3Fe_5O_{12}$
$ZrZn_2$	I - 284-285

INDEX BY SUBJECT

Abrikosov lattice	II - 248
Absorption	I - 172, 435
AC biasing	II - 308
Accelerometer	II - 345
Actuator	II - 29, 134, 216, 230
Additivity	II - 179
Adiabatic demagnetization	I - 347
AFM	see Atomic Force Microscopy
After effects	I - 98, 236
Aging	I - 236
Air gap	I - 44 • II - 5
Amorphous alloy	I - 369 • II - 256, 317
Ampère's law	I - 22
Amperian approach	I - 31
Amplitude modulation	I - 416
AMR effect	see Anisotropic Magnetoresistance
AMS	see Anisotropy of Magnetic Susceptibility
Analog recording	II - 307
Anderson transition	I - 453
Anderson's hamiltonian	I - 302
Anhysteretic magnetization curve	I - 98, 229 • II - 309
Animal magnetism	II - 429
Anisotropic exchange	I - 89
Anisotropic Magnetoresistance (AMR)	II - 284, 298, 329
Anisotropic magnetostriction	I - 352, 362 • II - 216
Anisotropy	II - 390, 399
Anisotropy constant	I - 90
Anisotropy field	I - 92, 94, 143, 191
Anisotropy of Magnetic Susceptibility (AMS)	II - 400, 426
Anisotropy of magnetization	I - 95
Antenna	II - 194
Antiferromagnetic coupling	II - 285, 288
Antiferromagnetic material	II - 291
Antiferromagnetic domains	I - 180
Antiferromagnetically coupled layers	II - 288, 296
Antiferromagnetism	I - 9, 81, 121
Apparent susceptibility	II - 391
Applied field	I - 37
Approach to saturation	I - 197
Arrott plot	I - 133-134
Asperomagnetism	I - 132

Atomic Force Microscopy (AFM)	II - 263
Auger spectroscopy	II - 263
Aurora australis	II - 412
Aurora borealis	II - 412
Ballistic behavior	II - 174
Band theory	I - 284
Barber pole	II - 330
Bearing	II - 30, 79
Bimorph	I - 375
Biot-Savart law	I - 20
Birefringence	I - 434 • II - 344
Bitter coil	II - 460
Bitter method	I - 158, 168 • II - 337
Bloch line	I - 159
Bloch wall	I - 87, 148
Blood Oxygenation Level Dependent (BOLD) Contrast	II - 372, 376, 379-380
Bohr magneton	I - 253
BOLD mechanism	see Blood Oxygenation Level Dependent Contrast
Boltzmann statistics	I - 111, 113
Bonded magnet	II - 20
Bound current	I - 31
Boundary conditions	I - 24
Bragg diffraction imaging	I - 170
Brillouin function	I - 114, 119
Brown's inequality	I - 204 • II - 40
Brown's paradox	I - 220 • II - 40
Brownian motion	II - 338
Bubble memory	II - 312
Buckling	I - 205
Canted state	I - 319
Cardiac valve	II - 433
Catheter	II - 435
Cerebral activation imaging	II - 371
CEF	see Crystalline Electric Field
Chemical shift	II - 360
Chemical Vapor Deposition (CVD)	II - 258
Chirality domains	I - 183
Circular magnetic dichroism	I - 172, 428
Circular permeability	II - 184
Circular susceptibility	II - 184
Circulator	II - 200
Climatic cycle	II - 398
Closure domains	I - 163, 223
Coaxial line	II - 166
Coercive field	I - 218 • II - 4
Coercivity	I - 198 • II - 12, 39
Cognition	II - 383
Coil methods	II - 164
Collapse field	II - 313

INDEX BY SUBJECT

Colossal magnetoresistance	I - 454
Compensation temperature	I - 83, 128 • II - 276
Competition	I - 146
Complex permeability	II - 155, 183
Complex susceptibility	II - 155
Compromise	I - 144
Configuration	I - 256-257
Continental drift	II - 417
Contrast	II - 370
Coordination	II - 267
Core field	II - 408, 415
Cotton-Mouton effect	I - 402
Coulomb integral	I - 312
Coulomb repulsion	I - 298
Coulombian approach	I - 29, 32
Coupling mechanisms	II - 272
Critical current	II - 237
Critical exponent	I - 330
Critical field	II - 237, 246
Critical radius	I - 166
Critical temperature	II - 235
CRM	see Crystallization Remanent Magnetization
Cross-tie	I - 159
Crustal magnetization	II - 409
Cryogenic magnetometer	II - 389
Crystal field	see Crystalline Electric Field
Crystalline Electric field (CEF)	I - 264-265, 272, 274, 299, 364 • II - 270
Crystallization Remanent Magnetization (CRM)	II - 404
Curie constant	I - 87, 110, 112
Curie law	I - 80, 110
Curie paramagnetism	I - 110
Curie temperature	I - 82, 120
Curie-Weiss law	I - 82, 87 • II - 343
Curling	I - 165, 205
Current density	I - 31
Current loop	I - 28
Current model	II - 281
CVD	see Chemical Vapor Deposition
Cyclotron effect	I - 448
Damper	II - 346
Damping	II - 169
DC biasing	II - 308
De Haas-Van Alphen effect	I - 455
Dead layer	II - 267
Declination	II - 389, 409-411
Defects	I - 211
ΔE Effect (Delta E effect)	I - 380, 386
Demagnetization curve	II - 4, 7
Demagnetized state	I - 225, 370

Demagnetizing field	I - 35, 145 • II - 160, 325
Demagnetizing field anisotropy	I - 167
Demagnetizing field coefficient	I - 36
Demagnetizing field energy	see Magnetostatic energy
Demagnetizing field line	I - 40, 184
Density of states	I - 287 • II - 264-265, 267, 281, 292
Detrital Remanent Magnetization (DRM)	II - 405
Diamagnetism	I - 79, 105
Dichroism	I - 172
Dielectric permittivity	I - 433
Differential actuator	II - 222
Differential susceptibility	I - 39
Diffusion after-effect	I - 99, 236, 243
Diffusion relaxation	II - 177
Digital Random Access Memories (DRAM)	II - 300
Digital recording	II - 310
Digitation	II - 348
Dimensional resonance	II - 161
Dimensionality	II - 266
Dipolar interaction	I - 55, 143, 145 • II - 270, 273, 339, 342
Dipolar magnetostriction	I - 376
Dipole field	I - 28, 36
Dipole interaction field	I - 36
Disaccommodation	I - 244 • II - 178
Domain configurations	I - 160
Domain propagation memory	II - 312
Domain structure	I - 143 • II - 102, 349
Domain wall	I - 147
Domain wall displacement	I - 184, 211
Domain wall pinning	I - 214
Döring mass	II - 174
Double coercivity multilayer	II - 289
Double exchange	I - 318
DRAM	see Digital Random Access Memories
DRM	see Detrital Remanent Magnetization
Dynamic susceptibility	I - 293
Dynamo model	II - 415, 417, 421
Dzyaloshinsky-Moriya interaction	I - 314
Earth materials	II - 387
Easy magnetization direction	I - 90, 93 • II - 269
Echo Planar Imaging (EPI)	II - 381
Eddy current	I - 236
Effective mass	II - 280
Effective medium	II - 181, 204
Effective moment	I - 114
Eigenmodes	I - 405, 414
Elastic constant	I - 101, 475 • II - 215, 487
Electrical resistivity	I - 100, 443, 445
Electromagnetic coupling	I - 45, 50

INDEX BY SUBJECT

Electromotive field	I - 48
Electromotive force	I - 47
Electron interference holography	I - 174
Electron microscopy	I - 174
Elinvar	I - 352 • II - 213
Ellipticity	I - 419
Energy	see Free energy
Energy dissipation	see Losses
Energy product	II - 8, 11
EPI	see Echo Planar Imaging
Epitaxy	II - 256, 262, 265
Epstein permeameter	II - 126, 444
Exchange coefficient	I - 145
Exchange energy	I - 117
Exchange integral	I - 312
Exchange interactions	I - 116, 143-144, 311
Exchange length	I - 146 • II - 257
Exchange magnetostriction	I - 101, 351, 359
External field	I - 37, 183
External susceptibility	I - 40 • II - 159, 309
Faraday effect	I - 163, 169, 400, 407, 415, 421 • II - 199
Faraday's experiment	I - 48
Fermi level	II - 264, 281, 286
Fermi surface	II - 280
Fermi velocity	II - 279
Fermi wavelength	II - 257
Fermi's golden rule	II - 281, 292
Ferrimagnetic macroscopic order	II - 276
Ferrimagnetism	I - 10, 83, 127
Ferrimagnets	I - 385
Ferrite magnet	II - 25
Ferritin	II - 427, 429
Ferrofluid	I - 168 • II - 337
Ferromagnetic materials	I - 475 • II - 487
Ferromagnetism	I - 82, 118
Fibrin	II - 428
Field coefficient	I - 24
Field detector	I - 427
Field gradient	II - 366
Field sensor	I - 430 • II - 298
Fluctuation after-effect	I - 98, 236-237
Flux channeling	II - 464
Flux conservation	I - 23
Flux gate	II - 392
Flux gate magnetometer	II - 432, 470
Flux leakage coefficient	II - 6
Flux quantization	II - 238
Flux quantum	II - 238
Fluxmeter	II - 444

fMRI sequence..II - 380
Force...I - 45, 52, 57
Force sensor...II - 227
Form effect..I - 354, 376
Foucault microscopy...I - 174
Fourier plane..II - 369
Frank-Van der Merwe growth..II - 263
Free current..I - 31
Free energy..I - 45, 57, 147
Free precession..II - 359
Fresnel coefficient...I - 435
Frustration..I - 133, 303
Fuchs-Sondheimer theory..II - 283
Functional Magnetic Resonance Imaging (fMRI)..II - 372
Garnet..I - 266 • II - 197, 318, 344
Gauss' theorem..I - 33
Generalized susceptibility..I - 292
Geomagnetic dating...II - 420
Geomagnetic field..II - 407
Geomagnetic field reversal..II - 413
Geomagnetism..II - 387
Giant Magnetoresistance (GMR)....................................I - 454 • II - 288, 298, 329
Gyromagnetic factor...I - 261 • II - 356
Gyromagnetic resonance..II - 170
Gyrotropy..II - 184
Haemoglobin..II - 374, 427
Haemosiderin..II - 427
Half-metal...I - 454 • II - 265
Hall effect...I - 443-444
Hall probe...II - 469
Hard drive..I - 14
Hard magnetization direction...I - 95
Hard material...II - 4, 490
Hartree approximation..I - 256, 297
Hartshorn bridge...II - 455
Heisenberg interaction..I - 116
Hele-Shaw cell...II - 348
Helimagnet...I - 183
Helimagnetism..I - 129-130
Hematite (α-Fe_2O_3)..I - 266 • II - 394-395
Hexaferrite..II - 193
High frequency electronics..II - 155
History...I - 3
Homogenization..II - 178
Hooke's law...I - 392
Hubbard's hamiltonian...I - 299
Hund's rules...I - 258
Hybridization...I - 302, 313 • II - 53, 292
Hysteresigraph..II - 453-454

Hysteresis	I - 209, 344
Hysteresis loop	I - 83 • II - 4, 92, 269, 284
Ideal system	I - 143
Inclination	II - 389, 402, 405, 409-411, 414, 417, 419
Inclinometer	II - 345
Incompatibility	I - 378
Induced anisotropy	I - 89, 99, 383
Induced magnetic moment	I - 54
Inductance	II - 194
Induction loop	II - 4
Inductive readout	II - 326
Initial magnetization curve	I - 82, 210, 225
Initial susceptibility	I - 39
Interdiffusion	II - 272
Interface anisotropy	II - 266
Interface dislocations	II - 271
Intermediate state	I - 427
Internal field	I - 32, 37, 41
Internal susceptibility	II - 159, 309
Interstitial compound	II - 63
Intrinsic propagation field	I - 189
Invar	I - 352 • II - 213
Inverse Wiedemann effect	I - 385 • II - 228
Iron-cobalt alloy	II - 108
Iron-nickel alloy	II - 108
Irreversibility	I - 199, 211
Isolator	II - 201
Itinerant magnetism	I - 262, 283, 365, 414
Johnson noise	II - 298, 432
Josephson effect	II - 239, 241
Josephson junction	II - 241
Joule magnetostriction	I - 352, 362, 364, 367, 372, 378
Kerr effects	I - 169, 403, 409, 415, 422 • II - 269, 296
Kondo effect	I - 303, 446
Kramers-Kronig relation	I - 158
Landau diamagnetism	I - 288
Landé factor	I - 113, 260
Langevin function	I - 112 • II - 343
Laplace's law	I - 21, 33
Larmor precession	I - 173, 448 • II - 360
LEED	see Low Energy Electron Diffraction
Legendre transformations	I - 322
Lenz's law	I - 48
Linear actuator	II - 221
Linear birefringence	I - 181
Linear elasticity	I - 391
Linear motor	II - 223
Linear response approximation	II - 363
Linearisation	II - 169

Load line	II - 6
Local density of states	I - 302
Localized electron model	I - 262, 414
Lodestone	see Magnetite
Longitudinal recording	II - 307
Lorentz field	I - 46
Lorentz force	I - 46
Lorentz microscopy	I - 174
Losses	II - 92-93
Loudspeaker	II - 29, 77
Low Energy Electron Diffraction (LEED)	II - 263
Lubrication	II - 345
Maghemite (γ-Fe_2O_3)	I - 4, 85, 266 • II - 314, 347, 394-395
Magic cylinder	I - 70 • II - 34, 468
Magnetic after-effect	I - 98, 236
Magnetic anisotropy	I - 89
Magnetic charge	I - 27-28, 32, 149
Magnetic circuit	I - 42 • II - 5
Magnetic dichroism	I - 172
Magnetic dipole	I - 27
Magnetic domain	I - 87, 173, 147 • II - 102, 349
Magnetic domain imaging	I - 168, 426
Magnetic field	I - 22, 32
Magnetic force	I - 29
Magnetic Force Microscopy (MFM)	I - 175
Magnetic head	II - 319
Magnetic image	I - 64
Magnetic induction	I - 20
Magnetic marking	II - 432
Magnetic mass	see magnetic charge
Magnetic moment	I - 27
Magnetic polarization	I - 32
Magnetic quantum number	I - 252
Magnetic Random Access Memory (MRAM)	II - 299
Magnetic recording	I - 14 • II - 305
Magnetic Resonance Imaging (MRI)	II - 355, 362
Magnetic semiconductor	I - 451
Magnetic separation	II - 399, 433
Magnetic structure	I - 138
Magnetic susceptibilities	I - 39 • II - 483
Magnetic torque	I - 29
Magnetic valence	I - 307-308
Magnetic X-ray diffraction	I - 404
Magnetite (Fe_3O_4)	I - 3-4, 266, 475-476 • II - 191, 217, 339, 394, 428, 487-488
Magnetization	I - 30, 339
Magnetization curling	I - 165, 205
Magnetization curve	I - 38
Magnetization loop	II - 4
Magnetization processes	I - 183, 211

INDEX BY SUBJECT

Magnetization reversal	I - 143, 198, 220 • II - 64, 67
Magnetization rotation	I - 190, 199 • II - 167
Magnetocaloric effect	I - 341
Magnetocardiography	II - 432
Magnetocrystalline anisotropy	I - 89, 143, 146 • II - 270
Magnetoelastic anisotropy	II - 271
Magnetoelastic effects	I - 101, 351
Magnetomechanical coupling coefficient	I - 388 • II - 225
Magnetometer	II - 432, 445, 446
Magnetometry	II - 228
Magnetomotive force	I - 43
Magnetomotive force loss coefficient	II - 6
Magneto-optical effects	I - 169, 399
Magneto-optical recording	I - 431 • II - 296, 311
Magneto-optical diffraction	I - 426
Magnetopause	II - 412
Magnetoresistance	I - 443, 448
Magnetoresistive materials	II - 298
Magnetoresistive readout	II - 329
Magnetosphere	II - 412
Magnetostatic energy	I - 58-59, 61, 185
Magnetostatic interaction	II - 323
Magnetostatic modes	II - 184, 187
Magnetostatics	I - 19
Magnetostriction	I - 327, 345, 359
Magnetostriction coefficient	II - 218
Magnetostrictive materials	II - 213, 297
Magnetostrictive transducer	I - 387
Magnetotactic bacteria	I - 169 • II - 428, 436
Magnetotherapy	II - 437
Major hysteresis loop	I - 227
Matteuci effect	I - 354, 385
Matthiessen's law	I - 445
Maxwell relations	I - 322, 325
Maxwell's equations	I - 50, 483
Mean field approximation	I - 116 • II - 267
Mean free path	II - 258, 260, 279
Media	II - 305
Meissner effect	II - 235-236
Metallic thin films	II - 316
Metamagnetism	I - 122, 125
Metglas	II - 226
MFM	see Magnetic Force Microscopy
Micromagnetism	I - 42
Microwave absorber	II - 203
Microwave applications	II - 298
Mineral	II - 392, 394
Minor loop	I - 228
Mixed state	I - 427

Modulated structure	I - 129-130
Molecular beam epitaxy (MBE)	II - 261
Molecular field	I - 10, 117, 145, 333
Molecular magnetism	I - 16 • II - 426
Morphic effect	I - 352
Motional narrowing	II - 379
Motor	II - 28
Motor function	II - 384
Mott transition	I - 452
Mott-Hubbard transition	I - 452
MRAM	see Magnetic Random Access Memory
Multilayers	I - 15 • II - 231, 255-256
Multiplet	I - 259
Mutual induction	I - 53
Nanocrystalline alloy	II - 114, 117-118, 121
Nanocrystalline material	II - 117
Nanostructures	I - 15
Narrow wall	I - 155
Natural abundance	II - 356, 393
Natural Remanent Magnetization (NRM)	II - 389, 403
Natural variables	I - 322
Near-field microscopy	see SNOM
Néel temperature	I - 81, 122
Néel wall	I - 158
Neutron diffraction	I - 137
Neutron interferometer	I - 173
Neutron refraction	I - 173
Neutron topography	I - 172, 181
Non-destructive testing	I - 431
Non-reciprocal device	II - 198
Non-reciprocity	I - 401, 429
NRM	see Natural Remanent Magnetization
Nuclear Magnetic Resonance (NMR)	II - 355
Nuclear magnetization	II - 357
Nuclear spin	II - 356
Nucleation	I - 199, 222 • II - 66
Nucleation coercivity	I - 235
Optical isolator	I - 429
Orange peel	II - 273
Orbital angular momentum	I - 252
Orbital magnetic moment	I - 251
Order parameter	I - 328
Palaeomagnetism	II - 388, 403, 413
Paramagnetic Curie temperature	I - 87, 120
Paramagnetism	I - 80
Particulate media	II - 314
Patterned media	II - 296
Pauli paramagnetism	I - 81, 287
Pauli susceptibility	I - 287

INDEX BY SUBJECT

Permalloy	II - 110, 282, 493
Permanent field source	II - 23
Permanent magnet	I - 13 • II - 3
Permanent magnet material	II - 17
Permeability	I - 39
Permeability of vacuum	I - 20
Permeance	I - 44
Perpendicular anisotropy	II - 268
Perpendicular recording	II - 296, 311
Petrofabric	II - 399
Phase modulation	I - 416
Phase rule	I - 92, 95, 194, 366
Phase transition	I - 425 • II - 266
Photoelectrons	I - 172
Physical Vapor Deposition (PVD)	II - 258
Piezomagnetism	I - 181, 327, 354
Piezoremanent Magnetization (PRM)	II - 405
Pinhole	II - 273
Pinning	II - 66
Pinning coercivity	I - 235
Planar anisotropy	I - 95
PLD	see Pulsed Laser Deposition
Poisson's equation	I - 33
Polarization	I - 418
Polder's tensor	II - 169
Pole piece	II - 5
Polishing	II - 346
Polycrystal	I - 196, 368, 372, 384
Position detector	II - 230
Precession	I - 108-109
Precipitation anisotropy	I - 175
Preisach's representation	I - 230
Principal quantum number	I - 252
Printing	II - 345
PRM	see Piezoremanent Magnetization
Propagation vector	I - 183
Pry and Bean model	II - 94
PVD	see Physical Vapor Deposition
Pulsed field	II - 461
Pulsed Laser Deposition (PLD)	II - 262
q domains	I - 180
Quantum number	I - 252
Quenching of the orbital moment	I - 265, 272
Radiofrequency pulse	II - 357
Rare earth	I - 261
Rare earth magnet	II - 26
Rayleigh's criterion	I - 437
Rayleigh's domain	I - 230, 232
Rayleigh's laws	I - 226

Readout	II - 305, 326
Reciprocal space	II - 368
Reciprocity theorem	I - 63 • II - 162, 441
Recoil loop	I - 228
Recording	I - 14 • II - 295, 305
Recording media	II - 295
Reflection High Energy Electron Diffraction (RHEED)	II - 263
Refractive index	I - 443
Refrigeration	I - 347
Relative permeability	I - 39
Reluctance	I - 43
Remanent magnetization	I - 39, 198 • II - 12
Reorientation	II - 269
Resonant magnetic X-ray scattering	I - 171
Resonator	II - 202
Reversibility	I - 199
Reversible initial susceptibility	II - 390
RHEED	see Reflection High Energy Electron Diffraction
Rock fabric	II - 399
Rotary motor	II - 224
Rotating machine	II - 127
Roughness	II - 272
Ruderman-Kittel-Kasuya-Yosida (RKKY) Interaction	I - 315 • II - 274
Russell-Saunders coupling	I - 258
S (spin) domains	I - 180
Satellite reflections	I - 183
Saturation	I - 134, 186, 198, 220
Scaling law	I - 330
Scanning Electron Microscopy with Polarization Analysis (SEMPA)	II - 264
Scanning Near-field Optical Microscopy (SNOM)	I - 170, 427, 437
Scanning Tunnelling Microscopy (STM)	II - 265
Seal	II - 345
Selective filter	II - 196
Selective pulse	II - 363
Self induction	I - 52
SEMPA	see Scanning Electron Microscopy with Polarization Analysis
Sensor	II - 32, 78, 225, 230, 433
Shape anisotropy	I - 89, 167 • II - 268, 270, 314
Shielding	I - 5, 69-70
Shock absorber	II - 346
Shot noise	II - 299
Shubnikov-de Haas effect	I - 455
Single domain particles	II - 394
Sintered magnet	II - 19
Skin effect	II - 161
Slater-Néel curve	I - 118
Slater-Pauling curve	I - 306
Slice selection	II - 363, 366
Small particles	I - 164

INDEX BY SUBJECT

Entry	Reference
Snoek's limit	II - 170
SNOM	see Scanning Near-field Optical Microscopy
Soft ferrite	II - 112
Soft materials	II - 89, 155, 297
Solar wind	II - 411
Solenoid	II - 458
Sonar	II - 225
Specific heat	I - 100, 339
Sperimagnetism	I - 132 • II - 317
Speromagnetism	I - 132
Spin	I - 10, 253
Spin dependent scattering	II - 281
Spin echo	II - 361, 379
Spin flip	I - 126 • II - 281, 294
Spin flop	I - 126
Spin flop transition	II - 276
Spin fluctuation	I - 446
Spin glass	I - 303
Spin lattice relaxation	II - 358
Spin magnetic moment	I - 253
Spin polarized electron tunnelling	II - 291
Spin Polarized Low Energy Electron Diffraction (SPLEED)	II - 264
Spin Polarized Scanning Tunnelling Microscopy (SPSTM)	I - 174
Spin quantum number	I – 252
Spin valve	II - 289, 298
Spin waves	II - 189
Spin-orbit coupling	I - 256 • II - 55
Spin-orbit interaction	I - 258 • II - 270
Spin-spin relaxation	II - 359
Spinel	II - 191
SPLEED	see Spin Polarized Low Energy Electron Diffraction
Spontaneous magnetic moment	I - 54
Spontaneous magnetization	I - 82, 333
SPSTM	see Spin Polarized Scanning Tunnelling Microscopy
Sputtering	II - 259
Squareness coefficient	II - 12
SQUID	I - 11 • II - 239, 242, 389, 432, 449
Stern-Gerlach experiment	I - 253
Stevens coefficient	I - 274, 278 • II - 56
Stevens operator	I - 274
STM	see Scanning Tunnelling Microscopy
Stoner's criterion	I - 289, 292, 299 • II - 267
Stoner's model	I - 288
Stoner-Wohlfarth model	I - 199 • II - 74, 314
Stransky-Krastanov growth	II - 263
Strong ferromagnetism	I - 290
Superconducting coil	II - 458
Superconductivity	I - 10 • II - 235
Superconductor	I - 427

Superexchange	I - 313
Superimposed susceptibility	I - 198
Superlattice	II - 256
Superparamagnetic limit	II - 295
Superparamagnetics	II - 374
Superparamagnetism	I - 115 • II - 315, 340
Surface anisotropy	II - 270
Surface magnetism	II - 265
Susceptibility	I - 471 • II - 374, 390
Synchrotron radiation	I - 171, 404, 428, 437 • II - 412
T (twin) domains	I - 180
T_1 contrast	II - 371
T_2 contrast	II - 371
Technical magnetization	II - 166
Tectonics	II - 419
Tectonic drift	II - 417
Telegraphone	II - 306
Terfenol-D	II - 216
Term	I - 258
Thermal activation	I - 209
Thermal expansion	I - 101 • II - 213
Thermodynamics	I - 321
Thermoremanent Magnetization (TRM)	II - 404
Thin film	I - 157 • II - 230-231, 255
Threshold field	I - 191
Tight-binding approximation	I - 295
Time reversal	I - 328
Torque	I - 45, 57
Torquemeter	II - 228
Total field	I - 32
Transformer	II - 123
Transition metal	I - 295
Transition metal magnet	II - 48
Transport	II - 278
Trilayer	II - 275-276
TRM	see Thermoremanent Magnetization
Tunnel magnetoresistance	II - 292
Two current model	II - 281
Type-I superconductor	II - 246
Type-II superconductor	II - 246
Ultrathin films	II - 265
Undulator	II - 35
Uniform magnetostatic resonances	II - 185
Unpinning	I - 218
Van der Waals force	II - 339
Van Vleck's paramagnetism	I - 81
Vector potential	I - 23
Very weak ferromagnet	I - 285
VGP	see Virtual Geomagnetic Pole

INDEX BY SUBJECT

Vibrating Sample Magnetometer (VSM)	II - 452
Virtual bound state	I - 301-302
Virtual Geomagnetic Pole (VGP)	II - 418
Viscosity	II - 343
Viscous Remanent Magnetization (VRM)	II - 405
Vision	II - 382
Voigt effect	I - 402, 408, 415
Voigt notation	I - 356, 393
Volmer-Weber growth	II - 263
Vortex lines	I - 174, 427 • II - 246
VRM	see Viscous Remanent Magnetization
VSM	see Vibrating Sample Magnetometer
Wall mobility	II - 176
Wall width	I - 150 • II - 258
Wave polarization	I - 432
Weak ferromagnet	II - 281
Weak ferromagnetism	I - 290
Weiss domains	see Magnetic domains
Wiedemann effect	I - 354, 379 • II - 222, 230
Wiedemann effect actuator	II - 222-223
Wiggler	II - 35
Working point	II - 6, 467
Writing	II - 305, 318
X-ray topography	I - 170, 181, 428
Zeeman energy	I - 61, 93, 124
ΔE Effect	see Delta E effect